Technical Physics

Technical Physics

P.J. Ouseph

University of Louisville

D. Van Nostrand Company

New York   Cincinnati   Toronto   London   Melbourne

Cover photograph: The MAC-8 microprocessor, courtesy of Western Electric.

D. Van Nostrand Company Regional Offices:
New York   Cincinnati

D. Van Nostrand Company International Offices:
London   Toronto   Melbourne

Library of Congress Catalog Card Number: 79-67376
ISBN: 0-442-26385-6

Published by D. Van Nostrand Company
135 West 50th Street, New York, N. Y. 10020

10 9 8 7 6 5 4 3 2 1

Preface

Technical Physics is designed to help students understand basic physical principles and apply them to their work as technologists. As technology advances, there is a need to prepare technical specialists to handle intelligently problems they may face. This new text achieves this goal with its practical world-of-work approach.

Over 200 solved problems, incorporated into each chapter with the basic principles, include applications to a wide variety of fields. Several chapters are devoted to important applied topics: Chapter 9 deals with properties of materials that interest civil and mechanical engineers; Chapters 23 and 26 with basic electrical properties; Chapter 11 with heat and insulating materials. Particularly noteworthy is the coverage of nuclear energy in Chapter 25 and the discussion of energy efficiency of light sources in Chapter 18.

The book is written in simple, understandable language and assumes the student has had no previous physics course or college-level mathematics course. Each chapter opens with an introduction to the chapter topics in order to point out their significance. Chapter-end summaries provide a helpful aid for study and review. The chapter-end problems are designed to be thought-provoking and comprehensive in quantity and variety. Since the problems vary in difficulty, instructors may choose problems at different levels suitable for their students.

The book is designed to give instructors flexibility in emphasizing certain topics in accord with their students' interests. For example, Chapters 24 and 25 can be left out for engineering technology students while these same chapters can be given emphasis for radiology, radiation, and medical technology students.

It is a pleasure to acknowledge the great amount of help I received in completing this book. Foremost among the people who helped me was Professor

Manuel Schwartz, University of Louisville, who patiently read the manuscript several times and made valuable suggestions. Professor Mary L. Harbold, Temple University, Professor Bob Martin, Tarrant County Junior College District, Texas, Professor David Mills, College of the Redwoods, California, and Professor Thomas Manakkil, Marshall University, reviewed the manuscript. Several of their suggestions are incorporated into the book, and the author is grateful to them for their help. I am also deeply indebted to Ms. Judy Anderson for her patience and care in typing the manuscript and to the members of my family for their patience and understanding during its preparation.

An Instructor's Manual containing solutions to all the chapter-end problems is available from the publisher. Also available is a manual of experiments appropriate to the laboratory associated with the course. This manual has been prepared by Manuel Schwartz and the author.

Contents

Technical Physics

Review of
Basic Mathematics

This book is written for students in various technologies who have had no college courses in mathematics and the level of high school mathematics the student has had is assumed to be minimal. Therefore a review of essential definitions and formulas in algebra, geometry, and trigonometry is given in this chapter. We start with relationships involving angles and lengths of the sides of a triangle. An introductory discussion of units is given at the end of the chapter.

1.1 TRIGONOMETRY

In Fig. 1.1 a circle of radius r ($= Oa$) is drawn with its center at the vertex of the angle defined by OA and OB. The ratio of the arc ab to the radius r defines the angle θ in radians, that is,

$$\theta \text{ (radians)} = ab/r. \tag{1.1}$$

Angles are also measured in degrees. To illustrate the relationship between degree and radian, consider the case where OA and OB [Fig. 1.1(b)] are lying in the same line (collinear). In this case angle θ is equal to $180°$ or π radians. In Fig. 1.1(c) the lines OA and OB are perpendicular to each other. The angle θ is equal to $90°$ or $\pi/2$ radians in this case. Since π is approximately equal to 3.14, we obtain

$$1 \text{ radian} = \frac{180°}{3.14} = 57.32°.$$

Triangle ABC, shown in Fig. 1.2, has neither its sides nor its angles equal. As is common practice, the vertices are marked by capital letters A, B, and

C and the sides opposite to them are marked by small letters. The side AC is called the base and the perpendicular distance from the base to the vertex is called the altitude designated by h in the figure. The area A of the triangle is given by the equation

$$A = bh/2, \qquad (1.2)$$

that is, the area is equal to half of the product of the base and the altitude.

The corresponding greek letters α(alpha), β(beta), and γ(gamma) stand for the three angles at the vertices A, B, and C. The sum of the three angles is

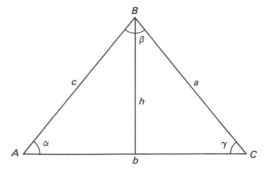

Figure 1.2. A triangle. A, B, and C are the vertices of the triangle and a, b, c indicate the lengths of the sides opposite vertices A, B, and C, respectively.

Figure 1.1. (a) Figure defines angle θ subtended by lines OA and OB; (b) two collinear lines subtend an angle of π radians (180°); (c) lines OA and OB are perpendicular to each other and the angle equals $\pi/2$ rad (90°).

(a)

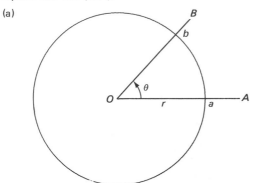

(b)

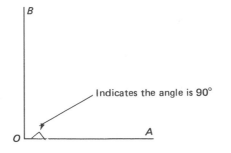

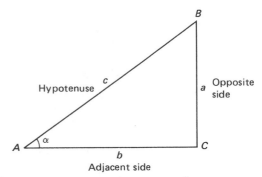

Figure 1.3. A right triangle. The side opposite to the 90° angle is called the hypotenuse; the side opposite to angle α is called the opposite side; and the third side close to angle α is called the adjacent side.

equal to 180°, which mathematically can be written as follows

$$\alpha + \beta + \gamma = 180°. \qquad (1.3)$$

A triangle for which the angle γ is equal to 90° is called a right triangle (Fig. 1.3). In this case the side BC is perpendicular to the base and, therefore, $h = a$. Therefore the area of a right triangle is

$$A = ab/2. \qquad (1.4)$$

The side opposite to the vertex C (opposite to the 90° angle) is called the hypotenuse. A relationship between the hypotenuse c and the sides a and b is

$$a^2 + b^2 = c^2. \tag{1.5}$$

This is known as the Pythagorean theorem which states that in a right triangle the square of the hypotenuse is equal to the sum of the squares of the other two sides.

Another set of relationships, extremely useful in solving physical problems, is known as the trigonometric relations. The definitions of the three basic functions in trigonometry, sine, cosine, and tangent, are given below:

$$\text{sine}\,\alpha = \frac{\text{opposite side (side } BC, \text{ opposite the angle)}}{\text{hypotenuse}}$$

$$\text{cosine}\,\alpha = \frac{\text{adjacent side (side } AC, \text{ close to angle)}}{\text{hypotenuse}},$$

$$\text{tangent}\,\alpha = \frac{\text{opposite side}}{\text{adjacent side}}.$$

These relations are usually written in short form as follows:

$$\sin\alpha = a/c, \tag{1.6}$$

$$\cos\alpha = b/c, \tag{1.7}$$

$$\tan\alpha = a/b. \tag{1.8}$$

We also have the relations (see Problem 1.3)

$$\sin^2\alpha + \cos^2\alpha = 1, \tag{1.9}$$

$$\tan\alpha = \frac{\sin\alpha}{\cos\alpha}. \tag{1.10}$$

Values of these functions may be found in tables (see Appendix). If one angle and one side of a right triangle are known, then, using these values, we can obtain the other sides and angles.

EXAMPLE 1.1

If the hypotenuse of a triangle is 2 ft and angle α is $30°$, obtain the lengths of sides AC and BC (Fig. 1.3) and the angle at the vertex B.

We use Eq. (1.6) first to obtain the length a,

$$\sin\alpha = a/c.$$

We can rearrange this equation in the form

$$a = c\sin\alpha.$$

Substituting values of c and α, we get

$$a = (2\,\text{ft})\sin 30°.$$

From the Table we get $\sin 30° = 0.5$

$$a = 2\,\text{ft} \times 0.5 = 1\,\text{ft}.$$

We can obtain b in two ways; both are shown below.

The first method is to use the Pythagorean theorem:

$$a^2 + b^2 = c^2$$

$$1^2 + b^2 = 2^2$$

or $b^2 = 2^2 - 1^2 = 3$

$$b = \sqrt{3} = 1.732\,\text{ft}.$$

The second method is to use the relationship for cosine which is

$$\cos\alpha = b/c \ \text{ or } \ b = c\cos\alpha.$$

Substituting the given values of c, we get

$$b = 2\cos 30° = 2 \times 0.86 = 1.732\,\text{ft}.$$

The angle at vertex B can be obtained using Eq. (1.3). Two of the three angles are given, $30°$ and $90°$. Therefore, for the sum to be $180°$ the angle at the vertex should be $60°$.

EXAMPLE 1.2

A 50 ft tree casts a 70 ft long shadow on a level ground. Obtain the angle made by the direction of the sun with the ground.

Since the adjacent and opposite sides of angle α (Fig. 1.4), which we have to determine, is given, we use Eq. (1.8) to calculate α.

$$\tan\alpha = \frac{50\,\text{ft}}{70\,\text{ft}} = 0.701.$$

From the Appendix we find the angle

$$\alpha = 70°.$$

1.2   GEOMETRIC FIGURES

Calculations of areas of surfaces and volumes of solids that appear often in applied physics are given below.

The area of a rectangle is the product of its length and width. The volume of a rectangular solid is the product of length, width, and height (Fig. 1.5).

Every point on the circle is equidistant from the center. The distance from the center to a point on the circle is called a radius (Fig. 1.6). The diameter d is the length of a straight line passing through the center with both its end points on the circle. The diameter is two times the radius. The circumference

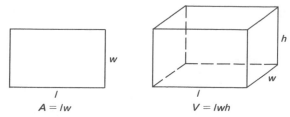

$$A = lw \qquad\qquad V = lwh$$

Figure 1.5.   A rectangle and rectangular solid.

of the circle is the distance of the curve measured once around (starting from say point 1 on the curve and ending at the same point). Circumference C is given by the following equation

$$C = 2\pi r = \pi d. \tag{1.11}$$

Note that this equation is consistent with the definition of angle in radians. The area of the circle is given by

$$A = \pi r^2 = \pi d^2/4, \tag{1.12}$$

where the greek letter π (pi) stands for a constant approximately equal to 22/7 or 3.14.

A sphere is a three-dimensional solid figure with every point on its surface equidistant from the

Figure 1.6.   A circle. Point O is the center of the circle; r, the radius; d, the diameter; C, the circumference.

Figure 1.4.   Example 1.2.

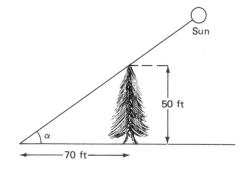

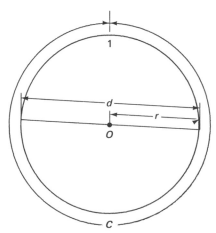

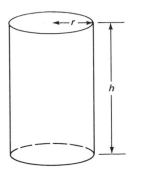

Figure 1.7. A cylinder. *h* is the height; *r*, the radius.

center. The distance from the center to the surface is the radius. Area of the spherical surface is given by

$$A = 4\pi r^2 \qquad (1.13)$$

and the volume of the sphere is given by

$$V = \frac{4}{3}\pi r^3. \qquad (1.14)$$

A cylinder having a circular cross section of radius *r* and height *h* is shown in Fig. 1.7. The area of a right cross-section and the end circular surfaces is the same (πr^2). The area of the lateral surface is $2\pi rh$ and hence the total surface area of a cylinder is given by

$$S = 2\pi r(r + h). \qquad (1.15)$$

The volume of the cylinder is given by

$$V = \pi r^2 h. \qquad (1.16)$$

1.3 VECTORS AND SCALARS

Some physical quantities can be completely specified by their magnitude. Such quantities are called scalar quantities and, for brevity, are also called scalars. Examples of scalar quantities are temperature, mass, density, and energy. On the other hand, magnitude alone will not sufficiently specify certain other quantities. They are called vector quantities and, for brevity, are also called vectors. Commonly encountered examples of vector quantities are force, velocity, and acceleration. Their directions are as important as their magnitude. For example, a velocity directed east will produce a different result from a velocity directed west even if their magnitudes are the same.

A scalar quantity is one that has a magnitude only. It is represented by a real number. This number is generally followed by a unit. (For example, 10°C, 10 is the number and °C is the unit).

A vector quantity is one that has magnitude and direction. It is represented by a directed line segment. The vector quantity is specified by a number and directional information and the number is generally followed by a unit. (For example, 50 km/hr, 30° south of east).

Let us consider a person driving from Louisville to Frankfort and from there to Cincinnati. The car, in going from L to F, has traveled a distance of 60 miles due east from the origin. In other words, the car has undergone a displacement of 60 miles east. One can also see that displacement is a vector since the magnitude distance of 60 miles and the direction east are both needed in describing the displacement. Let D_1 be this displacement, from L to F (Fig. 1.8). Similarly, let us say, the car is next driven from Frankfort to Cincinnati, a displacement designated by D_2 of 80 miles due north. What is the total displacement? The net effect of the two displacements is equivalent to a displacement from L to C. This new displacement **D** is equal to the vector "sum" of displacements D_1 and D_2. It is important to understand that the addition of vectors differs significantly from the addition of numbers. The magnitude of the vector sum **D** is not always equal to the algebraic sum of the magnitudes of D_1 and D_2. Similarly, the direction of **D** is different from the direction of D_1 and that of D_2 except in special cases.

In Fig. 1.8, vectors **D**, D_1, and D_2 are drawn in the conventional way. In Fig. 1.8, vector D_1 is represented by a line LF. The length of line LF is proportional to the magnitude of vector D_1 and the

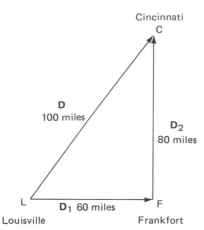

Figure 1.8. Addition of vectors. The vector **D** is equal to the sum of vectors **D**₁ and **D**₂.

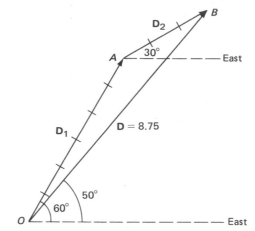

Figure 1.9. Graphical method of adding vectors.

arrow indicates its direction. Figure 1.8 also represents the equation $\mathbf{D}_1 + \mathbf{D}_2 = \mathbf{D}$. Vector **D** is the sum of **D**₁ and **D**₂. Thus Fig. 1.8 illustrates the addition of two vectors geometrically. The procedure we have followed is to place the second vector at the end of the first vector, and then join the beginning of the first vector and the end of the second vector to obtain the vector sum. The direction of the vector **D** is from L to C.

Consider two vectors **D**₁ and **D**₂ of magnitudes 6 units and 3 units making an angle of 60° and 30°, respectively, with the east direction. Their sum is obtained in Fig. 1.9. The procedure followed in obtaining the vector sum is outlined below.

1 The vector **D**₁ is drawn to scale. The length of the line *OA* representing vector **D**₁ is proportional to the magnitude of the vector **D**₁ and makes an angle of 60° with the east direction.

2 Vector **D**₂ is then drawn with its beginning at the end point of vector **D**₁. The length of line *AB* representing vector **D**₂ is proportional to the magnitude of **D**₂ and it makes an angle of 30° with the east direction.

3 The line *OB*, drawn from the beginning of **D**₁ to the end of **D**₂, represents the vector sum **D**. The magnitude of **D** can be obtained by meas-

uring the length of line *AB* and the angle can be read directly using a protractor from Fig. 1.9. The method described above is a graphical method of adding vectors.

EXAMPLE 1.3

A person walks 300 ft east, then 200 ft northeast (45° to the east direction), and then walks another 400 ft 30° south of west. Find the magnitude and the direction of the sum of the displacement vectors.

The displacement vectors are shown in Fig. 1.10. *OA* represents the displacement vector with magnitude 300 ft and directed east. *AB* and *BC* represents the other two displacements. The line *OC* represents the sum of the three vectors. The magnitude

Figure 1.10. Example 1.3.

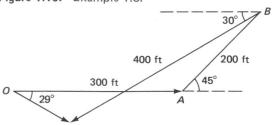

of $OC = 225$ ft and OC makes an angle of 29° south of east.

EXAMPLE 1.4

A mass located at the origin O has three forces acting on it. The forces are $\mathbf{F}_1 = 30$ lb east, $\mathbf{F}_2 = 40$ lb northwest, and $\mathbf{F}_3 = 50$ lb south [Fig. 1.11 (a)]. Find the resultant force acting on the mass.

Figure 1.11. (a) Example 1.4. (b) Graphical addition of vectors shown in (a).

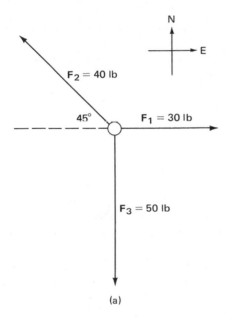

(a)

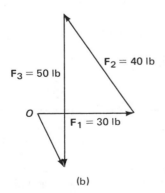

(b)

To solve this problem we note that a vector moved parallel to itself (magnitude remaining the same) is equal to the original vector. (Note, the principle of transmissibility of force applies to rigid bodies, but not to deformable bodies).

To add the vectors $\mathbf{F}_1$, $\mathbf{F}_2$, and $\mathbf{F}_3$, move $\mathbf{F}_2$ parallel to itself so that its initial point is located at the end of $\mathbf{F}_1$ [Fig. 1.11(b)]. Similarly, move $\mathbf{F}_3$ parallel to itself so that its initial point is located at the end of the vector $\mathbf{F}_2$ in its new position. The vector sum is represented by the line joining the origin O and the end point of the new position of the vector $\mathbf{F}_3$. The magnitude of the vector sum $= 21.7$ lb. The angle made by the sum vector with the east is 83° south of east.

In Example 1.4 we introduced the idea of equal vectors. Equal vectors are two vectors parallel to each other and having the same magnitude and direction (Fig. 1.12). A convention to remember regarding vectors is that, while a vector $\mathbf{A}$ is represented by a boldfaced letter $\mathbf{A}$, its magnitude is represented by the ordinary letter A.

1.4 COMPONENTS OF A VECTOR

Let us now consider a vector $\mathbf{A}$ at an angle θ to the line OX (Fig. 1.13). This vector can be regarded as the sum of two vectors $\mathbf{a}_x$ lying along OX and $\mathbf{a}_y$ lying in a direction perpendicular to OX. Note that

Figure 1.12. Vectors **a** and **b** are equal. Vectors **c**, **d**, and **e** are not equal because their directions are not the same, but note $\mathbf{c} = -\mathbf{d}$. Vectors **f** and **g** are not equal because their magnitudes are not equal.

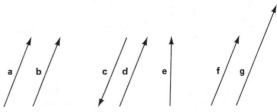

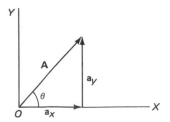

Figure 1.13. a_x and a_y are components of vector **A**.

a_x and a_y are projections of **A** along the OX and OY directions. These projections a_x and a_y are called components of the vector **A** along OX (x axis) and OY (y axis). In other words, we have resolved the vector **A** into its components along the two axes by drawing the projections. Also, from Fig. 1.13, the following relationships are evident:

$$a_x = A\cos\theta \qquad (1.17)$$

$$a_y = \sin\theta \qquad (1.18)$$

$$A^2 = a_x^2 + a_y^2 \qquad (1.19)$$

$$\tan\theta = a_y/a_x. \qquad (1.20)$$

In Fig. 1.13, the vector **A** is in the first quadrant of the coordinate system. The components are taken as positive in this case. If a vector is in any other quadrant (Fig. 1.14), its components may not be directed along the positive directions of the x and y axes. The components directed to the left

($-x$ direction) and downwards ($-y$ direction) are considered negative, whereas components to the right and upwards are considered positive. All possibilities in this regard are shown in Fig. 1.14.

EXAMPLE 1.5

The vector **A** in Fig. 1.13 has a magnitude of 3 m and its direction is at an angle of $37°$ to the x axis.

To obtain the components, we use Eqs. (1.17) and (1.18).

$$a_x = A\cos 37° = 3\,\text{m} \times 0.8 = 2.4\,\text{m}$$

$$a_y = \sin 37° = 3\,\text{m} \times 0.6 = 1.8\,\text{m}.$$

Note that the a_x and a_y we just calculated satisfy Eq. (1.19),

$$3^2 = 2.4^2 + 1.8^2.$$

EXAMPLE 1.6

The components of a displacement vector are $a_x = 3$ m, $a_y = 4$ m (Fig. 1.15). Calculate the magnitude of the vector and its direction. We use Eq. (1.19) to obtain the magnitude of the vector

$$A^2 = a_x^2 + a_y^2$$

$$= 3^2 + 4^2 = 25$$

$$A = 5\,\text{m}.$$

Figure 1.14. Components of vectors lying in different quadrants of the coordinate system.

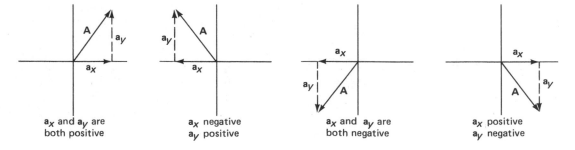

a_x and a_y are both positive

a_x negative a_y positive

a_x and a_y are both negative

a_x positive a_y negative

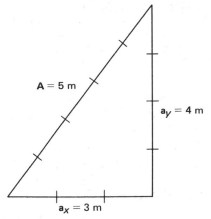

Figure 1.15. Example 1.6.

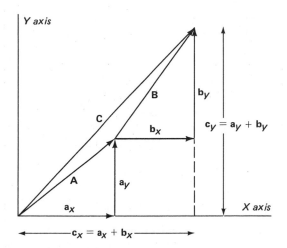

Figure 1.16. Addition of vectors by component method. The components of the sum vector **C** are equal to the sum of the components of the two vectors, $c_x = a_x + b_x$ and $c_y = a_y + b_y$.

To obtain the angle we could use Eq. (1.17) or (1.18). Let us use Eq. (1.17) so that

$$a_x = A \cos\theta$$

or $\cos\theta = a_y/A = 3/5 = 0.6$;

$\theta = 53°$ (from Appendix).

1.5 ADDING VECTORS USING COMPONENTS

Figure 1.16 illustrates addition of vectors **A** and **B** where the sum is vector **C**. The components a_x and a_y of **A** and b_x and b_y of **B** are also indicated in Fig. 1.16. It is evident that the sum of the x components of **A** and **B** is equal to the x component of **C**. Similarly the y components of **A** and **B** add up to the y components of **C**. Hence we have

$$a_x + b_x = c_x, \qquad (1.21a)$$

$$a_y + b_y = c_y. \qquad (1.21b)$$

The magnitude of vector **C** can be obtained using Eqs. (1.21a) and (1.21b) in terms of the components of **A** and **B** as follows:

$$C^2 = c_x^2 + c_y^2. \qquad (1.22)$$

Substituting for c_x and c_y, we get

$$C^2 = (a_x + b_x)^2 + (a_y + b_y)^2. \qquad (1.23)$$

The sum of the components in each direction gives us the component of the resulting vector in that direction. The direction of the new vector is given by

$$\tan\theta - \frac{a_y + b_y}{a_x + b_x}. \qquad (1.24)$$

EXAMPLE 1.7

Three forces acting on a mass are shown in Fig. 1.17. Find the resultant of the vectors.

To find the components, we first draw the x and y axes. Note that the two axes are perpendicular to each other. The angles made by forces are then determined.

$F_1 = 30$ lb force $0°$ with x axis to right

$F_2 = 40$ lb force $45°$ with x axis to left
(in the third quadrant)

$\mathbf{F}_3 = 50\,\text{lb force}\qquad 90°\ \text{with}\ x\,\text{axis}$

The components of the forces are then determined.

x components

$(30\ \text{lb})\cos 0° = 30\ \text{lb}$

$(-40\ \text{lb})\cos 45° = -28.28\ \text{lb}$

$(50\ \text{lb})\cos 270° = 0$
_____
sum of x components $= 2.72\ \text{lb}$

y components

$(30\ \text{lb})\sin 0° = 0$

$(40\ \text{lb})\sin 45° = 28.28\ \text{lb}$

$(50\ \text{lb})\sin 270° = -50\ \text{lb}$
_____
sum of y components $= -21.72\ \text{lb}$

The magnitude of the resultant vector is obtained by using Eq. (1.22)

$$c^2 = (2.72)^2 + (-21.72)^2 = 479.15$$

or $C = \sqrt{479.15} = 21.9\,\text{lb}.$

To obtain the angle of the resultant vector, Eq. (1.24) is used

$$\tan\theta = \frac{c_y}{c_x} = \frac{-21.72}{2.72} = -7.98$$

or $\theta = -83°$.

This example is the same as Example 1.4. Both methods, the graphical method and the component method, give us the same answer.

A summary of steps to add vectors by the component method is as follows: (a) find the x and y components of each vector; (b) add the x components and y components separately; (c) use Eq. (1.23) to obtain magnitudes of the new vector and Eq. (1.24) to obtain its direction.

In many practical problems, a coordinate system with three axes (OX, OY, and OZ) is generally used. These axes are mutually perpendicular, and they are arranged in such a way that, when a right-handed

Figure 1.18. Right-handed rectangular coordinate system. A right-hand screw, when rotated in the sense to X to Y, will advance along the Z axis.

Figure 1.17. Example 1.7.

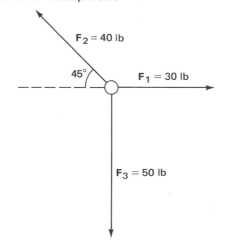

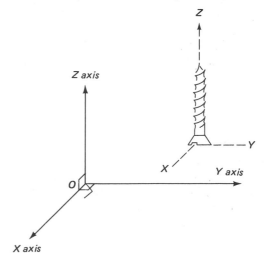

screw is rotated in the sense of X to Y, it will advance along the Z axis (Fig. 1.18). This coordinate system is called a right-handed rectangular coordinate system. A vector in such a coordinate system will have three components, a_x, a_y, and a_z and the magnitude of the vector is given by

$$A = \sqrt{a_x^2 + a_y^2 + a_z^2}. \qquad (1.25)$$

1.6 UNITS AND SYSTEMS OF MEASUREMENTS

As the reader is aware, distances are given in units such as miles or kilometers. For example, the distance between Chicago and New York is 855 miles or 1375 kilometers(km). This statement means that the distance between Chicago and New York is 855 times a standard distance called a mile or 1375 times a standard distance called a kilometer. The standard distance is a unit of length. Similarly, there are units for measurement of mass and time. There are two systems of units, widely used these days; one is known as the English system and the other one is known as the SI (Systeme International). The units used in these systems are given in Table 1.1.

The units giving the mass, length, and time are regarded as fundamental units. They differ from other units, for example, speed, in that the unit of speed is a derived unit. Since speed = (distance)/(time) and the distance is in feet (ft) and the time in seconds (s), it follows that a unit of speed is ft/s. Also we should note that the numbers on both sides of an equation and the units on both sides balance.

For example, if the distance traveled is 80 ft in 4 s then the speed is

$$20 \text{ ft/s} = \frac{80 \text{ ft}}{4 \text{ s}}.$$

The SI units are being used all over the world these days. One of the attractive features of this unit system is that it is a decimal system, that is, bigger or smaller units (secondary) are obtained by multiplying the base unit (meter, kilogram, or second) by powers of 10; for example, a kilometer is 1000 meters and a centimeter is 0.01 meter. Table 1.2 gives a list of commonly used multiples of base units and their prefixes (centi, in centimeter is the prefix). Table 1.3 lists units used in the SI and English system for measuring distance. It is evident that the relationship between the units in the English system is not as simple as in the SI units. Tables in the Appendix may be used to convert measurements from one unit system to the other.

In Table 1.2, 1000 is written as 10^3; the latter is known as exponential notation. The number 3 is the exponent, which tells us how many times the base number should be multiplied. The following examples illustrate this statement.

$$2^3 = 2 \times 2 \times 2,$$

$$5^4 = 5 \times 5 \times 5 \times 5,$$

$$10^6 = 10 \times 10 \times 10 \times 10 \times 10 \times 10 = 1,000,000.$$

In the decimal system the base number is 10. When

TABLE 1.1 Fundamental Units in the English System and the System International (SI).

QUANTITY	UNITS USED IN ENGLISH SYSTEMS	UNITS USED IN SI
length	foot	meter
mass	slugs	kilogram
time	seconds	seconds

TABLE 1.2 Powers of 10, Prefixes, and Symbols

FACTOR	PREFIX	SYMBOL	EXAMPLES
10^{-12} = 1/1,000,000,000,000	pico	p	1 pf = 1 picofarad
10^{-9} = 1/1,000,000,000	nano	n	1 nm = 1 nanometer
10^{-6} = 1/1,000,000	micro	μ	1 μm = 1 micrometer
10^{-3} = 1/1000	milli	m	1 mg = 1 milligram
10^{-2} = 1/100	centi	c	1 cm = 1 centimeter
10^3 = 1000	kilo	k	1 kg = 1 kilogram
10^6 = 1,000,000	mega	M	1 MW = 1 megawatt
10^9 = 1,000,000,000	giga	G	1 GW = 1 gigawatt

the exponent is negative, it means we have to take the reciprocal of the number with the same positive exponent, for example,

$$10^{-3} = 1/10^3 = 1/100 = 0.001.$$

Similarly we have

$$10^{-1} = 1/10 = 0.1 \, (\text{one-tenth})$$

TABLE 1.3 Units Generally Used in Measuring Distance

BRITISH UNITS
1 inch (in.) = $\frac{1}{12}$ ft
1 foot (ft) = 12 in.
1 yard (yd) = 3 ft
1 mile (mi) = 5280 ft = 1760 yd
SI UNITS
1 millimeter (mm) = 1/1000 m = 10^{-3} m
1 centimeter (cm) = 1/100 m = 10^{2} m
1 meter (m)   = 100 cm = 1000 mm
1 kilometer (km) = 1000 m

$$10^{-2} = 1/10^2 = 0.01 \, (\text{one-hundredth})$$

$$10^{-3} = 1/10^3 = 0.001 \, (\text{one-thousandth})$$

$$10^{-6} = 1/10^6 = 0.000001 \, (\text{one-millionth})$$

Multiplication of exponential numbers with the same base is shown below.

$$a^n \times a^m = a^{n+m},$$

$$a^n \times a^{-m} = a^{n-m},$$

that is, the exponent of the product is obtained by adding the exponents.

1.7 PROPORTIONS

As we have seen before, the circumference of a circle is equal to a constant π multiplied by its diameter. Circumference is, therefore, directly proportional to d, which we write as

$$C \propto d,$$

where $\propto$ signifies the proportionality between the two variables; therefore, π is called the proportionality constant. For any circle, the circumference divided by the diameter is equal to this constant. There are other physical situations where proportionalities between two or more variables are encountered. For example, the volume of a gas is proportional to temperature T, elongation of a spring, s, is proportional to the applied force F, etc. In these cases, we could write

$$V \propto T, \qquad V = k_T T; \qquad (1.26)$$

$$F \propto s, \qquad F = k_s s; \qquad (1.27)$$

where the subscripts indicate that the constants are not the same in each case. Instead of temperature, if the pressure is varied, the volume of a gas changes again too. In this case the volume is not directly proportional to pressure, but the volume is inversely proportional to pressure; that is, as the pressure increases volume decreases. Inverse proportionality is expressed by

$$V \propto 1/P \text{ or } V = k_p/P. \qquad (1.28)$$

The volume of a gas is thus directly proportional to T when pressure is held constant and inversely proportional to P when T is held constant. Both these proportionalities can be combined into one as shown below (proof is given later),

$$V \propto T/P \text{ or } PV = kT. \qquad (1.29)$$

We now prove that two proportionalities such as $V \propto T$ and $V \propto 1/P$ can be combined into one. Consider three quantities A, B, and C, where $A \propto B$ when C is constant and $A \propto C$ when B is constant. Let C be constant while B changes to b. Then A changes to a value a'. From the proportionality, we get

$$\frac{A}{a'} = \frac{B}{b}. \qquad (1.30)$$

Now let B be constant at the value b while C changes to c. Then A changes from a' to a. Hence

$$\frac{a'}{a} = \frac{C}{c}. \qquad (1.31)$$

From Eqs. (1.30) and (1.31) we obtain

$$\frac{A}{a'} \times \frac{a'}{a} = \frac{B}{b} \times \frac{C}{c}$$

$$A = \frac{a}{bc} BC \text{ or } A \propto BC. \qquad (1.32)$$

Thus the proportionality between several variables, whether direct or inverse, can be combined as a product.

SUMMARY   Equations of areas and volumes often used in physics are given below.

Triangle: $A = ah/2$;

Circle: $A = \pi r^2$;

Sphere: $A = 4\pi r^2$, $V = \frac{4}{3} \pi r^3$;

Cylinder: $A = 2\pi r(r + h)$, $V = \pi r^2 h$.

It is important to remember the following trigonometric relations:

$$\sin \alpha = \frac{\text{opposite side}}{\text{hypotenuse}};$$

$$\cos \alpha = \frac{\text{adjacent side}}{\text{hypotenuse}} \, ;$$

$$\tan \alpha = \frac{\text{opposite side}}{\text{adjacent side}} \, .$$

The Pythagorean theorem for a right-angle triangle is

$$a^2 + b^2 = c^2 .$$

Magnitude and direction have to be specified to describe a vector quantity while a scalar quantity can be specified by giving its magnitude alone.

When two vectors are added, we get a new vector. The magnitude of the new vector is not always equal to the algebraic sum of the magnitudes of the vectors. Similarly, the direction of the new vector is different from the original vectors.

Vectors can be added by graphical and component methods. When the vectors are added by the component method, special attention should be given to the signs of the components.

QUESTIONS AND PROBLEMS

1.1 From Tables, find the values of

$\sin 10°$	$\cos 80°$
$\sin 20°$	$\cos 70°$
$\sin 30°$	$\cos 60°$
$\sin 45°$	$\cos 45°$
$\sin 60°$	$\cos 30°$

1.2 Using the information obtained in Problem 1.1, can you make the statement

$$\cos\theta = \sin(90° - \theta) \text{ and } \sin\theta = \cos(90° - \theta)?$$

1.3 Prove $\cos^2\theta + \sin^2\theta = 1$. (Use the Pythagorean theorem and the definitions of $\sin\theta$ and $\cos\theta$ and $\sin\theta/\cos\theta = \tan\theta$.)

1.4 Obtain the height of the triangle given in Fig. 1.19. Calculate the length of its base. Next calculate the area of the triangle.

1.5 Calculate the values of the function $\sin\alpha$, $\cos\alpha$, and $\tan\alpha$ for angle α in Fig. 1.20.

1.6 Calculate the lengths of the hypotenuse and adjacent side for the triangle given in Fig. 1.21. Obtain the value of the angle at the vertex C.

1.7 The angle of elevation of a tree is $30°$ at a distance of 400 ft from it. Calculate the height of the tree.

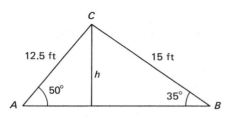

Figure 1.19. Problem 1.4.

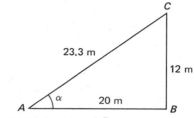

Figure 1.20. Problem 1.5.

1.8 A steep road has a 20% grade, which means it rises 20 ft for every 100 ft of horizontal distance. Find the angle of incline for the road.

1.9 The surface of a mountain road climbs at an angle of 30°. Calculate the rise in height for every 100 ft horizontal distance. What is the percent grade of the road?

1.10 A house has a simple gable roof pitched at 30°. If the width of the house is 40 ft, calculate the length of the rafters asssuming no overhang (see Fig. 1.22).

1.11 A steel cylinder tank, 10 ft high, has a radius of 2 ft. Calculate the area of the steel plate used to construct this tank.

1.12 Calculate the volume of the tank in Problem 1.11.

1.13 Pick out the vectors from the following quantities: displacement, distance, speed, velocity, temperature, momentum, energy, power and acceleration.

1.14 Under what condition will the sum of the magnitudes of vectors D_1 and D_2 be equal to the magnitude of the sum vector D?

1.15 Add the vectors D_1 and D_2 given below by graphical and component methods.

D_1		D_2	
MAGNITUDE	ANGLE	MAGNITUDE	ANGLE
6 units	east	5 units	north
10 ft	30° north of east	20 ft	40° north of west
40 lb	20° above x axis	50 lb	45° below x axis

Figure 1.21. Problem 1.6.

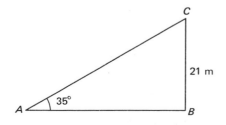

Figure 1.22. Problem 1.10.

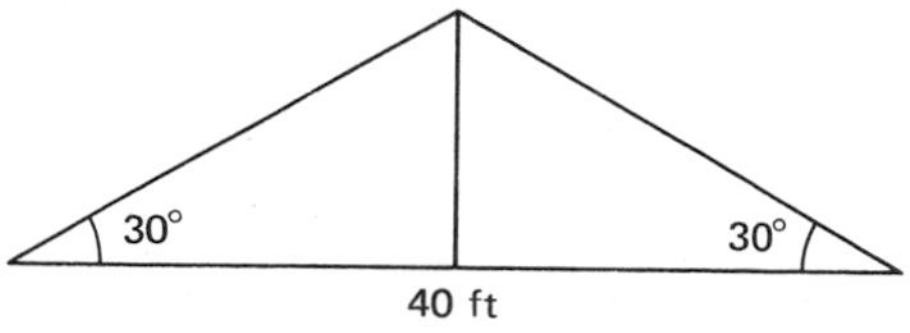

1.16   Resolve the vectors shown in Fig. 1.23.

1.17   Add the vectors shown in Fig. 1.23.

1.18   The velocity of a boat is 15 mi/h in still water. If the boat travels in a river with a velocity of 5 mi/h as shown in Fig. 1.24 what is the direction and magnitude of the resultant velocity?

1.19   Using the data in Fig. 1.24, if the boat starts at point A of one bank and wants to reach B, an opposite point on the second bank, in what direction should the boat be directed?

1.20   Two forces of 20 lb and 30 lb act on a mass as shown in Fig. 1.25. Find the direction and magnitude of a third force that will balance these forces.

Figure 1.24. Problems 1.18 and 1.19.

Figure 1.23. Problems 1.16 and 1.17.

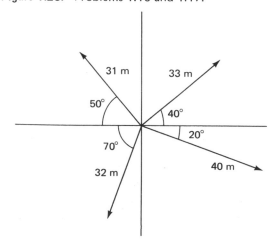

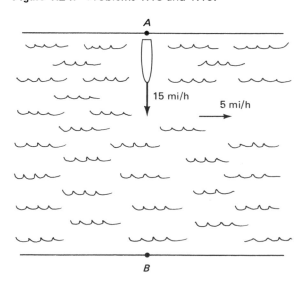

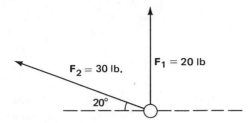

Figure 1.25. Problem 1.20.

2 One- and Two-Dimensional Motion

In this chapter we are concerned with motion of an object. Certain aspects of motion, such as the relationships of speed, distance, and time, are familiar to most people. For example, a driver can mentally compute the distance he will travel in 6 h when he is traveling at a speed of 50 mi/h (300 miles). The direction in which an object goes is also important. As we all know, when we press on a gas pedal, the speed of a car increases. The rate of such an increase is its acceleration. Thus there are several concepts with which we shall have to deal: distance, displacement, velocity, and acceleration. We shall, moreover, derive certain equations connecting these quantities, and apply them to problems involving one- and two-dimensional motion.

What is meant by motion? A continuous change in the position of a particle is called a *motion*. By the word motion we usually mean a translatory motion where an object moves from one point to another point. The other type of motion is known as rotary motion, where the object rotates about an axis. This will be discussed later.

2.1 SPEED AND ITS UNIT

Let us consider an object moving from point 1 to point 2 (see Fig. 2.1). In general, a moving object does not travel along a straight line path. If the distance traveled (not the straight-line distance, Fig.

2.1) and the time of travel are known, we can define the speed of the object as

$$\text{Speed} = \frac{\text{distance traveled}}{\text{time elapsed}}. \qquad (2.1)$$

This definition of speed does not involve direction or changes in direction of the path of the particle. Speed, therefore, is a scalar quantity.

Speed defined by Eq. (2.1) is an average quantity. For example, when a person travels 120 miles in 3 hours, his speed is

$$\text{Speed} = \frac{120 \text{ mi}}{3 \text{ h}} = 40 \text{ mi/h}.$$

This does not mean that the driver maintained a constant speed of 40 mi/h, but it means only that his average speed during his driving time was 40 mi/h. In this book, we shall generally use the word "speed" as synonymous with "average speed."

As we have seen in Chapter 1, distance is expressed in meters or multiples of meters in SI units, and in miles, feet, etc. in the English unit system. In both systems, time is expressed in seconds, minutes, or hours. Since speed is distance divided by time, we may see it expressed in any of the following units:

miles/hour (mi/h)	meters/second (km/h)
feet/second (ft/s)	kilometers/hour (km/h)
etc.	etc.

Since we encounter problems where we will have to convert from one unit to another, let us consider the procedure for such conversions. Suppose, for example, that we want to express a distance of 15 miles in feet. To achieve this conversion, we start with the known fact that 1 mile = 5280 feet. From this equality, we obtain

$$\frac{5280 \text{ ft}}{1 \text{ mi}} = 1 \quad \text{and the inverse} \quad \frac{1 \text{ mi}}{5280 \text{ ft}} = 1.$$

Now, we use the fact that multiplying by unity does not change the quantity and obtain

$$15 \text{ mi} = 15 \text{ mi} \times \frac{5280 \text{ ft}}{1 \text{ mi}} = 7.9 \times 10^4 \text{ ft},$$

and

$$15 \text{ mi} = 15 \text{ mi} \times \frac{1 \text{ mi}}{5280 \text{ ft}}.$$

In the first equation, the unit of mile cancels out on the right-hand side, and thus we obtain the conversion of the distance in feet. Note that the second equation, even though it is correct, does not give us the needed conversion. Hence, care should be taken in multiplying with the appropriate ratio. The following examples further illustrate this procedure.

Express 60 miles in kilometers.

$$1 \text{ mi} = 1.62 \text{ km} \quad \text{or} \quad 1 = \frac{1.62 \text{ km}}{1 \text{ mi}}.$$

Therefore, we get

$$60 \text{ mi} = 60 \text{ mi} \times \frac{1.62 \text{ km}}{1 \text{ mi}}$$

$$= 97.2 \text{ km}.$$

Figure 2.1. Motion of an object from position 1 at t_i to position 2 at t_f. The straight line represents the net displacement.

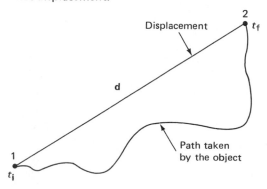

Express 1 mi/h in ft/s.

$$1 \text{ mi} = 5280 \text{ ft}, \quad 1 = \frac{5280 \text{ ft}}{1 \text{ mi}};$$

$$1 \text{ h} = 3600 \text{ s}, \quad 1 = \frac{3600 \text{ s}}{1 \text{h}};$$

$$1 \text{ mi/h} = 1 \text{ mi/h} \times \frac{5280 \text{ ft}}{1 \text{ mi}} \times \frac{1 \text{ h}}{3600 \text{ s}}$$

$$= 1.46 \text{ ft/s}.$$

A convenient conversion to remember is 30 mi/h = 44 ft/s, which is obtained from

$$30 \frac{\text{mi}}{\text{h}} = 30 \frac{\text{mi}}{\text{h}} \times \frac{5280 \text{ ft}}{1 \text{ mi}} \times \frac{1 \text{ h}}{3600 \text{ s}} = 44 \frac{\text{ft}}{\text{s}}$$

EXAMPLE 2.1

An airplane travels at a constant speed of 700 mi/h. Express its speed in ft/s and calculate the time it will take to travel from New York to London (3466 miles):
We know

$$30 \text{ mi/h} = 44 \text{ ft/s};$$

$$700 \text{ mi/h} = 700 \frac{\text{mi}}{\text{h}} \times \frac{44 \text{ ft/s}}{30 \text{ mi/h}}$$

$$= 1026.66 \text{ ft/s}.$$

Since speed × time = distance traveled, we get

$$\text{time} = \frac{\text{distance traveled}}{\text{speed}} = \frac{3455 \text{ mi}}{700 \text{ mi/h}} = 4.95 \text{ h}.$$

2.2 DISPLACEMENT AND VELOCITY

In Fig. 2.1 the object is at point 1 at time t_i and at point 2 at t_f. The displacement in time interval $t = t_f - t_i$ is the straight line drawn between points

1 and 2. We define average velocity by the equation

$$\overline{V} = \frac{d}{t_f - t_i} = \frac{d}{t} \qquad (2.2)$$

that is, *average velocity is the displacement per unit time.* Since displacement is a vector, velocity—obtained by dividing a vector by a scalar—is also a vector.

The line drawn above the letter **V** indicates an average. Velocity calculated for very short periods of time is called instantaneous velocity. Since velocity is displacement (directed distance between two points) divided by time, its units are the same as the speed.

To further illustrate the difference between speed and velocity, consider a runner in a circular track (Fig. 2.2). Let us assume that he runs at a constant speed once around the 100 m track in 10 s. The speed of the runner = (100 m)/(10 s) = 10 m/s. His net displacement on completion of one round of running is zero. Hence, his average velocity is zero. However, if one considers intermediate stages of his running, one can calculate average velocities which are continuously changing in magnitude and direction. Let us calculate his average velocity for travel from S to A (quarter way) and for S to B (half

Figure 2.2. In a circular path the average velocity calculated depends on the time interval, even if the speed is constant.

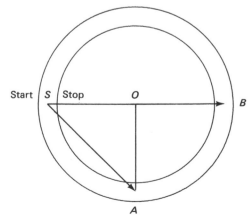

way).   To make these calculations we need the
radius of the track.

Circumference $= 2\pi r$, where r is the radius

$$r = \frac{\text{circumference}}{2\pi}$$

$$= \frac{100 \text{ m}}{2\pi} = 15.9 \text{ m}.$$

Magnitude of vector $\mathbf{SA} = \sqrt{OS^2 + OA^2}$

$$= \sqrt{(15.9)^2 + (15.9)^2}$$

$$= 22.5 \text{ m}.$$

Average velocity from S to A $= \dfrac{22.48 \text{ m}}{2.5 \text{ s}}$

$$= 9.00 \text{ m/s}, 45°$$
$$\text{south of east}.$$

Average velocity from S to B $= \dfrac{2r}{t} = \dfrac{2 \times 15.9 \text{ m}}{5 \text{ s}}$

$$= 6.36 \text{ m/s east}.$$

In dealing with practical problems, we encounter
two different cases. They are as follows:
   (a) Instantaneous velocity of the object remains
constant as illustrated in Fig. 2.3(a). In this case,
the average velocity is the same as the instantaneous
velocity. Distance traveled in this case linearly increases at a constant rate [Fig. 2.3(b)] and can be
calculated by the equation $d = Vt$.
   (b) Velocity changes with time [Fig. 2.4(a)].
We restrict ourselves to cases where the magnitude
of the velocity increases (decreases) at a constant
rate. At time t_i Fig. 2.4(a), the object has a velocity
$\mathbf{V_i}$; and at t_f its velocity is $\mathbf{V_f}$ where i and f refer,
respectively, to the initial and final values. The
average velocity for the time period t_i to t_f is given
by

$$\overline{\mathbf{V}} = (\mathbf{V_i} + \mathbf{V_f})/2. \tag{2.3}$$

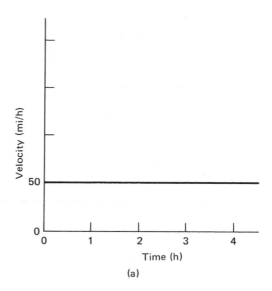

(a)

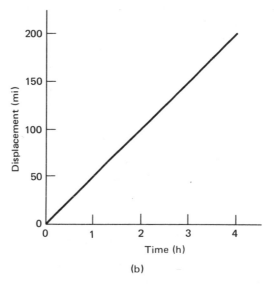

(b)

Figure 2.3.   (a) Velocity remains constant. (b) Displacement as a function of time for constant
velocity.

The total displacement at the end of the period
$t = t_f - t_i$ can be obtained by

$$\mathbf{d} = \overline{\mathbf{V}}t = \tfrac{1}{2}(\mathbf{V_i} + \mathbf{V_f})(t_f - t_i). \tag{2.4}$$

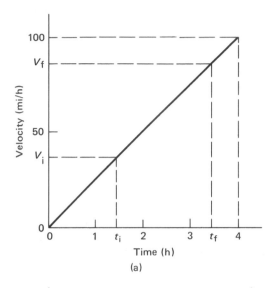

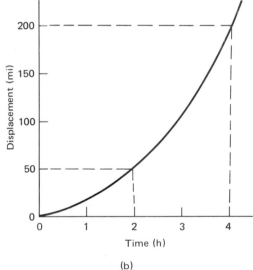

Figure 2.4. (a) Velocity increases at a constant rate with time. (b) Displacement as a function of time when velocity increases at constant rate.

Compare Fig. 2.3 with Fig. 2.4. The displacement at the end of the fourth hour in both cases is the same. This is because the average velocity in this time period in the second case is the same as the constant velocity. However, the displacement at the end of 2 h is not the same in both cases. The

average velocity in the two-hour period, 25 m/h in the example of Fig. 2.4, is smaller than the constant velocity (50 mi/h) in the example of Fig. 2.3.

EXAMPLE 2.2

The velocity of a car increases at a constant rate from 0 to 50 mi/h in 6 s. Calculate the distance it travels in 6 s.

Average velocity of the car is

$$\overline{V} = \frac{V_i + V_f}{2} = \frac{0 + 50}{2} = 25 \frac{mi}{h} \times \frac{1.46 \text{ ft/s}}{1 \text{ mi/h}}$$

$$= 36.7 \text{ ft/s}.$$

Distance traveled $= \overline{V}t = 36.7 \text{ ft/s} \times 6 \text{ s} = 220 \text{ ft.}$

2.3 ACCELERATION

When the velocity of an object is increasing (or decreasing) the object is accelerating (or decelerating). Acceleration is defined as the change in velocity per unit time, or

$$\mathbf{a} = \frac{\mathbf{V}_f - \mathbf{V}_i}{t_f - t_i} = \frac{\mathbf{V}_f - \mathbf{V}_i}{t}. \qquad (2.5)$$

The letter **a** stands for acceleration and $\mathbf{V}_i$ and $\mathbf{V}_f$ are the instantaneous velocities at t_i and t_f, respectively. From the definition it is evident that **a** is a vector. As in the definition of velocity, if t is finite, Eq. (2.5) defines average acceleration and, if t is very small, instantaneous acceleration. In problems discussed here, a is assumed constant and, therefore, average acceleration and instantaneous acceleration are the same.

Since acceleration is given by

$$a = \frac{\text{change in velocity}}{\text{time}}$$

the units for a are as follows:

SI units $\qquad \dfrac{m/s}{s} = \dfrac{m}{s \times s}$ or $\dfrac{m}{s^2}$;

English units $\dfrac{ft/s}{s} = \dfrac{ft}{s \times s}$ or $\dfrac{ft}{s^2}$.

EXAMPLE 2.3

Calculate the acceleration of the car in Example 2.2.

The initial velocity, $\mathbf{V_i} = 0$; the velocity at the end of 6 s = 50 mi/h = 36.7 ft/s; the change in velocity = 36.7 − 0 = 36.7 ft/s.

$$\text{Acceleration} = \frac{36.7 \text{ ft/s}}{6 \text{ s}}$$

$$= 6.11 \text{ ft/s}^2.$$

Note that an acceleration of 6.11 ft/s^2 means that the velocity increases by 6.11 ft/s each second.

2.4 ONE-DIMENSIONAL MOTION

The three quantities discussed above—displacement, velocity, and acceleration—are vector quantities. Velocity is obtained by dividing a vector, displacement, by a scalar, time; acceleration is obtained by dividing a vector, velocity, by a scalar, time. In one-dimensional problems all these vectors are directed along one direction only. These problems are easier to solve and they are discussed in this section.

The vector equations, (2.2), (2.3), and (2.5) can now be written as scalar equations, taking the direction of motion as the x axis.

$$x = \overline{V}t, \tag{2.6}$$

$$a = (V_f - V_i)/t, \tag{2.7}$$

$$\overline{V} = (V_f + V_i)/2. \tag{2.8}$$

If the velocity decreases with time, that is, if the final velocity V_f is less than the initial velocity V_i, it is customary to say that the object is undergoing deceleration.

In Eq. (2.6), x is the distance in the x direction, the direction of motion. Also, in all succeeding discussions, it is assumed that a remains constant. Using the above equations, we now proceed to derive certain useful relationships involving x, V, a, and t.

Equation (2.7) can be rearranged in the form

$$t = (V_f - V_i)/a. \tag{2.9}$$

Substituting the values of $\overline{V}$ from Eq. (2.8) and t from Eq. (2.9) in Eq. (2.6), we get

$$x = \frac{V_f + V_i}{2} \times \frac{V_f - V_i}{a}$$

$$= \frac{V_f^2 - V_i^2}{2a}$$

or

$$V_f^2 = V_i^2 + 2ax. \tag{2.10}$$

Equation (2.7) also can be written in the form

$$V_f = V_i + at$$

and hence

$$\overline{V} = (V_f + V_i)/2 = V_i + \frac{1}{2} at. \tag{2.11}$$

Substituting this value of $\overline{V}$ into Eq. (2.6), we get

$$x = V_i t + \frac{1}{2} at^2. \tag{2.12}$$

One-dimensional problems can be solved by using one or more of the equations (2.6), (2.7), (2.8), (2.10), and (2.12). For convenience they are collected together in Table 2.1.

TABLE 2.1 Equations for One-dimensional Motion

EQUATION	EQUATION NUMBER IN TEXT
$x = \bar{V}t$	(2.6)
$a = (V_f - V_i)/t$	(2.7)
$\bar{V} = (V_f + V_i)/2$	(2.8)
$V_f^2 = V_i^2 + 2ax$	(2.10)
$x = V_i t + \frac{1}{2}at^2$	(2.12)

In solving a problem, the student will find it convenient to follow some simple steps: write down all the given information with units; write down the quantities we are looking for; pick the appropriate equations from Table 2.1. This procedure is followed in the numerical example given below.

EXAMPLE 2.4

A car accelerating from rest travels a distance of 440 ft in 10 s. Calculate its final velocity and acceleration.

We are given

$$V_i = 0,$$

$$t = 10 \text{ s},$$

$$x = 440 \text{ ft}.$$

We have to find the values of a and V_f. One can solve this problem in two different ways. In the first and simpler method, we use Eqs. (2.6) and (2.7)

$$x = \bar{V}t$$

$$440 \text{ ft} = \bar{V} \times 10 \text{ s}, \quad \bar{V} = 44 \text{ ft/s};$$

$$\bar{V} = \frac{V_i + V_f}{2} = \frac{0 + V_f}{2}, \quad V_f = 88 \text{ ft/s};$$

$$a = \frac{V_f - V_i}{t} = \frac{88 \text{ ft/s} - 0}{10 \text{ s}} = 8.8 \text{ ft/s}^2.$$

The second method uses Eqs. (2.12) and (2.7)

$$x = V_i t + \frac{1}{2}at^2$$

$$440 \text{ ft} = 0 \times 10 \text{ s} + \frac{1}{2}a(10 \text{ s})^2;$$

$$a = (440 \text{ ft})\left(\frac{2}{100 \text{ s}^2}\right) = 8.8 \text{ ft/s}^2.$$

$$V_f = V_i + at$$

$$V_f = 0 + (8.8 \text{ ft/s}^2)(10 \text{ s}) = 88 \text{ ft/s}.$$

EXAMPLE 2.5

An object accelerating from rest attains a final velocity of 980 m/s, while it travels a distance of 49,000 m. Calculate the acceleration of the object and the time of travel.

We are given

$$V_i = 0, \quad V_f = 980 \text{ m/s}, \text{ and } x = 49,000 \text{ m}.$$

Acceleration can be obtained in two different ways.

1 The average velocity of the object is

$$\bar{V} = \frac{V_i + V_f}{2} = \frac{0 + 980 \text{ m/s}}{2} = 490 \text{ m/s};$$

the time of travel is

$$\frac{x}{\bar{V}} = \frac{49,000 \text{ m}}{490 \text{ m/s}} = 100 \text{ s}.$$

Acceleration can be calculated using Eq. (2.7):

$$V_f = V_i + at$$

$$980 \text{ m/s} = 0 + a \times 100 \text{ s}$$

$$a = 9.8 \text{ m/s}^2.$$

2 Using Eq. (2.10) we can first calculate *a:*

$$V_f^2 = V_i^2 + 2ax$$

$$(980 \text{ m/s})^2 = 0 + 2a \, (49{,}000 \text{ m})$$

$$a = 9.8 \text{ m/s}^2.$$

Now we use Eq. (2.7) to obtain the time:

$$V_f = V_i + at$$

$$980 = 0 + 9.8 \, t$$

$$t = 100 \text{ s}.$$

2.5 FREE FALL

The velocity of an object falling freely to the ground increases at a constant rate. The acceleration of a freely falling object is called the acceleration due to gravity. The magnitude of the acceleration due to gravity at any place on the earth is approximately a constant. It has small variations from place to place, depending on its height above or below sea level. However, in simple calculations, these variations can be neglected. The acceleration at sea level is usually taken as the standard. The reason for this variation, along with a general discussion of gravity, is given in Chapter 4.

The acceleration due to gravity is generally taken as equal to 32 ft/s^2 or 9.8 m/s^2. Every object, irrespective of its mass, size, and shape, experiences the same acceleration. Galileo Galilei, in his famous experiments from the tower of Pisa, proved this fact. (When one tries to do such an experiment, proper account should be taken of the frictional and buoyant effects of the air.) The letter g stands for acceleration due to gravity, namely,

$$g = 32 \text{ ft/s}^2 = 9.8 \text{ m/s}^2.$$

Figure 2.5. Velocity, average velocity, and position of a free-falling object in gravitational field.

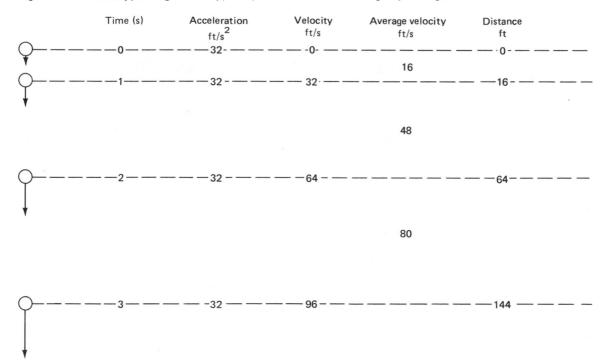

The velocity of a free-falling object increases every second by 32 ft/s (9.8 m/s). The distance it travels in each second also increases as it falls to the ground. As an illustration, let us look at a ball falling to the ground (Fig. 2.5). Its velocity and its position at the end of every second can be calculated by using these equations:

$$V_f = V_i + at,$$

$$y = \overline{V}t,$$

$$y = V_i t + \frac{1}{2} at^2.$$

(Letter y is used for distance instead of x here. It is a common practice to use x for horizontal distance and y for vertical distance).

At the instant the particle is released, i.e. at $t = 0$, its velocity is zero and its position, indicated by the dotted line, is at $y = 0$. The velocity of the ball at the end of the 1st, 2nd, and 3rd second, etc. can be calculated as follows:

$$V_i = 0,$$

$$V_1 = 0 + 32 \text{ ft/s}^2 \times 1 \text{ s} = 32 \text{ ft/s},$$

$$V_2 = 0 + 32 \text{ ft/s}^2 \times 2 \text{ s} = 64 \text{ ft/s},$$

$$V_3 = 0 + 32 \text{ ft/s}^2 \times 3 \text{ s} = 96 \text{ ft/s}.$$

The average velocities in each 1 s time period are

$$\overline{V}(\text{1st second}) = 16 \text{ ft/s},$$

$$\overline{V}(\text{2nd second}) = 48 \text{ ft/s},$$

$$\overline{V}(\text{3rd second}) = 80 \text{ ft/s}.$$

The distances traveled during these 1 s periods are

$$\overline{V}_1 \times t = 16 \text{ ft},$$

$$\overline{V}_2 \times t = 48 \text{ ft},$$

$$\overline{V}_3 \times t = 80 \text{ ft}.$$

By adding these distances, the positions of the particle can be obtained. Direct application of Eq. (2.12) also will give the distances. Verify this. Figure 2.5 shows the velocity, average velocity, and position of the particle with time.

Graphs, showing changes in one quantity as a second quantity varies, are a powerful means of visually demonstrating the functional relationship between two quantities (Figs. 2.3 and 2.4). As an example, Fig. 2.6 shows the variation of the velocity of a free-falling object as a function of time. The graph is a straight line, indicating a linear relationship between V and t.

The graph (Fig. 2.6) can also be used to calculate the acceleration. The velocities at $t = 0.5$ s and $t = 1$ s are 16 ft/s and 32 ft/s, respectively.

$$\text{Acceleration} = \frac{\text{change in velocity}}{\text{time}}$$

$$= \frac{32 \text{ ft/s} - 16 \text{ ft/s}}{0.5 \text{ s}}$$

$$= 32 \text{ ft/s}^2.$$

Figure 2.6. Variation of velocity as a function of time for a constant acceleration case.

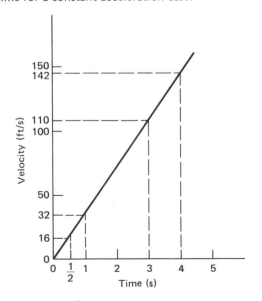

A similar calculation will show that the acceleration due to gravity, during the time period between 3 s and 4 s, is also 32 ft/s^2.

The method we have used above to calculate acceleration is used to determine the slope of the curve. The slope of a curve is defined by

$$\text{Slope} = \frac{\text{change in quantity along the } y \text{ axis} (\Delta y)}{\text{change in quantity along the } x \text{ axis} (\Delta x)}.$$

$$(2.13)$$

Both changes are estimated between the same two points on the curve. The change along the vertical axis is often called the rise and along the horizontal axis, the run. Hence, slope is also defined by

$$\text{Slope} = \frac{\text{rise}}{\text{run}}.$$

Figure 2.7 shows the change in position as a function of time for the free-falling object. The graph representing the relationship between distance and time is not a straight line, indicating that the distance is not proportional to time. In fact, the distance is proportional to the square of the time $(y = \frac{1}{2} gt^2)$, where g is the acceleration due to gravity. A distance versus time curve, similar to the one shown in Fig. 2.7, can be used to calculate average velocity. Two examples are shown in Fig. 2.7. In one case we are calculating the average velocity in time periods between 1 and 2 s and the second case between 4 and 5 s. The change in the position Δx in each case is, respectively, 48 ft and 144 ft. (Δ, delta, stands for changes in the quantity following it.) Δt, change in time, in both cases is 1 s. The average velocity between 1 and 2 s is

$$\frac{\Delta x}{\Delta t} = \frac{48 \text{ ft}}{1 \text{ s}} = 48 \text{ ft/s}.$$

The average velocity between 4 and 5 s is

$$\frac{\Delta x}{\Delta t} = 144 \text{ ft/s}.$$

When the time period is very small, i.e., when it approaches zero, the velocity we calculate by the method above is called the instantaneous velocity. When $\Delta t \to 0$, we are talking about a point on the curve. The slope or the rate of change of position with time in this case can be obtained from the slope of a tangent drawn to the curve at the point under consideration. A tangent (Fig. 2.7) is a line which touches the curve only at one point. The instantaneous velocity at 5.6 s is obtained from the slope of the tangent drawn to the curve at 5.6 s and its value is 182 ft/s (200/1.1).

EXAMPLE 2.6

A ball is kicked vertically up with a velocity of 20 m/s (Fig. 2.8). Assume there is no frictional force acting on the ball. How high will the ball rise? How much time will it take to reach the maximum point? What will be its velocity when it falls back to its original position?

As the ball moves up, acceleration due to gravity is acting to decrease its velocity. Hence, its velocity decreases from 20 m/s to zero. From the point where its velocity has been reduced to zero, the velocity starts to increase in the direction of the acceleration. For the first part of the problem, acceleration is in the opposite direction to the direction of the velocity, and for the second part they are both in the same direction.

For problems like this, it is convenient to assign signs to different quantities as follows:

Distance, velocity, and acceleration are positive if they are directed up.

They are negative if their direction is down.

Let us now proceed to carry out the first part of the problem. We are given $V_i = 20$ m/s, $a = -9.8$ m/s^2. The velocity $V_f = 0$. The average velocity $= (20 \text{ m/s} + 0)/2 = 10$ m/s. The easiest way to

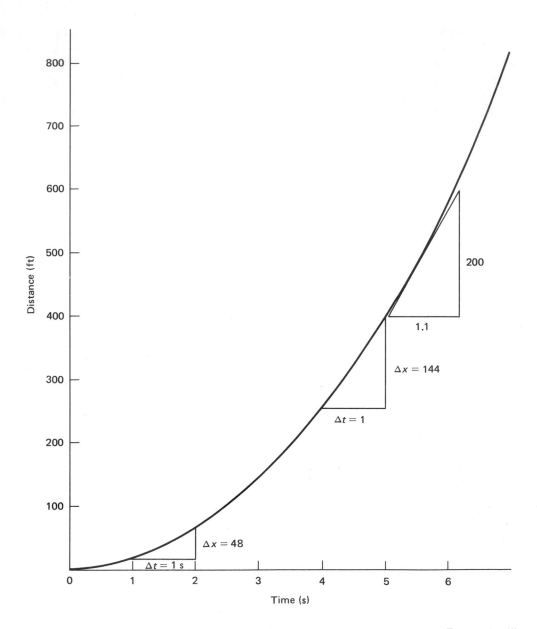

Figure 2.7. Change in position as a function of time for a constant acceleration case. Figure also illustrates the method for calculation of average velocity and instantaneous velocity.

obtain the height h is to first calculate the time using Eq. (2.7):

$$a = (V_f - V_i)/t$$

or

$$V_f = V_i + at.$$

Substituting the known values, we get

$$0 = 20 \text{ m/s} - (9.8 \text{ m/s}^2)t$$

or
$$t = \frac{20 \text{ m/s}}{9.8 \text{ m/s}^2} = 2 \text{ s}.$$

The height h can now be obtained by using the equation

$$h = \overline{V} \times t$$

$$= 10 \text{ m/s} \times 2 \text{ s} = 20 \text{ m}.$$

For the second half of the problem when the ball falls back from its peak position, its velocity increases from 0. Therefore, we have $V_i = 0$, $a = -9.8 \text{ m/s}^2$ and distance traveled $= -20 \text{ m}$. Note that the distance is negative with respect to the starting point. To calculate the final velocity, we use Eq. (2.10):

$$V_f^2 = V_i^2 + 2ay.$$

Figure 2.8. Path of a ball thrown vertically up with a velocity of 20 m/s. The magnitude of its velocity is the same at the same height going up or down. Velocity is zero at the maximum height.

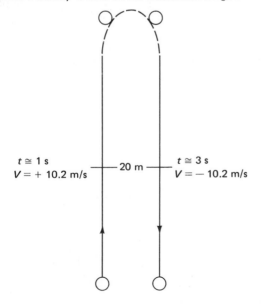

$t \cong 1$ s
$V = +10.2$ m/s

— 20 m —

$t \cong 3$ s
$V = -10.2$ m/s

Substituting the given values, we get

$$V_f^2 = 0 + 2(-9.8 \text{ m/s}^2)(-20 \text{ m})$$

$$V_f = \sqrt{392} = \pm 19.8 \text{ m/s}$$

The answer we get has $+$ and $-$ signs in it. Since we know the velocity is directed down, we choose the correct sign, which is negative

$$V_f = -19.8 \text{ m/s}.$$

We can also show that the time taken for the ball to reach the original position from the peak position is the same as the time taken to reach the peak. As shown in Fig. 2.7, magnitudes of velocities at the same height from the ground, going up and coming down, are the same.

EXAMPLE 2.7

A package is dropped from a hot-air balloon at a height of 1000 ft. Assuming there are no frictional forces acting on it, calculate the time it will take for the package to reach the ground.

We are given $V_i =$ initial velocity in the vertical direction $= 0$; $a = -32 \text{ ft/s}^2$; $y = -1000 \text{ ft}$.

To calculate the time, we use the equation

$$y = V_i t + \frac{1}{2} at^2$$

$$-1000 \text{ ft} = 0 + \frac{1}{2} (-32 \text{ ft/s}^2)t^2.$$

Rearranging the terms, we get

$$t^2 = 1000 \text{ ft} \times \frac{2}{32 \text{ ft/s}^2} = 62.5 \text{ s}^2$$

$$t = 7.9 \text{ s}.$$

2.6 TWO-DIMENSIONAL MOTION

With careful selection of appropriate equations from Table 2.1, problems where motion is confined to one dimension can be solved. However, most of the problems we encounter involve more than one dimension. In this section we discuss two-dimensional problems.

As an example of two-dimensional motion, let us consider a ball rolling off a table (Fig. 2.9). As it leaves the table, its velocity V_x is in the x direction. The path of the ball is represented by the dashed line. As it falls to the ground, it travels a distance X in the horizontal direction in a time t. As soon as the ball leaves the table, the gravitational field starts to accelerate the ball downward. Since the gravitational field is directed downward, the acceleration of the ball is in the vertical direction and there is no horizontal component to the acceleration. The acceleration in the vertical direction changes the vertical component of the velocity, while it does not affect the horizontal velocity. The motion of the ball can be fully solved by treating it as two separate one-dimensional problems with the time t as the only common quantity.

In general, we have to use two sets of equations to solve two-dimensional problems. Table 2.2 lists both sets of equations; subscript x is used for quantities in the x direction and subscript y for those in the y direction.

In the problem of the ball rolling off the table, there is no acceleration in the x direction and hence,

V_x remains constant. We simply have

$$x = V_x t. \tag{2.14}$$

For the y direction, we note that

$$V_{yi} = 0, \quad a_y = -g, \quad y = -h.$$

Using Eq. (2.16), we get

$$-h = 0 \times t \ + \frac{1}{2}(-g)t^2$$

or

$$h = + \frac{1}{2}gt^2. \tag{2.19}$$

Equation (2.19) can be rewritten by substituting for t from Eq. (2.14) in the form

$$h = \frac{1}{2}g\, x^2 / V_x^2. \tag{2.20}$$

If h and x are known, V_x can be calculated; or, if the ball is projected with a known velocity V_x and if h is known, the horizontal distance can be calculated from Eq. (2.20). If more than one quantity needs to be calculated, Eq. (2.19) will have to be used. The components of the velocity as the ball

Figure 2.9. Path of a ball rolling off a table. Horizontal component of the velocity remains constant. Vertical component increases continuously.

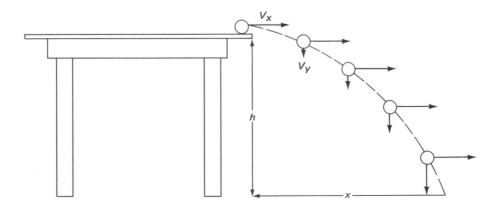

TABLE 2.2 **Equations For Two-dimensional Motion**

x axis	y axis	
$x = \bar{V}_x t$	$y = \bar{V}_y t$	(2.14)
$a_x = \dfrac{V_{xf} - V_{xi}}{t}$	$a_y = \dfrac{V_{yf} - V_{yi}}{t}$	(2.15)
$\bar{V}_x = \dfrac{V_{xf} + V_{xi}}{2}$	$\bar{V}_y = \dfrac{V_{yf} + V_{yi}}{2}$	(2.16)
$x = V_{xi}t + \dfrac{1}{2}a_x t^2$	$y = V_{yi}t + \dfrac{1}{2}a_y t^2$	(2.17)
$V_{xf}^2 = V_{xi}^2 + 2a_x x$	$V_{yf}^2 = V_{yi}^2 + 2a_y y$	(2.18)

touches the ground are V_x and gt. A numerical example is given below.

EXAMPLE 2.8

A ball leaves the top of a table from a height of 4 ft with a horizontal velocity of 20 ft/s. How much time does it take for the ball to reach the ground? How far does it travel in the horizontal direction? What are the x and y components of the velocity as it touches the ground?

To obtain the time we use Eq. (2.17)

$$y = V_{yi}t + \frac{1}{2}a_y t^2.$$

Substituting the given quantities, we have

$$y = -4 \text{ ft},$$

$$V_{yi} = 0,$$

$$a_y = -g = -32 \text{ ft/s}^2.$$

From these values, we get

$$-4 \text{ ft} = 0 \times t + \frac{1}{2}(-32 \text{ ft/s}^2)t^2$$

$$t = \sqrt{\tfrac{1}{4}} = \tfrac{1}{2} \text{ s}.$$

The distance traveled along the horizontal direction is

$$x = V_x t = 20 \text{ ft} \times \frac{1}{2} \text{ s} = 10 \text{ ft}.$$

Components of the velocity of the ball as it touches the ground are

$$V_x = 20 \text{ ft/s},$$

$$V_y = V_{yi} + at$$

$$= 0 + (-32 \text{ ft/s}^2) \times \frac{1}{2} \text{ s}$$

$$= -16 \text{ ft/s}.$$

Another group of problems of interest concerns projectile motion. A football kicked away at an angle is an example of projectile motion. The range R of the ball (distance traveled along the horizontal direction) and the maximum height h are determined by the velocity of the ball, as well as the direction of the velocity vector at the instant of kicking (Fig. 2.10). The x and y components of the initial velocity are $V\cos\theta$ and $V\sin\theta$, respectively. Since there is no acceleration in the horizontal direction, the x component remains constant. However, the y component decreases continuously and is zero at the maximum point of the trajectory. From that point, the y component of the velocity increases. The values of these components, as the ball reaches the ground, are equal to the initial values, but the y component will be in the opposite direction.

First, let us calculate the height h of the maximum point of the trajectory. Equations for the maximum height h and range R are derived below in terms of velocity V and angle θ. The initial value of the y component of velocity $= V\sin\theta$. The value of the y component at maximum height $= 0$. Using the equation for the y direction, namely,

$$V_{yf} = V_{yi} + at,$$

we obtain

$$0 = V\sin\theta + (-g)t, \tag{2.21}$$

$$t = (V\sin\theta)/g.$$

The time we have calculated is the time for half of the flight of the ball. The average y component of the velocity during this half of the flight is given by

$$\overline{V}_y = \frac{V_{yi} + V_{yf}}{2} = \frac{V\sin\theta}{2}.$$

Hence we obtain the height h from

$$h = \overline{V}_y t = \frac{V\sin\theta}{2} \times \frac{V\sin\theta}{g},$$

$$h = \frac{V^2 \sin^2\theta}{2g}. \tag{2.22}$$

The total time of flight is

$$2t = 2(V\sin\theta)/g.$$

Therefore, the range is given by

$$R = V_x \times \frac{2V\sin\theta}{g} = \frac{V\cos\theta \times 2V\sin\theta}{g}$$

$$= \frac{V^2 2\cos\theta \sin\theta}{g}$$

$$R = \frac{V^2 \sin 2\theta}{g}. \tag{2.23}$$

If $\theta = 45°$, $\sin 2\theta = \sin 90° = 1$. Since the sine of the angle is a maximum, it follows that at $45°$, the range R is a maximum (Fig. 2.10).

EXAMPLE 2.9

A motorcycle daredevil tries to cross a gap as shown in Fig. 2.11. His velocity is 60 mi/h. Calculate the maximum width of the gap.

In the problem it is convenient to convert the velocity to ft/s, i.e., 60 mi/h = 88 ft/s. The components of the velocity in the horizontal and vertical directions are as follows:

$$V_x = (88 \text{ ft/s}) \cos 45° = (88 \text{ ft/s}) \times 0.707$$

$$= 62.2 \text{ ft/s},$$

$$V_y = (88 \text{ ft/s}) \sin 45° = (88 \text{ ft/s}) \times 0.707$$

$$= 62.2 \text{ ft/s}.$$

For the maximum gap, the trajectory should be like the trajectory shown in Fig. 2.11 where the motorcycle comes down to the original height as it crosses the gap. That means, in the time it travels a distance x in the horizontal direction, it returns to the same height, that is, $y = 0$.

Figure 2.10. Projectile motion. Trajectory of a ball kicked at an angle θ to the horizontal direction.

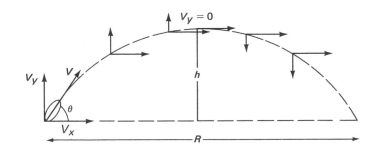

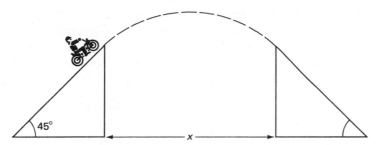

Figure 2.11. Example 2.8.

We are given

$$V_{xi} = 62.2 \text{ ft/s}, \quad V_{yi} = 62.2 \text{ ft/s};$$

$$a_x = 0, \quad a_y = -32 \text{ ft/s}^2;$$

$$x = ? \text{ to be calculated}, \quad y = 0;$$

$$t = t, \text{ to be calculated}$$

Let us first calculate the time by solving the y com-

ponent part of the problem. For this purpose, we use the equation

$$y = V_{yi}t + \frac{1}{2}at^2.$$

Substitution yields

$$0 = (62.2 \text{ ft/s})t + \frac{1}{2}(-32 \text{ ft/s}^2)t^2, \quad t = 3.88 \text{ s}.$$

From $R = x = V_x t$, the gap distance is 62.2 ft/s $\times$ 3.88 s = 241 ft.

SUMMARY

In this chapter we defined quantities such as speed, velocity, and acceleration. For solving practical problems, the following equations are used. Some of these equations are derived from other equations that give definitions.

$$x = \overline{V}_x t, \quad y = \overline{V}_y t \qquad \text{Defines average velocity.}$$

$$\overline{V}_x = \frac{V_{xi} + V_{xf}}{2}, \quad \overline{V}_y = \frac{V_{yi} + V_{yf}}{2} \qquad \text{Suitable for cases where velocity increases at a constant rate or acceleration remains a constant.}$$

$$a_x = \frac{V_{xi} - V_{xf}}{t}, \quad a_y = \frac{V_{yf} - V_{yi}}{t} \qquad \text{Defines acceleration.}$$

$$X = V_{xi}t + \frac{1}{2}a_x t^2, \quad Y = V_{yi}t + \frac{1}{2}a_y t^2$$

$$V_{xf}^2 = V_{xi}^2 + 2a_x x, \quad V_{yf}^2 = V_{yi}^2 + 2a_y y \qquad \text{Mathematically derived from the first three equations.}$$

For one-dimensional problems we use equations without x or y subscripts. When we try to solve two-dimensional problems, remember that only time t is common to both sets of equations, and a_x does not change the velocity in the y direction or vice versa. For projectile problems, a two-dimensional problem, we derived the following equation for range:

$$R = (V^2 \sin 2\theta)/g,$$

where V is the initial velocity at angle θ to the horizontal.

Acceleration due to gravity g affects the velocity of all free-falling objects.

QUESTIONS AND PROBLEMS

2.1 Some baseball pitchers throw balls at 90 mi/h. Convert this speed into ft/s. If the distance between the mound and the hitter is 60 ft, how long does it take the ball to reach the hitter?

2.2 In the Kentucky Derby horse race, the horse Seattle Slew ran the track of $1\frac{1}{8}$ miles in 2 min $2\frac{1}{5}$ s. Calculate the horse's average speed in that race. If the starting gate and the finish line are at the same point, calculate the horse's velocity.

2.3 A driver takes 1½ hr to travel from A to B and 1½ hr from B to C (see Fig. 2.12). Calculate his average speed from A to B, B to C, and A to C. Also calculate his average velocity from A to B, B to C, and A to C.

2.4 Recently the expressway speed limit was changed from 70 mi/h to 55 mi/h. How much more time does it take to travel from Cincinnati to New York (650 mi) because of this change?

2.5 A plane following a straight course at 500 mi/hr maintains a height of 1 mi. A gun on the plane fires a shot. Where will the plane be when an observer just below the plane hears this shot? (Speed of sound is 1100 ft/s).

2.6 Displacement of an object as a function of time is given by the curve in

Figure 2.12. Problem 2.3.

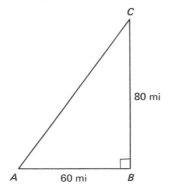

Fig. 2.13. Is the velocity constant from 1 to 5 s? What is the velocity of the object from 5 to 10 s? How will the curve look if the velocity is continuously changing with time?

2.7   A car accelerates from rest to 50 mi/h in 5 s. Assuming the acceleration is constant, calculate its average velocity and the distance it travels in that time.

2.8   A car traveling at 70 mi/h was noticed by a policeman. The police car accelerated from rest at the instant the car passed him. If the police car maintains an acceleration of 20 ft/s^2, when will it catch up with the car? How far do the cars travel in this time?

2.9   A bullet moving horizontally is stopped in a tree. The velocity of the bullet just before it enters the tree is 30 m/s and it stops in 0.12 s. Calculate the acceleration of the bullet in the wood, and determine the distance traveled.

2.10 A driver driving at a speed of 56 km/h notices the traffic light changing to red 50 m from the light. He brakes the car to slow down. Estimate the deceleration of the car (assuming constant) so that the car stops at the light. How long did it take to stop the car?

2.11 A stone is dropped from a height of 96 ft. Calculate the time it will take to reach the ground. Also, obtain its velocity when it reaches the ground.

2.12 A stone is thrown downward from a height of 96 ft. If the stone reaches the ground in 2 s, calculate its initial velocity.

2.13 A ball kicked straight up reaches a height of 50 m. Calculate its initial velocity and the time taken by the ball to reach the maximum height.

2.14 Pile drivers are often used in construction. If the hammer (Fig. 2.14) falls

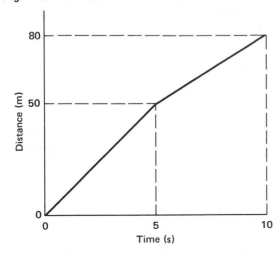

Figure 2.13. Problem 2.6.

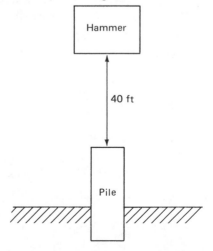

Figure 2.14. Problem 2.14. In a pile driver a heavy hammer is lifted and then dropped on the head of a pile to drive it into the ground.

through a height of 40 ft, what is its velocity as it hits the piling?

2.15 A stone is thrown vertically upward with a velocity 60 ft/s. How high does it go? How much time does it take to reach the maximum height? In what time does it return to its starting point?

2.16 A ball rolls over the top of a table (4 ft from the ground). How far does it fall from the base of the table on the ground if its velocity is 30 ft/s as it leaves the table? How much time does it take to fall to the ground?

2.17 A plane traveling at a speed of 200 mi/h drops a bomb from a height of 1000 ft. If the bomb is to hit a target, at what horizontal distance from the target should it be dropped?

2.18 A shell is fired at an elevation of 30° with a velocity of 500 m/s. Calculate the vertical component of the velocity, the horizontal component of the velocity, the maximum height, and the range.

2.19 A man tries to hit a target 500 ft away (Fig. 2.15). The target is 10 ft above the ground. What should be the velocity of the bullet, and at what angle should it be aimed?

2.20 A football kicked at 45° elevation has a horizontal range of 40 yd. Calculate its velocity. What is its maximum height?

2.21 If the football of Problem 2.20 is kicked at a 35° elevation, calculate its velocity. Repeat the calculation for a 60° elevation.

2.22 A river is 120 m wide. A stone is thrown horizontally. If it is thrown from a height 3 m above the bank and it falls on the opposite bank, calculate the minimum velocity of the stone.

2.23 If the stone in the above problem is thrown at an angle of 30° with the horizontal from a height 3 m from the ground level of the bank, what should be its initial velocity if it falls on the opposite bank?

2.24 A ball thrown by a quarterback at an angle of 35° with the horizontal is caught at a distance of 50 m by the receiver at the same height. What was the ball's velocity as it left the passer's hand? How high above the passer's and receiver's hand did the ball reach?

Figure 2.15. Problem 2.19.

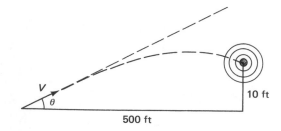

Force

What causes objects to move? Objects in the heavens, like the stars and the planets, are continuously moving without any apparent slowing down. Ancient Greeks thought the gods driving their chariots were responsible for this motion.

In our experience on earth, all moving objects slow down unless precautions are taken to sustain the motion. We will see later that in most of these cases frictional forces are responsible for the slowing down. In order to move an object at rest, force has to be applied to the object. Hence, we could say from our experience that force produces acceleration (increase in velocity) or deceleration (decrease in velocity) and consequent changes in the velocity of the objects.

3.1 NEWTON'S FIRST LAW

Newton summarized all these observations into three laws known as Newton's laws. The first law deals with the general experience that force is needed to change the velocity. Newton's first law states *an object at rest or moving with a constant velocity continues in its state of motion unless acted upon by an unbalanced external force.*

The property of an object to continue in its state of motion is called inertia and hence, the first law is sometimes referred to as the law of inertia. The meaning of the word inertia is similar to the popular conception of the word. A heavier object at rest is harder to move compared to a lighter object; hence, we say the heavy object has a larger inertia.

Newton's first law states that, if the sum of forces acting on the object is zero, then (i) a stationary object remains stationary, and (ii) the velocity of the object remains constant; its velocity does not increase or decrease in magnitude or change its direction.

3.2 NEWTON'S SECOND LAW

Newton's second law is concerned with the change in velocity that results from the application of a force. Newton's second law states, *if a net force is acting on an object, it will accelerate in the direction of the force, and the magnitude of the acceleration is proportional to the magnitude of the force.*

The second law is often written in the following mathematical form:

$$\mathbf{F} = m\,\mathbf{a} \tag{3.1}$$

Force = mass × acceleration.

The m which appears in Eq. (3.1) is the mass of the object. Force and acceleration are vectors, while mass is a scalar.

Equation (3.1) states that (i) acceleration is proportional to applied force, and (ii) acceleration is inversely proportional to the mass; or, for a given force, as the mass of the object increases, the acceleration decreases. The second law is often called the law of acceleration.

Equation (3.1) can now be used to define the units of force in English units and SI units. In the English system, the unit of force is pounds and mass is expressed in slug:

$$F(\text{lb}) = m(\text{slug}) \times a(\text{ft/s}^2).$$

In this equation, if $m = 1$ slug and $a = 1$ ft/s^2, we get $F = 1$ lb, that is, a force of 1 lb acting on a mass of 1 slug produces an acceleration of 1 ft/s^2.

The pound is a unit commonly used to express the weight of an object. Weight of an object is the gravitational force acting on it. Since the acceleration due to gravity is 32 ft/s^2, we have

Weight in pounds = mass in slugs

$$\times \text{ gravitational acceleration in ft/s}^2$$

or

$$w(\text{lb}) = m(\text{slug})\,g(\text{ft/s}^2), \tag{3.2}$$

where w stands for weight.

The mass in slugs can be obtained from the weight in pounds, for example, the mass of an object weighing 64 lb is $(\frac{64}{32}) = 2$ slugs. Since g, the acceleration due to the gravity, varies from place to place, the weight of an object as measured at these points will vary, even though the mass in Newtonian mechanics is a constant. As an illustration, a mass of 1 slug has a weight of 32 lb at a standard point on the earth, and it has a weight averaging 5.12 lb on the moon (Fig. 3.1).

In the SI system, mass is expressed in kilograms and the force in newtons (N):

$$F(\text{N}) = m(\text{kg})a(\text{m/s}^2).$$

Hence, we have

$$1\,\text{N} = 1\,\text{kg} \times 1\,\text{m/s}^2,$$

that is, when a force of 1 N acts on an object having a mass of 1 kg, it accelerates the object by 1 m/s^2.

Weight in SI system is expressed in newtons (N),

$$w(\text{N}) = m(\text{kg})g\,(\text{m/s}^2).$$

A weight of 1 N is equal to 0.25 lb.

EXAMPLE 3.1

An object weighing 20 lb is suspended by a string from a ceiling. Calculate the tension in the string (Fig. 3.2).

Since the object remains stationary, its acceleration is zero. The two forces acting on the object are the tension, acting up, and the gravitational force

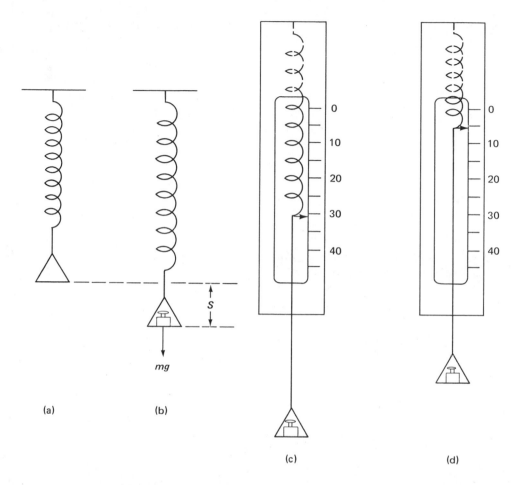

Figure 3.1. A spring balance is often used to determine the weight of an object: (a) spring with no force; (b) spring stretches when force is applied to it. Extension is proportional to the force *mg*. Therefore, a calibrated spring, i.e., a spring for which elongation for unit weight is determined, can be used to obtain the weight of an object. (c) An object of 1 slug weighs 32 lb on the earth. (d) The same object weighs only 5.12 lb on the moon.

mg, acting down. Using Eq. (3.1), we get

$$\underbrace{T - mg}_{\text{net force}} = \underbrace{m \times 0}_{\text{mass} \times \text{acceleration}}$$

or

$T = mg = 20$ lb.

In Fig. 3.2(b), the object is isolated from the system (ceiling and string) and the forces acting on the object are indicated. Such a diagram, called a free-body diagram, is often used to solve problems involving complicated systems.

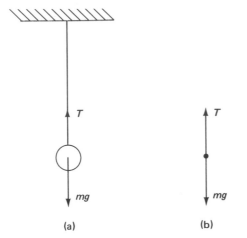

(a)   (b)

Figure 3.2. (a) Example 3.1. (b) Free-body diagram showing the force acting on the object.

EXAMPLE 3.2

A 30 kg mass is suspended by two strings, as shown in Fig. 3.3. Calculate the tensions in the strings.

The forces acting on the mass are the tensions T_1 and T_2 and the gravitational force mg. Since the object is stationary, the vector sum of the forces is zero, that is, the components of vectors in the x and y directions are separately equal to zero. The sum of these components is

$$-T_1 \cos 45° + T_2 \cos 30° = 0, \qquad (3.3)$$

$$T_1 \sin 45° + T_2 \sin 30° - mg = 0. \qquad (3.4)$$

(Note that in the above equations, negative signs are used for components directed along the left and downwards, and positive signs for the components directed to the right and upwards.)

From Eq. (3.3), we get

$$T_1 = T_2 \frac{\cos 30°}{\cos 45°} = 1.2\, T_2. \qquad (3.5)$$

Substituting the value of T_1 in Eq. (3.4) we get

$$1.2\, T_2 \sin 45° + T_2 \sin 30° = mg$$

$$T_2 (1.2 \sin 45° + \sin 30°) = (30\ \text{kg})\,(9.8\ \text{m/s}^2)$$

or

$$T_2(1.35) = 294\ \text{N}$$

$$T_2 = 217\ \text{N}.$$

From Eq. (3.5) we obtain

$$T_1 = 1.2 T_2 = 261\ \text{N}.$$

EXAMPLE 3.3

An object weighing 20 lb is kept on a horizontal surface (Fig. 3.4). A rope is attached to the block, and a force 60 lb is applied to the other end of the

Figure 3.3. (a) Example 3.2. (b) Free-body diagram showing the forces acting on the object.

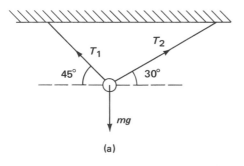

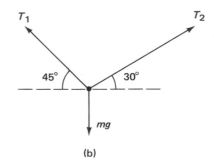

(a)   (b)

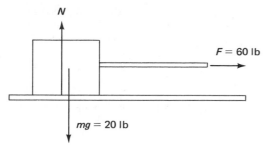

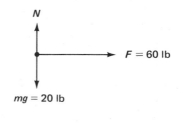

Figure 3.4. Example 3.3.

rope. Calculate the acceleration of the block. Assume no friction between the surfaces.

The rope is used to transmit the force from one point to the other—in this case, from the hand to the block. We usually take the rope to be massless. In such cases we can say the force is transmitted undiminished to the block. The force acting at any point in the rope is called tension, and the tension remains a constant in a continuous massless (never a reality) rope.

Force acting on the block $= F = 60$ lb,

$$\text{Mass of the block} = \frac{\text{weight of the block}}{\text{acceleration due to gravity}}$$

$$= \frac{20 \text{ lb}}{32 \text{ ft/s}^2} = 0.625 \text{ slug},$$

$$\frac{\text{Acceleration of the block}}{\text{in the direction of the force}} = \frac{\text{force}}{\text{mass}}$$

$$= \frac{60 \text{ lb}}{0.625 \text{ slug}}$$

$$= 96 \text{ ft/s}^2.$$

Gravitational force is acting on the block as indicated in Fig. 3.4. However, there is an equal force supplied by the surface in the opposite direction (N). The net force in the vertical direction is, therefore, zero and hence, there is no motion in that direction.

EXAMPLE 3.4

Atwood's Machine

Two masses m_1 and m_2 are tied together by a string which goes around a pulley as shown in Fig. 3.5. Assuming $m_1 > m_2$ describe the motion of the masses. Obtain the acceleration of the masses and tension of $m_1 = 20$ kg and $m_2 = 15$ kg.

We assume that there is no frictional force at the

Figure 3.5. Atwood's machine. Example 3.4.

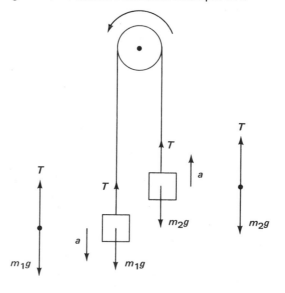

pulley and its mass is zero. Since the string is continuous and assumed massless, the tension acting at
the two masses are equal. Also, since the string does
not stretch, the magnitudes of the accelerations of
the two masses are equal.

Considering the forces acting on m_1 we obtain its
acceleration from Newton's second law $F = ma$

$$\underbrace{T - m_1 g}_{\text{net force}} = \underbrace{-m_1 a}_{\text{mass} \times \text{acceleration}} \qquad (3.6)$$

Note that we are still following the convention that
all quantities directed down are negative. The equation of motion for the second mass is

$$T - m_2 g = m_2 a. \qquad (3.7)$$

Multiplying Eq. (3.6) by -1, we get

$$-T + m_1 g = m_1 a. \qquad (3.8)$$

Adding Eqs. (3.7) and (3.8), we obtain

$$m_1 g - m_2 g = (m_1 + m_2) a$$

$$a = \frac{(m_1 - m_2) g}{m_1 + m_2}. \qquad (3.9)$$

Substituting the given values for m_1 and m_2 in
Eq. (3.8), we get

$$a = \left(\frac{20 \text{ kg} - 15 \text{ kg}}{20 \text{ kg} + 15 \text{ kg}}\right)(9.8 \text{ m/s}^2) = 1.4 \text{ m/s}^2.$$

Tension in the string for this example can be obtained from Eq. (3.7) so that

$$T - (15 \text{ kg})(9.8 \text{ m/s}^2) = (20 \text{ kg})(1.4 \text{ m/s}^2)$$

$$T = 168 \text{ N}.$$

EXAMPLE 3.5

Inclined Plane

An object is sliding down an inclined plane as
shown in Fig. 3.6. Assuming there is no friction,
obtain the acceleration of the block down the
inclined plane.

Angle θ indicated in Fig. 3.6 is called the angle of
the incline or the angle of inclination. The motion
of the block is parallel to the plane, suggesting that
the net force is along the inclined plane. The forces
perpendicular to the plane balance each other.
Therefore, it is natural, in order to solve this problem,

Figure 3.6. Inclined plane. Example 3.5.

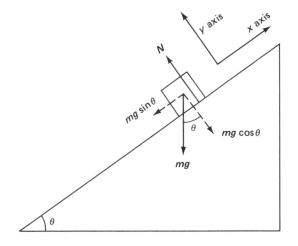

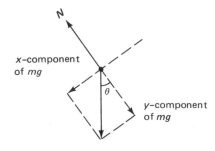

to draw a coordinate system with one axis along the plane and the other axis perpendicular to the plane. The forces acting on the block are the normal force N and the weight of the block mg.

Resolving the forces along the x and y axis, we get

force	x axis	y axis
N	0	N
mg	$-mg\sin\theta$	$-mg\cos\theta$

Since there is no motion along the y axis, the sum of the force components along this axis should add up to zero, that is,

$$N - mg\cos\theta = 0$$

or

$$N = mg\cos\theta. \tag{3.10}$$

Along the x axis, the sum of the components is $mg\sin\theta$. This net force along the x axis produces the acceleration of the block down the plane, given by

$$a = (mg\sin\theta)/m = g\sin\theta. \tag{3.11}$$

The acceleration a is independent of the mass and is a function of the angle θ. If $\theta = 0$, that is, the plane is horizontal, the acceleration is zero as to be expected. For $\theta = 90°$, the plane is perpendicular to the horizontal. In this case, the block falls down like a free-falling object with $a = g$. Equation (3.11) gives us $a = g$ for $\theta = 90°$.

3.3 NEWTON'S THIRD LAW

In the discussion of Example 3.3, we mentioned that the gravitational force acting on a mass placed on a table is balanced by a reaction force from the table. Every time an object exerts a force (action force) on another object, the second object exerts a reaction force in the opposite direction on the first object. In Example 3.3, the block exerts a downward force (action) on the table and, therefore, the table exerts an upward force (reaction) on the block.

Newton's third law states *action and reaction are equal in magnitude and opposite in direction.*

A few examples are shown in Fig. 3.7. Action and reaction forces always occur in pairs. The force on one body has a reaction on a second body.

3.4 FRICTION

We are always made aware of frictional forces in our daily life, sometimes in a dramatic fashion, like the case of a driver who had run out of gasoline and thus became incapable of overcoming frictional forces. In a moving automobile, close to a quarter of the energy generated by the engine is used to overcome frictional forces. However, there are beneficial aspects to frictional forces. Because of these forces, we are able to perform simple and natural acts such as walking, holding an object like a pencil in our hands, sticking a stamp to an envelope, etc.

The forces of friction are classified into two groups: *Static friction* is friction between surfaces of objects in contact, and not moving relative to each other. *Kinetic friction* is friction between surfaces of objects, one of which is moving relative to the other.

By using a simple experimental arrangement shown in Fig. 3.8, we can study frictional forces.

Figure 3.7. Examples of Newton's third law.

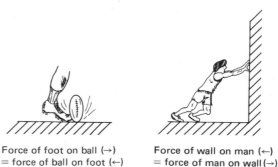

Force of foot on ball (→)
= force of ball on foot (←)

Force of wall on man (←)
= force of man on wall (→)

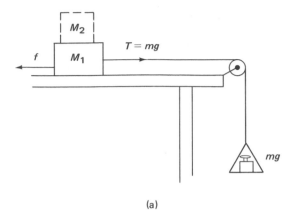

(a)

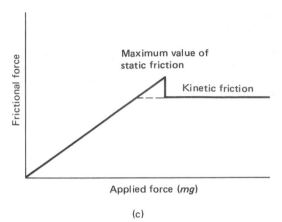

(c)

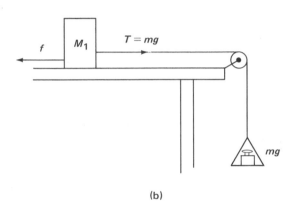

(b)

Figure 3.8. Experimental setup to determine the frictional force acting between the surfaces of the object and the table. In both (a) and (b) frictional force f is acting in the opposite direction to the applied force, in this case tension in the string. Tension is equal to the weight in the pan. (c) Variation of friction as a function of applied force. Friction between two surfaces, when there is relative motion between the two surfaces, is less than the maximum value of the static friction.

The block M_1 is in contact with the table surface. A force of mg, where m is the mass of the object in the pan, acts on the block M_1, and f indicates the frictional force acting in a direction opposite to the applied force. The mass m is slowly increased from zero, but initially we do not observe any motion of the block. This means that the frictional force in these instances equals the increasing applied force. When the applied force increases slightly above a certain value, the block starts to move. This limiting value of the static frictional force, f_s, is the maximum value the static frictional force can attain.

When a force, just slightly bigger than the maximum static force, is applied to the block, the block starts to accelerate. This indicates that the kinetic friction f_k is less than the maximum value of the static friction. The results are plotted in Fig. 3.8(c).

If the above experiment is repeated by placing additional blocks, like M_2, on M_1 we notice that both f_s and f_k increase proportional to the total weight of the blocks, that is, they are proportional to the normal force between the surfaces. Similarly if we keep block M_1, as shown in Fig. 3.8(b), with less contact area, we will observe that the frictional forces are still the same and that they are approximately independent of the area of contact. Also, f_k is approximately independent of the velocity.

Summarizing the above results, we note the following:

1 Kinetic friction is less than the maximum static friction. (The force required to keep an object moving at constant velocity is less than the force required to start it moving.)

2 Both kinetic and static friction are proportional to the normal force acting between the surfaces.

3 Both kinetic and static friction are independent of the area of the contact.

TABLE 3.1 Coefficient[a] of Friction for Selected Surfaces

SURFACES IN CONTACT	KINETIC μ_k	STATIC μ_s
Wood on wood	0.4	0.7
Steel on steel	0.09	0.15
Rubber on concrete (dry)	0.7	0.9
Rubber on concrete (wet)	0.6	0.7

[a] Approximate values.

4 The kinetic friction is independent of the relative velocity of the surfaces.

The proportionality between friction and normal force (statement 2 above) can be expressed in mathematical form as

$$f_s = \mu_s N, \tag{3.12}$$

$$f_k = \mu_k N, \tag{3.13}$$

where N is the normal force, μ_s and μ_k are, respectively, the coefficients of friction for the two cases given by Eqs. (3.12) and (3.13). For a given set of surfaces, kinetic friction is less than the maximum value of static friction, $\mu_k < \mu_s$, and usually both μ_s and μ_k are less than 1 (Table 3.1). The coefficients μ_s and μ_k have no units since they are ratios of two forces. Note that the normal force is the upward reaction force of the surface on which the object rests. As the name indicates, it is always perpendicular to the surface on which the object rests. The normal force is not necessarily equal to the weight of the object, as illustrated in Fig. 3.9.

What causes frictional forces? Surfaces, even the smoothest to our eyes, on magnification appear rough and irregular as shown in Fig. 3.10(a). There are several points of contact where tiny parts of surface A project into surface B and vice versa, as illustrated in Fig. 3.10. These projections hinder the relative motion of the surfaces. From Fig. 3.10(b) one can see that the effects of lubricants are to separate the two surfaces and thus to reduce the friction.

EXAMPLE 3.6

A block weighing 50 lb is kept on a table (Fig. 3.11). The coefficients of friction are $\mu_s = 0.4$ and $\mu_k = 0.33$. Calculate the maximum force one can apply without moving the block, and the maximum force needed to maintain a constant velocity when the block is moving. Repeat the calculation for the case shown in Fig. 3.11(b), where an additional block weighing 20 lb is kept over the first block.

The maximum force that can be applied without moving the block = maximum value of the static friction = $\mu_s N$ = 0.4 × 50 lb = 20 lb. When the object is moving, if the frictional force is equal to the applied force, since they are in opposite directions, there will be no acceleration. The maximum force without acceleration = $\mu_k N$ = 0.33 × 50 lb = 16.5 lb.

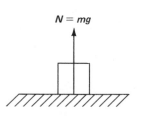

$N = mg$

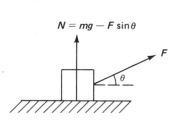

$N = mg - F\sin\theta$

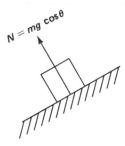

$N = mg\cos\theta$

Figure 3.10 (a) Relative motion between two objects is hindered by microprojections on the surfaces. The frictional force results from this. (b) Lubricant separates the surfaces and thus reduces the frictional force.

In the second case, the normal force has increased. The maximum force without moving the object = 0.4 × 70 lb = 28 lb. The maximum force without acceleration = 0.33 × 70 lb = 23.1 lb.

The coefficient of friction between two surfaces can be determined by the "inclined plane" method. This simple method is illustrated in the following example.

EXAMPLE 3.7

A block weighing 50 lb is kept on an inclined plane and the angle θ is slowly increased (Fig. 3.12). The block starts sliding down for an angle of $40°$. Calculate the coefficient of static friction between the block and the inclined plane.

For an angle θ, the normal force and the component of the gravitational force along the inclined plane are given by

$$N = mg\cos\theta,$$

$$F \text{ (along the plane)} = mg\sin\theta.$$

The frictional force $=\mu_s N = \mu_s mg\cos\theta$. At $\theta = 40°$, F is slightly higher than the frictional force. Hence, we have on the verge of sliding down

$$mg\sin\theta = \mu_s mg\cos\theta$$

or

$$\mu_s = \frac{\sin\theta}{\cos\theta} = \tan\theta. \qquad (3.14)$$

Since $\theta = 40°$,

$$\mu_s = \tan 40° = 0.84.$$

Thus Eq. (3.14) gives us the value of the coefficient of static friction μ_s in experiments using the "inclined plane" method. We should remember that the value of θ is the angle of the inclined plane for which the object is on the verge of sliding down.

EXAMPLE 3.8

In an experimental setup as in Fig. 3.12, it is found that as the block starts moving, it moves at

Figure 3.11. (a) and (b): Example 3.6.

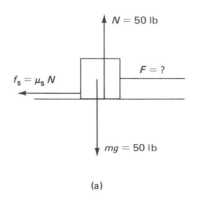

(a)

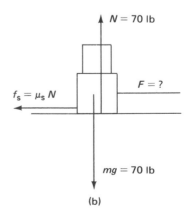

(b)

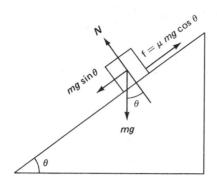

Figure 3.12. Examples 3.7 and 3.8.

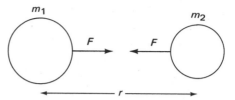

Figure 3.13. Gravitational force acting between two objects is proportional to their masses and inversely proportional to the square of the distance between the masses. Also note that the magnitude of the force acting on m_1 is equal to the magnitude of force acting on m_2.

a constant speed when the angle of the inclined plane is 35°. Calculate the coefficient of kinetic friction.

The block slides down without acceleration, which means that the net force is equal to zero, that is, the component of the gravitational force along the plane is equal to the frictional force acting in the opposite direction. Therefore, we have

$$mg \sin 35° = \mu_k \, mg \cos 35°$$

or

$$\mu_k = \tan 35° = 0.7.$$

3.5 UNIVERSAL GRAVITATIONAL FORCE

We have seen that objects falling freely to the ground accelerate, because of the earth's gravitational force. This is a force that attracts objects to the earth. Newton suggested that this attractive force, experienced on earth, is only a special case of universal attractive forces which exist among all matter. The force which makes the apple fall down and which keeps the earth in an orbit around the sun are demonstrations of this universal force.

The force between any two point objects or spherical objects of masses m_1 and m_2 (Fig. 3.13) is given by the equation

$$F = G \, m_1 m_2 / r^2, \tag{3.15}$$

where r is the distance between the centers of the two objects, and G is the universal gravitational constant. The value of G is given by

$$G = 6.67 \times 10^{-11} \text{ N} \cdot \text{m}^2/\text{kg}^2,$$

$$G = 3.44 \times 10^{-8} \text{ lb} \cdot \text{ft}^2/\text{slug}^2.$$

Equation (3.15), known as Newton's Law of Universal Gravitation, can be stated as follows:

The gravitational attractive force acting between two point objects is proportional to the product of their masses and is inversely proportional to the square of the distance between the objects. The gravitational force is a mutually attractive force, that is, the force on m_1 due to m_2 is exactly equal to the force on m_2 due to m_1, but they are in opposite directions. This relation was extended by Newton, using the integral calculus, to two spherical objects with their masses regarded as concentrated at their respective centers.

Relation Between g and G

One can easily obtain the value of g from Eq. (3.15). Assume m_1 is the earth and m_2 an object near the surface of the earth. The force acting on m_2 is

$$F = G \, m_1 m_2 / r^2.$$

The distance between the centers is given by

$$r = R_E,$$

where R_E is the radius of the earth.

Hence we obtain

$$F = G \, m_1 m_2 / R_E^2 = m_2 g. \qquad (3.16)$$

Therefore, we get

$$g = G \, m_1 / R_E^2. \qquad (3.17)$$

Since G the gravitational constant, m_1 the mass of the earth, and R_E the radius of the earth are constants, g is also a constant.

SUMMARY Newton's laws form the basis of classical mechanics. The first law states *every object continues in its state of motion unless acted upon by an external force.* The third law states *action and reaction are equal and opposite.* The second law states *an object accelerates only when there is a net force acting on it.*

Units associated with $F = m \times a$

$$N = kg \times m/s^2 \ \text{(SI system)}$$

$$lb = slug \times ft/s^2 \ \text{(English system)}.$$

In English units the mass is in slugs, and force is in pounds. When we say that the weight of an object is 10 lb, we mean that the gravitational force of the earth on the object is 10 lb.

Static and kinetic friction are both proportional to the normal force, and independent of the area of contact. Equations expressing this relationship are

$$f_s = \mu_s N \ \text{ and } \ f_k = \mu_k N.$$

The quantities μ_s and μ_k are coefficients of friction.

The gravitational force between two point masses or two spherical masses is

$$F = G \, m_1 m_2 / r^2$$

where G is the universal gravitational constant and r is the distance between centers.

QUESTIONS AND PROBLEMS

3.1 An object is moving with constant velocity. What is the force acting on it?

3.2 A force is applied to a block lying on a table, but it is not moving. Is this against Newton's second law? Explain.

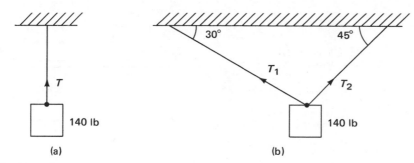

Figure 3.14. Problem 3.4.

3.3 Complete the following table.

Weight of the object	Mass of the object	Force acting on the object	Acceleration
20 lb	?	40 lb	?
?	5 slugs	?	4 ft/s^2
100 lb	?	?	20 ft/s^2
?	5 kg	30 N	?
980 N	?	250 N	?
?	30 kg	?	5 m/s^2
?	?	50 N	6 m/s^2
?	?	360 lb	30 ft/s^2

3.4 An object weighing 140 lb is supported as shown in Fig. 3.14. Calculate the tension in the string for case (a) and the tensions in the two strings for case (b).

3.5 Force is applied on an object kept on a plane surface as shown in Fig. 3.15. Calculate the acceleration in each case. Assume there is no friction between the surfaces.

3.6 An electrically charged particle is in an electric field (Fig. 3.16). It has a mass of 0.1 kg and the electric force acting on it due to the electric field is 1 N. What is the direction of the resultant force (vector sum of the gravitational force and the electric force)? Obtain the acceleration of the particle.

Figure 3.15. Problem 3.5.

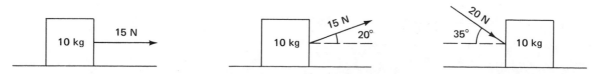

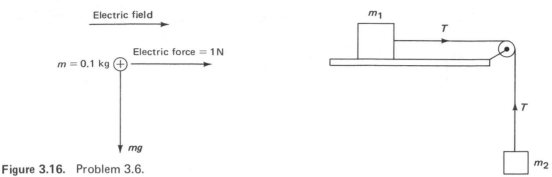

Electric field

Electric force = 1 N

$m = 0.1$ kg $\oplus$

mg

Figure 3.16. Problem 3.6.

Figure 3.17. Problem 3.9.

3.7 A car weighing 2000 lb is traveling at 60 mi/hr. If it is to be stopped in 1000 ft, what is the average force that should be applied? Assume a constant deceleration.

3.8 An 1800 kg car accelerates to 50 km/hr in 10 s. Assuming uniform acceleration, calculate the force acting on the car.

3.9 In Fig. 3.17 $m_1 = 2.5$ kg and $m_2 = 0.5$ kg. Calculate (a) the acceleration of m_1, (b) the acceleration of m_2, and (c) the tension T on the string. Assume no frictional forces.

3.10 In Atwood's machine shown in Fig. 3.18, $m_1 = 2$ kg and $m_2 = 2.2$ kg. Calculate the acceleration of m_1 and m_2 and the tension on the string.

3.11 If the Atwood's machine of Problem 3.10 is taken to the moon, what will be the acceleration of m_1 and m_2? Acceleration due to gravity on the moon is 1.6 m/s^2.

Figure 3.18. Problem 3.10.

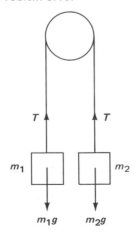

3.12 A block with a mass of 2 kg is sliding down an inclined plane (Fig. 3.12). Angle θ of the inclined plane is 30°. What is its acceleration? Assume no friction between the block and the inclined plane.

3.13 If the coefficient of friction between the surfaces is 0.1, will the block of Problem 3.12 move down the inclined plane? If it does, calculate its acceleration.

3.14 In Fig. 3.17 the weight of m_1 is 30 lb and that of m_2 25 lb. Calculate the acceleration of m_1 and m_2 (a) assuming no friction, and (b) assuming a coefficient of kinetic friction of 0.05 between m_1 and the plane.

3.15 Calculate the frictional force and acceleration of the block for cases shown in Fig. 3.15. The coefficient of kinetic friction is 0.1.

3.16 A box weighing 50 lb is at rest on a level floor. What is the minimum force needed to move it if the coefficient of static friction is 0.3? What is the minimum force needed to keep it moving at a constant velocity if its kinetic friction is 0.2?

3.17 A sled has a 30 kg mass. It is being pulled over a snow-covered flat field. If $\mu_s = 0.3$ and $\mu_k = 0.2$, calculate the force needed to (a) start the sled moving, and (b) keep it moving at a constant velocity.

3.18 A man weighing 200 lb is standing on a scale in an elevator. How much will he weigh when the elevator is (a) moving up with an acceleration of 8 ft/s^2; (b) moving up with a constant velocity of 4 ft/s; (c) moving down with a constant velocity of 4 ft/s?

3.19 Using Eq. (3.17), obtain the value of g on the earth (radius of the earth $= 6.38 \times 10^6$ m; mass of the earth $= 5.98 \times 10^{24}$ kg).

3.20 Using Eq. (3.17), calculate the acceleration due to gravity on the surface of the moon (radius of the moon $= 1.74 \times 10^6$ m; mass of the moon $= 7.35 \times 10^{22}$ kg).

3.21 If the distance between the earth and the moon is 3.84×10^8 m, calculate the gravitational force between the earth and moon.

3.22 Translate the statement "28.3 g of prevention is worth 0.453 kg of cure" by converting the masses from SI units to English units and expressing them in their equivalent weights.

Work, Energy, and Power

In everyday usage, work is associated with exertion. Thus mental as well as physical labor would constitute work. In physics, however, work is associated with force; as we shall see later, we mean by work the application of a force to move an object through a distance. However, confusion exists in our understanding of the differences involved in the words, work, energy, and power. In this chapter we define these terms and discuss their units.

4.1 WORK

Work performed by a force on an object is the product of the component of the force in the direction of the displacement and the magnitude of the displacement.

This definition involves three factors: (a) the magnitude of the force, (b) the direction of the force, and (c) the displacement of the object. Application of force does not necessarily result in work. Work generally results only if the object moves as a result of the application of force [Fig. 4.1(a)]. In

the example shown in Fig. 4.1(a) the work done is zero. In Fig. 4.1(b), the direction of the force is different from that of the motion of the object. In such cases only part of the force, the component along the direction of motion, is effective in moving the object. Hence work is equal to the product of the distance and the component of the force along the direction of motion.

A simple example is illustrated in Fig. 4.2(a). A force F is applied to an object resting on the surface

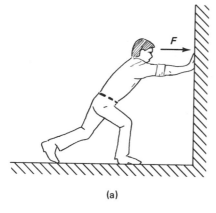

(a)

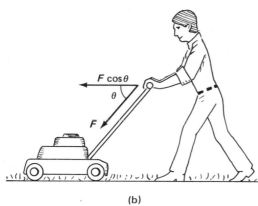

(b)

Figure 4.1. (a) Force does not move the object; hence work done is zero. (b) Work done, as the object moves through a distance d, is equal to the product of the effective force ($F \cos\theta$) and the distance.

of a table. The force is just enough to overcome the frictional force. Therefore, the force produces no increase in velocity. The work done by the force F against friction is given by

$$W = Fd. \qquad (4.1)$$

If the force is not in the direction of the displacement [Figs. 4.2(b) and 4.2(c)], work is given by the equation

$$W = (F\cos\theta)d. \qquad (4.2)$$

Since θ is the angle between the force and the displacement, $F\cos\theta$ is the component of the force along d or it is the effective part of the force that moves the object along d. It is interesting to note that the amount of work done in the above three cases varies because of the variations in the frictional force. (See the numerical examples below.)

Units of Work

In the SI unit system, the unit of work is a joule (J). A joule is the amount of work done by a force of 1 newton when the displacement is 1 meter.

$$W\text{(joules)} = F\text{(newton)} \times d\text{(meter)}. \qquad (4.3)$$

In the English units foot-pound (ft · lb) is used as the unit for work

$$W\text{(foot-pound)} = F\text{(pound)} \times d\text{(foot)}. \qquad (4.4)$$

EXAMPLE 4.1

The mass on the table in Fig. 4.2(a), (b), and (c) is 10 kg. The coefficient of kinetic friction between the surfaces is 0.1. Angle $\theta = 45°$. Calculate the minimum force applied and amount of work done in moving the mass over a distance of 0.5 m in each case.

Case I [Fig. 4.26(a)]

$$f_k = \mu_k\, mg = (0.1)(10 \text{ kg})(9.8 \text{ m/s}^2).$$

Force to be applied $= F = f_k = 9.8$ N.

Work done in moving object $= Fd$

$$= (9.8 \text{ N})(0.5 \text{ m})$$

$$= 4.9 \text{ J}.$$

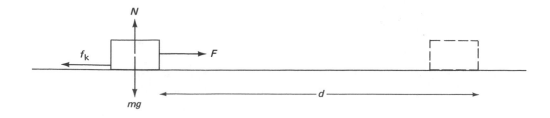

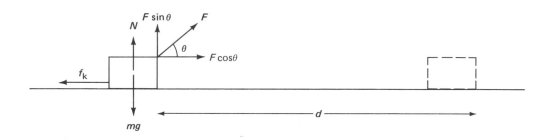

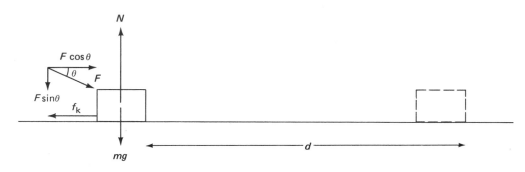

Figure 4.2. (a) The applied force balances the frictional force. Work done is given by $W = Fd$. (b) $W = (F\cos\theta)d$. (c) $W = (F\cos\theta)d$. Note that in (b) and (c) the normal force is not equal to the weight of the object.

Case II [Fig. 4.2(b)]

In the vertical direction, in addition to mg and the normal we have a component of F and they satisfy

$$N + F\sin\theta - mg = 0$$

$$N = mg - F\sin\theta$$

Frictional force $= \mu N = \mu(mg - F\sin\theta)$

$$= 0.1\,(10\text{ kg} \times 9.8\text{ m/s}^2$$

$$- F \times 0.707)$$

$$= 9.8\text{ N} - 0.0707$$

The component of F in the horizontal direction,

$F\cos\theta$ should be at least equal to f_k.

$$F\cos 45° = 9.8 \text{ N} - 0.0707F$$

$$0.707F = 9.8 \text{ N} - 0.0707F$$

$$0.7777F = 9.8 \text{ N}$$

$$F = \frac{9.8 \text{ N}}{0.777} = 12.72 \text{ N}$$

Work done $= (F\cos\theta)\,d$

$$= (12.72 \text{ N})(0.707)(0.05 \text{ m})$$

$$= 4.45 \text{ J.}$$

Case III [Fig. 4.2(c)]

In this case the vertical component of F is in the direction of mg thereby increasing the normal N between the surfaces; we obtain for the normal component N

$$N = mg + F\sin\theta$$

$$= (10 \text{ kg})(9.8 \text{ m/s}^2) + (F)(0.707)$$

$$= 98 \text{ N} + 0.707F$$

Frictional force $= \mu N = 0.1(98 \text{ N} + 0.707F)$

$$= 9.8 \text{ N} + 0.0707F.$$

The component of force required to overcome the frictional force is given by

$$F\cos\theta = 9.8 \text{ N} + 0.0707F$$

$$F(0.707 - 0.0707) = 9.8 \text{ N}$$

$$F = \frac{9.8}{0.6363} \text{ N} = 15.4 \text{ N}$$

Work done $= (F\cos\theta)d$

$$= (15.4\text{N})(0.707)(0.5 \text{ m})$$

$$= 5.44 \text{ J.}$$

Even though the end result, moving the object through a distance of 0.6 m, is the same in all three cases discussed above, the force applied and the work done are not the same. The example also shows that by properly choosing the direction of the force, whenever possible, the amount of work done can be minimized. In the third case more work is done than in the other two cases. Why?

4.2   KINETIC ENERGY

If a force F is the only force acting on an object of mass m, its velocity increases in the direction of the force. The increase in velocity as it travels a distance d is given by the equation (derived on page 23).

$$V_f^2 - V_i^2 = 2ad,$$

where V_i and V_f are the initial and final velocities, and a is the acceleration. Since $a = F/m$,

$$V_f^2 - V_i^2 = 2(F/m)d.$$

Rearranging the above equation, we obtain

$$\frac{1}{2}mV_f^2 - \frac{1}{2}mV_i^2 = Fd. \tag{4.5}$$

The work done Fd is equal to the change in the quantity, $\frac{1}{2}mV^2$. This quantity is called the kinetic energy of the object, that is, energy due to its motion. Kinetic energy, KE, is given by

$$\text{KE} = \frac{1}{2}mV^2. \tag{4.6}$$

According to Eq. (4.6) *an object of mass m and speed V has a kinetic energy ½ mV²*. Also Eq. (4.5) tells us that *the result of doing work on a free particle (a particle on which no force other than the force doing the work is acting) is to increase its kinetic energy.* Units of work and kinetic energy, therefore, are the same.

EXAMPLE 4.2

Calculate (a) the increase in kinetic energy and (b) the final velocity when work of 40 ft · lb is done on an object weighing 20 lb. The initial velocity is zero.

$$\text{Mass of the object} = \frac{\text{weight}}{\text{acceleration due to gravity}}$$

$$= \frac{20 \text{ lb}}{32 \text{ ft/s}^2} = 0.625 \text{ slug}.$$

Work done = Increase in kinetic energy

or

Work done = Final KE − Initial KE

that is,

$$40 \text{ ft} \cdot \text{lb} = \frac{1}{2}(0.625 \text{ slug})(V_f^2) - \frac{1}{2}(0.625 \text{ slug})(V_i^2)$$

Since $V_i = 0$, initial KE = 0. Therefore,

$$\text{Final KE} = \frac{1}{2}(0.625 \text{ slug})V_f^2 = 40 \text{ ft} \cdot \text{lb}$$

$$\text{Final velocity} = V_f = \sqrt{\frac{(40 \text{ ft} \cdot \text{lb})2}{0.625 \text{ slug}}} = 11.3 \text{ ft/s}.$$

4.3 POTENTIAL ENERGY

Consider an object in the earth's gravitational field (Fig. 4.3). To raise this object against the gravitational force mg, a second force at least equal to it in magnitude should be applied. The work done lifting the particle through a height h is mgh. The effect of this work is simply to raise the object, its velocity does not change and no other form of energy is generated, like heat when friction is involved.

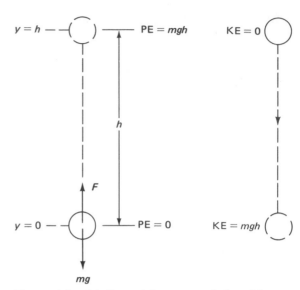

Figure 4.3. (a) Potential energy of the object at height h above the ground level ($y = 0$) is *mgh*. Potential energy is equal to the work done to raise the object. (b) The velocity of the object increases as it falls down from a height h. At ground level ($y = 0$), its kinetic energy equals its original potential energy.

What happened to the work? It is stored in the system containing the particle and the field. The system contains higher energy as the particle height increases. This energy which is a function of the position of the particle is called gravitational potential energy. Gravitational potential energy, *mgh*, in our example, is the work done to raise the object through a height h above the ground. In most problems, the earth's surface is taken as the reference level and potential energy is the amount of work needed to raise an object from this level to the desired height. Hence, *gravitational potential energy of an object at a height h from the ground is given by*

$$PE = mgh. \tag{4.7}$$

Can we recover this stored energy? Yes, the word potential implies it is available. If the object at a height h is released, its potential energy decreases and its kinetic energy increases. When it reaches the ground, all the potential energy is converted into kinetic energy,

$$\frac{1}{2}mV^2 = mgh.$$

Hence we have

$$V^2 = 2gh.$$

Note that this equation can be obtained from Eq. (2.10) by making the following substitutions:

$$V_i = 0, \quad X = -h, \quad \text{and} \quad a = -g.$$

At any point between h and ground, the decrease in potential energy is equal to the increase in kinetic energy, which in mathematical form is

$$-\Delta PE = \Delta KE,$$

$$\Delta PE + \Delta KE = 0. \tag{4.8}$$

From Eq. (4.8) we also obtain that the sum of the PE and KE remains a constant. This is a very restricted form of the principle of conservation of energy. Conservation of energy says that it is not possible to create or destroy energy. Of course, when this principle is applied universally, other forms of energy should be included. The statement that the sum of PE and KE is a constant can be written mathematically as

$$PE + KE = \text{constant}$$

or

$$(PE + KE)_i = (PE + KE)_f, \tag{4.9}$$

where i and f stand for initial and final values.

EXAMPLE 4.3

A crane lifted a car weighing 3500 lb to a height of 20 ft from the ground without acceleration and then released it. How much work is done by the crane to lift the car? What is the potential energy of the car when it was 20 ft high? What is the velocity of the car just before it touched the ground?

Weight of the car $= mg = 3500$ lb;

Mass of the car $= \dfrac{w}{g} = \dfrac{3500 \text{ lb}}{32 \text{ ft/s}^2} = 109.37$ slugs;

Work done in lifting the car $= mgh$

$$= 3500 \text{ lb} \times 20 \text{ ft}$$

$$= 70{,}000 \text{ ft} \cdot \text{lb};$$

Potential energy with respect to ground

$$= \text{work done} = 70{,}000 \text{ ft} \cdot \text{lb}.$$

To calculate the velocity we use Eq. (4.9). The initial state is just before the car is released from the crane, and the final state just before it touches the ground. Therefore, $KE_i = 0$ and $PE_f = 0$.
Hence

$$PE_i = KE_f \text{ (potential energy converts to kinetic energy)}$$

$$70{,}000 \text{ ft} \cdot \text{lb} = \frac{1}{2}(109.37 \text{ slugs})(V_f^2)$$

$$V_f = 35.8 \text{ ft/s}.$$

EXAMPLE 4.4

A mass of 10 kg is placed 1 m high on an inclined plane (Fig. 4.4). If this mass is released to slide down the inclined plane, what will be its velocity when it reaches the bottom? (Assume no friction). We apply a method for solving this problem, different from the method we used earlier (page 43). This method is simpler than the previous method. The potential energy of the mass $= mgh = (10 \text{ kg})(9.9 \text{ m/s}^2)(1 \text{ m}) = 98$ J. This potential energy converts to kinetic energy when it reaches the ground

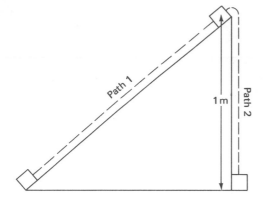

Figure 4.4. As the object slides down the inclined plane, its potential energy decreases and its kinetic energy increases. Kinetic energy of the object at ground level is equal to the potential energy at the top of the inclined plane.

level. Therefore,

$$\frac{1}{2}mV^2 = \frac{1}{2}(1 \text{ kg})(V^2) = 98 \text{ J}$$

$$V = 14 \text{ m/s}.$$

Also note that the velocity of the object at ground level will be the same for the object traversing either path 1 or path 2.

How do we include friction in problems such as the example we discussed immediately above? When work is done against friction, internal energy, a form of energy stored internally, of the system comprising the sliding object and the inclined plane, increases. The increase in internal energy may manifest itself as an increase in temperature of the system. Hence, the mechanical energy (KE + PE) decreases by an amount equal to the work done against friction. Taking this into account, we may rewrite Eq. (4.9) in the following form, namely,

$$(PE + KE)_i = (PE + KE)_f + W_{fr}, \qquad (4.10)$$

where W_{fr} is the work done against friction. The following example illustrates the use of Eq. (4.10).

EXAMPLE 4.5

Calculate the velocity of the object in Fig. 4.5 when it reaches the bottom of the ramp. In this case, assume the coefficient of friction between the surface of the inclined plane and the object is 0.15.

The length of the inclined plane is 15 m, its height 10 m and the mass of the object is 10 kg. Initial KE and final PE are zero. Therefore Eq. (4.10) becomes

$$PE_i = KE_f + W_{fr}.$$

Frictional work $W_{fr} = f_k d = \mu N d = \mu mg(\cos\theta)d$

$$= (0.15)(10 \text{ kg})(9.8 \text{ m/s}^2)$$

$$\times (\frac{11.18}{15})(15 \text{ m})$$

$$= 164 \text{ J}.$$

$$mgh = \frac{1}{2}mV^2 + \text{frictional work}$$

$$(10 \text{ kg})(9.8 \text{ m/s}^2)(10 \text{ m}) = \frac{1}{2}(10 \text{ kg})V^2 + 164 \text{ J}$$

$$V = 12.8 \text{ m/s}.$$

Figure 4.5. Example 4.5. In this case the kinetic energy at the ground level is equal to the potential energy minus the work done against friction along the inclined plane.

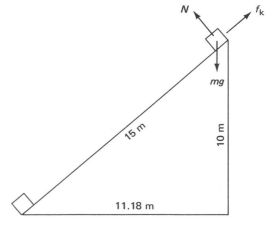

In Section 4.2, we have seen that work increases kinetic energy of an object and, in Section 4.3, we have seen that work increases potential energy. In Section 4.3, we have also seen that energy produces work. Thus work and energy are equivalent.

Energy can, therefore, be considered as the ability to do work.

EXAMPLE 4.6

A force of 20 lb directed vertically up is applied on an object weighing 15 lb. Calculate the velocity of the object if the force is applied through a distance of 10 ft.

The work done increases the KE and the PE of the object.

$$\text{Work done} = (20 \text{ lb})(10 \text{ ft}) = 200 \text{ ft} \cdot \text{lb};$$

$$\text{Increase in PE} = mgh = (15 \text{ lb})(10 \text{ ft}) = 150 \text{ ft} \cdot \text{lb};$$

$$\text{Increase in KE} = W - mgh = 50 \text{ ft} \cdot \text{lb}.$$

Assuming the initial KE = 0, we get

$$\text{KE}_f = \frac{1}{2} m V_f^2 = 50 \text{ ft} \cdot \text{lb}$$

$$\frac{1}{2}\left(\frac{15 \text{ lb}}{32 \text{ ft/s}^2}\right) V^2 = 50 \text{ ft} \cdot \text{lb}.$$

or

$$V = \sqrt{\frac{2(50 \text{ ft} \cdot \text{lb})32 \text{ ft/s}^2}{15 \text{ lb}}} = 14.6 \text{ ft/s}.$$

4.4 POWER AND ITS UNITS

One of the advantages of machines over men and beasts of burden is that they can do work at a faster rate. This superiority of the machines can be measured in terms of a quantity called power. *Power is the rate at which work is done.* This definition is expressed in the equation

$$P = W/t, \tag{4.11}$$

where P stands for power and W is the work done in a time t.

In the English unit system, work done per unit time has the unit of ft $\cdot$ lb/s. However, a commonly used unit is horsepower, which is equivalent to 550 ft $\cdot$ lb/s or 33,000 ft $\cdot$ lb/min. In the SI system, the watt, which is equal to the J/s, is used as the unit of power.

A slightly different but useful equation is derived below for power:

$$P = \frac{W}{t} = \frac{Fd}{t} = F\frac{d}{t} = FV. \tag{4.12}$$

EXAMPLE 4.7

An electric motor is used to pump water from a well 40 ft deep. The amount of water pumped per hour to the ground level is 237,000 gallons. What is the power of the motor assuming there is no power loss?

$$\text{Amount of water pumped} = 237,000 \text{ gal/h}$$

$$= \frac{237,000 \text{ gal/h}}{3600 \text{ s/h}}$$

$$= 65.8 \text{ gal/s};$$

$$\text{Weight of water lifted} = (65.8 \text{ gal/s}) \times (8.34 \text{ lb/gal})$$

$$= 548.8 \text{ lb/s};$$

$$\text{Work done} = (548.8 \text{ lb/s})(40 \text{ ft})$$

$$= 21,952 \text{ ft} \cdot \text{lb/s};$$

$$\text{Power} = \frac{\text{Work done}}{\text{second}}$$

$$= \frac{21,952 \text{ ft} \cdot \text{lb/s}}{550 \text{ (ft} \cdot \text{lb/s)/hp}}$$

$$= 40 \text{ hp}.$$

4.5 EFFICIENCY

In the example we discussed above, it was assumed that the power of the motor is transmitted undiminished to the pump. This is never true. The power available at the pump is always less than the electrical power supplied to the motor. Loss in power occurs mainly in overcoming the friction in the moving parts. In our example the moving parts of the motor and the pump share in the losses.

The efficiency of a machine can be defined as the ratio of its output power to its input power. The efficiency as defined above is always less than unity. Efficiency is always expressed in percent by multiplying the above ratio by 100:

Percent efficiency = $\eta\%$

$$= \frac{\text{power output}}{\text{power input}} \times 100. \quad (4.13)$$

In Eq. (4.13) the symbol η stands for efficiency. Since power is work per second, Eq. (4.13) can also be written in the following form:

$$\eta\% = \frac{\text{Useful work output}}{\text{Work input}} \times 100. \quad (4.14)$$

EXAMPLE 4.8

If the pump in Example 4.6 has only 75% efficiency, calculate the power input.

Useful power output (Ex. 4.6) = 40 hp

Efficiency ($\eta\%$) = 75%

$$\text{Power input} = \frac{\text{Power output}}{\text{Efficiency}}$$

$$= 40 \div \frac{75}{100}$$

$$= 53.3 \text{ hp.}$$

SUMMARY   In this chapter we defined work, energy, and power. These definitions are given by the following equations

$$W = (F\cos\theta)d$$

$$\text{KE} = \frac{1}{2}mV^2,$$

$$P = W/t.$$

We also defined potential energy, which is the energy associated with the position of a particle in a field. In a gravitational field, it is given by

$$\text{PE} = mgh,$$

that is, potential energy is equal to the work done in lifting the mass through a height h. The potential energy at the ground is arbitrarily taken equal to zero.

The sum of potential and kinetic energy is a constant, when no friction is involved and the mechanical energy is not converted to other forms of energy. In the presence of friction, the work done against friction should be included in the conservation of energy equation.

QUESTIONS AND PROBLEMS

4.1 A 10 kg block is on a table (Fig. 4.6). A force of 50 N acts on it. Assuming no friction, calculate the work done by the force if the block moved 25 m.

4.2 Did the work performed in Problem 4.1 appear as (a) potential energy, (b) kinetic energy, or (c) heat?

4.3 If the coefficient of kinetic friction is 0.3 between the block and the table of Problem 4.1, how much of the work done will appear as heat (work done against friction)? In what form does the rest appear?

4.4 A 4000 lb car is traveling at a speed of 45 mi/h. What is its kinetic energy?

4.5 A 250 kg car traveling at 75 km/h is stopped in a distance of 500 m. How much work is done in stopping the car? Assuming a constant frictional force was responsible for stopping the car, calculate the frictional force.

4.6 A beer drum, weighing 200 lb, is on a truck 4 ft above the ground. Calculate (a) its potential energy, (b) its velocity if it drops to the ground, and (c) its velocity if it slides down a 6 ft long frictionless inclined plane.

4.7 A hammer of a pile driver weighs 600 lb and it is 15 ft above the pile. The pile is driven 6 in. deep into the ground by the falling hammer. What is the work done by the hammer? What is the average force driving the pile?

4.8 A volume of water equal to 400 m^3 is pumped into a tank 80 m high every second. (a) Calculate the work done by the pump in 1 h. (b) What is the power of the pump?

4.9 In the waterfalls of Niagara, 200,000 ft^3 of water drop through a height of 160 ft every second. (a) Calculate the kinetic energy of the water falling in 1 s when it reaches the bottom. (b) If the energy of the water can be converted to electricity with 20% efficiency what is the power generated.

4.10 A conveyor belt of length 15 m at a 35° angle is shown in Fig. 4.7. It can hold 4000 kg of ore over its total length and it travels the distance of its length in 30 s. Calculate the power rating of the motor that drives the conveyor belt.

Figure 4.6. Problem 4.1.

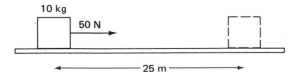

Figure 4.7. Problem 4.10.

4.11 Ready mix concrete weighing 500 lb is lifted to a construction site 50 ft above the ground by an elevator in 10 s. What is the power needed?

4.12 A 20 g bullet is traveling at a speed of 200 m/s. What is its kinetic energy? How much work has to be done to stop it?

4.13 A ball is thrown up with a velocity 20 m/s. What is its velocity when it is 10 m above the ground?

4.14 A pendulum, shown in Fig. 4.8, is swinging without any loss in energy. (a) Calculate its potential energy at points, 1, 2, and 3 (take point 2 as the reference point). (b) Calculate the velocities at points 1, 2, and 3. (c) Is the tension in the string doing any work?

4.15 An object slides down a chute (Fig. 4.9) from a height of 1 m. If the coefficient of friction is 0.2 on the flat plane, calculate how far it will travel along the flat plane. Assume no friction on the inclined plane.

Figure 4.8. Problem 4.14.

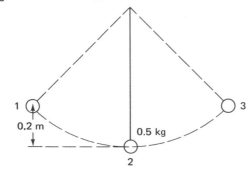

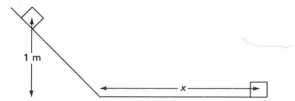

Figure 4.9. Problem 4.15.

4.16 A mechanical jack is used to lift a 300 kg piece of aluminum plate through a distance of 20 cm. (a) What is the minimum force required to lift the aluminum plate. (b) How much work is done in lifting the plate? (c) What is the potential energy of the aluminum after it has been lifted?

4.17 An elevator weighing 100 lb is coming down with a velocity of 10 ft/s. The cable breaks at a height of 100 ft. (a) Assuming there is no friction, what will be its velocity as it reaches the ground? (b) If there is a frictional force of 50 lb between the elevator and its guides, what will be its velocity as it reaches the ground?

4.18 A 2.5 lb hammer head is moving 15 ft/s at the instant it strikes a nail. The nail is driven into the wood ¼ in. What average force is exerted by the hammer?

4.19 A sled of 5 kg mass is pulled by applying a force at an angle of 30° to the horizontal. (a) If the coefficient of kinetic friction is 0.1, what is the minimum force to be applied to move the sled at a constant speed? (b) How much work is done when it is moved through a distance of 100 m?

4.20 An astronaut lifts a 50 kg packet on the moon through a height of 2 m. (a) How much work did he do? (b) What is the potential energy gained by the packet? (c) If the same packet is lifted on earth through the same distance, how much work is to be done?

5

Impulse, Momentum, and Collisions

In the last two chapters we discussed problems in which forces act on an object for long periods of time. In such problems, calculation of acceleration from the force and then the velocity at the end of a given time period can be done with the help of Newton's second law and Eqs. (2.14) to (2.18). However, we often encounter problems where the force acts only for very short periods of time. There are numerous examples of such situations, hitting a baseball with a bat, kicking a football, punching with your hand, etc. How do we calculate velocity changes in such cases? This chapter deals with problems of this kind.

5.1 IMPULSE

Consider a batter hitting a baseball. During the short time of contact, of the order of milliseconds, between the bat and the ball, the bat exerts a force on the ball producing a change in its velocity. To obtain a relationship between these quantities, force, time of interaction, and change in velocity, let us start with Newton's second law, namely,

$$F = ma$$

or

$$F = m \frac{V_f - V_i}{t}.$$

Multiplying both sides by t, we get

$$Ft = mV_f - mV_i. \tag{5.1}$$

We note that the product Ft becomes a useful quantity to calculate velocity changes when large forces act for short periods of time. The force in Eq. (5.1) is not necessarily constant during time t. For example, the force F during a collision between a bat and a ball is variable as shown in Fig. 5.1. During the initial period of the collision, the force increases and during the later period, as the ball separates from the bat, the force decreases. In such cases we replace the variable force by an average force $\overline{F}$ that produces the same velocity change. We

define impulse I as

$$I = \overline{F}t, \tag{5.2}$$

that is, the impulse is the product of the average force and the time during which the force acts.

Since the SI unit of force is the newton and that of time is the second the unit of impulse is newton · second. Similarly in the English system, the unit of impulse is pound · second. We note that kg · m/s and slugs · ft/s are also the units for impulse in the SI and English units, respectively.

EXAMPLE 5.1

A baseball with a velocity of 20 m/s is hit by a bat (Fig. 5.2). After the collision the ball travels in the opposite direction with a velocity of 40 m/s. The contact time is 10^{-3} s and the mass of the ball is 0.15 kg. Calculate the impulse and the average force the ball experienced during collision.

The initial velocity $= -20$ m/s. (Note the $-$ sign since the ball is traveling to the left, following the sign conventions we discussed in previous chapters.) The final velocity $= 40$ m/s.

Using Eq. (5.1) we obtain

$$\text{Impulse} = mV_f - mV_i$$

$$= 0.15 \text{ kg} \times 40 \text{ m/s} - 0.15 \text{ kg} \times (-20 \text{ m/s})$$

$$= 9 \text{ kg} \cdot \text{m/s or N} \cdot \text{s}.$$

Impulse by definition is also given by

$$I = \overline{F}t$$

$$t = 0.001 \text{ s}.$$

Hence we get for the average force

$$\overline{F} = \frac{I}{t} = \frac{9 \text{ N} \cdot \text{s}}{0.001 \text{ s}} = 9000 \text{ N}.$$

Figure 5.1. Variation of force during collision between a bat and a ball. The force acts only for a short period of time during which the two objects are in contact. It is customary to replace the variable force by an average force which produces the same momentum change in calculations.

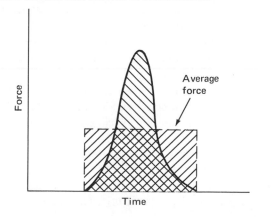

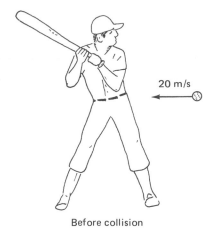

20 m/s

Before collision

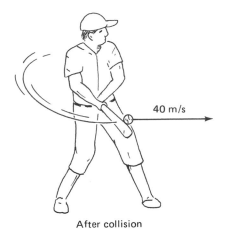

40 m/s

After collision

Figure 5.2. Example 5.1. Change in velocity of the ball after collision.

5.2   MOMENTUM

The product of mass and velocity appearing on the right-hand side of Eq. (5.1) is called momentum. The mathematical form of this definition is

$$P = mV, \tag{5.3}$$

where P stands for momentum. Since momentum is

obtained by multiplying a vector V by a scalar m, the momentum is a vector. Equation (5.1) can be rewritten in the form

$$\bar{F}t = P_f - P_i, \tag{5.4}$$

that is, impulse is equal to change in momentum.

EXAMPLE 5.2

A 0.45 kg soccer ball is kicked to give it a velocity of 30 m/s. Calculate its momentum.

Momentum of the ball $= mV = (0.45 \text{ kg})(30 \text{ m/s})$

$$= 13.5 \text{ kg} \cdot \text{m/s}.$$

Conservation of Momentum

If there is no net force acting on an object, according to Eq. (5.4),

$$P_f - P_i = 0$$

or

$$P_f = P_i,$$

that is, its momentum remains a constant or there is no change in momentum. This principle, known as the conservation of momentum, states *if no net external force is acting on an object, its momentum remains a constant.*

This principle can be extended to a system of particles which experiences no net external force. The presence of internal forces, forces within the system between the particles, does not alter the above statement of conservation of momentum. To illustrate this point let us consider the collision between two balls shown in Fig. 5.3. The velocities of the balls before collision are V_{1i} and V_{2i} and after collision are V_{1f} and V_{2f}. During collision ball m_1 exerts a force F_{12} on m_2, and an equal and opposite force F_{21} acts on m_1. Therefore, we have

Figure 5.3. Collision between two balls. Momentum is conserved during collisions.

from Newton's third law

$$F_{12} = -F_{21}. \qquad (5.5)$$

The change in momentum for both particles is given by

$$F_{21}t = m_1 V_{1f} - m_1 V_{1i}, \qquad (5.6)$$

$$F_{12}t = m_2 V_{2f} - m_2 V_{2i}, \qquad (5.7)$$

where t is the time of interaction. Addition of Eqs. (5.6) and (5.7) gives us the following equation

$$(F_{21} + F_{12})t = m_1 V_{1f} - m_1 V_{1i} + m_2 V_{2f} - m_2 V_{2i}$$

Since $F_{12} = -F_{21}$ the left-hand side is zero.

$$0 = m_1 V_{1f} - m_1 V_{1i} + m_2 V_{2f} - m_2 V_{2i}.$$

Rearranging the terms, we get

$$m_1 V_{1i} + m_2 V_{2i} = m_1 V_{1f} + m_2 V_{2f}. \qquad (5.8)$$

The above equation states that the sum of the momenta of the particles before collision is equal to the sum of momenta after collision. Remember

that there are no external forces acting on the particles, but there are internal forces between the particles during collisions. Note that the net internal force of the system is zero.

5.3 ELASTIC AND INELASTIC COLLISIONS

As we have seen in the previous paragraphs, if no external force is acting, conservation of momentum is always satisfied. The other conservation principle we have seen before is the conservation of energy. More than one type of energy may be involved in collisions one of them always being the kinetic energy. If no other forms of energy, such as potential energy or internal energy, is changed, kinetic energy will be conserved during collisions. Collisions are classified into elastic and inelastic based on whether the kinetic energy remains constant or not. *In elastic collisions, kinetic energy and momentum are conserved. In inelastic collision, only momentum is conserved.*

The following example illustrates the difference between elastic and inelastic collisions.

EXAMPLE 5.3

A 1 kg ball traveling with a velocity of 0.01 m/s hits head-on with another 1.5 kg ball at rest (Fig. 5.4). (a) Calculate the velocities of the balls after collision assuming it is an elastic collision. (b) If we place a small amount of glue on the second ball so that the balls stick to each other, we will have an inelastic collision. Calculate the velocity after such a completely inelastic collision.

In elastic collision, kinetic energy and momentum are conserved. Equations representing these conservation principles are

$$\frac{1}{2}m_1 V_{1i}^2 + \frac{1}{2}m_2 V_{2i}^2 = \frac{1}{2}m_1 V_{1f}^2$$

$$+ \frac{1}{2}m_2 V_{2f}^2, \qquad (5.9)$$

Elastic collision

Inelastic collision

Figure 5.4. Example 5.3.

$$m_1 V_{1i} + m_2 V_{2i} = m_1 V_{1f} + m_2 V_{2f}. \qquad (5.10)$$

The subscripts i stand for before collision and f for after collision. Substituting the given values in the above equation we get

$$\tfrac{1}{2}(1 \text{ kg})(0.01 \text{ m/s})^2 + \tfrac{1}{2}(1.5 \text{ kg})(0 \text{ m/s})^2$$

$$= \tfrac{1}{2}(1 \text{ kg}) V_{1f}^2 + \tfrac{1}{2}(1.5 \text{ kg}) V_{2f}^2, \qquad (5.11)$$

$$(1 \text{ kg})(0.01 \text{ m/s}) + (1.5 \text{ kg})(0)$$

$$= (1 \text{ kg}) V_{1f} + (1.5 \text{ kg}) V_{2f}. \qquad (5.12)$$

Simplifying these equations we obtain

$$0.001 = V_{1f}^2 + 1.5 V_{2f}^2 \qquad (5.13)$$

$$0.01 = V_{1f} + 1.5 V_{2f} \qquad (5.14)$$

From Eq. (5.14) we have

$$V_{1f} = 0.01 - 1.5 V_{2f}. \qquad (5.15)$$

Substituting the value of V_{1f} into Eq. (5.13), we have

$$0.001 = (0.01 - 1.5 V_{2f})^2 + 1.5 V_{2f}^2. \qquad (5.16)$$

Algebraic simplification of Eq. (5.16) yields

$$3.75 V_{2f}^2 = 0.03 V_{2f}$$

or

$$V_{2f} = \frac{0.03}{3.75} = 0.008 \text{ m/s}.$$

Substituting this value of V_2 into Eq. (5.15), we obtain

$$V_{1f} = 0.01 - 1.5 \times 0.008$$

$$= -0.002 \text{ m/s}.$$

The negative value for V_{1f} means that the particle is moving in the negative direction which is opposite to the direction of its velocity before collision (Fig. 5.4).

Now let us consider the second part of the problem where we have a completely inelastic collision. Only momentum is conserved in this case. The equation for conservation of momentum is

$$m_1 V_{1i} + m_2 V_{2i} = (m_1 + m_2)V_f.$$

On the right-hand side of the equation we have only one term since the two masses stick together and move with a velocity V_f. Substituting the given values, we have

$$(1 \text{ kg})(0.01 \text{ m/s}) + (1.5 \text{ kg})(0 \text{ m/s})$$

$$= (1 \text{ kg} + 1.5 \text{ kg})V_f$$

or

$$V_f = \frac{0.01 \text{ kg} \cdot \text{m/s}}{2.5 \text{ kg}} = 0.004 \text{ m/s}.$$

EXAMPLE 5.4

In the inelastic collision of Example 5.3, show that kinetic energy is not conserved. How much kinetic energy is lost?

Kinetic energy before collision

$$= \frac{1}{2}(1 \text{ kg})(0.01 \text{ m/s})^2 + \frac{1}{2}(1.5 \text{ kg})(0)^2$$

$$= 0.00005 \text{ J}.$$

Kinetic energy after collision

$$= \frac{1}{2}(2.5 \text{ kg}) \times (0.004)^2 \text{ m}^2/\text{s}^2$$

$$= 0.00002 \text{ J}.$$

Kinetic energy lost

$$= \text{kinetic energy before} - \text{kinetic energy after}$$

$$= 0.00003 \text{ J}.$$

In this problem more than half of the kinetic energy is lost. The lost kinetic energy is used in spreading and redistributing the glue between the balls. It eventually appears as the internal energy of the system (heat).

5.4 ONE-DIMENSIONAL ELASTIC COLLISIONS

In the above example we discussed the method of obtaining the velocities after collision. Now we extend the discussion to certain useful general ideas. As indicated before, the two conservation principles in a one-dimensional collision are

$$\underbrace{\frac{1}{2}m_1 V_{1i}^2 + \frac{1}{2}m_2 V_{2i}^2}_{\text{kinetic energy before}} = \underbrace{\frac{1}{2}m_1 V_{1f}^2 + \frac{1}{2}m_2 V_{2f}^2}_{\text{kinetic energy after}}$$

$$(5.17)$$

$$\underbrace{m_1 V_{1i} + m_2 V_{2i}}_{\text{total momentum before}} = \underbrace{m_1 V_{1f} + m_2 V_{2f}}_{\text{total momentum after}}$$

$$(5.18)$$

where V_{1i} and V_{2i} are the initial velocities of the masses m_1 and m_2, respectively, and V_{1f} and V_{2f} the velocities after collision.

These equations can be rearranged in the form

$$m_1(V_{1i}^2 - V_{1f}^2) = m_2(V_{2f}^2 - V_{2i}^2), \qquad (5.19)$$

$$m_1(V_{1i} - V_{1f}) = m_2(V_{2f} - V_{2i}). \qquad (5.20)$$

Dividing Eq. (5.19) by Eq. (5.20), we obtain

$$V_{1i} + V_{1f} = V_{2f} + V_{2i}$$

which can be rearranged in the form

$$V_{1i} - V_{2i} = -(V_{1f} - V_{2f}). \qquad (5.21)$$

The term $V_{1i} - V_{2i}$ is the relative velocity of m_1 with respect to m_2 before collision and similarly $V_{1f} - V_{2f}$ is the relative velocity after collision.

Equation (5.21) states that the *relative velocity of one of the masses with respect to the other before collision and after collision are equal in magnitude but opposite in sign in a perfectly elastic one-dimensional collision.*

In a perfectly inelastic collision $V_{1f} - V_{2f} = 0$ since the objects stick to each other (Examples 5.3 and 5.4).

The ratio of relative velocities after and before collision is used as a measure of "elasticity" in a collision. This ratio is called the coefficient of restitution (e):

$$e = \frac{V_{2f} - V_{1f}}{V_{1i} - V_{2i}}. \qquad (5.22)$$

For a perfectly elastic collision, $e = 1$ and for a perfectly inelastic collision, $e = 0$. The value of e lies between 0 and 1 for collisions which do not belong to these two extreme categories. A collision that does not fall in the two limiting groups is the most common type of collision. It is inelastic, but not completely so.

Using Eqs. (5.19), (5.20), and (5.21) we now derive the final velocities of the masses, m_1 and m_2, in terms of their initial velocities. Rearranging Eq. (5.21) we have

$$V_{2f} = V_{1i} + V_{1f} - V_{2i}. \qquad (5.23)$$

Substituting this into Eq. (5.20) and solving for V_{1f} we obtain

$$V_{1f} = (\frac{m_1 - m_2}{m_1 + m_2})V_{1i}$$

$$+ (\frac{2m_2}{m_1 + m_2})V_{2i}. \qquad (5.24)$$

Similarly we get for V_{2f},

$$V_{2f} = (\frac{2m_1}{m_1 + m_2})V_{1i}$$

$$+ (\frac{m_2 - m_1}{m_1 + m_2})V_{2i}. \qquad (5.25)$$

The following are some special cases:

(a) $m_1 = m_2$. In this case Eqs. (5.24) and (5.25) give us

$$V_{1f} = V_{2i} \quad \text{and} \quad V_{2f} = V_{1i}. \qquad (5.26)$$

The two particles exchange their velocities in this case.

(b) $m_1 = m_2$ and $V_{2i} = 0$. In this case we get $V_{1f} = 0$ and $V_{2f} = V_{1i}$, that is, the incoming particle stops and the second particle proceeds with a velocity equal to the original velocity of the incoming particle.

(c) $m_2 \gg m_1$ and $V_{2i} = 0$. In this case Eqs. (5.24) and (5.25) give us

$$V_{1f} \simeq -V_{1i} \quad \text{and} \quad V_{2i} = 0, \qquad (5.27)$$

that is, the incoming light particle reverses its velocity and the massive particle remains at rest. This is exactly what happens when a ball is thrown against a wall.

(d) $m_1 \gg m_2$. This approximation yields

$$V_{1f} \simeq V_{1i} \quad \text{and} \quad V_{2f} \simeq 2V_1 - V_2, \qquad (5.28)$$

that is, the heavy particle continues to travel with very little change in its velocity as is to be expected.

5.5   ELASTIC COLLISIONS IN TWO-DIMENSIONS

In an elastic collision kinetic energy and momentum are conserved. Momentum is a vector quantity and hence in collisions involving more than one dimension the sum of the components of momentum before and after the collision along each axis (x, y, and z axes) remain unchanged. As an example, consider the two-dimensional collision illustrated in Fig. 5.5. This collision is not a head-on collision and the final directions of the particles are not along the x axis but they lie in the x-y plane. The equaton for kinetic energy can be easily written in this

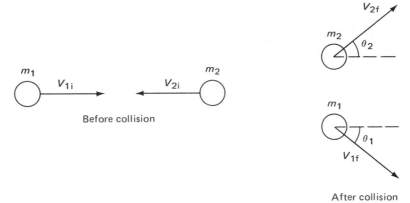

Figure 5.5. Two-dimensional elastic collision.

case (remember kinetic energy is only a scalar) as follows:

$$\frac{1}{2}m_1 V_{1i}^2 + m_2 V_{2i}^2$$

$$= \frac{1}{2}m_1 V_{1f}^2 + \frac{1}{2}m_2 V_{2f}^2. \qquad (5.29)$$

For the conservation of momentum there are two equations. For the x direction,

$$m_1 V_{1i} + m_2 V_{2i}$$

$$= m_1 V_{1f}\cos\theta_1 + m_2 V_{2f}\cos\theta_2. \qquad (5.30)$$

For the y direction

$$0 = m_1 V_{1f}\sin\theta_1 - m_2 V_{2f}\sin\theta_2. \qquad (5.31)$$

These three equations are used to solve problems of collisions in two dimensions as shown in the following example.

EXAMPLE 5.5

Two billiard balls collide with each other as shown in Fig. 5.6. One of them was at rest before collision and the other one had a velocity of 20 ft/s.

After collision one goes at an angle of $30°$. Calculate the velocities of the balls after collision and the angle the second ball makes with the x direction.

Conservation of kinetic energy gives us

$$\frac{1}{2}m(20\text{ ft/s})^2 = \frac{1}{2}mV_{1f}^2 + \frac{1}{2}mV_{2f}^2.$$

Simplifying this equation, we obtain

$$V_{1f}^2 + V_{2f}^2 = 400\text{ ft}^2/\text{s}^2. \qquad (5.32)$$

Momentum conservation gives us

$$m(20\text{ ft/s}) = mV_{1f}\cos30° + mV_{2f}\cos\theta_2$$

$$0 = mV_{1f}\sin30° - mV_{2f}\sin\theta_2.$$

Simplifying these equations, we get

$$20\text{ ft/s} = 0.866V_1 + V_2\cos\theta_2 \qquad (5.33)$$

$$0.5V_{1f} = V_{2f}\sin\theta_2. \qquad (5.34)$$

Rearranging Eqs. (5.33) and (5.34), we obtain

$$V_{2f}\cos\theta_2 = 20\text{ ft/s}, \qquad (5.35)$$

$$V_{2f}\sin\theta_2 = 0.5V_{1f}. \qquad (5.36)$$

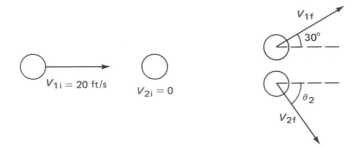

Figure 5.6.   Example 5.5.

Squaring Eqs. (5.35) and (5.36) and adding, and using Eq. (5.32) yields

$$V_{1f}^2 + 400 - 34.64V_{1f}^2 + V_{1f}^2 = 400$$

or

$$V_{1f}(2V_{1f} - 34.64) = 0$$

which yields

$$V_{1f} = 17.52 \text{ ft/s.}$$

Substituting the value of V_{1f} in Eq. (5.32), we get

$$V_{2f} = \sqrt{400 - V_{1f}^2} = 10 \text{ ft/s.}$$

Also we can obtain the value of θ_2 from Eq. (5.35) or (5.36) as

$$\theta_2 = 60°.$$

In collisions between two equal masses the sum of the angles θ_1 and θ_2 is always equal to $90°$ as we have obtained in the above example.

5.6   ENERGY TRANSFER IN COLLISIONS

Collisions are used to slow down or reduce the energy of a moving particle. A well-known example is the slowing down of high energy fission neutrons by letting them collide with atoms of light elements such as hydrogen. It is easy to see that the maximum energy transfer occurs for a head-on collision. Also we have seen that if the masses of the colliding particles are equal, complete energy transfer occurs because the incoming particle stops while the struck particle takes off with the original velocity. On the other extreme, if $m_1 \gg m_2$ negligible energy transfer occurs and the velocity of the incoming particle just reverses [Eq. (5.28)]. Therefore, as the ratio of masses (m_2/m_1) changes from 1 to a large number, the fraction of energy transferred decreases from a maximum to zero. This is the reason why in nuclear reactors light atoms are preferred as moderators (materials to slow down neutrons). Of course, the amount of energy transferred is always less for a collision which is not a head-on collision.

EXAMPLE 5.6

Hydrogen (water) and carbon (graphite) are used as moderators to slow down neutrons in a reactor. What fraction of energy is lost in a neutron–hydrogen and neutron–carbon head-on collision. Assume $m_n \cong m_H$ and $m_C = 12m_n$ and assume the moderator atoms are at rest.

In a neutron–hydrogen collision practically all the energy will be transferred to the hydrogen atom [special Case (b) discussed in Section 5.4]. Therefore, the energy transfer is 100%.

To calculate the energy transfer in a neutron–carbon collision, let us first calculate the velocity of carbon after the collision by using Eq. (5.25), namely,

$$V_{2f} = \left(\frac{2m_1}{m_1 + m_2}\right)V_{1i} + \left(\frac{m_2 - m_1}{m_1 + m_2}\right)V_{21}.$$

We are given

$$m_1 = m_n, \quad m_2 = 12m_n, \quad V_{2i} = 0.$$

Therefore,

$$V_{2f} = \left(\frac{2m_n}{m_n + 13m_n}\right)V_{1i}$$

$$= \frac{2}{13}V_{1i}.$$

Kinetic energy of the neutron before collision

$$= \frac{1}{2}m_n V_{1i}^2.$$

Kinetic energy of carbon atom after collision

$$= \frac{1}{2}(12m_n)\left(\frac{2}{13}V_{1i}\right)^2 = 0.28\left(\frac{1}{2}m_n V_{1i}^2\right).$$

The fraction of kinetic energy transferred is

$$\frac{\text{KE of carbon atom after collision}}{\text{KE of neutron before collision}}$$

$$= \frac{0.28\left(\frac{1}{2}m_n V_{1i}^2\right)}{\frac{1}{2}m_n V_{1i}^2} = 0.28 = 28\%.$$

5.7 SOME OTHER ASPECTS OF CONSERVATION OF MOMENTUM

Consider the two masses with a compressed spring between them (Fig. 5.7). When the masses are released, the smaller mass attains a higher velocity than the heavier mass. How is this related to conservation of momentum? Before the masses are released, the total momentum is zero since they are at rest. Therefore, the final momentum should also be zero,

$$-MV + mv = 0.$$

Therefore, the velocity of the small object is given by

$$v = V M/m.$$

Since $M > m$, $v > V$, that is, the velocity of the ligher object is larger than that of the heavier object. Examples of conservation of momentum can be seen all around us, such as recoil of a gun when a bullet is fired, the spectacular spherical distribution of luminous particles from an exploding firecracker and the thrust on a rocket engine due to the release of propellant.

Let us now consider the rocket engine. When the propellant shoots out, the rocket moves forward to conserve momentum. If the propellant is ejected at a constant rate, a constant thrust in the opposite direction will develop on the rocket. Let us derive a

Figure 5.7 Conservation of momentum: the lighter block travels with a higher velocity than the heavier block. The momenta of the two blocks are numerically equal but in opposite directions.

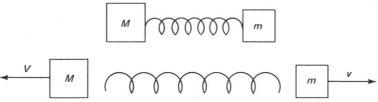

simple equation for this thrust force as a function of the rate of propellant released and its velocity. Let V be the velocity of propellant gases relative to the rocket and m be the mass of gases released in time t. The change in momentum of the gases is mV. An equal amount of momentum change in the opposite direction is experienced by the rocket. The force of thrust acting on the rocket can therefore be calculated using Eq. (5.1):

$$F_{thrust} = mV/t.$$

Chemical reaction provides the internal force to separate the rocket and the propellants just like the compressed spring separates the masses M and m in Fig. 5.7.

EXAMPLE 5.7

A rocket weighing 6,000,000 lb is being launched. The velocity of exhaust gases is 7000 ft/s relative to the rocket. The ejection rate of gases is 1000 slug/s. Calculate the thrust.

$$F_{thrust} = (7000 \text{ ft/s})(\frac{1000}{1} \text{ slug/s})$$

$$= 7,000,000 \text{ lb.}$$

SUMMARY   New definitions in this chapter are

$$\text{Impulse} = Ft$$

$$\text{Momentum } P = mV.$$

Impulse produces change in momentum, i.e.,

$$\overline{F}t = mV_f - mV_i$$

Conservation of momentum is always true. The total momentum of a particle or a system of particles does not change if no external force is acting on it.

In an elastic collision, momentum and kinetic energy are conserved. In a two-particle dimensional collision, we get two equations, namely,

$$\frac{1}{2}m_1 V_{1i}^2 + \frac{1}{2}m_2 V_{2i}^2 = \frac{1}{2}m_1 V_{1f}^2 + \frac{1}{2}m_2 V_{2f}^2$$

$$m_1 V_{1i} + m_2 V_{2i} = m_1 V_{1f} + m_2 V_{2f}.$$

In a two-dimensional collision, there are two momentum equations, one for each direction

$$m_1 V_{1i} + m_2 V_{2i} = m_1 V_{1f}\cos\theta_1 + m_2 V_{2f}\cos\theta_2$$

$$0 = m_1 V_{1f}\sin\theta_1 - m_2 V_{2f}\sin\theta_2$$

In an inelastic collision, only momentum is conserved. Students should also be familiar with other aspects of conservation of momentum and its consequences such as rocket propulsion.

QUESTIONS AND PROBLEMS

5.1   A rotating water spray is shown in Fig. 5.8. Explain how it works.

5.2   Which way is it easier to hit a home run, striking a fast ball or a slow ball?

5.3   In bowling a straight ball hitting the head pin head-on results in a split, while a ball in the pocket most of the time results in a strike. Can you explain this.

5.4   An "executive's" toy is shown in Fig. 5.9. All the masses are equal. If one ball is lifted and released (Fig. 5.9), what do you expect to happen? If one ball from each end is lifted to the same height and released, what will happen?

5.5   In a popular toy a ball is kept floating by water rushing through a tube (Fig. 5.10). Explain how it works.

5.6   A 4000 lb car is traveling at a speed of 60 mi/h. (a) Calculate its momentum. (b) This car strikes a tree and comes to rest in 0.05 s. Calculate the average force acting on the car.

5.7   A golf ball (47 g) at rest is struck to impart to it a velocity of 60 m/s. Assuming the impact time is 1.5 ms, obtain (a) the momentum of the ball, (b) impulse imparted to the ball, and (c) the average force exerted on the ball by the club.

5.8   A steel ball weighing 1000 lb is suspended by a 50 ft long cable. The ball is allowed to hit a wall from an angular position of 45° (Fig. 5.11). If the ball stops in 0.1 s, what is the average force acting on the wall during this time?

5.9   A 2 lb hammer moving at a velocity of 20 ft/s hits a nail. If it is brought to rest in 0.01 s, calculate the average driving force on the nail.

5.10   A uranium nucleus splits into two parts (fission). The mass of one part is 3 times higher than the other. Calculate (a) the ratio of their velocities and (b) the ratio of their kinetic energies.

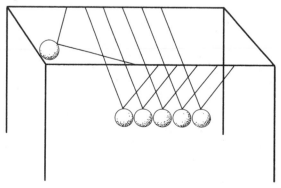

Figure 5.9. Problem 5.4.

Figure 5.8. Problem 5.1.

Figure 5.10. Problem 5.5.

Figure 5.12. Problem 5.11.

5.11 A person sitting in a boat (Fig. 5.12) fires 10 g bullets at the rate of 100 per second. If the man, boat, and gun have a combined mass of 500 kg, calculate the velocity of the boat at the end of the first second. Assume no friction between boat and water. The velocity of the bullets with respect to the boat is 750 m/s.

5.12 A truck weighing 10,000 lb traveling at a speed of 60 mi/h hits a compact car weighing 1000 lb traveling at 40 mi/h from behind. During the collision the two vehicles get attached to each other and move forward. What type of collision is this? What is their velocity after collision?

5.13 A 5 g bullet traveling at a velocity of 1000 m/s hits a 1 kg wooden block on a frictionless table. If the bullet gets embedded in the block, what is their final velocity?

5.14 A ball weighing 1 lb hits another ball weighing 3 lb head-on. The velocity of the first ball is 10 ft/s and that of the second ball is 15 ft/s as shown in Fig. 5.13. Assuming a perfectly elastic collision, calculate the velocity of the balls after collision.

Figure 5.11. Problem 5.8.

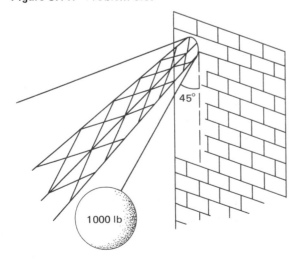

45°

1000 lb

Figure 5.13. Problem 5.14.

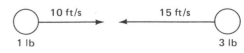

10 ft/s   15 ft/s

1 lb   3 lb

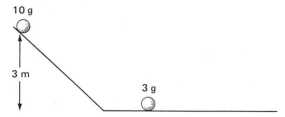

Figure 5.14. Problem 5.15.

5.15 A 10 g ball, released 3 m above the ground is sliding down a frictionless chute (Fig. 5.14). It strikes another ball at rest at the bottom of the incline. The mass of the second ball is 5 g. Calculate (a) the velocity of the first ball as it reaches the bottom of the inclined plane and (b) the velocity of both balls after collision.

5.16 A ball weighing 1 lb traveling with a velocity of 100 ft/s hits a block of 500 lb resting on a frictionless table. Calculate the velocity of the ball after collision. Could you have used the approximation Eq. (5.27)?

5.17 A billiard ball with a velocity of 5 m/s collides with another ball at rest. After collision, one goes at $30°$ and the other goes at $60°$ to the original direction. Assuming the balls are of equal mass, calculate their velocities.

5.18 Prove that the collision of Problem 5.17 is elastic.

5.19 Calculate the velocities V_{1f} and V_{2f} and the angle θ for the collision between two masses (20 kg and 30 kg) shown in Fig. 5.15. The velocity of the 30 kg mass before collision is 10 m/s.

5.20 A rocket weighing 35,000 lb ejects gas at a rate of 10 slug/s. The velocity of the exhaust gas relative to the plane is 4000 ft/s. Obtain the thrust on the rocket.

Figure 5.15. Problem 5.19.

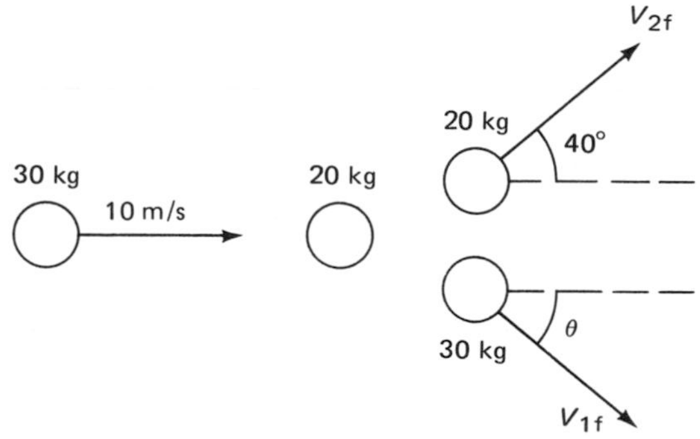

Circular Motion and Rotation of Rigid Bodies

In the previous chapters, we discussed velocity, acceleration, and displacement in translational motion. We have also seen how forces affect these quantities. Energy, work, and power associated with such motion were also previously discussed. In this chapter we define similar quantities associated with circular motion.

6.1 ANGULAR DISPLACEMENT, VELOCITY, AND ACCELERATION

In Fig. 6.1, consider an object at point 1. As it moves along the circle of radius r, its angular position with respect to the radius, 01, changes. As it reaches point 2, its angular displacement is θ.

As previously shown in Chapter 1, the angular displacement θ (in radians) is given by

$$s/r = \theta, \qquad (6.1)$$

where s is the arc length (Fig. 6.1).

Let the time taken by the particle in traveling from point 1 to 2 along the arc s be t. If we divide both sides of Eq. (6.1) by t, we get

$$\frac{1}{r}\frac{s}{t} = \frac{\theta}{t}. \qquad (6.2)$$

In Eq. (6.2), s/t is the speed of the particle. Let

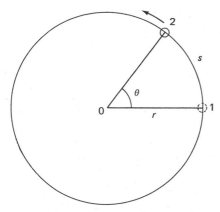

Figure 6.1. Motion of an object in a circular path. Object moves from position 1 to 2 in time *t*. Its angular velocity, $\omega = \theta/t$.

us designate it by V. The quantity on the right-hand side is the angular displacement divided by time and it is called the angular velocity. (Note the similarity with the definition of linear velocity which is linear displacement divided by time.) The greek letter ω (omega) is used to represent angular velocity. Now Eq. (6.2) can be rewritten in the following way:

$$V/r = \omega. \tag{6.3}$$

Our starting point in deriving Eq. (6.3) is Eq. (6.1). In Eq. (6.1) the angle θ is in radians. Therefore, the

unit of angular velocity is radians/second. The quantity ω may also be expressed in other units such as degrees/second or revolutions/second. However, when Eq. (6.3) is used, for example, to calculate V from ω, the angular velocity, if given in other units, must first be converted to units of radians/second.

Let us now restate the definition of angular velocity: *Average angular velocity is defined as angular displacement divided by elapsed time or*

$$\bar\omega = \theta/t. \tag{6.4}$$

If angular velocity changes with time, an angular acceleration α can be defined as the change in angular velocity divided by time, that is,

$$\alpha = (\omega_f - \omega_i)/t, \tag{6.5}$$

where ω_i is the initial angular velocity and ω_f is the final angular velocity. Just as in linear motion we can derive a set of equations among the variables θ, ω, and t. They are shown in Table 6.1 along with the corresponding equations in linear motion.

EXAMPLE 6.1

A record is revolving at an angular velocity of 33 revolutions per minute. Calculate (a) the angular velocity in radians/second and (b) the speed of a

TABLE 6.1 Similar Equations in Linear Motion and Circular Motion

linear displacement, d	angular displacement, θ	
velocity, $\bar V = d/t$	$\bar\omega = \theta/t$	(6.4)
acceleration, $a = (V_f - V_i)t$	$\alpha = (\omega_f - \omega_i)/t$	(6.5)
$d = \bar V t = \frac{1}{2}(V_f + V_i)t$	$\theta = \bar\omega t = \frac{1}{2}(\omega_f + \omega_i)t$	(6.6)
$d = V_i t + \frac{1}{2}at^2$	$\theta = \omega_i t + \frac{1}{2}\alpha t^2$	(6.7)
$V_f^2 = V_i^2 + 2ad$	$\omega_f^2 = \omega_i^2 + 2\alpha\theta$	(6.8)

point on the record 5 cm from the center.

1 rev = 2π rad

Angular velocity = $33 \dfrac{\text{rev}}{\text{min}} = \dfrac{33 \text{ rev} \times 2\pi \dfrac{\text{rad}}{\text{rev}}}{1 \text{ min} \times \dfrac{60 \text{ s}}{\text{min}}}$

$= 3.45$ rad/s.

The speed of a point can be obtained by using Eq. (6.3), namely,

$$V = r\omega,$$

where r is the distance to the point from the center. Hence we get

$$V = (5 \text{ cm})(3.45 \text{ rad/s})$$

$$= 17.25 \text{ cm/s}.$$

EXAMPLE 6.2

A grinding wheel rotating at an angular velocity of 1500 rev/min is turned off. It is found that it comes to rest in 20 min. Calculate (a) the angular deceleration of the wheel and (b) the total angular displacement in this time.

Initial angular velocity, $\omega_i = 1500$ rev/min;

Final angular velocity, $\omega_f = 0$;

Angular acceleration $= \dfrac{\omega_f - \omega_i}{t}$

$= \dfrac{0 - 1500 \text{ rev/min}}{20 \text{ min}}$

$= -75 \dfrac{\text{rev}}{\text{min}^2}$;

Total angular displacement $= \bar{\omega}t = \dfrac{1}{2}(\omega_f + \omega_i)t$

$= \dfrac{1}{2}(0 + 1500 \text{ rev/min})$

$\times 20 \text{ min}$

$= 15{,}000$ rev.

6.2 UNIFORM CIRCULAR MOTION

An object rotating with a constant angular velocity is said to be in a uniform circular motion. According to Newton's first law, unless a force acts on an object, it will continue with a constant speed in the same direction. In a circular motion, however, the direction is continuously changing. Thus there must be a force producing an acceleration which is responsible for this change in direction. Before we prove that the necessary force and the resulting acceleration are directed to the center of the circle, we discuss the instantaneous velocity of the object in circular motion.

Consider the mass rotating in Fig. 6.2(a). The instantaneous velocity of this object is always along the tangent of the circle; and hence is called the tangential velocity. If the string holding the object is broken, one observes that the object flies away tangential to the circle [Fig. 6.2(b)]. The magnitude of the tangential velocity V_T is equal to the constant speed V of the object along the circle. This can be proved by an arrangement shown in Fig. 6.3. The weight Mg produces a tangential force acting on the wheel. Since there is no slippage between the cord and the wheel, the linear distance traveled by the mass downwards in a short time t equals the arc

Figure 6.2. (a) Instantaneous velocity of an object undergoing uniform circular motion is tangential to its path. (b) If the string keeping the object in the circular path is broken, the object will fly away in the direction of the instantaneous velocity.

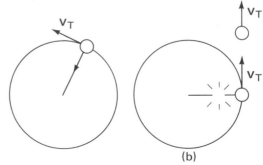
(b)

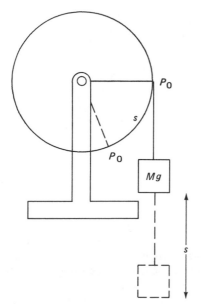

Figure 6.3. As the object on the string travels a distance downwards, the point P_0 moves an equal distance s. The point P_0 also experiences a tangential acceleration equal to the acceleration of the object.

length s of the circumference of the wheel. The point P_0 on the wheel traveled through a distance s in the same time. Therefore, the average speed $V(= s/t)$ is equal to the average tangential velocity $V_T(= s/t)$, so that

$$V_T = V. \tag{6.9}$$

Since we have already proved $V = r\omega$, it follows that

$$V_T = r\omega. \tag{6.10}$$

In Fig. 6.3 the mass M starts from rest. Its velocity is increasing with time. Therefore the point P_0 experiences an acceleration directed downwards along the string, that is, it experiences a tangential acceleration. Note that in addition to this acceleration, there is another acceleration directed to the center acting on P_0 which will be discussed later. If V_i and V_f are the respective velocities at times

t_i and t_f and ω_i and ω_f are the corresponding angular velocities we obtain the tangential acceleration a_T as

$$a_T = \frac{V_f - V_i}{t_f - t_i}.$$

From Eq. (6.10) it follows that

$$a_T = \frac{r\omega_f - r\omega_1}{t_f - t_i} = r\alpha. \tag{6.11}$$

Every point on the rim of the wheel experiences a tangential acceleration of the same amount. The tangential acceleration is present only when the speed of rotation is changing, that is, when α is not equal to zero.

Now we proceed to show that the acceleration responsible for the change in the direction of the object in uniform circular motion ($a_T = 0$ and $\alpha = 0$) is directed towards the center of the circle. To demonstrate this, consider the rotating object in Fig. 6.4. Its instantaneous velocities at t_i and t_f are V_{Ti} and V_{Tf}, respectively. The magnitudes of these velocities are the same, but their directions are not the same. The difference, $V_{Tf} - V_{Ti}$, is obtained by the vector addition method. Vectors V_{Ti} and V_{Tf} are moved parallel to themselves as shown in Fig. 6.4(b). The vector marked $V_{Tf} - V_{Ti}$ is the difference between V_{Tf} and V_{Ti}. Note that the three vectors now satisfy the vector addition, $V_{Ti} + (V_{Tf} - V_{Ti}) = V_{Tf}$. The difference vector $V_{Tf} - V_{Ti}$ is almost directed to the center. As we take the velocities V_{Tf} and V_{Ti} at points closer to each other than the points A and B shown in Fig. 6.4(a), we will see that the direction of $V_{Tf} - V_{Ti}$ is closer to that of a vector directed to the center along the radius. Hence in the limit as t_f approaches t_i, the acceleration $(V_{Tf} - V_{Ti})/(t_f - t_i)$ is directed to the center.

In Fig. 6.4, the angle between OA and OB is equal to the angle between V_{Ti} and V_{Tf} since V_{Ti} is perpendicular to OA and V_{Tf} is perpendicular to OB. Also two sides of triangle AOB are equal ($OA = OB$) and the two corresponding sides of the

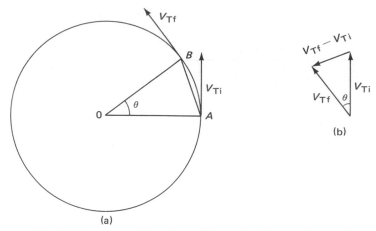

Figure 6.4. The tangential velocity of an object in uniform circular motion continuously changes. The change in velocity ($V_{Tf} - V_{Ti}$) is approximately directed toward the center. The change in velocity will point directly to the center if the two points A and B become coincident. Hence the acceleration of an object in uniform circular motion is directed to the center.

triangle formed by the vectors [Fig. 6.4(b)] are equal since the magnitudes of V_{Ti} and V_{Tf} are the same. Therefore the two triangles are similar. Hence we obtain

$$\frac{AB}{OA} = \frac{V_{Tf} - V_{Ti}}{V_{Tf}}$$

or $\qquad V_{Tf} - V_{Ti} = \frac{AB}{OA} V_{Tf}.$

Dividing both sides by $t_f - t_i$, we get

$$\frac{V_{Tf} - V_{Ti}}{t_f - t_i} = \frac{AB}{OA} \times \frac{V_{Tf}}{t_f - t_i}. \qquad (6.12)$$

The left-hand side by definition is the acceleration, and since $V_{Tf} - V_{Ti}$ is directed to the center of the circle it follows that this acceleration is directed to the center of the circle. If $t_f - t_i$ is very small, $AB \simeq r\theta$. Also $OA = r$, $V_{Tf} = V$ (speed of object) and $\theta/(t_f - t_i) = \omega$. Substituting these in Eq. (6.12) we obtain

$$a_c = \frac{r\theta}{r} \times \frac{V}{t_f - t_i} = \frac{\theta V}{t_f - t_i} = \omega V,$$

where a_c, the acceleration towards the center is called centripetal acceleration. From $V = r\omega$, a_c can also be expressed in the forms

$$a_c = V^2/r \qquad (6.13)$$

and

$$a_c = r\omega^2 \qquad (6.14)$$

The force that produces this acceleration is, therefore, given by

$$F_c = ma_c = mV^2/r = mr\omega^2. \qquad (6.15)$$

We note that the force F_c, like the acceleration, is directed towards the center and, for this reason, it is called the centripetal force. The centripetal force keeps the object in circular motion.

In the discussion above, we obtain two accelerations, tangential acceleration and centripetal acceleration. The tangential acceleration is responsible for an increase or decrease in the magnitude of

TABLE 6.2 Basic Definitions in Circular Motion

QUANTITY	USEFUL RELATIONS IN CIRCULAR MOTION	EQUATION NUMBER
Angular displacement θ	$s = r\theta$	(6.1)
Angular velocity $\omega = \theta/t$	$V = r\omega$	(6.3)
Angular acceleration $\alpha = (\omega_f - \omega_i)/t$	$a_T = r\alpha$	(6.11)
Centripetal acceleration a_c	$a_c = V^2/r$	(6.13)
Centripetal force F_c	$F_c = mV^2/r$	(6.15)

the tangential velocity and hence the angular velocity. The centripetal acceleration is responsible for continuously changing the direction of the object to keep it in the circular path and hence the direction of the tangential velocity. Thus centripetal acceleration is always present if the object is in a circular path, while tangential acceleration is present only if the speed of the object is changing. Table 6.2 lists equations related to the basic definitions in circular motion.

EXAMPLE 6.3

In Fig. 6.2 the mass of the rotating object is 10 g, the radius of the circular path (length of the cord) is 5 cm, and the velocity of the object is 15 cm/s. Calculate (a) the angular velocity of the object, (b) the centripetal acceleration, and (c) the tension of the string.

Angular velocity of the object

$$\omega = \frac{V}{r} = \frac{15 \text{ cm/s}}{5 \text{ cm}} = 3 \text{ rad/s.}$$

Centripetal acceleration $= \dfrac{V^2}{r} = \dfrac{(0.15 \text{ m/s})^2}{0.05 \text{ m}}$

$$= 0.45 \text{ m/s}^2.$$

The tension in the string supplies the centripetal force. Therefore,

$$T = F_c = \frac{mV^2}{r} = (\frac{10}{1000} \text{ kg}) (0.45 \text{ m/s}^2)$$

$$= 0.0045 \text{ N.}$$

EXAMPLE 6.4

A 150 kg earth satellite is in a circular orbit 1000 km above the surface of the earth (Fig. 6.5). Calculate (a) the gravitational force acting on the satellite, (b) the velocity of the satellite in the

Figure 6.5. Example 6.4. Satellite in a circular orbit around the earth.

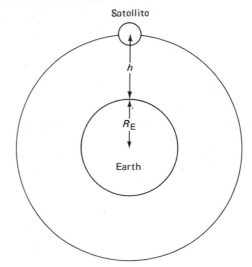

Satellite

h

R_E

Earth

orbit, (c) the centripetal acceleration, and (d) the time taken by the satellite to complete the orbit.

The force between the satellite and the earth can be calculated using the equation

$$F = \frac{GM_E M_S}{(R_E + h)^2},$$ (6.16)

where G is the gravitational constant; M_E, mass of the earth; M_S, mass of the satellite; R_E, radius of the earth; and h the height of the orbit above the earth's surface. The quantity $GM_E/(R_E + h)^2$ is the acceleration due to the earth's gravitational field at a height h. When $h = 0$, this quantity becomes g, the acceleration due to gravity at the surface of the earth, that is,

$$g = GM_E/R_E^2.$$ (6.17)

Hence the acceleration at a height h designated by g_h in terms of g is

$$g_h = \frac{GM_E}{R_E^2} \times \frac{R_E^2}{(R_E + h)^2} = g \times \frac{R_E^2}{(R_E + h)^2}.$$ (6.18)

We are given

$$R_E = 6370 \text{ km},$$

$$h = 1000 \text{ km},$$

$$g = 9.8 \text{ m/s}^2.$$

Substituting these values in Eq. (6.18) we obtain

$$g_h = (9.8 \frac{\text{m}}{\text{s}^2})\frac{(6370 \text{ km})^2}{(7370 \text{ km})^2} = 7.3 \text{ m/s}^2.$$

(a)   Gravitational force acting on the satellite

$$= M_S g_h = 150 \text{ kg} \times 7.3 \text{ m/s}^2 = 1095 \text{ N}.$$

(b) The force that keeps the satellite in its orbit is the gravitational force, that is, the gravitational force is the centripetal force in this case. Thus

$$M_S g_h = \frac{M_S V^2}{r} = \frac{M_S V^2}{R_E + h}$$

or

$$V^2 = g_h(R_E + h).$$

Hence the velocity of the satellite is

$$V = \sqrt{g_h (R_E + h)}$$

$$= \sqrt{(7.3 \text{ m/s}^2)(6,370,000 \text{ m} + 1,000,000 \text{ m})}$$

$$= 7335 \text{ m/s}.$$

(c) As stated earlier, the centripetal force is the gravitational force which we calculated earlier.

$$F_c = M_S g_h = 1095 \text{ N};$$

$$a_c = \frac{F_c}{M_S} = \frac{1095 \text{ N}}{150 \text{ kg}} = 7.3 \text{ m/s}^2 = g_h.$$

(d) The time taken by the satellite is equal to the circumference of the orbit divided by the speed.

$$t = \frac{2\pi(R_E + h)}{V} = \frac{2\pi(7,370,000 \text{ m})}{7335 \text{ m/s}}$$

$$= 6310 \text{ s or } 1.75 \text{ h}.$$

EXAMPLE 6.5

Conical Pendulum

A 5 kg mass is attached to a 2 m long string. The mass rotates in a circular orbit lying in a plane as shown in Fig. 6.6. The radius of the circular orbit is 1 m. Calculate (a) the value of θ, (b) the velocity of the mass, and (c) the tension in the string.

The two forces acting on the mass m are the tension T and its weight mg. Since there is no

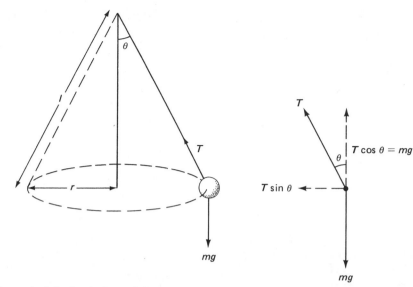

Figure 6.6. Example 6.5. Conical pendulum.

motion perpendicular to the plane of the orbit, the vertical component of T, $T\cos\theta$, should balance mg. Hence we have

$$T \cos \theta = mg \qquad (6.19)$$

The horizontal component $T \sin\theta$, which is always directed to the center of the circle, supplies the necessary centripetal force to maintain the circular motion. Hence

$$T \sin\theta = mV^2/r. \qquad (6.20)$$

Dividing Eq. (6.20) by Eq. (6.19), we obtain

$$\tan\theta = V^2/rg. \qquad (6.21)$$

(a) $\sin\theta = r/l = (1 \text{ m})(2 \text{ m}) = 0.5$ which yields $\theta = 30°$.

(b) Velocity of the mass can be obtained using Eq. (6.21)

$$\tan 30° = \frac{V^2}{1 \text{ m} \times 9.8 \text{ m/s}^2}$$

$$V = \sqrt{0.577 \times 1 \text{ m} \times 9.8 \text{ m/s}^2} = 2.36 \text{ m/s}.$$

(c) The tension can be obtained using Eq. (6.19) or (6.20). Using Eq. (6.19), we obtain

$$T\cos 30° = 5 \text{ kg} \times 9.8 \text{ m/s}^2 = 49 \text{ N}$$

$$T = 49 \text{ N}/0.866 = 56.6 \text{ N}.$$

Two conical pendulums connected together as shown in Fig. 6.7 are used to regulate the speed of rotation of the engines. When used for this purpose, they are called the governors. As the velocity changes the position of the masses A and the cylinder C changes. Change in the position of C is in turn used to control the fuel injection into the engine by connecting a valve (not shown in Fig. 6.7) to cylinder C.

The problem of the conical pendulum and the problem of the car traveling on a banked highway are similar. When a car travels on a curved section of a highway, it is necessary that the centripetal force be supplied to keep the car in a circular path. Friction between the tire and the car is not always enough to keep the car in a circular path. "Banking"

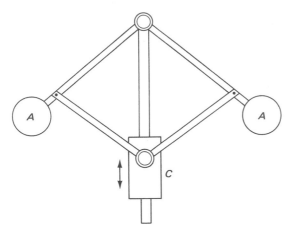

Figure 6.7. A governor: as the speed of rotation changes, the position of masses *A* and the cylinder *C* changes. As the speed increases, *C* moves up; as the speed decreases, *C* moves down. The motion of *C* in turn controls fuel injection into the engine to regulate the speed.

of highways helps in this regard.

Banking raises the outer edge of the highway above the level of the inner edge. The angle of banking (θ in Fig. 6.8) is adjusted for the sharpness of the curve and the speed limit.

Comparing the conical pendulum and the banked highway, one sees that the normal N takes the place of tension T of the pendulum. One component of the normal balances the weight and the other component supplies the centripetal force. Hence we have

$$N\cos\theta = mg, \qquad (6.22)$$

$$N\sin\theta = mV^2/r. \qquad (6.23)$$

Dividing Eq. (6.23) by Eq. (6.22) we get

$$\tan\theta = V^2/rg. \qquad (6.24)$$

Equation (6.24) tells us that the banking angle θ should be higher for higher velocities and for lower values of r, i.e., as the sharpness of the curve increases. If the speed is less than V given by Eq. (6.24), the car will slide down and if it is more than

V, it will slide up along the incline. However, because of frictional forces there is a range of velocities on either side of V for which the car will not slide.

EXAMPLE 6.6

A highway has a curved section for which the radius of curvature is 400 ft. If the highway is designed for a maximum speed of 60 mi/h, obtain the angle of banking. 60 mi/h = 88 ft/s.

Using Eq. (6.21), we obtain

$$\tan\theta = \frac{V^2}{rg} = \frac{(88 \text{ ft/s})^2}{400 \text{ ft} \times 32 \text{ ft/s}^2} = 0.605,$$

that is, $\theta = 31°$.

6.3 TORQUE AND MOMENT OF INERTIA

What is meant by rotation? For an extended rigid body, rotation means that every point in the object has the same angular velocity about an axis of rotation. For example, if the rigid body in Fig. 6.9 is rotating about the axis *aa*, points 1 and 2 have the same angular velocity. However, the linear velocity of point 2 ($r_2\omega$) is higher than that of point 1 ($r_1\omega$) because $r_2 > r_1$.

What causes rotation? Application of force on an object may not always result in the rotation of the object. It is an everyday experience for us that a force F_1 applied parallel to the plane of a door (Fig. 6.10) will not produce rotation. On the other hand, a force F_2 applied perpendicular to the door produces rotation. Also the applied force F_3 equal in magnitude to F_2 but nearer to the axis than F_2, produces a rotational effect smaller than that of F_2.

From the above observations we see that (a) the rotational effect is proportional to the magnitude of the force, and (b) it is proportional to the length of a line drawn from the axis of rotation perpendicular to the line of action of the force. In the example above, since the force F_1 passes through the axis,

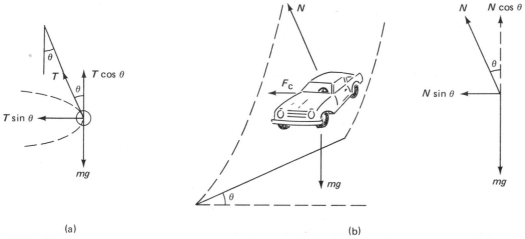

(a)   (b)

Figure 6.8. Forces acting on a car on a banked highway and forces acting on a mass of a conical pendulum are similar [compare (a) and (b)]. The centripetal force which keeps the mass of the conical pendulum in circular path is $T \sin\theta$ and the centripetal force which keeps the car in a circular path on the highway is $N \sin\theta$.

the perpendicular distance is zero. In the other two cases, the perpendicular distance r_3 is smaller than r_2, hence the rotational effect produced by F_3 is smaller than F_2.

We now define a quantity called torque, that measures the rotational effects of a force, by the equation

$$T = r_\perp F, \qquad (6.25)$$

that is, *the torque is the product of the force and length $r_\perp$ of the perpendicular dropped from the axis to the line of action of the force.*

Figure 6.11 shows $r_\perp$ and the line of action of force F. To obtain another form of Eq. (6.25), we again refer to Fig. 6.11. We note that

$$r_\perp = r \sin\theta.$$

Figure 6.9. A rigid body rotates about an axis *aa*. Every point in the rigid body has the same angular velocity, but every point does not necessarily have the same linear velocity since $V = r\omega$.

Figure 6.10. Force F_1, with line of action through the axis, does not rotate the door. Forces F_2 and F_3 perpendicular to the plane of the door produce rotation.

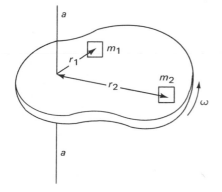

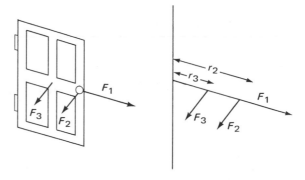

Hence, the torque is given by

$$T = (r\sin\theta)F. \qquad (6.26)$$

From Fig. 6.11 it is also clear that $F\sin\theta$ is the component of force perpendicular to the line joining the axis of rotation and the point of application of force. Therefore, torque is also given by

$$T = rF_\perp. \qquad (6.27)$$

where $F_\perp = F\sin\theta$.

Any one of the equations (6.25), (6.26), and (6.27) may be used to obtain the torque due to the force F about a point O. The torque about a point O means the torque about an axis through O.

Since torque is equal to the product of force and distance, torque has the units of newton-meter and pound-foot in SI and English unit systems, respectively.

The line drawn perpendicular to the force from the axis is called the moment arm. (Note: The torque produced by a force is also called moment of force.)

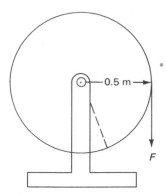

Figure 6.12. Example 6.7.

EXAMPLE 6.7

A force of 2N acts tangentially on a wheel (Fig. 6.12). If the radius of the wheel is 0.5 m, calculate the torque.

In this case the perpendicular distance from the axis to the force is the radius. Torque $= r \times F$ $= 0.5$ m $\times$ 2N $= 1$ N $\cdot$ m.

EXAMPLE 6.8

If the force acting on the wheel of the above example is not tangential as shown in Fig. 6.13, calculate the torque due to the force.

In this case the perpendicular distance is $r_\perp$

$$r_\perp = r\sin 30° = (0.5 \text{ m})(0.5) = 0.25 \text{ m}$$

$$T = r_\perp F = (0.25 \text{ m})(2 \text{ N}) = 0.5 \text{ N} \cdot \text{m}.$$

Force produces a proportional acceleration on an object (Newton's second law). We will now show that a torque produces a proportional angular acceleration in a rotational motion. In Fig. 6.14 a force acts on an object, producing a tangential acceleration a_T given by

$$F = ma_T.$$

Figure 6.11. Torque is equal to the product of force and perpendicular distance from the axis to the force. Also, it is equal to the product of the perpendicular component of the force and distance from the axis to the point of application of the force.

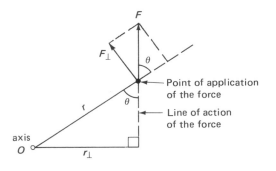

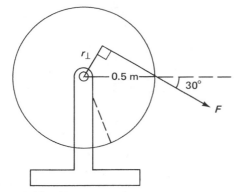

Figure 6.13. Example 6.8.

Since $a_T = r\,\alpha$, we get

$$F = mr\alpha.$$

Multiplication of both sides by r yields

$$rF = mr^2\alpha$$

or $T = (mr^2)\alpha.$ (6.28)

The mass times the square of the distance to the mass, assumed to be a point mass, from the axis is called the moment of inertia, usually represented by

the letter I. Hence we have

$$T = I\alpha \qquad (6.29)$$

which is an equation similar to $F = ma$. The correspondence between T, I, and α in rotational motion and F, m, and a in linear motion is evident. Equation (6.29) tells us the following:

1 Angular acceleration is proportional to torque.
2 Resistance to angular acceleration is provided by the moment of inertia, like mass offers resistance to linear acceleration.

TABLE 6.3 Moments of Inertia of Various Geometrical Bodies Along Indicated Axis. Each Object Has a Total Mass M.

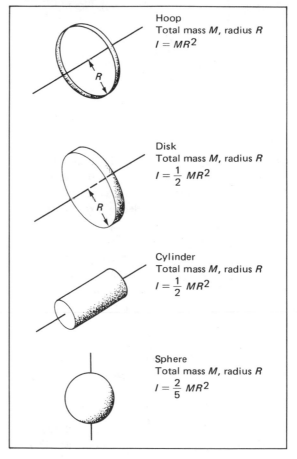

Figure 6.14. Force F produces tangential acceleration of the mass in circular motion.

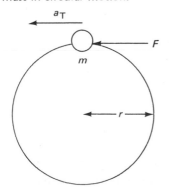

The calculation of moment of inertia I about an axis for an extended object is not that simple. A solid uniform disk is shown in Fig. 6.15. The axis about which the moment of inertia is calculated is perpendicular to the disk and passing through its center. To calculate the moment of inertia about this axis, we first divide the disk into smaller sections of masses $m_1, m_2, m_3, \ldots$, etc. at distances r_1, $r_2, r_3, \ldots$, etc. from the axis. The moment of inertia is then obtained as

$$I = m_1 r_1^2 + m_2 r_2^2 + m_3 r_3^2 + \cdots$$

$$I = \sum_i m_i r_i^2 \qquad (6.30)$$

($\sum\limits_i$ stands for summation of all terms like $m_i r_i^2$.)

In the case of a disk the summation can be reduced to a simple form, namely,

$$I = \tfrac{1}{2} M r^2, \qquad (6.31)$$

where M is the total mass of the disk and r is the radius of the disk. Table 6.3 gives the equations of the moment of inertia of some of the common geometrical solids about selected axes of rotation. Since moment of inertia is $\sum_i m_i r_i^2$. its units are kg $\cdot$ m^2 and slug $\cdot$ ft^2.

It was pointed out earlier that force produces linear acceleration and that torque produces angular acceleration. A comparison of the two equations may be appropriate at this point

$$
\begin{array}{ccccc}
F & = & m & \times & a \\
\downarrow & & \downarrow & & \downarrow \\
\text{causes linear} & & \text{offers resistance} & & \\
\text{acceleration} & & \text{to acceleration} & & \text{acceleration}
\end{array}
$$

$$
\begin{array}{ccccc}
T & = & I & \times & \alpha \\
\downarrow & & \downarrow & & \downarrow \\
\text{causes angular} & & \text{offers resistance to} & & \text{angular} \\
\text{acceleration} & & \text{angular acceleration} & & \text{acceleration}
\end{array}
$$

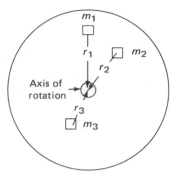

Figure 6.15. The disk is divided into smaller sections to calculate its moment of inertia. Moment of inertia is given by the sum of the products of mass of each elemental section and the square of the distance of that section from the axis.

Also note that m is an independent quantity which depends on how much material there is, while I is a dependent quantity which depends on the distribution of mass and the choice of the axis.

EXAMPLE 6.9

A 15 kg solid disk has a radius of 2 m. Calculate (a) the moment of inertia of the disk and (b) its angular acceleration if a force of 30 N is applied tangentially to the disk (Fig. 6.14).

(a) For a disk, $I = \tfrac{1}{2} M r^2 = \tfrac{1}{2} (15 \text{ kg}) (2 \text{ m})^2 = 30 \text{ kg} \cdot \text{m}^2$

(b) Torque acting is

$$T = rF = (2 \text{ m})(30 \text{ N}) = 60 \text{ N} \cdot \text{m}.$$

Since $T = I\alpha$, we have

$$\alpha = \frac{T}{I} = \frac{60 \text{ N} \cdot \text{m}}{30 \text{ kg} \cdot \text{m}^2} = 2 \text{ rad/s}.$$

EXAMPLE 6.10

Two masses are tied to a string which goes over a heavy solid disk pulley without slipping (Fig.

6.16). If the pulley weighs 40 lb and has a radius of 2 ft, calculate (a) the moment of inertia of the pulley, (b) the acceleration of the masses, (c) the angular acceleration in the pulley, and (d) the tensions in the string on both sides of the pulley.

In Fig. 6.16 tensions on both sides of the pulley are marked T_1 and T_2. The sense of rotation of the pulley produced by torque due to T_1 is in the clockwise direction, while the sense of rotation due to T_2 is in the counterclockwise direction. If the two tensions are equal, the effects of the two torques cancel each other. Therefore, only if T_1 and T_2 are unequal will the net torque produce a rotation. We are assuming that there is no slippage between the string and the pulley. (Note the difference between this problem and Example 3.4 in Chapter 3 where we assumed an ideal massless pulley, which did not require a net torque to rotate it. In the problem the tensions on both sides of the string were the same.)

Figure 6.16. Example 6.10.

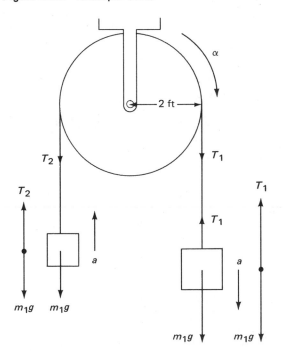

Moment of inertia of the pulley

$$= \frac{1}{2}Mr^2 = \frac{1}{2}(\frac{40}{32} \text{ slug}) (2 \text{ ft})^2$$

$$= 2.5 \text{ slug} \cdot \text{ft}^2;$$

Torque due to $T_1 = 2T_1$ (lb $\cdot$ ft);

Torque due to $T_2 = 2T_2$ (lb $\cdot$ ft).

The net torque in the direction of rotation (rotation is in the clockwise direction since $m_1 > m_2$) = $2T_1$ (lb·ft) $- 2T_2$ (lb·ft). Hence we get

$$2T_1 - 2T_2 = I\alpha = (2.5 \text{ slug} \cdot \text{ft}^2)\alpha. \quad (6.32)$$

Considering the motion of two masses, we get

$$T_1 - 35 \text{ lb} = \frac{35}{32} \text{ slug} \times (-a), \quad (6.33)$$

$$T_2 - 20 \text{ lb} = \frac{20}{32} \text{ slug} \times (a). \quad (6.34)$$

Subtracting Eq. (6.34) from Eq. (6.33) yields

$$T_1 - T_2 = -\frac{55}{32} a + 15. \quad (6.35)$$

Since $\alpha = a/r = a/2$, Eq. (6.32) can be rewritten as

$$T - T_2 = \frac{2.5}{4} a. \quad (6.36)$$

Combining Eqs. (6.35) and (6.36), we obtain

$$-\frac{55}{32} a + 15 = \frac{2.5}{4} a$$

$$a = 6.4 \text{ ft/s}^2.$$

Substituting the value of a into Eqs. (6.33) and (6.34), we obtain T_1 and T_2, namely,

$$T_1 = (\frac{35}{32} \text{ slug})(-6.4 \text{ ft/s}^2) + 35 \text{ lb} = 28 \text{ lb};$$

$$T_2 = (\frac{20}{32} \text{ slug})(6.4 \text{ ft/s}^2) + 20 \text{ lb} = 24 \text{ lb}.$$

In the above example we calculated the net torque by taking the difference between the torques that produce rotations in the clockwise and counterclockwise directions. However, it is customary to take torques that produce clockwise rotation as negative and torques that produce counterclockwise direction as positive.

6.4   KINETIC ENERGY, WORK, AND POWER IN ROTATIONAL MOTION

An extended object rotating at an angular velocity of ω is shown in Fig. 6.17. Every point in the object has the same angular velocity, but their linear speeds are different. Hence to calculate the kinetic energy we divide the object into a large number of masses m_1, m_2, ..., etc. at distance r_1, r_2, ..., etc. from the axis. The total kinetic energy is therefore given by

$$KE = \frac{1}{2}m_1 V_1^2 + \frac{1}{2}m_2 V_2^2 + \frac{1}{2}m_3 V_3^2 + \cdots .$$

Since $V_1 = r_1\omega$, $V_2 = r_2\omega$, and $V_3 = r_3\omega$ we have

$$KE = \frac{1}{2}m_1 r_1^2 \omega^2 + \frac{1}{2}m_2 r_2^2 \omega^2 + \frac{1}{2}m_3 r_3^2 \omega^2$$

$$= \frac{1}{2}(m_1 r_1^2 + m_2 r_2^2 + \cdots)\omega^2.$$

Figure 6.17. Every point in the rigid body rotates with the same angular velocity. Total kinetic energy $= \frac{1}{2}(\Sigma m_i r_i^2 \omega^2) = \frac{1}{2}I\omega^2$.

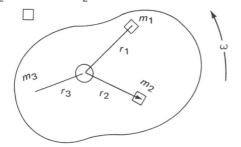

The quantity in the parentheses is the moment of inertia of the object. Therefore, kinetic energy of a rotating object is given by the equation

$$KE = \frac{1}{2}I\omega^2. \tag{6.37}$$

It is instructional at this point to verify that the units are balanced on both sides of Eq. (6.37):

$$KE = \frac{1}{2}I\omega^2$$

Units of $KE = $ joules $= $ (newton)(meter)

$$= (kilogram)(\frac{meter}{second^2})(meter)$$

Units of $I\omega^2 = (kg \cdot m^2)(\frac{1}{s})^2 = \frac{kg \cdot m^2}{s^2}$

$$= \frac{kg \cdot m^2}{s^2}$$

Thus both sides have the same units $kg \cdot m^2/s^2$. The reader is urged to do the same using units in the English system.

EXAMPLE 6.11

A flywheel, assumed to be a uniform disk weighing 500 lb and having a diameter of 4 ft, is rotating about an axis through its center. If its angular velocity is 1800 rev/min, calculate its kinetic energy.

Moment of inertia of the flywheel

$$= \frac{1}{2}Mr^2$$

$$= \frac{1}{2}(\frac{500}{32} slug) \times (2 ft)^2$$

$$= 31.25 \ slug \cdot ft^2;$$

Angular velocity = 1800 rev/min

$$= \frac{1800 \text{ rev/min}}{60 \text{ s/min}} \times \frac{2\pi \text{ rad}}{\text{rev}} = 188.57 \frac{\text{rad}}{\text{s}};$$

Kinetic energy $= \frac{1}{2} I \omega^2$

$$= \frac{1}{2}(31.25 \text{ slug} \cdot \text{ft}^2)(188.57 \text{ rad/s})^2$$

$$= 5.5 \times 10^5 \text{ ft} \cdot \text{lb}.$$

In this chapter so far we have limited our discussion to problems where an object is rotating about a fixed axis. However, in many problems, the objects possess both translational motion and rotational motion. Consider the wheel rolling, without slipping, as illustrated in Fig. 6.18. Originally the point P_0 is in contact with the surface and as the wheel rolls the center of the wheel moves along a straight line and the point P_0 rotates about the center of the wheel. As the point P_0 completes one revolution, the center travels a distance of $2\pi r$. In

Figure 6.18. When the disk rolls, the point P_0 rotates through an angle θ about its center and the center travels through a linear distance $s = r\theta$ where r is the radius of the disk.

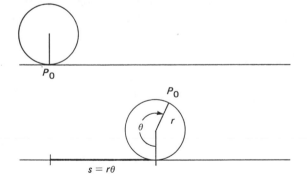

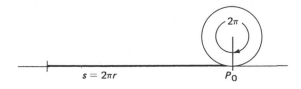

general, for any angular displacement θ, we have the equation

$$s = r\theta, \tag{6.39}$$

and $$V = r\omega, \tag{6.39}$$

where V is the velocity of the center. If ω changes with time, we also get the following relation between the linear acceleration of the center and the angular acceleration.

$$a = r\alpha. \tag{6.40}$$

Energy of a rolling object has two parts, the kinetic energy of the linear motion and the kinetic energy of the rotation about the center. Hence we have

$$(KE)_{total} = \frac{1}{2}MV^2 + \frac{1}{2}I\omega^2. \tag{6.41}$$

EXAMPLE 6.12

A rolling ball of radius 4.5 ft and weighing 16 lb is rolling along a lane without slipping. Its average linear velocity is 20 ft/s. Calculate (a) the moment of inertia, (b) the angular velocity, and (c) the total kinetic energy.

(a) Moment of inertia $= \frac{2}{5}Mr^2$

$$= \frac{2}{5}(16 \text{ lb})(4.5 \text{ ft})^2 = 130 \text{ lb} \cdot \text{ft}.$$

(b) $V = r\omega$, hence $\omega = V/r = \dfrac{(20 \text{ ft/s})}{(4.5 \text{ ft})} = 4.4 \text{ rad/s}.$

(c) $(KE)_{total} = \frac{1}{2}MV^2 + \frac{1}{2}I\omega^2$

$$= \frac{1}{2}MV^2 + \frac{1}{2}(\frac{2}{5}Mr^2 V^2 / r^2) = \frac{7}{10}MV^2$$

$$= 4.5 \times 10^3 \text{ J}.$$

EXAMPLE 6.13

A 10 kg cylinder starts from rest and rolls down an inclined plane from a height of 1 m. The radius of the cylinder is 0.2 m. Calculate the linear velocity of the cylinder as it reaches the ground (Fig. 6.19).

As discussed in the previous example, the total kinetic energy is given by

$$(KE)_{total} = \frac{1}{2}mV^2 + \frac{1}{2}I\omega^2.$$

The first term on the right-hand side is the translational kinetic energy and the second term is the rotational kinetic energy. The linear velocity V and the angular velocity ω are related by the equation

$$V = r\omega \quad \text{and} \quad I(\text{cylinder}) = \frac{1}{2}Mr^2.$$

Hence $\quad (KE)_{total} = \frac{1}{2}mV^2 + \frac{1}{2}(\frac{1}{2}Mr^2)V^2/r^2$

$$= \frac{3}{4}mV^2.$$

In this example the PE transformed to KE. Therefore,

$$mgh_A = (KE)_B = \frac{3}{4}mV^2$$

$$(10 \text{ kg})(9.8 \text{ m/s}^2)(1 \text{ m}) = (\tfrac{3}{4})(10 \text{ kg})V^2$$

Figure 6.19. Example 6.13.

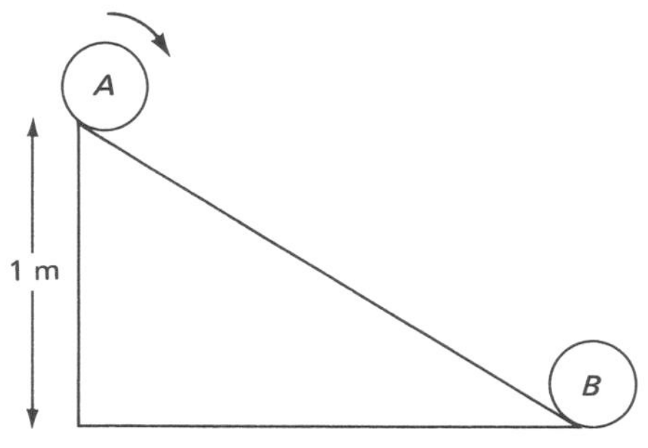

$$V^2 = 13 \text{ m}^2/\text{s}^2$$

or $\qquad V = \sqrt{13} = 3.6 \text{ m/s}.$

To obtain an equation for work in rotational motion, consider the disk free to rotate about an axis shown in Fig. 6.20. A tangential force acts over a distance $s (= r\theta)$.

Work done in this case is

$$W = Fs = Fr\theta.$$

Since $Fr = T$

$$W = T\theta. \tag{6.42}$$

Work done by the applied torque is equal to torque multiplied by the angular displacement.

Equation (6.42) is similar to $W = Fd$.

Since power is work done per unit time, dividing Eq. (6.42) by time we obtain

$$P = \frac{W}{t} = T\frac{\theta}{t} = T\omega. \tag{6.43}$$

This equation is also similar to the corresponding equation in linear motion, $P = FV$.

Table 6.4 lists equations in rotational motion along with their corresponding linear equations. Also listed are similar quantities in both cases.

EXAMPLE 6.14

An automobile engine delivers a torque of 250 lb·ft. How much work is done in 10 revolutions?

$$W = T\theta,$$

where θ should be expressed in radians

$$\theta = 10 \text{ rev} \times 2\pi \text{ rad/rev} = 62.8 \text{ rad},$$

$$W = (250 \text{ lb} \cdot \text{ft})(62.8 \text{ rad})$$

$$= 15,700 \text{ ft} \cdot \text{lb}.$$

TABLE 6.4 Relationships Between Quantities in Circular Motion. Corresponding Relationships in Linear Motion are Given for Comparison.

LINEAR MOTION	ROTATIONAL MOTION
Displacement d	Angular displacement θ $(= s/r)$
Velocity V	Angular velocity ω $(= V/r)$
Acceleration a	Angular acceleration α $(= a/r)$
$d = \bar{V}t$	$\theta = \bar{\omega}t$
$\bar{V} = \frac{1}{2}(V_i + V_f)$	$\bar{\omega} = \frac{1}{2}(\omega_i + \omega_f)$
$V_f = V_i + at$	$\omega_f = \omega_i + \alpha t$
$V_f^2 = V_i^2 + 2\,\text{rad}$	$\omega_f^2 = \omega_i^2 + 2\alpha\theta$
$d = V_i t + \frac{1}{2}at^2$	$\theta = \omega_1 t + \frac{1}{2}\alpha t^2$
Force F	Torque $T = rF$
Mass m	Moment of inertia, $I\ (= \Sigma_i m_i r_i^2)$
$F = ma$	$T = I\alpha$
$KE = \frac{1}{2}mV^2$	$KE = \frac{1}{2}I\omega^2$
Work $= Fd$	Work $= T\theta$
Power $= FV$	Power $= T\omega$
Momentum $P = mV$	Angular momentum* $L = I\omega$

*Angular momentum is discussed on p. 97.

EXAMPLE 6.15

A 15 lb grinding wheel takes 1 min to come to a stop after it is turned off. It is a uniform disk of radius 6 in. and its angular velocity before turning off was 600 rev/min. Calculate (a) the angular acceleration, (b) the total angular displacement during this 1 min period, (c) the average torque acting on the wheel, and (d) the total work done by frictional forces during slowing down.

(a) We are given

$$\omega_i = 600\ \text{rev/min} = \frac{600\ \text{rev/min}}{60\ \text{s/min}} \times 2\pi\ \frac{\text{rad}}{\text{rev}}$$

$$= 628\ \text{rad/s};$$

$$\omega_f = 0;\ t = 1\ \text{min} = 60\ \text{s}.$$

From Eq. (6.5)

$$\omega_f = \omega_i + \alpha t.$$

We obtain

$$\alpha = \frac{\omega_f - \omega_i}{t} = \frac{0 - 628\ \text{rad/s}}{60\ \text{s}} = -10.5\ \text{rad/s}^2.$$

(b) The angular displacement is

$$\frac{1}{2}(\omega_1 + \omega_2)t = \frac{1}{2}\ (0 + 628\ \text{rad/s}) \times 60\ \text{s}$$

$$= 18{,}840\ \text{rad}.$$

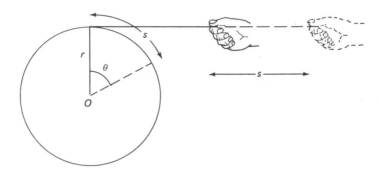

Figure 6.20.   Work done by the force F as the string is pulled out through a distance s is $Fs = Fr\theta = T\theta$.

(c) Since torque $T = I\alpha$, we have to first calculate I to obtain the torque

$$I = \frac{1}{2}Mr^2 = \frac{1}{2}(\frac{15}{32}\text{ slug})(\frac{1}{2}\text{ ft})^2 = 0.0586 \text{ slug} \cdot \text{ft}^2$$

$$T = I\alpha = (0.0586 \text{ slug} \cdot \text{ft}^2)(10.5 \text{ rad/s}^2)$$

$$= 0.615 \text{ lb} \cdot \text{ft}.$$

(d) Total work done $= T\theta = (0.615\,\text{lb}\cdot\text{ft})$ $\times (18,840 \text{ rad}) = 1,15 \times 10^4$ ft $\cdot$ lb. The frictional force reduces the kinetic energy of the wheel. Change in kinetic energy, therefore, should also give us the amount of work done by the frictional forces.

Initial kinetic energy $= \frac{1}{2}I\omega^2$

$$= \frac{1}{2}(0.0586 \text{ slug} \cdot \text{ft}^2)(628 \text{ rad/s})^2$$

$$= 11,555 \text{ ft} \cdot \text{lb};$$

Final kinetic energy $= 0$;

Work done $=$ change in kinetic energy

$$= 1.15 \times 10^4 \text{ ft} \cdot \text{lb}.$$

EXAMPLE 6.16

An electric motor delivers 750 watts power to a water pump (Fig. 6.21). The pulley of the pump is connected by a V-belt. The pump pulley is rotating at a speed of 1200 rev/min. Assuming no slippage between the belt and the pulleys, calculate the net torque acting on the pulley.

As we have seen earlier, the torques on both sides of the belt are not equal. The difference in torques produces the rotation and hence the work.

$$\omega = 1200 \text{ rev/min} = \frac{1200 \text{ rev/min}}{60 \text{ s/min}} \times 2\pi \frac{\text{rad}}{\text{rev}}$$

$$= 126 \text{ rad/s}.$$

Angular displacement in 1 s $= 126$ rad.

Work done in 1 s $= T\theta = T \times 126$ J.

Figure 6.21.   Example 6.16.

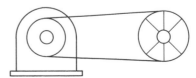

Work done per second is the power. Hence we get

$$T \times 126 \text{ J/s} = 750 \text{ W}$$

$$T = \frac{750 \text{ W}}{126 \text{ rad}} = 5.9 \text{ N} \cdot \text{m}.$$

EXAMPLE 6.17

A pulley is driven by a belt as shown in Fig. 6.22. The tension on one section of the belt is 30 lb and the other section is 5 lb. If the angular velocity of the driving wheel is 300 rev/min, calculate the power delivered to the wheel by the belt. The radius of the wheel is 6 in.

The net torque on the wheel = (30 lb) (½ ft) − (5 lb) (½ ft) = 12.5 lb · ft.

$$\omega = 300 \frac{\text{rev}}{\text{min}} = 300 \frac{\text{rev}}{\text{min}} \times 2\pi \frac{\text{rad}}{\text{s}} \times \frac{1 \text{ min}}{60 \text{ s}}$$

$$= 31.4 \frac{\text{rad}}{\text{s}}.$$

Power = $T\omega$ = 12.5 lb · ft × 31.4 rad/s

$$= 392.5 \text{ ft} \cdot \text{lb/s}.$$

6.5 ANGULAR MOMENTUM

The last equation included in Table 6.4 defined angular momentum. The definition is similar to that of linear momentum.

Figure 6.22. Example 6.17.

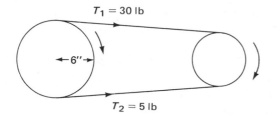

$T_1 = 30$ lb

←6″→

$T_2 = 5$ lb

Angular momentum is defined as the product of the moment of inertia of an object and its angular velocity.

$$L = I\omega, \qquad\qquad (6.44)$$

where L stands for angular momentum

Equation (6.24) if rewritten in the form

$$T = I\alpha = I(\frac{\omega_f - \omega_i}{t})$$

$$= \frac{I\omega_f - I\omega_i}{t} = \frac{L_f - L_i}{t}$$

leads to the principle of the conservation of angular momentum. If $T = 0$, $L_f = L_i$, that is, if the torque acting on an object is zero its angular momentum remains constant.

The conservation of angular momentum states that if no external torque acts on an object, its angular momentum remains a constant about a fixed axis or about the center of mass.

A figure skater [Fig. 6.23(a)] by changing her moment of inertia (by changing the mass distribution, remember $I = \Sigma \, m_i r_i^2$) can change her angular velocity. Figure skaters achieve the change in moment of inertia by spreading their arms or by bringing them closer to their body. A popular demonstration of this phenomenon is illustrated in Fig. 6.23(b). As the man standing on the rotating platform brings the weights closer to his body his moment of inertia decreases and angular speed increases.

The three conservation principles we have seen so far, conservation of energy, conservation of momentum, and conservation of angular momentum, apply equally in all branches of physics from nuclear physics to astrophysics.

EXAMPLE 6.18

A man standing on a rotating platform has his hands stretched out with 10 lb weights in each of his hands [Fig. 6.23(b)]. His moment of inertia in

(a)   (b)

Figure 6.23. (a) A figure skater changes her angular velocity by changing her moment of inertia. (b) As the masses are brought closer to the central axis of the body, the moment of inertia about this axis decreases. Consequently, a faster rotation results.

this position is 10 slug·ft^2 and it reduces to 3.5 slug·ft^2 when the weights are brought closer to his body. If the angular velocity is 25 rev/min for the hands in the out-stretched position, calculate his angular velocity when the weights are brought closer to his body.

Conservation of angular momentum may be expressed by the equation

$$I_i\omega_i = I_f\omega_f,$$

where i and f stands for initial and final values. We are given

$$I_i = 10 \text{ slug} \cdot \text{ft}^2,$$

$$\omega_i = 25 \text{ rev/min},$$

$$I_f = 3.5 \text{ slug} \cdot \text{ft}^2.$$

Final angular velocity $\omega_f = \dfrac{I_i\omega_i}{I_f}$

$$= \frac{(10 \text{ slug} \cdot \text{ft}^2)(25 \text{ rev/min})}{3.5 \text{ slug} \cdot \text{ft}^2}$$

$$= 71.4 \text{ rev/min}.$$

SUMMARY   In this chapter all the parameters in circular motion, angular displacement, velocity, and acceleration are defined. Moment of inertia takes the place of mass in rotational motion. Similarly T and L correspond, respectively, to F and P. Table 6.4 summarizes the definitions and lists the relations between the different parameters in rotational motion. Students should be thoroughly familiar with Table 6.4 and should know how to use the equations in Table 6.4 to calculate kinetic energy, work, angular momentum, etc.

QUESTIONS AND PROBLEMS

6.1 Indicate which of the following statements are true and which ones are false.

(a) Net force acting on a particle traveling at a constant speed is zero.

(b) Net force acting on a particle traveling with a constant velocity is zero.

(c) There is no force on a particle traveling with a constant speed in a circular orbit.

(d) The instantaneous velocity of a particle in a circular path is tangential to the path.

(e) The acceleration acting on a particle with a constant speed in a circular path is tangential to the path.

(f) A tangential acceleration increases the instantaneous velocity of a particle in a circular orbit.

(g) If tangential acceleration is present, angular acceleration is also present.

(h) Force always produces torque.

(i) No angular acceleration means no force acting.

(j) Kinetic energy of a rolling wheel is given by $\frac{1}{2} MV^2$.

6.2 Bicyclists and motorcyclists lean to one side when they navigate a turn. To which side do they lean? Why? Draw a diagram and explain.

6.3 What is the force that keeps the earth in an orbit around the sun?

6.4 Some flywheels have their mass concentrated near their rim. What is the advantage?

6.5 Two cans having equal mass and equal outer radii roll down an inclined plane. In one of the cans, part of its mass is concentrated along the axis and in the other the mass is concentrated near the outer rim (Fig. 6.24). Which one will reach the bottom faster?

6.6 If conservation of angular momentum is always true, why does a spinning ball slow down?

6.7 A spinning mass explodes into two. As the pieces move out, what will happen to their angular velocity?

Figure 6.24. Problem 6.5.

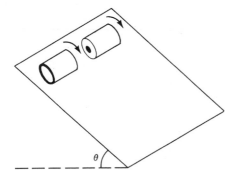

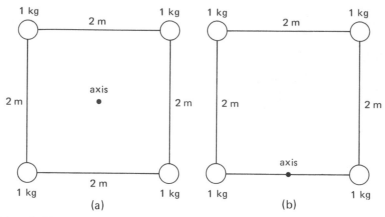

Figure 6.25. Problem 6.15.

6.8 A 12 in. record rotates at an angular velocity of 45 rev/min. Calculate its angular velocity:
(a) in rad/min;
(b) in rad/s;
(c) in rad/h;
(d) in deg/min;
(e) in deg/s;
(f) deg/h.

6.9 Calculate the speed of a point on the rim of the above record.

6.10 A merry-go-round revolves 10 times in a minute. If you are sitting 5 ft from the center, calculate your speed.

6.11 An armature of a motor picks up its rated angular velocity of 1800 rev/min in 15 sec after turning on the motor. Calculate its angular accelerations and the number of revolutions it made in this time.

6.12 A car is traveling at a speed of 60 mi/h. The radius of its tires is 36 in. The car is brought to a stop in 10 min. Calculate (a) the initial angular velocity of the tire, (b) the angular acceleration during slowing down,

Figure 6.26. Problem 6.18.

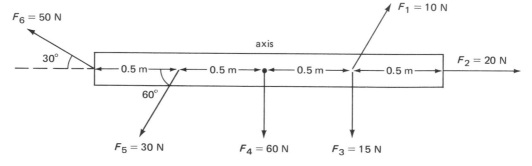

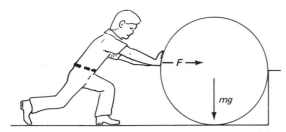

Figure 6.27. Problem 6.19.

(c) the angular displacement of the tire during slowing down, and (d) the distance traveled by the car during slowing down.

6.13 Derive Eq. (6.8).

6.14 Derive Eq. (6.9).

6.15 Four 1 kg masses are fixed to the corners of a metal frame of negligible mass. Calculate the moment of inertia of the system when (a) the axis is at the center of the square [Fig. 6.25(a)] and (b) the axis is at the center of one of the edges [Fig. 6.25(b)]. (Note: Moment of inertia of a system varies with the axis of rotation.)

6.16 Calculate the moment of inertia of a 40 lb cylinder about its central axis. Its diameter is 6 in. and its length is 1 ft.

6.17 A ball weighs 16 lb. Its radius is 6 in. Calculate its moment of inertia about a diameter.

6.18 Calculate the torques produced by each force in Fig. 6.26 about the axis at the center of the rod. Obtain the sum of the torques. Will the beam rotate about its axis? If so, in what direction?

6.19 The child does not look happy because the wheel is not rolling over the step (Fig. 6.27). Why is it not rolling?

6.20 What is the minimum force to be applied at point *A* (Fig. 6.28) so the wheel can be rolled over?

6.21 A 10 g ball tied to a string is whirled around in a horizontal circular path of 1 m radius. Calculate the tension in the string if the ball makes 10 rev/s.

Figure 6.28. Problem 6.20.

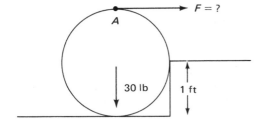

6.22 A girl weighing 60 lb is sitting in a 6 ft swing. If the swing is released from a position where it makes an angle of 30° with the vertical, calculate its velocity as it reaches the bottom (potential energy changing to kinetic energy). Calculate the tension on the ropes at this point.

6.23 An automobile makes a U-turn in a 20 ft radius circular path. If the coefficient of friction is 0.4, what is the maximum safe speed with which the U-turn can be made (centripetal force should be equal to the frictional force)?

6.24 A bucket tied to a 4 ft long rope is whirled in a vertical circle (Fig. 6.29). What is the minimum speed the bucket should have so that the water will not spill when it is at the top of the circle? If the bucket and water together weigh 20 lb, calculate the tension in the rope when the bucket is at the top and bottom. Assume its speed is maintained at the minimum speed.

6.25 A 400 lb earth satellite is in an orbit 200 mi above the surface of the earth. Calculate the speed and the period of the satellite assuming it is making a circular orbit.

6.26 If the satellite in Problem 6.25 is to be raised to an orbit 300 mi above the earth, what change should be made in its velocity. If the satellite is to be brought back to the earth, what should be done?

6.27 The radius of curvature of a curved section of an expressway is 200 ft. Calculate the angle of banking for a 60 mi/h speed limit.

6.28 The end sections of a race track have a radius of 300 ft and an angle of banking of 50°. What is the maximum safe speed limit?

6.29 A force acts on a uniform solid disk as shown in Fig. 6.30. Calculate (a) the torque acting on the disk, (b) the angular acceleration of the disk, and (c) the angular velocity of the disk after 5 s, assuming it starts from rest.

Figure 6.29. Problem 6.24.

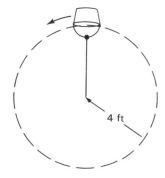

4 ft

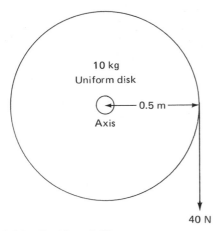

10 kg
Uniform disk

|←— 0.5 m —→|

Axis

40 N

Figure 6.30.   Problem 6.29.

6.30 The armature of a motor weighs 50 lb and has a radius of 4 in. Assuming it is a uniform cylinder, calculate its moment of inertia. If the armature speed increases to 18,000 rev/min from zero in 20 s, what is the net torque acting on the armature? How much work is done during this time?

6.31 A grinding wheel with a 12 in. diameter is driven at a constant speed of 120 rev/min. A metal piece to be polished is pressed against it with 30 lb for 10 min. The coefficient of friction between the wheel and the metal is 0.3. How much work is done during this time? Calculate the power output.

7

Static Equilibrium

As we climb up a ladder leaning against a wall, we sometimes wonder whether the ladder is stable against sliding down. Engineers worry about the stability of bridges. Can they stand all the forces that may possibly act on them, such as the weight of traffic and the force of wind? Both of these examples are problems of equilibrium, against linear and rotational motion. Such problems are dealt with in this chapter.

7.1 CONDITIONS OF EQUILIBRIUM

When is an object in static equilibrium?

1 When the sum of the forces acting on an object is zero, the object is in equilibrium against translational motion.

2 When the sum of the torques acting on an object is zero, the object is in equilibrium against rotational motion.

When condition 1 is satisfied, it is not necessarily true that condition 2 is automatically satisfied and vice versa. As examples consider the situations illustrated in Fig. 7.1(a) and 7.1(b). In Fig. 7.1(b) the two forces, whose magnitudes are the same but directions are opposite, act on an object lying on a table. These forces add up to zero. However, the torque about any arbitrary point is not zero. The object thus will rotate, but will have no translational motion. In Fig. 7.1(a), the two forces produce no net torque and hence no rotation about its center, but there is a net force resulting in a translational motion.

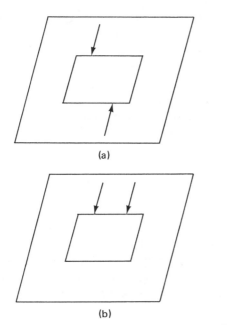

(a)

(b)

Figure 7.1. (a) The two forces produce rotation. (b) The two forces produce translational motion but no rotational motion.

7.2 EQUILIBRIUM OF PARTICLES

The equilibrium of a small particle can be treated in a very simple manner since the small particle can be considered as a point object. In such cases the line of action of all the forces passes through the point under consideration and hence torques due to the forces are zero (Fig. 7.2). Thus only condition 1 above needs to be satisfied, namely,

$$\Sigma F = 0. \tag{7.1}$$

When more than one direction is involved, it is convenient to resolve the forces and condition 1 then can be written in component form as

$$\Sigma F_x = 0, \tag{7.2}$$

$$\Sigma F_y = 0, \tag{7.3}$$

$$\Sigma F_z = 0. \tag{7.4}$$

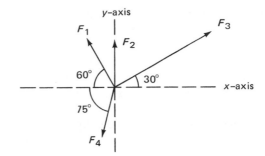

Figure 7.2. Example 7.1.

EXAMPLE 7.1

Four forces, shown in Fig. 7.2, are acting on a point object. Show that the object is in equilibrium.

If the object is in equilibrium, the sum of the x components and the y components of the forces add to zero separately. For forces $F_1 = 60$ lb, $F_2 = 29$ lb, $F_3 = 70$ lb, and $F_4 = 120$ lb.

x Components		y Components	
$-F_1 \cos 60°$	$= -30$ lb	$F_1 \sin 60°$	$= 52$ lb
	0	F_2	$= 29$ lb
$F_3 \cos 30°$	$= 61$ lb	$F_3 \sin 30°$	$= 35$ lb
$F_4 \cos 75°$	$= -31$ lb	$F_4 \sin 75°$	$= -116$ lb
ΣF_x	$= 0$	ΣF_y	$= 0$

In Example 7.1 all forces are acting on a point object. Such forces are called concurrent forces. Also, in the above example, three of the forces are balanced by the fourth force. For example, F_4 balances the combined effect of F_1, F_2, and F_3. The force F_4 is called the equilibrant of the other three forces.

7.3 CENTER OF GRAVITY

Before we discuss the problems of equilibrium of extended bodies, we have to find out what is the effect of gravitational attraction on such an object. Consider an extended object shown in Fig. 7.3 which is divided into a number of sections. The gravitational force on the first section is m_1g, on the second section m_2g, etc. The sum of the gravitational forces acting is $m_1g + m_2g + m_3g + \cdots$.

Can we replace all these forces by one force? If so, where should it be located? The gravitational forces which were assumed here to be parallel, can be replaced by a force Mg $(= m_1g + m_2g + \cdots)$. The point of application of this resultant force is called the center of gravity (c.g. for short). The point, center of gravity, is the same irrespective of the orientation of the body.

Figure 7.3. (a) Gravitational forces acting on each section of an object can be replaced by a single force acting through its center of gravity. (b) A force equal to the weight acting at the center of gravity balances the gravitational forces. (c) If the force is not applied at the center of gravity, the object will not be in equilibrium. It rotates.

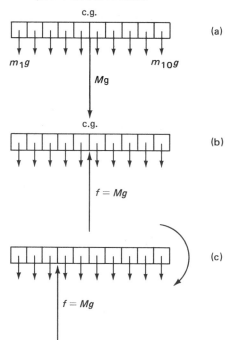

For bodies having simple shapes and uniform mass distribution (constant density), the center of gravity is the geometrical center of the body. For example, for the bar shown in Fig. 7.3 the center of gravity is located at its center. If Mg acting at c.g. replaces all the forces, m_1g, m_2g, $\cdots$, an equal force in the opposite direction should be able to balance these forces [Fig. 7.3(b)]. Inspection of Fig. 7.3(b) shows that such a force balances the gravitational forces and the net torque of all the forces (e.g., torque due to pairs m_1g and $m_{10}g$ are equal and of opposite sense) about c.g. is zero. Therefore, the weight Mg acting at the center of gravity truly replaces the distributed gravitational forces. Also note that Mg applied at any other point will result in zero force but not in zero torque [Fig. 7.3(c)]. *The center of gravity of an object is the point through which its weight (sum of gravitational forces) acts.*

For objects having nonuniform shape and/or mass distribution the center of gravity may be determined experimentally. When an object is suspended from a point, we note that it will come to rest with its center of gravity below the suspension point.

Draw a vertical line from the point of suspension (Fig. 7.4). Repeat with a second point of suspension The intersection of the two lines thus obtained is the center of gravity.

7.4 EQUILIBRIUM OF EXTENDED OBJECTS

For an extended object to be in equilibrium, both conditions 1 and 2, namely, sum of forces=0 and sum of torques=0 have to be satisfied. Before we discuss problems involving extended objects let us make sure we understand the calculation of torques. For this purpose some examples are illustrated in Fig. 7.5.

In Fig. 7.5(a), $T = 0$ since the perpendicular distance from the axis to the line of force is zero. In Fig. 7.5(b), $T = rF$, since F and r are perpendicular. In Fig. 7.5(c), $T = xF$ since the perpendicular distance is x not r. Also note that $x = r\sin\theta$. In Fig. 7.5(d), $T = rF$ since F and r are perpendicular.

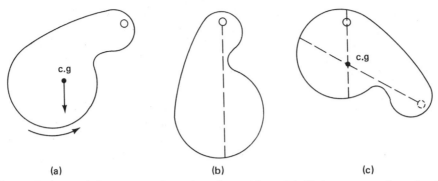

(a) (b) (c)

Figure 7.4. Determination of the center of gravity of an object. (a) If the center of gravity is not below the point of suspension, the object will rotate. (b) Object is in equilibrium when its center of gravity is below its point of suspension. Draw a vertical line through the point of suspension. The line passes through the center of gravity. (c) Repeat the procedure of (b) for another point of suspension. The point where the two lines meet is the center of gravity.

Let us consider Fig. 7.5(b) and 7.5(d) again. The rotation produced by force F in Fig. 7.5(b) is clockwise, while in Fig. 7.5(d) it is counterclockwise. Hence they are assigned different signs as follows: Torque producing rotation in an anticlockwise direction is positive (+); Torque producing rotation in a clockwise direction is negative (−).

Now since we have all the tools to solve problems, it is a good idea to have a plan of orderly "attack." Some problems may appear very complicated, but such a plan can avoid confusion and error. The procedure one may follow is as follows.

Figure 7.5. (a) Torque is zero; (b) clockwise (−) torque; (c) clockwise (−) torque; (d) counterclockwise (+) torque.

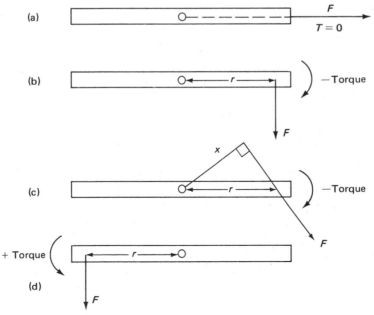

1 Make a line drawing of the problem under study.

2 Mark all the forces and other quantities (distances and angles) on the sketch.

3 Isolate a point or a body of the system and draw a free-body diagram of it.

4 Write the equations for equilibrium. (When calculating torques, a judicial choice of the axis will sometimes simplify the problems. For example, select an axis through a point at which several unknown forces meet. When a body is in equilibrium, any point may be considered to be an axis of rotation since no rotation actually occurs.)

5 Solve for unknown quantities.

6 If the problem is not completely solved go to the next point or the next body.

Several examples following this procedure are given below.

EXAMPLE 7.2

A uniform 4 ft long aluminum rod weighs 2 lb. Two 6 lb weights are attached to the rod, one 0.5 ft from one end and the other 2.5 ft from the same end. The rod is in equilibrium at a pivot as shown in Fig. 7.6. Find the position of the pivot and the force acting at the pivot. Where is the center of gravity of the system of the rod and the weights?

The weight of the rod is acting through its center of gravity which is at its geometrical center located 2 ft from the ends. Let the pivot be located at a distance x from the center of gravity of the rod and the force acting up at the pivot be N. Since all the forces acting are parallel or antiparallel (all lying along one axis), we need only two equations for the equilibrium conditions.

Referring to the free-body diagram, we get the following equations for the forces and torques:

$$N - 6\text{ lb} - 2\text{ lb} - 6\text{ lb} = 0$$

$$N = 14\text{ lb}.$$

Torques are calculated about the pivot:

$$(6\text{ lb})(1.5\text{ ft} - x) = (2\text{ lb})(x) + (6\text{ lb})(0.5\text{ ft} + x)$$

$$9 - 6x = 2x + 3 + 6x.$$

Rearranging the terms in the above equation yields

$$10x = 6$$

$$x = 0.6\text{ ft}.$$

Since all the forces are balanced (no net torque) the pivot is at the center of gravity of the combined system of rod and weights.

Figure 7.6.   Example 7.2.

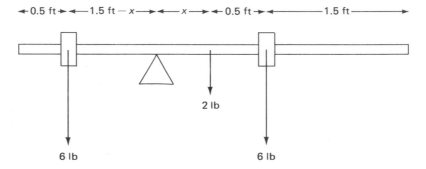

2 lb

6 lb   6 lb

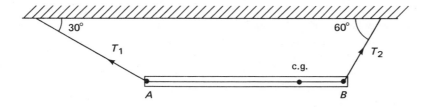

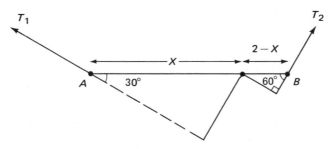

Figure 7.7. Example 7.3.

EXAMPLE 7.3

The nonuniform bar shown in Fig. 7.7 is 2 m long and has a mass of 60 kg. The bar is suspended so that it stays horizontal. Calculate the tensions of the strings and the position of the center of gravity of the bar. Let us start by assuming that the center of gravity is at a distance X from one end of the bar. We have to determine T_1, T_2, and X. First we use the first condition, $\Sigma F_x = 0$, $\Sigma F_y = 0$.

x Components

$$-T_1 \cos 30° = -0.866T_1$$

$$T_2 \cos 60° = 0.6T_2$$

$$0$$

$$\overline{\rule{5cm}{0.4pt}}$$

$$\Sigma F_x = -0.866T_1 + 5T_2 = 0$$

or $T_2 = \dfrac{0.866}{0.5}\,T_1$

y Components

$$T_1 \sin 30° = 0.5T_1$$

$$T_2 \sin 60° = 0.866T_2$$

$$-60\ \text{kg} \times 9.8\ \text{m/s}^2 = -588\ \text{N}$$

$$\overline{\rule{6cm}{0.4pt}}$$

$$\Sigma F_y = 0.5T_1 + 0.866T_2 - 588\ \text{N} = 0$$

Substituting for T_2 we obtain

$$0.5T_1 + 0.866(\tfrac{0.866}{0.5})T_1 = 588\ \text{N}$$

or

$$T_1(0.5 + \frac{0.866 \times 0.866}{0.5}) = 588\ \text{N}$$

$$T_1 = 294\ \text{N}.$$

Substituting the value of T_1 into the equation for T_2, we obtain

$$T_2 = \frac{0.866}{0.5} \times 294\ \text{N} = 509\ \text{N}.$$

To obtain the distance X, we use the torque equation, torques being calculated about the center of gravity. The perpendicular distances to the tensions are not X and $2 - X$, but they are r_1 and r_2,

$$r_1 = X \sin 30° = 0.5X$$

$$r_2 = (2 - X)\sin 60° = 0.866(2 - X).$$

Hence,

Torque due to tension T_1

$$= T_1 r_1 = -(294 \text{ N})(0.5X);$$

Torque due to tension T_2

$$= T_2 r_2 = (509 \text{ N})(0.866)(2 - X);$$

The sum of the torques should be zero, so

$$-(294 \text{ N})(0.5X) + (0.866)(2 - X)(509 \text{ N}) = 0$$

$$X = \frac{882}{588} = 1.5 \text{ m.}$$

EXAMPLE 7.4

A 16 ft long ladder weighing 60 lb is leaning against a frictionless wall (Fig. 7.8). Calculate the

Figure 7.8. Example 7.4.

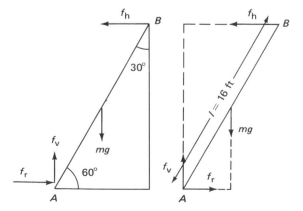

forces acting at the points of contact, on the ground and on the wall.

In addition to the weight of the ladder, the forces acting on the ladder are f_h, normal force at the ladder wall contact; f_v, vertical reaction force at the ground; f_r, frictional force at the ladder ground contact point.

From the conditions of equilibrium we get for the x direction forces:

$$f_r - f_h = 0 \text{ or } f_r = f_h;$$

for the y direction forces:

$$f_v - mg = 0 \text{ or } f_r = 60 \text{ lb.}$$

For the calculation of the torque, we select point A as the axis (eliminates 2 forces from consideration), so that

$$f_h \times \ell \sin 60° - mg \times \frac{1}{2} \ell \cos 60° = 0.$$

The above equation simplifies to

$$0.866 f_h = 0.5 \times 60 \text{ lb}$$

$$f_h = \frac{0.5}{0.866} \times 60 \text{ lb} = 34.6 \text{ lb.}$$

Therefore, the unknown forces are

$$f_r = f_h = 34.6 \text{ lb};$$

$$f_v = 60 \text{ lb.}$$

EXAMPLE 7.5

A derrick, often used to hoist heavy loads, is shown in Fig. 7.9. The uniform boom is 20 ft long and weighs 500 lb. The angles made by the boom and the cable with the mast are indicated in Fig. 7.9. The weight being lifted is 7000 lb. Calculate the tension in the cable connecting the boom and the mast and the forces acting at the other end of the boom.

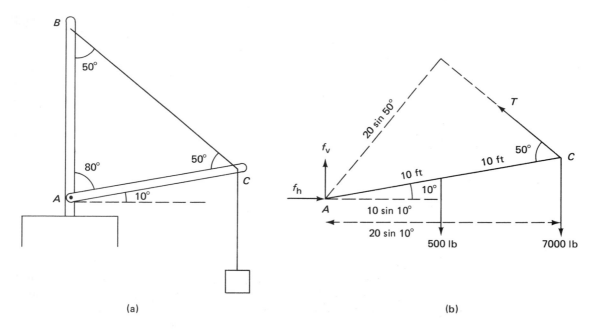

Figure 7.9. Example 7.5. (a) derrick; (b) the boom is isolated and the forces acting on it are shown for convenience in order to solve the problem.

The free-body diagram indicates the forces acting at the various parts of the boom. The weight of the boom is acting at its center.

Considering the vertical forces, we obtain

$$f_v = T\sin 40° = 500 \text{ lb} + 7000 \text{ lb}$$

or

$$f_v + 0.643T = 7500 \text{ lb}.$$

Considering the horizontal forces, we obtain

$$f_h - T\cos 40° = 0 \text{ or } f_h = 0.766T. \qquad (7.5)$$

Refer to Fig. 7.9(b) for the calculation of torques.

We select point A, where there are two unknown forces, as the axis.

$$T \times 20\sin 50° - 500 \text{ lb} \times 10\cos 10°$$

$$- 7000 \text{ lb} \times 20\cos 10° = 0$$

$$T = 9320 \text{ lb}.$$

From Eq. (7.5) we obtain

$$f_h = 0.766T = 7139 \text{ lb};$$

$$f_v = 7500 \text{ lb} - 0.643T$$

$$= 1507 \text{ lb}.$$

SUMMARY   An object is in equilibrium when the vector sum of the forces acting is zero and the vector sum of the torques due to these forces about any arbitrary point is zero. These conditions can be written as

$$\Sigma F_x = 0, \text{ sum of the } x \text{ components} = 0.$$

$$\Sigma F_y = 0, \text{ sum of the } y \text{ components} = 0.$$

$$\Sigma T = 0, \text{ sum of the torques} = 0.$$

In adding torques, remember that counterclockwise torques are taken positive and clockwise negative. Note the sum of the torques is zero about any point.

QUESTIONS AND PROBLEMS

7.1   An object is suspended by two strings as shown in Fig. 7.10. Calculate the tension in the string.

7.2   The object in Problem 7.1 is suspended with the strings making a smaller angle with the horizontal than in Problem 7.1 (Fig. 7.11). Calculate the tension.

7.3   Using the results of Problems 7.1 and 7.2, can you make a comment on the following statement. "If a hammock is tied tight it might break easier than when it is tied loosely."

Figure 7.11. Problem 7.2.

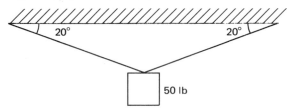

Figure 7.10. Problem 7.1.

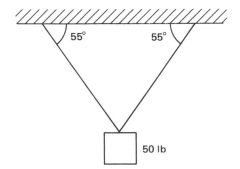

Figure 7.12. Problem 7.4.

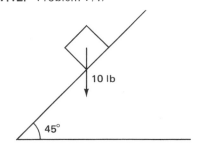

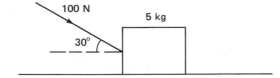

Figure 7.13. Problem 7.5.

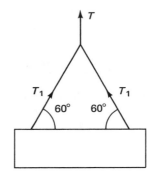

Figure 7.14. Problem 7.6.

7.4 The block on an inclined plane is stationary (Fig. 7.12). Calculate the frictional force acting on the block.

7.5 A force of 100 N acts on a block as shown in Fig. 7.13. The block is stationary. Calculate the normal force, the frictional force, and the coefficient of friction.

7.6 A billboard weighing 60 lb is suspended as shown in Fig. 7.14. Calculate the tension in each cord.

7.7 Calculate the forces T and F indicated in Fig. 7.15(a) and (b). Assume the beam to be of negligible mass.

Figure 7.15. Problem 7.7.

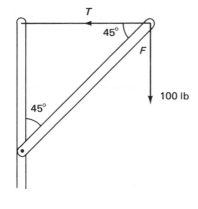

(a)

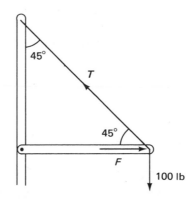

(b)

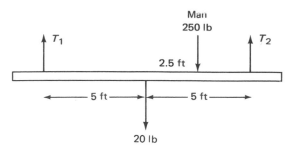

Figure 7.16. Problem 7.8.

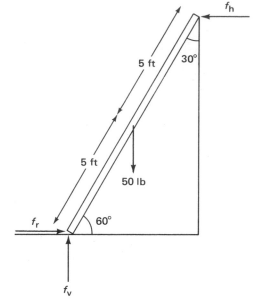

Figure 7.18. Problem 7.10.

7.8   A painter's platform is supported by two ropes as shown in Fig. 7.16. If the painter weighs 250 lb, what are the tensions in each cord?

7.9   Some "brave" physics professor tries to prove the point of force distribution at the support points by lying on a bed of nails (Fig. 7.17). If nails are spaced 1 in. apart, about 1000 nails will be supporting him. If the professor weighs 200 lb, what is the reaction force at each nail.

7.10 A ladder weighing 50 lb is leaning against a frictionless wall (Fig. 7.18). Calculate the reaction force at the wall and the forces acting at the point where it touches the ground.

Figure 7.17. Bed of nails. Problem 7.9.

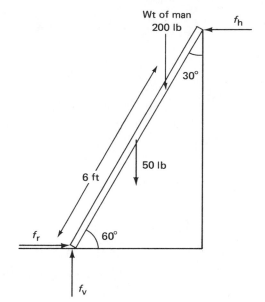

Wt of man
200 lb

f_h

30°

50 lb

6 ft

f_r 60°

f_v

Figure 7.19. Problem 7.11.

45°

T

45°

f_h 20 lb   600 lb

f_v

Figure 7.20. Problem 7.12.

7.11 If a 200 lb man stands on the ladder of Problem 7.10 6 ft from the end which touches the ground, the ladder becomes unstable (Fig. 7.19). Calculate the coefficient of friction between the ladder and the ground.

7.12 A hoist is kept at a 45° angle position with the vertical mast (Fig. 7.20). Calculate the forces T, f_v, and f_h for this case.

7.13 Calculate the center of gravity of the masses shown in Fig. 7.21.

7.14 Calculate the tension on the string, when the hoist is pulled up by a rope (Fig. 7.22).

7.15 A uniform beam 2 m long has a mass of 10 kg. The beam is suspended as shown in Fig. 7.23. Calculate tensions T_1 and T_2 and the angle θ_2.

7.16 A 4000 lb car is 2/3 the way across the bridge (Fig. 7.24). The bridge is uniform and weighs 1000 lb. What are the reaction forces by the supports at each end of the bridge?

Figure 7.21. Problem 7.13.

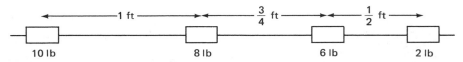

|←———— 1 ft ————→|←— $\frac{3}{4}$ ft —→|←— $\frac{1}{2}$ ft —→|

10 lb   8 lb   6 lb   2 lb

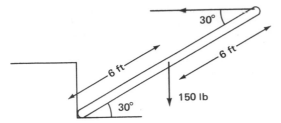

Figure 7.22. Problem 7.14.

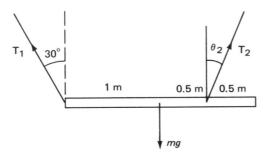

Figure 7.23. Problem 7.15.

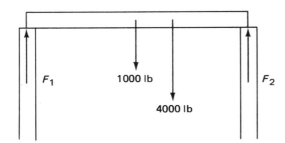

Figure 7.24. Problem 7.16.

Simple Machines

Machines are used to ease the burden of work, sometimes by reducing the necessary force required to do work and at other times by simply altering the direction of the applied force. Inclined planes and some levers belong to the first group while a single simple pulley belongs to the second group. We are now familiar with several aspects of inclined planes (see Chapters 4 and 5). Therefore, we start the discussion of the general characteristics of a simple machine in terms of those of an inclined plane.

8.1 INCLINED PLANE

An inclined plane is often used to lift heavy objects. The force, F_i, which one has to apply along the inclined plane in order to push the object is $mg \sin \theta$ (see Fig. 8.1). This force is smaller than the force mg, which one applies to lift the object without the help of the inclined plane. We are now discussing an ideal case where no friction is assumed. The force F_i is called the input force and the force $F_o = mg$ is called the output force. For this ideal case where no friction is involved, *the ratio F_o/F_i is called ideal mechanical advantage (IMA), namely,*

$$IMA = F_o/F_i. \tag{8.1}$$

By using the inclined plane we gained the advantage of applying a smaller force. However, the amount of work with and without the inclined plane is the same (remember we are still talking about the ideal

117

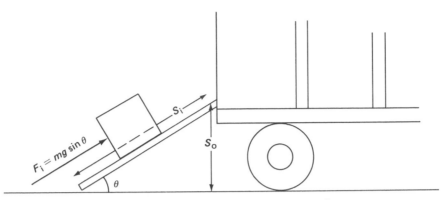

Figure 8.1.   Inclined plane. S_i is the input distance and S_o the output distance.

case), as shown below.   The work output is the amount of work to lift the weight mg through a distance S_o.   The work output is $F_o S_o$ ($= mgS_o$) and the work input is $F_i S_i$ [$= mg(\sin\theta)S_i$].   Since $S_o = S_i \sin\theta$, we obtain

$$F_i S_i = mg(\sin\theta)S_i = mgS_o = F_o S_o. \quad (8.2)$$

Equation (8.2) is a restatement of the conservation of energy.

From Eq. (8.2), Eq. (8.1) for ideal mechanical advantage can be written as

$$\text{IMA} = \frac{F_o}{F_i} = \frac{S_i}{S_o} \quad \text{(ideal)}. \quad (8.3)$$

Since the input distance S_i and the output distance S_o are dimensions of the machine which do not change, IMA is often defined as follows: *the ideal mechanical advantage of a machine is the ratio of input distance to output distance.*

As we are aware, ideal situations are never achieved in practice.   In the case of the inclined plane F_i is always greater than $mg\sin\theta$ because of friction.   Hence the ratio F_o/F_i is no longer the ideal mechanical advantage.   It is called the actual mechanical advantage such that

$$\text{AMA} = F_o/F_i. \quad (8.4)$$

The quantity F_i in Eq. (8.4) is given by

$$F_i = mg\sin\theta + \text{frictional force}.$$

Since the force to be applied, F_i, is larger than the ideal force, $mg\sin\theta$, AMA is always less than IMA.   IMA can always be calculated by taking the ratio of S_i, the distance through which the input force is applied, to S_o, the distance through which the output force moves while the calculation of AMA requires exact knowledge of hidden forces such as friction.

The efficiency of a machine was defined in Chapter 4 as

$$\text{Efficiency} = \frac{\text{Work output}}{\text{Work input}} = \frac{\text{Power output}}{\text{Power input}}.$$

In the case of simple machines, we can rewrite the equation in the form

$$\text{Efficiency} = \eta = \frac{F_o S_o}{F_i S_i} = \frac{F_o/F_i}{S_i/S_o} = \frac{\text{AMA}}{\text{IMA}}. \quad (8.5)$$

[The greek letter η (eta) is often used to represent efficiency.]   In the case of an ideal machine, AMA = IMA and consequently efficiency will be equal to 100%.   In a real machine, AMA < IMA and efficiency is less than 100%.

EXAMPLE 8.1

An inclined plane (Fig. 8.1) is used to raise an 800 lb crate to a truck. The length of the inclined plane is 5 ft and the angle is 30°. The coefficient of friction between the crate and the plane is 0.2. Calculate the IMA, AMA, and the efficiency of the inclined plane.

Input force distance S_i = 5 ft;

Output force distance S_o = 5 ft $(\sin 30°)$

$$= 5 \text{ ft} \times 0.5 = 2.5 \text{ ft};$$

Input force = Component of the weight along the inclined plane + frictional force

$$= mg \sin 30° + \mu mg \cos 30°$$

$$= (800 \text{ lb})(0.5) + (0.2)(800 \text{ lb})(0.866)$$

$$= 539 \text{ lb};$$

Output force = 800 lb;

$$\text{AMA} = \frac{F_o}{F_i} = \frac{800 \text{ lb}}{539 \text{ lb}} = 1.5;$$

$$\text{IMA} = \frac{S_i}{S_o} = \frac{5 \text{ ft}}{2.5 \text{ ft}} = 2;$$

$$\text{Efficiency, } \eta\% = \frac{\text{AMA}}{\text{IMA}} \times 100\%$$

$$= \frac{1.5}{2} \times 100\% = 75\%.$$

Modifications of Inclined Plane

The wedge (Fig. 8.2) is often used to lift heavy objects through very small distances. As the wedge is driven in with a force F_i through a distance S_i, the object is raised through a height S_0. The IMA of the wedge is S_i/S_o.

The screw jack is one of several forms of a screw which is a modification of an inclined plane. A screw may be considered as an inclined plane curved around a central cylinder (Fig. 8.3). The force is applied on a handle of length r. When the handle is rotated once, that is, for an input force distance of $2\pi r$, the screw moves up a distance p, called the pitch, lifting the weight (w) through the distance p. Therefore, the IMA is $2\pi r/p$.

The cam shown in Fig. 8.4 is also a modification of the inclined plane.

8.2 LEVER

A simple lever is a rigid rod (Fig. 8.5) free to turn about a fixed point called the fulcrum. The fulcrum is a convenient point to which distances are measured.

Figure 8.2. A wedge.

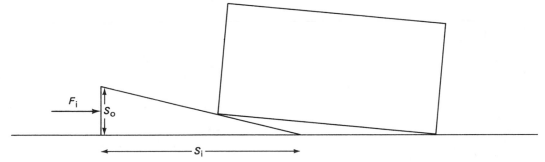

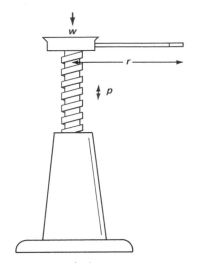

Figure 8.3.   A screw jack.

The applied force is called the effort and the load being lifted is called the resistance. The distance from fulcrum to the effort ℓ_e is called the effort arm, and the distance to the resistance ℓ_r is called the resistance arm. As the effort moves through a distance of S_o, the resistance moves through a distance of S_i. The triangles AFC and BFD are similar. Hence we have

$$\text{IMA} = \frac{S_i}{S_o} = \frac{\ell_e}{\ell_r}. \qquad (8.6)$$

Figure 8.4.   A cam. The input force distance is L and the output distance is $r_2 - r_1$ in this case.

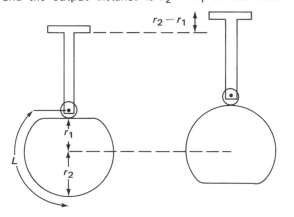

There are three classes of levers which are illustrated in Fig. 8.5 and some commonly found examples which are shown in Fig. 8.7. Our hand belongs to the third class lever which is the least efficient of the three types, in fact, the IMA is less than 1, because $\ell_e < \ell_r$ (Fig. 8.5), whereas most machines have an IMA > 1.

Figure 8.5.   (a) Lever; (b) Lever of the first class: IMA $= \ell_e/\ell_r$; (c) lever of the second class; (d) lever of the third class.

(a)

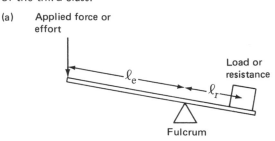

(b)

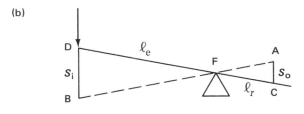

(c)

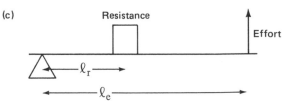

(d)

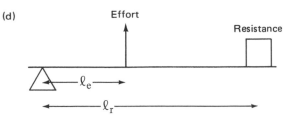

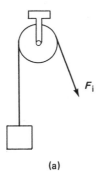

(a)

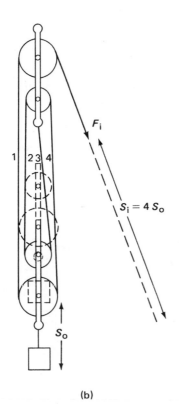

1 23 4

$S_i = 4S_o$

S_o

(b)

Figure 8.6 (a) Single pulley. (b) Block and tackle. When the load moves through distance S_o, the applied force moves through a distance of $4S_o$. IMA = 4 = number of strings holding the load.

The wheel and the axle is a commonly used modification of lever which consists of a wheel of radius R fixed to an axle of radius r, as shown in

Fig. 8.8. The output work is done by the tension of the cord attached to the axle and the input work by the input force applied through the cord around the wheel. As the wheel is rotated through an angle θ, the axle also rotates through the same angle θ. The distance the load moved during this rotation, S_o, is therefore $r\theta$ and the point of application of the input force moves through a distance $S_i\ (=R\theta)$. The IMA of the wheel and the axle is

$$\text{IMA} = \frac{S_i}{S_o} = \frac{R}{r}. \qquad (8.7)$$

Ideal mechanical advantage is thus the ratio of the two radii.

8.3 THE PULLEY

The pulley is a commonly used simple machine and has a large variety of modifications. A simple pulley, illustrated in Fig. 8.6(a), changes the direction of the force. Its IMA is 1. Another pulley system, illustrated in Fig. 8.6(b), is called a block and tackle. In order to raise the load through a height of S_o, each strand of rope (marked 1,2,3, and 4) should be shortened through S_o. Consequently, the rope at the operator's end moves through a distance of $S_i = 4S_o$. Hence IMA in this case is given by

$$\text{IMA} = \frac{S_i}{S_o} = 4 \qquad (8.8)$$

= number of ropes supporting the load

which is equal to the number of strands that support the movable block.

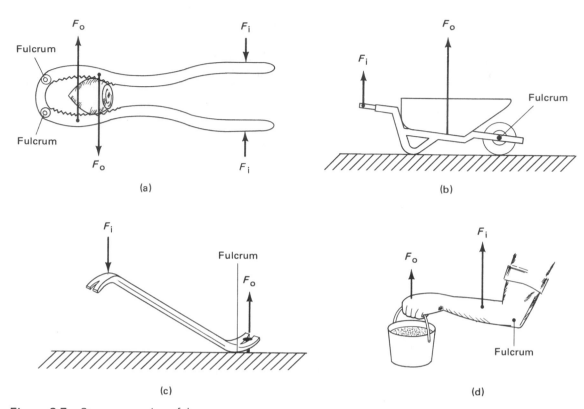

Figure 8.7. Some examples of levers.

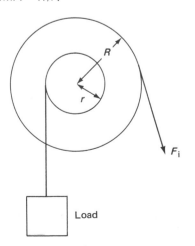

Figure 8.8. Wheel and axle. A modification of levers: IMA $= R/r$.

8.4 TRANSMISSION OF TORQUE

In rotational motion, work is transmitted by pulleys and wheels, gear systems, etc. (Fig. 8.9). As in simple machines discussed earlier, sometimes we get a larger output torque than the input torque. On the other hand, some arrangements produce a higher angular velocity at the output delivering a reduced torque.

In rotational motion, work is given by

$$W = T\theta,$$

where the angle θ takes the place of the linear distance and T the place of force in translatory motion. If T_o is the output torque and T_i the input torque and θ_o is the angular displacement of the output torque and θ_i the corresponding displacement of the

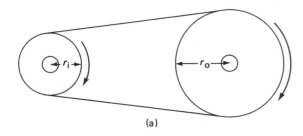

(a)

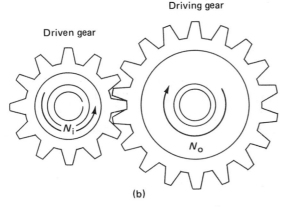

Driving gear

Driven gear

N_i

N_o

(b)

Figure 8.9. (a) The belt drive. IMA $= \dfrac{r_o}{r_i}$. (b) Spur gears: IMA $= \dfrac{N_o}{N_i}$

input torque, the IMA can be defined as

$$\text{IMA} = \theta_i/\theta_o; \qquad (8.9)$$

AMA can be defined by

$$\text{AMA} = T_o/T_i; \qquad (8.10)$$

and the efficiency, defined as the ratio of output work to input work, is given by

$$\eta\% = \frac{T_o\theta_o}{T_i\theta_i} \times 100 = \frac{\text{AMA}}{\text{IMA}} \times 100. \quad (8.11)$$

Since angular displacement is directly equal to the angular velocity multiplied by the time, IMA can also be written in the form

$$\text{IMA} = \frac{\omega_i t}{\omega_o t} = \frac{\omega_i}{\omega_o}. \qquad (8.12)$$

Table 8.1 lists the corresponding equations and quantities of significance to the discussion of simple machines.

For a given efficiency, if the IMA increases, the AMA also increases, that is,

$$\frac{T_o}{T_i} \propto \frac{\omega_i}{\omega_o}. \qquad (8.13)$$

It follows that for a given efficiency if the transmission device reduces the output speed, the torque output increases or vice versa.

Two of the common devices for transmission of torques are shown in Fig. 8.9. In both these cases, the linear distance traveled by a point on the edge of both wheels or both interconnecting gears has to be equal. Therefore, we get

$$\omega_i r_i = \omega_o r_o$$

or

$$\frac{\omega_i}{\omega_o} = \frac{r_o}{r_i}. \qquad (8.14)$$

Substituting this ratio into Eq. (8.11) we get

$$\text{IMA} = r_o/r_i. \qquad (8.15)$$

Since the circumference of the gears is proportional to their radii, the number of teeth is proportional to

TABLE 8.1 A Comparison of Equations and Quantities of Significance in the Discussion of Simple Machines in Linear and Rotational Motion.

ROTATIONAL MOTION	LINEAR MOTION
Displacement, θ	Displacement, S
Torque, T	Force, F
IMA $= \theta_i/\theta_o$	IMA $= S_i/S_o$
AMA $= T_o/T_i$	AMA $= F_o/F_i$
Efficiency, $\eta\% = \dfrac{\text{AMA}}{\text{IMA}} \times 100$	

the radii. Hence we also get

$$\text{IMA} = \frac{\text{number of teeth on the output wheel } (N_o)}{\text{number of teeth on the input wheel } (N_i)}.$$

$$(8.16)$$

Equation (8.15) is applicable to wheels interconnected by a nonslipping V-belt and Eq. (8.16) is applicable to gear systems. Examples illustrating the use of the above derived equations are given below.

EXAMPLE 8.2

A V-belt and two pulleys are used to transmit power from a 10 hp motor to a machine. The radius of the driving wheel is 6 in. and that of the driven wheel is 1 ft. The motor rotates the driving wheel at a speed of 240 rev/min. The efficiency of transmission is 75%. Calculate the angular velocity and the torque delivered to the wheel.

Since the radius of the driven wheel is larger than the driving wheel, the angular velocity of the driven wheel will be smaller. Using Eq. (8.14), we have

$$\frac{\omega_i}{\omega_o} = \frac{r_o}{r_i} \quad \text{or} \quad \omega_i r_i = \omega_o r_o.$$

Since $\omega_i = 240$ rev/min, $r_i = 6$ in. and $r_o = 1$ ft,

$$(240 \text{ rev/min})(\tfrac{1}{2} \text{ ft}) = \omega_o (1 \text{ ft})$$

$$\omega_o = 120 \text{ rev/min}.$$

$$\text{IMA} = \frac{\omega_i}{\omega_o} = \frac{240 \text{ rev/min}}{120 \text{ rev/min}} = 2$$

$$\text{AMA} = \eta \times \text{IMA}$$

$$= \frac{75}{100} \times 2 = 1.5$$

$$\text{AMA} = T_o/T_i.$$

Since the relation between torque and power is

$P = T\omega$, we get

$$T_i = \frac{P_i}{\omega_i} = \frac{10 \text{ hp} \times 550(\text{ft} \cdot \text{lb/s})/\text{hp}}{240 \frac{\text{rev}}{\text{min}} \times \frac{1 \text{ min}}{60 \text{ s}} \times 2\pi \frac{\text{rad}}{\text{rev}}}$$

$$= 219 \text{ ft} \cdot \text{lb}$$

$$T_o = T_i \times \text{AMA} = 219 \times 1.5 = 328 \text{ ft} \cdot \text{lb}.$$

EXAMPLE 8.3

A spur gear system is used to increase the output speed. The gear connected to the motor has 50 teeth and the smaller gear has 10 teeth. Calculate the ratio ω_i/ω_o of the gear system and its IMA

$$\frac{\omega_o}{\omega_i} = \frac{N_i}{N_o} = \frac{50}{10} = 5.$$

Using Eq. (8.16) we have

$$\text{IMA} = N_o/N_i = 0.2.$$

EXAMPLE 8.4

The gear system in the above example is connected to a 20 kW electric motor, driving the gear at 300 rev/min. The efficiency of the spur gears is 80%. Calculate (a) ω_o, (b) torque output, and (c) the torque at the second gear.

$$\omega_o = 5 \, \omega_i = 5 \times 300 \text{ rev/min} = 1500 \text{ rev/min}$$

$$\omega_i = \frac{300 \text{ rev/min}}{60 \text{ s/min}} \times 2\pi \frac{\text{rad}}{\text{rev}} = 31.4 \text{ rad/s}.$$

Since power is given by $P = T\omega$, we have

$$\text{Torque input} = \frac{20 \times 10^3 \text{ W}}{31.4 \text{ rad/s}} = 637 \text{ N} \cdot \text{m};$$

$$\text{Efficiency} = \frac{\text{AMA}}{\text{IMA}} = 0.8;$$

$$AMA = 0.8 \times 2 = 0.16.$$

To obtain the torque output we use Eq. (8.11):

$$AMA = T_o/T_i$$

$$T_o = AMA \times T_i = 0.16 \times 637 = 102 \text{ N} \cdot \text{m}.$$

In all the examples of simple machines above, we have derived equations for IMA in terms of the physical dimensions, such as the length of effort arm, resistance arm, and radii of wheels. In general AMA, given by the ratio of output force (torque) to input force (torque), is difficult to calculate because of our lack of understanding of the frictional forces which dissipate part of the input force. It is also interesting to point out that most of the complicated machinery we see today is a combination of simple machines and their modifications.

SUMMARY Simple machines, for a smaller input force F_i, produce a larger output force F_o. However, there is no savings in work since the work at the input end is done over a longer distance compared to the work done without the machine. Work done at the input end $F_i S_i$ is larger than the output work $F_o S_o$ because of frictional force. We defined the following quantities:

$$IMA = S_i/S_o,$$

$$AMA = F_o/F_i,$$

$$\text{Efficiency} = AMA/IMA.$$

The simple machines discussed above were the inclined plane, levers, screw jack, and pulleys. One should be familiar with the application of the above equations for these machines.

For devices that transmit torque, pulleys and wheels and gear systems, we have the following equations:

$$IMA = \frac{\theta_i}{\theta_o} = \frac{\omega_i}{\omega_o} = \frac{r_o}{r_i} = \frac{N_o}{N_i}.$$

The ratio N_o/N_i is for gear systems, where N_o stands for the number of teeth on the output wheel and N_i, the number of teeth on the input wheel.

$$AMA = T_o/T_i$$

$$\text{Efficiency} = AMA/IMA.$$

QUESTIONS   *8.1*   Why is AMA always less than IMA?

AND   *8.2*   When a gear system is used to increase the output speed, the output
PROBLEMS   torque is less. Why?

                8.3   An inclined plane is used to lift a cubic box. Instead of pushing the box
                        up it is tipped from one edge to the other edge as shown in Fig. 8.10. Is
                        there any advantage in using the inclined plane like this?

                8.4   An inclined plane with a 20° slope is used to lift a 200 lb weight through
                        a height of 10 ft. The coefficient of friction is 0.3. Calculate (a) IMA,
                        (b) the minimum force to be applied, (c) AMA, (d) total work done, and
                        (e) work done against friction.

                8.5   An inclined plane 2 m long is used to lift objects through a height of 1 m.
                        A 100 kg object is lifted along this inclined plane. The minimum force
                        required to keep the object moving is found to be 600 N. Calculate
                        (a) the coefficient of friction between the object and the inclined plane;
                        (b) the IMA; (c) the AMA; (d) the efficiency of the inclined plane.

Figure 8.11.   Problem 8.7.

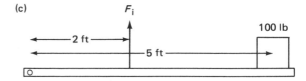

Figure 8.10.   Problem 8.3.

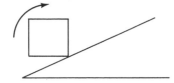

8.6   Find three practical examples each for the three types of levers.

8.7   Considering the levers shown in Fig. 8.11 to be ideal, calculate (a) F_i, (b) the **AMA**, and (c) the **IMA** for each case.

8.8   A 4000 lb car is being lifted by a jack by applying a 40 lb force. In each stroke when the input handle moves through a distance of 3 ft the car lifts up through $\frac{3}{10}$ in. Calculate (a) the **IMA**; (b) the **AMA**; (c) the efficiency; (d) the work done in each stroke.

8.9   Obtain the IMA for the pulley arrangement shown in Fig. 8.12.

8.10   For a screw jack the input force is applied through a handle of length 40 cm and the pitch is 1 mm. It has an efficiency of 25%. Obtain the required input force to lift a 1000 kg object.

8.11   Step pulleys (Fig. 8.13) are often used to change the output speed (example in lathes and drills). The pulley diameters are 10, 15, and 20 cm. If an electric motor turns the input pulley at a speed of 2400 rev/min obtain the possible output speeds.

Figure 8.12.   Problems 8.9 and 8.14.

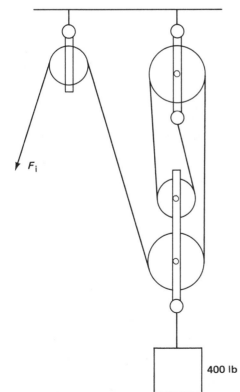

F_i

400 lb

Figure 8.13.   Problem 8.11.

8.12   Two gears with 15 and 255 teeth, respectively, are used to increase the output speed. What is the ratio of the output to input speeds? If the system is ideal, obtain the ratio of output torque to input torque obtainable in this case.

8.13   If the gear system in Problem 8.12 is 55% efficient, calculate the output torque if the input gear is run by a 500 W motor at 1000 rev/min.

8.14   The block and tackle arrangement shown in Fig. 8.12 is used to lift a 4000 lb weight. Assuming ideal conditions, calculate (a) the minimum force to be applied to lift the weight and (b) the tension on each strand of rope.

8.15   A pulley system (90% efficiency) is used to transmit torque from a motor to a lathe. The motor has a power of 1 hp and it is running at a speed of 1745 rev/min. The output wheel rotates at a speed of 200 rev/min. Calculate the net torque acting on the output wheel.

Properties of Matter

Often we are faced with the problem of selecting materials for a specific purpose. For example, when a cable is chosen for a crane, then for the maximum rated load the tension should be below the breaking limit, known as ultimate strength. Similarly, when a cable is selected to hang traffic signals, we consider their thermal expansion properties so that it would not hang too low in the summer. In this chapter some of the mechanical properties of matter are considered. Thermal, electric, and magnetic properties are discussed in later chapters.

Matter is classified into three groups: solids, liquids, and gases:

1 Solids have definite shape and definite volume.
2 Liquids have definite volume, but indefinite shape.
3 Gases have neither definite shape nor definite volume. Gases take the shape and expand into the volume of the container. Figure 9.1 illustrates the differences.

Liquids and gases are together called fluids, the implication being that they are capable of flowing. Because of this ability to flow, the fluids exert buoyant force on an immersed body, transmit an applied pressure undiminished to other parts of an enclosed fluid, etc.

Before we discuss the properties of matter in the three groups separately, let us consider a property common to all, that is, the density.

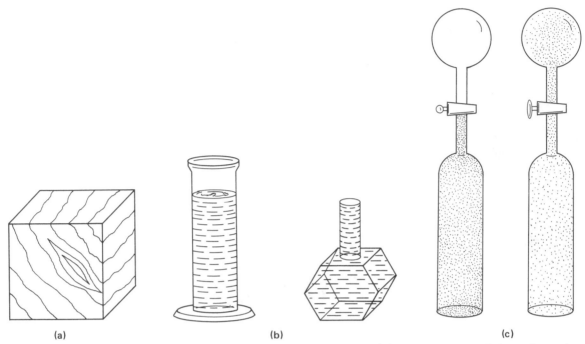

(a)   (b)   (c)

Figure 9.1. (a) Solids have definite shapes and definite volumes. (b) Liquids have definite volumes but they take the shapes of the containers. (c) Gases take the shape and occupy the volumes of the containers.

9.1 MASS DENSITY AND WEIGHT DENSITY

A "denser" material has more mass per unit volume. A material like lead is denser than a lighter material such as aluminum. The mass density of a material is defined as the mass of that material contained in a unit volume. Hence

$$d = m/V, \qquad (9.1)$$

where d stands for mass density, m for mass, and V for volume.

The unit of density in SI system is kg/m^3 although g/cm^3 is in common use because they are smaller numbers by a factor of 1000 and hence easier to use. In the English system the unit is $slug/ft^3$. However, density is often quoted in units of weight per unit volume or weight density in the English system.

$$\text{Weight density}, D = mg/V = dg, \qquad (9.2)$$

that is, weight density is equal to mass density multiplied by the acceleration due to gravity. In SI units weight density is expressed in N/m^3.

Densities of a few selected materials are listed in Table 9.1.

The ratio of the density of a material to the density of water is called specific gravity of the material.

$$\text{Specific gravity} = \frac{d_m}{d_w} = \frac{D_m}{D_w}. \qquad (9.3)$$

where subscript m stands for the material and w for water. Since the density of water is close to $1 \ g/cm^3$,

TABLE 9.1 Density of Materials at Room Temperature, 20°C.

SUBSTANCE	d, kg/m³	d, g/cm³	D, lb/ft³
Solids:			
Aluminum	2,700	2.7	169
Brass	8,700	8.7	540
Copper	8,890	8.89	555
Glass	2,600	2.6	162
Gold	19,300	19.3	1,204
Ice	920	0.92	57
Iron	7,850	7.85	490
Lead	11,300	11.3	705
Silver	10,500	10.5	654
Steel	7,800	7.8	487
Wood			
Maple (sugar)	620–750	0.62–0.75	39–47
Oak	610–730	0.61–0.73	38–45
Poplar	350–500	0.35–0.5	22–31
Walnut	640–700	0.64–0.70	40–43
Liquids:			
Alcohol	790	0.79	49
Benzine	880	0.88	547
Gasoline	680	0.68	42
Mercury	13,600	13.6	850
Water (3.98°C)	1,000	1.0	62.4
Gases (0°C):			
Air	1.29	0.00129	0.0807
Hydrogen	0.090	0.00009	0.0058
Helium	0.178	0.000178	0.0110
Oxygen	1.43	0.00143	1.43

$$\text{Specific gravity} = \frac{D_m V}{D_w V}$$

$$= \frac{\text{Weight of a volume of a substance}}{\text{Weight of an equal volume of water}}. \quad (9.4)$$

EXAMPLE 9.1

A cylindrical iron rod is 2 m long and 5 cm in radius. Its mass is 120 kg. Calculate the mass density, weight density, and specific gravity of iron.

$$\text{Volume of the rod} = \pi r^2 \ell$$

$$= \pi \left(\frac{5 \text{ cm}}{100 \text{ cm/m}}\right)^2 (2 \text{ m}) = 0.016 \text{ m}^3.$$

$$\text{Mass density } d = \frac{\text{mass}}{\text{volume}}$$

$$= \frac{120 \text{ kg}}{0.016 \text{ m}^3} = 7.5 \times 10^3 \text{ kg/m}^3.$$

$$\text{Weight density } D = dg$$

$$= (7.5 \times 10^3 \text{ kg/m}^3)(9.8 \text{ m/s}^2)$$

$$= 73.5 \times 10^3 \text{ N/m}^3.$$

$$\text{Specific gravity} = \frac{\text{Density of material}}{\text{Density of water}}$$

$$= \frac{7.5 \times 10^3 \text{ kg/m}^3}{1 \times 10^3 \text{ kg/m}^3} = 7.5.$$

9.2 DETERMINATION OF DENSITY

specific gravity of a material is equal numerically to the density expressed in g/cm³. Specific gravity of course has no units, since it is a ratio. Multiplying the denominator and numerator of the right-hand side of Eq. (9.3) by V, we get a convenient form of the definition of specific gravity, namely,

As illustrated in the above example, it is easy to determine the density of a solid object which has a regular shape. How do we determine the density of liquids and of irregularly shaped objects? We can readily determine the density of liquids based on density = mass/volume, by accurately determining the mass (by weighing) of a measured volume of liquid. Another method, where measurement of volume is not needed uses specific gravity bottles (Fig. 9.2). A surprisingly simple procedure is as follows: (a) Clean and dry the bottle carefully. Weigh it with the stopper in its place. (b) Fill the bottle with water and close the stopper. Excess water flows out through the capillary tube in the stopper. Dry the outside and weigh the bottle. (c) After removing the water, fill it with the liquid and obtain the weight. Specific gravity is then calculated by using Eq. (9.4).

Specific gravity

$$= \frac{\text{Wt. (liquid + bottle)} - \text{Wt. (bottle)}}{\text{Wt. (water + bottle)} - \text{Wt. (bottle)}} \qquad (9.5)$$

From the specific gravity, density can be calculated. Note that in this method we do not have to determine the volume of the liquid. However, special care should be taken to maintain a constant temperature of the bottle to avoid volume changes.

EXAMPLE 9.2

In an experiment to determine the density of alcohol using a specific gravity bottle the following

Figure 9.2. Specific gravity bottle.

results were obtained: mass of bottle 479 g; mass of bottle and water, 148.2 g; mass of bottle and alcohol 125.4 g. Calculate the specific gravity and density of alcohol.

We use Eq. (9.5) to calculate the specific gravity of alcohol.

$$\text{Specific gravity} = \frac{125.4 \text{ g} - 47.9 \text{ g}}{148.2 \text{ g} - 47.9 \text{ g}}$$

$$= \frac{77.5 \text{ g}}{100.3 \text{ g}} = 0.77.$$

Density of alcohol

= specific gravity × density of water

$$= 0.77 \times 1 \text{ g/cm}^3 = 0.77 \text{ g/cm}^3.$$

9.3 PRESSURE IN A FLUID

Pressure is defined as the normal force acting on a surface divided by its area.

Pressure = Force/Area

$$P = F/A. \qquad (9.6)$$

Pressure P is expressed in N/m^2, lb/ft^2 or lb/in^2. Pressure defined by Eq. (9.6) is a scalar, being the magnitude of the component of the force in a direction normal to the area divided by the area. Equation (9.6) makes no distinction whether the force is produced by solid materials or fluids.

To illustrate the difference between pressure and force, let us consider Fig. 9.3(a) and 9.3(b). The same number of bricks are stacked up in both cases and the forces acting on areas A_1 and A_2 are the same, being the weight of six bricks.

However, the pressure on area A_1 will be twice as much as the pressure of A_2 since the force in the second case is distributed over an area two times bigger than the area in the first case.

In an analogous situation of the liquid it is clear that the pressure at the bottom surface of container

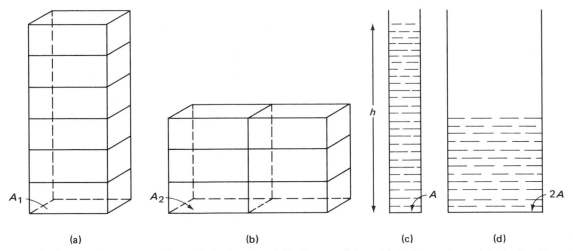

(a)   (b)   (c)   (d)

Figure 9.3. Weight acting on the table is the same in both cases (a) and (b). Since the weight is distributed over a larger area in the second case (b), pressure acting on the surface in the second case is less than that in the first case (a). Similarly, pressure due to the liquid column in (c) is larger than in (d).

1 is twice as much as the pressure at the bottom surface of container 2 [Fig. 9.3(c) and 9.3(d)].

To derive an equation for the pressure, consider Fig. 9.3(c). Let A be the area, h the depth of the liquid column, and d its density. The force acting on area A is given by

Force on area A

$= $ Volume of liquid $\times$ density $\times g$

$= Ahdg.$

Therefore, for pressure we get

$$P = Ahdg/A = hdg. \qquad (9.7)$$

Even though we derived Eq. (9.7) for a surface at the bottom of the fluid, Eq. (9.7) can be applied to obtain pressure at any point in the fluid. Pressure at a point depends only on the depth of the fluid at that point. The pressure in a fluid increases in proportion to the depth in a linear way. Pressure at a point in a fluid does not depend on the size or shape of the fluid column above that point. In addition, pressure at any point is the same in all directions and, irrespective of the orientation of the surface, pressure acting on it is normal to the surface. The force produced on a surface in contact with the fluid is also at right angles to the surface. All these points are illustrated in Fig. 9.4.

EXAMPLE 9.3

Each brick in Fig. 9.3(a) and 9.3(b) weighs 2 lb and its dimensions are 12 in. $\times$ 6 in. $\times$ 3 in. Calculate the pressure exerted on the surface supporting the bricks in each case.

Total force acting on the surface in both cases

$$= 6 \text{ lb} \times 2 = 12 \text{ lb};$$

$$\text{Area } A_1 = 12 \text{ in.} \times 6 \text{ in.} = 72 \text{ in.}^2;$$

$$\text{Pressure on area } A_1 = \frac{12 \text{ lb}}{72 \text{ in.}^2} = \tfrac{1}{6} \text{ lb/in.}^2$$

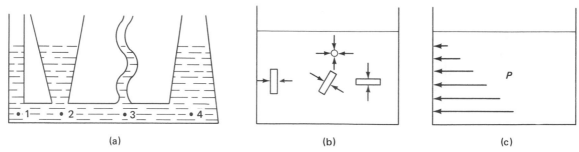

Figure 9.4. (a) Pressure at points, 1, 2, 3, and 4 is the same. Pressure at any point in a liquid depends only on the height of the liquid column above that point, not on the shape of the liquid column. (b) At any point inside a liquid, the pressure is the same in all directions. Pressure acts perpendicular to the surface. If the surfaces are at the same height, pressures acting on them have the same magnitude. (c) Pressure increases proportional to the depth in the liquid.

EXAMPLE 9.4

The water tank (Fig. 9.5) is half filled. Calculate the water pressure at the bottom of the tank and in the pipe on the ground level.

Pressure at any depth is given by $P = hdg$. Pressure at the bottom of the tank is

$$5 \text{ m} \times (1 \times 10^3 \text{ kg/m}^3) \times 9.8 \text{ m/s}^2$$

$$= 4.9 \times 10^4 \text{ N/m}^2.$$

Pressure at the ground level is

$$30 \text{ m} \times (1 \times 10^3 \text{ kg/m}^3) \times 9.8 \text{ m/s}^2$$

$$= 2.94 \times 10^5 \text{ N/m}^2.$$

9.4 ARCHIMEDES' PRINCIPLE

How do we determine the density of irregularly shaped objects? This question was posed to Archimedes in the second century B.C. The question was actually asked by a king in a different form: "Can you detect the presence of base metals in the gold crown?" Fortunately while taking his bath and observing that his body displaced water he discovered a method and the rest is history (Fig. 9.6).

Density of gold is 19 g/cm^3 and that of copper is

8.9 g/cm^3, only half that of gold. If a sufficient amount of copper is added to the gold crown, the density measurement should indicate a lowering of the density. It is interesting to point out that this is the first nondestructive testing method developed. Even though there are more powerful methods like the fluorescence method (Chapter 22) available now, the density measurements are still used because of their simplicity.

Archimedes' principle states that a body wholly or partially submerged in a fluid experiences a

Figure 9.5. Example 9.4.

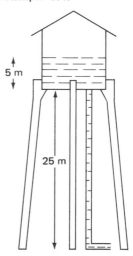

Archimedes

Figure 9.6.

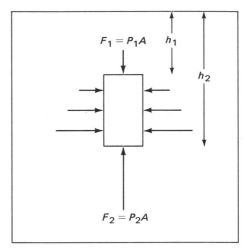

Figure 9.7. Buoyant force acting on an object in a liquid (or fluid) results from the differences in the pressures at the top and bottom surfaces.

buoyant force equal to the weight of the displaced fluid.

We now give a proof of Archimedes' principle. Consider the cylindrical object immersed in a fluid as shown in Fig. 9.7. The forces acting on the cylindrical surface cancel each other, but the force acting upon the bottom is larger than the force down on the top. The pressures at depth h_1 and h_2 are $h_1 dg$ and $h_2 dg$, respectively, and hence the forces acting on the two surfaces are

Force F_1 acting down on the top surface $= Ah_1 dg$.

Force F_2 acting up on the bottom surface $= Ah_2 dg$.

The buoyant force is the difference between the two forces, namely,

$$F_2 - F_1 = Ah_2 dg - Ah_1 dg$$

$$= A(h_2 - h_1)\, dg = A\ell dg$$

$$= Vdg = VD = W,$$

where W is the weight of a volume of liquid equal to the volume of the immersed solid, $A\ell\,(=V)$. Hence the buoyant force is equal to the weight of the displaced liquid.

Now let us see how we can use the Archimedes' principle and what it says.

Figure 9.8 illustrates the effect of the buoyant force on the weight when the object is immersed in a fluid. In Fig. 9.8(b) the object is fully immersed in water. The forces acting on the object are the gravitational force mg down and the buoyant force up equal to the weight of the displaced water. What is read on the balance is the tension in the string which is equal to the difference between these two forces. Let us designate

weight of the object in air, W_a

weight of the object in water, W_w

volume of the object, V

density of the object, d

density of water, d_w.

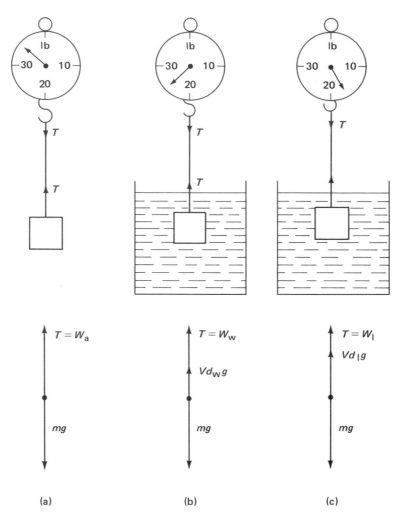

Figure 9.8. When an object is weighed in water (b) and liquid (c), its apparent weight is less than its real weight in air. The difference between the real weight and the apparent weight is the buoyant force which is equal to the weight of the displaced water or liquid.

Referring to the free-body diagram, since the object is in equilibrium, we get

Buoyant force (Vd_wg) + Weight in water (W_w)

= Weight of the object (W_a),

that is,

$$Vd_wg + W_w = W_a. \tag{9.8}$$

By rearranging this equation we may obtain the following equation for the volume of the object, the weight density of the object, and its specific gravity.

$$\text{Volume } V = \frac{W_a - W_w}{d_wg}; \tag{9.9}$$

Weight density of the object $D = \dfrac{W_a}{V}$;   (9.10)

Weight density of water $D_w = d_w g$

$$= \frac{W_a - W_w}{V};$$ (9.11)

Specific gravity of the object $= \dfrac{D}{D_w}$

$$= \frac{W_a}{W_a - W_w}.$$ (9.12)

The term $W_a - W_w$ is known as the apparent loss of weight in water. According to Eq. (9.12), the specific gravity of an object is the ratio of the weight in air to the apparent loss of weight in water.

If the specific gravity of a liquid of unknown density is to be determined, immerse the object in the liquid and note the weight W_1 in the liquid [Fig. 9.8(c)]. We have an equation similar to Eq. (9.8), namely,

$$V d_1 g + W_1 = W_a,$$ (9.13)

where d_1 is the density of the liquid. From Eqs. (9.8) and (9.13), we have

Specific gravity of the liquid

$$= \frac{W_a - W_1}{W_a - W_w}.$$ (9.14)

Note that $W_a - W_1$ is the apparent loss of weight in liquid.

EXAMPLE 9.5

The mass of an object as determined in air, water, and alcohol is 18 kg, 16 kg, and 17 kg, respectively. Determine the volume of the object, its density and the density of the alcohol.

Buoyant force acting on the object in water

$$= 2 \text{ kg} \times g;$$

Mass of water displaced by the object = 2 kg;

Volume of water displaced = Volume of the object

$$= \frac{2 \text{ kg}}{d_w} = \frac{2 \text{ kg}}{1000 \text{ kg/m}^3} = 2 \times 10^{-3} \text{ m}^3;$$

Mass density of the object $= \dfrac{m}{V} = \dfrac{18 \text{ kg}}{2 \times 10^{-3} \text{ m}^3}$

$$= 9 \times 10^{-3} \text{ m}^3;$$

Mass of displaced liquid $= 18 \text{ kg} - 17 \text{ kg} = 1 \text{ kg};$

Mass density of the liquid $= \dfrac{1}{2 \times 10^{-3} \text{ m}^3}$

$$= 5 \times 10^2 \text{ kg/m}^3.$$

If the weight of an object is less than the weight of an equal volume of a liquid, the object floats with only part of it immersed in the liquid. What fraction of the object will be immersed in the liquid? When an object is dropped into a liquid, two forces act on it: its weight and the buoyant force (Fig. 9.9). The latter increases as the object sinks deeper into

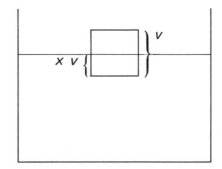

Figure 9.9. When an object floats in a liquid, its weight is equal to the weight of the displaced liquid. Volume of the object is V and XV is the fraction of the volume immersed in the liquid.

the liquid and displaces more liquid, and eventually it becomes equal to the weight of the object. The object does not sink further.

Let us denote the fraction of the object immersed in the liquid by X. For a floating object, the fact that its weight is equal to the weight of the displaced liquid gives us the equation

$$Vdg = XVd_1g$$

or

$$d = Xd_1, \qquad (9.15)$$

where V is the total volume of the object, d its density, and d_1 the density of the liquid. For a floating object X is equal to or less than 1. Hence, from Eq. (9.15) we get the conditions for floating as $d \leqslant d_1$, that is, a less dense object always floats in a denser liquid. Thus ice floats in water because its density is less than that of water and cream floats in milk for the same reason. Ships made of steel, a material of higher density, float in water too (Why?)

Figure 9.10. Example 9.7.

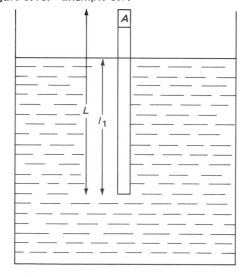

Three points to remember are as follows: (a) volume of a displaced liquid is equal to the volume of the part of the object immersed in the liquid; (b) buoyant force is equal to the weight of the displaced liquid whether fully immersed or floating; (c) buoyant force for a floating object is equal to the weight of the object.

EXAMPLE 9.6

A uniform wooden stick with a specific gravity of 0.13 floats in water, alcohol, and gasoline. Calculate the fraction of the length of the stick in each liquid (Fig. 9.10).

Let A be the area of the cross-section and L the length of the stick. Let ℓ_1, ℓ_2, ℓ_3 indicate the lengths of the rod submerged in water, alcohol, and gasoline, respectively.

Weight of the rod = Volume × Weight density

$$= LAD$$

Weight of water displaced = $\ell_1 AD_w$

Weight of alcohol displaced = $\ell_2 AD_a$

Weight of gasoline displaced = $\ell_3 AD_g$

These weights are equal to the weight of wood. Therefore, we have

$$\ell_1 AD_w = \ell_2 AD_a = \ell_3 AD_g = LAD$$

or

$$\ell_1 = L \times \frac{D}{D_w}; \qquad (9.16)$$

$$\ell_2 = L \times \frac{D}{D_a}; \qquad (9.17)$$

$$\ell_3 = L \times \frac{D}{D_g}. \qquad (9.18)$$

The length of the immersed section is inversely proportional to the density of the liquid.

For the example above it is clear that the length of the part of the stick under water is related to the density of liquid. This fact has been used in the design of an instrument called the hydrometer which measures the density of liquids. From our experience we know that it is hard to keep a long object floating vertically in a liquid. To make hydrometers float vertically, they are loaded at the bottom (Fig. 9.11). Hydrometers have vertical scales fixed to them to read the densities directly.

Battery acid testers are a common form of the hydrometer. In a battery, specific gravity of the electrolyte varies from 1.3 in a fully charged state to 1.15 in a fully discharged state of the battery. By reading the specific gravity by means of a hydrometer, the condition of the battery can be determined.

Figure 9.11. Hydrometer. The weight at the bottom keeps the hydrometer vertical.

Scale

Weight

9.5 MEASUREMENT OF PRESSURE

In all the discussions above we neglected the effect of atmospheric pressure. What causes atmospheric pressure? The earth is surrounded by a layer of air, of decreasing density with height. Like all other fluids air exerts a force equal to the weight of air everywhere on the surface of the earth. Pressure due to this force, that is, the weight of air per unit area is the atmospheric pressure. The standard atmospheric pressure at sea level is equal to 14.7 lb/in.^{2} or 1.013×10^5 N/m^2.

EXAMPLE 9.7

A table top is 2 m long and 1 m wide. Calculate the force acting on the table due to atmospheric pressure.

$$\text{Atmospheric pressure} = 1.013 \times 10^5 \text{ N/m}^2$$

$$\text{Area of the table top} = 2 \text{ m} \times 1 \text{ m} = 2 \text{ m}^2$$

Total force acting on the table

$$= PA = (1.013 \times 10^5 \text{ N/m}^2)(2 \text{ m}^2)$$

$$= 2.026 \times 10^5 \text{ N}.$$

This force is equivalent to keeping a weight of 20,000 kg on the table, yet the table experiences no ill effect. Is the table that strong or is there some other reason?

Let us come back to the topic of this section, the measurement of pressure. Consider a U-tube filled with a liquid [Fig. 9.12(a)] of mass density d.

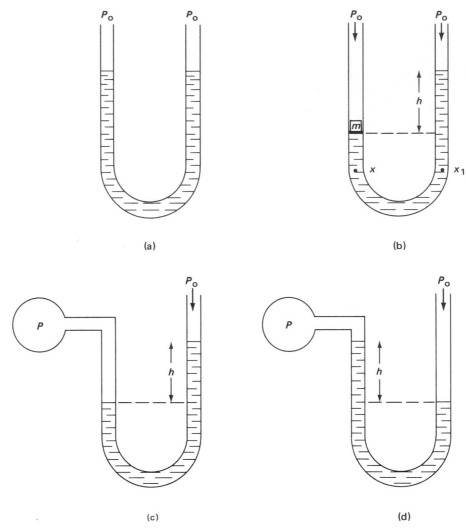

Figure 9.12. (a) U-tubes filled with liquid are often used for measuring pressure. (b) When force (due to the weight) is applied on one end of the tube levels become uneven so that the pressure at points x and x_1 are of equal height. (c) and (d) Liquid levels change if the pressure inside a system connected to the U-tube is different from atmospheric pressure.

Atmospheric pressure is acting on both sides, and the levels are the same on both sides. By placing a mass m on a tight fitting but movable piston, apply a force $F\,(=mg)$ on one side. An additional pressure F/A acts on this side [Fig. 9.12(b)]. The liquid on this side drops while the liquid on the other side rises. For equilibrium, the pressures at two points such as x and x_1 at the same level are equal. Hence

it follows that

$$F/A = hdg. \tag{9.19}$$

Thus the increase in pressure on one side can be measured in terms of the difference in height h of the liquid columns.

Instead of adding weight to the left-hand side,

suppose it is connected to a chamber of higher pressure [Fig. 9.12(c)]. In this case the pressure inside the chamber is given by

$$P = hdg + P_0,\qquad(9.20)$$

where h is the difference in the heights of the two columns and P_0 is the atmospheric pressure.

If the chamber pressure is lower than P_0 [Fig. 9.12(d)], the liquid column in the open tube goes down. The pressure in this case is given by

$$P + hdg = P_0.\qquad(9.21)$$

Equations (9.20) and (9.21) may be combined into one equation, namely,

$$P = P_0 \pm hdg,\qquad(9.22)$$

where the $+$ indicates the measured pressure is higher than the atmospheric pressure, while $-$ indicates the measured pressure is lower than the atmospheric pressure.

Liquid filled U-tubes described above are generally known as manometers. In this and other common instruments what is determined is the difference between the actual pressure and the atmospheric pressure. This difference, like the quantity hdg in Eq. (9.22), is called the gauge pressure. Hence we obtain the absolute pressure from

Absolute pressure = Atmospheric pressure

+ Gauge pressure   (9.23)

Barometers are used to measure atmospheric pressure. They are also called Torricelli barometers after its inventor Evangelista Torricelli. A barometer consists of a long glass tube which after being filled with mercury is inverted in a dish of mercury (Fig. 9.13). The height of the mercury column adjusts so that the pressure due to the mercury column (hdg) is equal to the atmospheric pressure. The space above the mercury column resulted because the

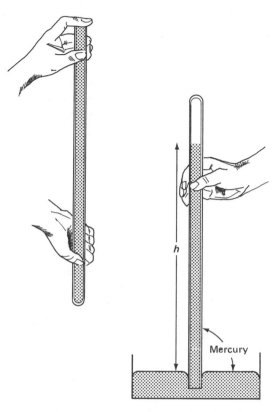

Figure 9.13. Mercury barometer. Height of mercury column h is proportional to the atmospheric pressure.

mercury moved out of that region. This region contains only a very small amount of mercury vapor. The pressure due to the mercury vapor can be neglected at room temperature. Hence by reading the height of the mercury column above the level in the open dish, one can obtain the atmospheric pressure in absolute units as

$$P_0 = hdg.\qquad(9.24)$$

However, pressure is often read and quoted in terms of the height of the column, that is, in cm of Hg (mercury), mm of Hg or inches of Hg. The following example illustrates the relationship between cm of Hg and the basic unit.

EXAMPLE 9.8

What is the pressure in absolute units when pressure is 76 cm of Hg?

When we say pressure is 76 cm of mercury, we mean that the atmospheric pressure is equivalent to a column of mercury of 76 cm length. Therefore, the pressure corresponding to 76 cm of mercury or 0.76 m of mercury is

$$hdg = (0.76 \text{ m})(13.6 \times 10^3 \text{ kg/m}^3)(9.8 \text{ m/s}^2)$$

$$= 1.01 \times 10^5 \text{ N/m}^2.$$

Mechanical Barometers

The liquid barometers are cumbersome to use, while the mechanical barometers are compact and provide direct reading of the pressure on a scale. One of the latter types, known as the aneroid barometer, is illustrated in Fig. 9.14. The chamber C made of corrugated metal is partially evacuated. The metal spring S holds the chamber without collapsing due to the pressure difference inside and outside the chamber. As the outside pressure changes, the corrugated surface and the attached spring move. This motion in turn is amplified by the lever L which causes an indicator to move on a scale. The aneroid barometer is not an absolute barometer and, therefore, it should be calibrated against a mercury barometer.

As we have seen, there are several units in use for pressure in addition to the basic units of N/m^2 and lb/ft^2. We list all currently used units in Table 9.2.

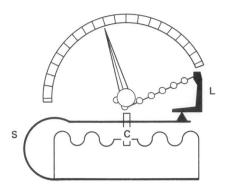

Figure 9.14. Aneroid barometer. As the pressure changes, the chamber expands or contracts, causing the lever and the needle to deflect. The pressure is read on the calibrated scale.

9.6 TRANSMISSION OF PRESSURE—PASCAL'S LAW

Pressure inside a fluid is directly proportional to the depth. For example, the pressure experienced by the wall of the container in Fig. 9.15 at point 1 is $h_1 dg$ and at point 2 it's $h_2 dg$, etc. Suppose an additional pressure $P (= F/A)$ is applied to the top of the liquid through a tight fitting piston. With the piston in position we have a completely enclosed system confining the static liquid. The pressure everywhere inside the liquid increases by an amount equal to P; pressure at point 1 is now $P + h_1 dg$ and at point 2 is now $P + h_2 dg$. This fact was first observed by Pascal and the statement summarizing these observations is now known as Pascal's Law.

TABLE 9.2 Units Used in Pressure Measurements

UNITS	RELATION TO BASIC UNITS
Pascal, Pa	1 Pa = 1 N/m^2
1 lb/ft^2 and lb/in.^{2}	1 lb/ft^2 = 144 lb/in.^{2}
cm of mercury	1 cm Hg = 1330 N/m^2
mm of mercury called torr	1 torr = 133 N/m^2
bar, standard atmospheric pressure	1 bar = 1.013 × 10^5 N/m^2
millibar (often used in meteorology)	1 mb = 1.013 × 10^2 N/m^2
Atmosphere	1 atm = 1.013 × 10^5 N/m^2 = 14.7 lb/in.^{2}

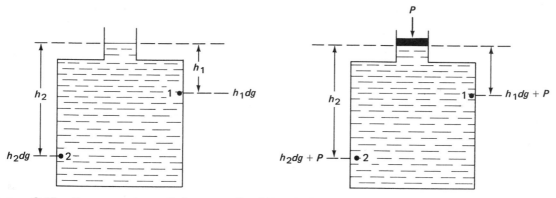

Figure 9.15. If pressure is applied to a confined liquid, the pressure at every point inside the liquid increases by the same amount.

Pascal's Law

Whenever pressure in a confined, incompressible fluid is increased or decreased at any point, this change in pressure is transmitted undiminished through the entire fluid and the inner walls of the container.

An important application of Pascal's law is found in the hydraulic press, which contains two interconnected cylinders (Fig. 9.16). Assume the area of one of the cylinders A_i, to be smaller than the area A_o of the other one. Consider a force F_i applied to the piston in the smaller cylinder. The resulting increase in pressure $P (=F_i/A_i)$ is transmitted to the piston in the larger cylinder. It experiences a force F_o which is a product of P and A_o. Hence

$$F_o = PA_o = \frac{F_i}{A_i} \times A_o$$

or

$$F_o/F_i = A_o/A_i. \tag{9.25}$$

Thus we obtain a larger output force F_o by applying a smaller input force F_i. The output force can be made as large as possible by increasing the ratio of A_o/A_i of the areas. As mentioned in the case of a simple machine, it is important to remember that there is no savings in work. The distance d_i, the

force F_i travels, is much larger than the distance d_o, the force F_o travels. When the piston in the smaller cylinder moves through a distance d_i, the volume of the fluid that moves into the bigger cylinder is $A_i d_i$. To accommodate this amount of fluid, the piston in the larger cylinder moves through a distance d_o, so that $d_o A_o = d_i A_i$ or $d_o = d_i A_i/A_o$. The work input at the small piston is

$$W_i = F_i \times d_i$$

and the work output at the bigger piston is

$$W_o = F_o \times d_o.$$

Figure 9.16. Hydraulic press. For a smaller input force F_i a larger output force F_o is obtained. The ratio of the forces F_o/F_i is equal to the ratio of the areas A_o/A_i.

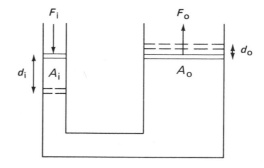

By substituting for F_o and d_o, we get

$$W_o = \frac{F_i A_o}{A_i} \times \frac{d_i A_i}{A_o} = F_i d_i.$$

Hence the work output is the same as the input under ideal conditions. However, actual work output is usually smaller because of energy losses due to friction, etc.

The ideal mechanical advantage of the hydraulic press is the ratio of the input distance to output distance.

$$\text{IMA} = d_i/d_o. \tag{9.26}$$

Since $d_i A_i = d_o A_o$, Eq. (9.26) may also be written as

$$\text{IMA} = A_o/A_i, \tag{9.27}$$

that is, the ideal mechanical advantage is equal to the ratio of the areas of the two cylinders. The actual mechanical advantage, as in the case of other machines, is given by

$$\text{AMA} = F_i/F_o. \tag{9.28}$$

There are innumerable applications of this principle governing the hydraulic press such as a barber's chair, automobile lifts in garages, and stamping presses in industrial plants.

EXAMPLE 9.9

The input and output pistons of a hydraulic jack have diameters 2 in. and 8 in., respectively. (a) Calculate the IMA of the hydraulic press. (b) What input force is required to deliver an output force of 1000 lb assuming no losses. (c) Obtain the distance the input piston moves when the output piston moves by 2 in.

Area of the input piston

$$= \pi (\frac{2 \text{ in.}}{2})^2 = 3.14 \text{ in.}^2;$$

Area of the output piston

$$= \pi (\frac{8 \text{ in.}}{2})^2 = 50.24 \text{ in.}^2;$$

Ideal mechanical advantage

$$= \frac{A_o}{A_i} = \frac{50.24 \text{ in.}^2}{3.14 \text{ in.}^2} = 16.$$

If there is no frictional loss, IMA = AMA = 16.

$$\text{AMA} = F_o/F_i$$

Hence we get

$$16 = (1000 \text{ lb})/F_i.$$

Input force $F_i = (1000 \text{ lb})/16 = 62.5 \text{ lb}.$

We use Eq. (9.26) to obtain the distance traveled by the input piston.

$$\text{IMA} = d_i/d_o$$

$$\text{IMA} = 16, \ d_o = 2 \text{ in.}$$

Distance traveled by the input piston is

$$d_i = \text{IMA} \times d_o$$

$$= 16 \times 2 \text{ in.} = 32 \text{ in.}$$

9.7 ELASTICITY

With little effort the shape of gases can be changed, but large amounts of force are needed to produce distortions in solids. The common types of distortions, namely, elongation, compression, bending, twisting, and shear are shown in Fig. 9.17. Materials that return to their original shape on withdrawal of the distorting force are called elastic while those that do not return to the original shape are called inelastic. Steel wires are perfectly elastic if the applied force does not exceed a certain limit,

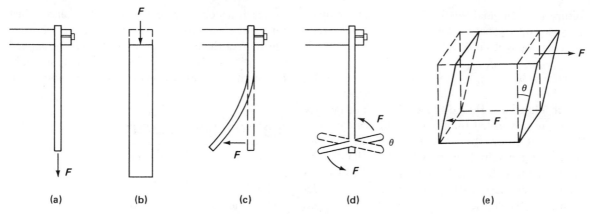

Figure 9.17. Common types of distortions: (a) tension, (b) compression, (c) bending, (d) twisting, and (e) shear.

while a piece of gum is inelastic. Thus elasticity is a measure of the ability of a material to return to the original size and shape when the distorting force is withdrawn.

In all types of deformations mentioned earlier, distortion is linearly proportional to the applied force if the applied force does not exceed a certain limit. Beyond this limit, known as the elastic limit, the linear relationship breaks down. To understand these common features, let us perform an experiment where gradually increasing amounts of force

are applied to a wire and the resulting increases in length are measured using an equipment similar to the one illustrated in Fig. 9.18. Let us also plot the results (Fig. 9.19).

The results plotted in Fig. 9.19 show that there are two well-defined regions of the curve: (I) The elastic region, where the elongation is proportional to the force and the material is perfectly elastic, that is, on removal of the weights the wire will

Figure 9.18 An arrangement to measure elongation as a function of applied force.

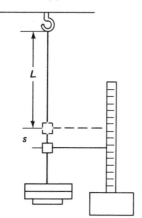

Figure 9.19. Elongation of a wire as a function of the tension. In the first region, elongation is proportional to the tension.

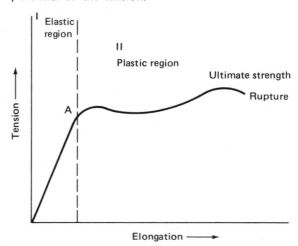

return to its original length. In this case, the elongation is proportional to the applied force. This relation may be written as

$$F = ks \qquad (9.29)$$

which is known as Hooke's Law. The proportionality constant k is called the force or stiffness constant. The point A on the curve which divides the two regions is generally known as the elastic limit. The elastic limit indicates the maximum elongation a material can have without being permanently deformed. (II) Plastic region, where the curve is no longer a straight line, indicating there is no linear relationship between elongation and applied force. In other words, Hooke's law does not apply here. Also once material is in this region, removal of the weights does not restore its original form. In the section of the curve where the curve slopes towards the horizontal axis, elongation proceeds even when the tension decreases. This behavior is known as plastic flow. Also note that the rate of elongation is faster in this region than in the elastic region. Finally

we reach a point where rupture takes place. Just before the rupture point the curve has a peak, which gives the maximum force that can be applied to the wire. This force is called the ultimate strength. This is the maximum tension a wire can withstand.

9.8 YOUNG'S MODULUS

We will now confine our discussion to the elastic region.

Let us repeat the above experiment this time to understand the relationship between elongation, applied force, length of the wire, and area of the wire (Fig. 9.20). Take a wire (copper wires or steel wires are used in laboratories for these experiments) of length L, apply a force F and observe the elongation ΔL. Take another wire twice the length. We observe an elongation which is two times bigger for the same applied force. If the experiment is repeated with another wire of the same material, but having twice the cross-sectional area we will observe that to get the same elongation ΔL, we have to apply

Figure 9.20. Elongation per unit length is proportional to force per unit area, i.e., strain is proportional to stress.

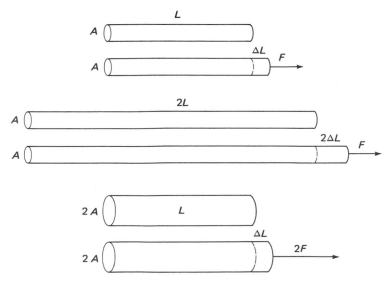

twice the amount of force. The results of our experiments are the following:

1 ΔL is proportional to F (in the elastic limit);
2 ΔL is proportional to L;
3 ΔL is inversely proportional to A.

Upon combining these proportionalities, we obtain

$$\Delta L \propto \frac{F}{A} L.$$

Dividing both sides by L, we obtain the convenient proportionality

$$\frac{\Delta L}{L} \propto \frac{F}{A},$$

that is, *the fractional increase in length is proportional to the force per unit area. The fractional increase $\Delta L/L$ is called the strain and force per unit area is called the stress:*

$$\text{Strain} = \Delta L/L, \tag{9.30}$$

$$\text{Stress} = F/A. \tag{9.31}$$

The proportionality $\Delta L/L \propto F/A$ may now be stated as strain is proportional to stress. This is another form of Hooke's law.

TABLE 9.3 Approximate Values of Young's Modulus for Selected Materials.

MATERIAL	YOUNG'S MODULUS	
	N/m^2	lb/in.^{2}
Aluminum	7×10^{10}	10×10^6
Brass	9.1×10^{10}	13×10^6
Copper	12.5×10^{10}	18×10^6
Iron, Cast	9.1×10^{10}	13×10^6
Steel, drawn	20×10^{10}	2.3×10^6
Rubber, vulcanized	14×10^4	29×10^6
Tungsten, drawn	35×10^{10}	51×10^6

If we introduce a proportionality constant Y, we obtain the equation

$$\frac{\Delta L}{L} Y = \frac{F}{A}. \tag{9.32}$$

The constant Y, which is a constant for a given material, is called Young's modulus. Rearranging Eq. (9.32) we have

$$Y = \frac{F/A}{\Delta L/L} = \frac{\text{Stress}}{\text{Strain}}, \tag{9.33}$$

that is, Young's modulus is equal to stress divided by strain. The stress F/A has the units N/m^2, lb/ft^2, lb/in.^{2} The strain has no units. Therefore, Y and stress have the same units. Values of Y are often quoted in N/m^2 or lb/in.^{2}

Typical values of Young's modulus are given in Table 9.3. Note that Eqs. (9.32) and (9.33) are equally valid for compression, in which case ΔL stands for the decrease in length.

EXAMPLE 9.10

A force of 90 lb is applied to a copper wire which is 10 ft long and 0.05 in. diameter. How much does it stretch.

Area of cross-section of the wire

$$= \pi (\frac{0.05 \text{ in.}}{2})^2 = 0.002 \text{ in.}^2$$

To calculate ΔL we use Eq. (9.33), which on rearranging gives us

$$\Delta L = FL/AY.$$

We are given

$$A = 0.002 \text{ in.}^2$$

$$F = 90 \text{ lb}$$

$$L = 10 \text{ ft}$$

TABLE 9.4 Typical Values of Elastic Limits and Ultimate Strengths of Materials Under Tension

MATERIAL	ELASTIC LIMIT 10^8 N/m^2	10^3 lb/in.^{2}	ULTIMATE TENSILE STRENGTH 10^8 N/m^2	10^3 lb/in.^{2}
Aluminum	1.3	19	1.4	21
Brass	3.8	55	4.7	67
Copper	1.5	22	3.4	49
Iron, Wrought	1.6	24	3.2	27
Steel, hard	2.7	40	5.5	80
Steel, spring tempered	11.7	170	13.8	200

$$Y = 18 \times 10^6 \text{ lb/in.}^2 \quad \text{(from Table 9.3)}$$

$$\Delta L = \frac{FL}{AY} = \frac{(90 \text{ lb})(10 \text{ ft})}{(0.002 \text{ in.}^2)(18 \times 10^6 \text{ lb/in.}^2)}$$

$$= 2.5 \times 10^{-2} \text{ ft} = 0.3 \text{ in.}$$

EXAMPLE 9.11

Obtain the maximum load that can be supported by the copper wire of Example 9.10. The ultimate strength of copper (see Table 9.4) is 4.9×10^4 lb/in.^{2}

The cross-sectional area of the wire = 0.004 in.^{2} The maximum load divided by the area should be equal to the ultimate strength. Therefore, we have

$$\text{Maximum load} = 4.9 \times 10^4 \text{ lb/in.}^2 \times 0.002 \text{ in.}^2$$

$$= 98 \text{ lb.}$$

EXAMPLE 9.12

How much will an aluminum bar 6 by 12 cm in cross-section and 5 m long shorten under a compresson load of 3000 newtons.

Area of cross-section of the bar

$$= \frac{6 \text{ cm}}{100 \text{ cm/m}} \times \frac{12 \text{ cm}}{100 \text{ cm/m}} = 0.0072 \text{ m}^2$$

Young's modulus of aluminum = 7×10^{10} N/m^2.

Substituting the given quantities into Eq. (9.33) we get

$$Y = \frac{F/A}{\Delta L/L}$$

$$7 \times 10^{10} \text{ N/m}^2 = \frac{3000 \text{ N}/0.0072 \text{ m}^2}{\Delta L/5 \text{ m}}$$

or

$$\Delta L = 3 \times 10^{-5} \text{ m} = 3 \times 10^{-2} \text{ mm.}$$

For safety considerations maximum load applied is always less than the ultimate strength. The ratio of the ultimate strength to maximum load is known as the factor of safety. The factor of safety varies from material to material. A factor of safety of 4 is used for steel and 10 for brick structures.

9.9 SHEAR MODULUS

Two equal forces applied in opposite directions and on opposite sides (top and bottom) produce a distortion known as shear distortion (Fig. 9.21). A similar distortion can be easily demonstrated by applying forces on a thick book (Fig. 9.22). Shear effect in a solid results from the slippage of layers of atoms from each other like the pages in the book. The force is applied on the area A and hence the

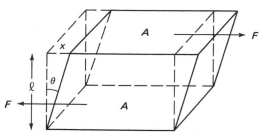

Figure 9.21. Shear distortion is produced by two equal forces applied in opposite directions and on opposite sides.

TABLE 9.5 Shear Modulus of Selected Materials

| MATERIAL | SHEAR MODULUS | |
	10^{10} N/m^2	10^6 lb/in.^{2}
Aluminum	2.4	3.4
Brass	3.6	5.1
Copper	4.2	6.0
Iron	7	10
Lead	0.56	0.8
Steel	8.4	12

stress in this case is given by

$$\text{Stress} = F/A. \quad (9.34)$$

Strain is measured in terms of the angle ϕ in Fig. 9.20. Since the angles are ordinarily very small, ϕ can be replaced by x/ℓ. Within the elastic limits, shear stress is proportional to shear strain. The shear modulus S is defined as the shear stress divided by the strain

$$S = \frac{F/A}{x/\ell}. \quad (9.35)$$

The units of shear modulus are the same as those of Young's modulus. From the values given in Table 9.5, it is clear that shear modulus of a material is smaller than Young's modulus, the obvious conclusion being that it is easier to slide atoms past one another than to pull them apart.

Twisting of rigid rods is another demonstration of shear distortion. One end of the rod in Fig. 9.23 is fixed and a torque is applied to the other end.

Figure 9.22. Shear type distortion is produced on a book by applying a force on the top cover of the book. In shear distortion layers of atoms slide past each other like pages in this book.

The result of the torque is a twisting of the rod, the top layer experiencing an angular displacement θ from the original position. The fact that layers of atoms do slip in this case is illustrated in Fig. 9.23 just like they do when shear stress is applied on a block.

The amount of twisting θ, measured as the displacement of the extreme layer will depend on the torque, length of the rod, and the shear modulus. It is given by

$$\theta = 2TL/\pi SR^4, \quad (9.36)$$

Figure 9.23. A torque applied to one end of a rod, the other end being fixed, produces a twisting. A vertical line drawn on the surface of the rod (a) twists as shown in (b) as the rod twists due to the torque. The displacement of the line is a maximum at the top where the twisting is a maximum. Slippage between layers of atoms in twisting is shown in (c).

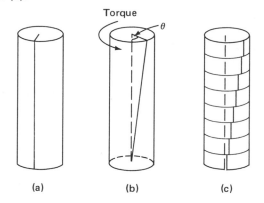

where θ is the angle of twisting in radians, T the torque, L length of the rod, R the radius, and S the shear modulus. In SI units torque is in N · m, L in m, R in m, and S in N/m^2. In English units torque is usually given in lb · ft, L in ft, R in in., and S in lb/in.^{2} Since for length both ft and in. are used, converting all quantities so that they contain only ft or in. should be the first step in solving problems as illustrated in the following example.

EXAMPLE 9.13

A 10 ft long 2 in. diameter drive shaft is used in an engine to deliver 200 hp at 800 rev/min. Calculate the angle through which the shaft is twisted. The rod is made of steel ($S = 12 \times 10^6$ lb/in.^{2}).

We first calculate the torque acting on the shaft using

$$P = T\omega$$

$$P = 200 \text{ hp} \times \frac{550 \text{ ft} \cdot \text{lb/s}}{\text{hp}}$$

$$= 1.1 \times 10^5 \text{ ft} \cdot \text{lb/s};$$

$$\omega = \frac{(800 \text{ rev/min}) \times (2\pi \text{ rad/rev})}{60 \text{ s/min}} = 83.8 \text{ rad/s};$$

$$T = \frac{1.1 \times 10^5 \text{ ft} \cdot \text{lb/s}}{83.8 \text{ rad/s}} = 1312 \text{ lb} \cdot \text{ft}.$$

Since S is given in lb/in.^{2}, it is convenient to convert all quantities in units of inches:

$$T = 1312 \text{ lb} \cdot \text{ft} \times 12 \text{ in./ft} = 1.57 \times 10^4 \text{ lb} \cdot \text{in}.$$

$$L = 10 \text{ ft} \times 12 \text{ in./ft} = 120 \text{ in}.$$

$$R = 1 \text{ in}.$$

$$S = 12 \times 10^6 \text{ lb/in.}^2$$

$$\theta = \frac{2TL}{\pi SR^4} = \frac{2(1.57 \times 10^4 \text{ lb} \cdot \text{in.})(120 \text{ in.})}{3.14(12 \times 10^6 \text{ lb/in.}^2)(\text{in.})^4}$$

$$= \frac{1}{10} \text{rad} = 5.7°.$$

9.10 BULK MODULUS

When forces act uniformly over the whole surface of a body (Fig. 9.24), its volume changes. The stress in this case is the force per unit area (only forces normal to the surface contribute to the volume change; therefore, only normal forces are taken into account). Hence

$$\text{Stress} = F/A = P \qquad (9.37)$$

$$\text{Strain} = \Delta V/V. \qquad (9.38)$$

In Eq. (9.37), P stands for pressure, force per unit area. The ratio of stress is called bulk modulus and is expressed as

$$B = \frac{F/A}{\Delta V/V} = \frac{P}{\Delta V/V}. \qquad (9.39)$$

Figure 9.24. Change in volume per unit volume is proportional to the force per unit area.

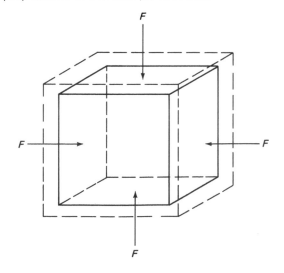

Bulk modulus also has the same units as Young's modulus. We note further that the units are the same as the units of pressure. Table 9.6 gives the modulus of selected bulk materials.

EXAMPLE 9.14

A pressure of 1×10^5 Pa is applied to 2 m^3 of water. Calculate the change in volume.

Change in volume can be calculated using Eq. (9.27), namely,

$$B = \frac{P}{\Delta V / V}$$

or

$$\Delta V = \frac{PV}{B} = \frac{1 \times 10^5 \times 2}{2.3 \times 10^9 \text{ Pa}} = 8.7 \times 10^{-5} \text{ m}^3.$$

9.11 DISTORTIONS IN FLUIDS

Fluids in general do not offer resistance to tension and shear stresses. However, liquids can stand the compressive stress of pressure. Even though bulk moduli of liquids are smaller than those of solids, substantial pressures have to be applied to produce measurable volume changes.

Unlike solids and liquids, gases are easily compressible. At atmospheric pressure if a gas occupies a volume of 2 m^3, increasing the pressure by an amount of 1×10^5 N/m^2 (another atmospheric pressure) decreases it to 1 m^3. Compare it to the small change in volume of 2 m^3 of water (Problem 9.14); the volume changes by 8.7×10^{-5} m^3 for an increase in pressure of 1×10^5 N/m^2.

The volume change of a given amount of gas at constant temperature can be studied by using equipment similar to that shown in Fig. 9.25. The syringe is sealed off at the bottom, and a wooden platform attached to the moving piston. The pressure of the gas inside can be increased or decreased by adding to or taking away weights kept on the platform.

The volume of a gas is read at various pressures obtained by adding weights and results are plotted in Fig. 9.26. When the pressure doubles, volume reduces to half and when pressure triples, volume reduces by $\frac{1}{3}$. The relationship between P and V can now be deduced. At point A on the graph, the product of pressure and volume is $2P \times V/2 = PV$ and at point B the product is again PV. Similarly, at every point on the curve, we get the product of

TABLE 9.6 Bulk Modulus of Selected Materials

MATERIAL	N/m^2	lb/in.^{2}
Solids		
Aluminum	7×10^{10}	10×10^6
Brass	6.1×10^{10}	8.5×10^6
Copper	14×10^{10}	20×10^6
Iron	10×10^{10}	14×10^6
Lead	0.77×10^{10}	1.1×10^6
Steel	16×10^{10}	23×10^6
Liquids		
Alcohol, ethyl	0.9×10^9	1.3×10^5
Benzene	1.05×10^9	1.5×10^5
Kerosene	1.3×10^9	1.0×10^5
Mercury	28×10^9	40×10^5
Water	2.1×10^9	3.1×10^5

Figure 9.25. A syringe adapted to study pressure-volume relationship of a gas.

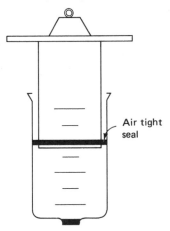

Air tight seal

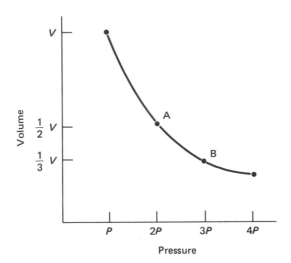

Figure 9.26. Variation of volume of a gas as a function of pressure at constant temperature.

pressure and volume to be a constant equal to the original value of the product.

Robert Boyle (1627-1691) originally discovered this pressure-volume relationship of the gases which is known as Boyle's law and which states *the volume of a confined body of gas varies inversely as the pressure, provided the temperature remains constant.* The law can also be expressed as

$$P_1 V_1 = P_2 V_2. \qquad (9.40)$$

The equation tells us that, as pressure changes, volume also changes, but that their product is a constant.

EXAMPLE 9.15

An air bubble having a radius of 1 cm is 10 m below the water surface. Calculate its volume when it rises to the water level.

As the air bubble rises, its pressure decreases. The volume change can be calculated using Boyle's law.

Pressure 10 m below water level

= Atmospheric pressure

+ pressure due to the water column.

Atmospheric pressure = 1.01×10^5 N/m^2.

Pressure due to water column

$$= h d_{\mathrm{w}} g$$

$$= (10 \text{ m})(1 \times 10^3 \text{ kg/m}^3)(9.8 \text{ m/s}^2)$$

$$= 9.8 \times 10^4 \text{ N/m}^2.$$

Pressure 10 m below water level

$$= (1.01 \times 10^5 \text{ N/m}^2) + 9.8 \times 10^4 \text{ N/m}^2$$

$$= 1.99 \times 10^5 \text{ N/m}^2.$$

Volume at 10 m under water

$$= \frac{4}{3}\pi r^3 = 4.2 \times 10^{-6} \text{ m}^3.$$

Volume at the surface of water

$$= \frac{P_1 V_1}{P_2}$$

$$= \frac{(1.99 \times 10^5 \text{ N/m}^2)(4.2 \times 10^{-6} \text{ m}^3)}{1.01 \times 10^5 \text{ N/m}^2}$$

$$= 8.3 \times 10^{-6} \text{ m}^3.$$

Radius at the surface of the water = 2.7 cm.

9.12   DENSITY OF GASES

The definition of the mass density still remains the same for gases, mass per unit volume. However, because of the drastic changes in volume due to pressure and temperature changes, the density is measured at a fixed temperature and pressure. The standard temperature often used is $0°C$ and the standard pressure of the atmosphere (76 cm Hg).

The mass density of gases varies from gas to gas, but the number density, number of molecules per unit volume, is the same for all gases provided the pressure and temperature are the same. Amadeo Avogadro (1776–1856) for the first time noted this. Now it is known as Avogadro's law.

At the same temperature and pressure equal volumes of all gases contain equal numbers of molecules.

An amount of gas containing 6.02×10^{23} molecules is known as a mole, and the number 6.02×10^{23} is called Avogadro's number. A mole of a gas occupies a volume of 22.4 liters at $0°C$ and 76 cm Hg.

EXAMPLE 9.16

Calculate the number of oxygen (O_2) molecules in $1\ m^3$ of oxygen.

$$1\ m^3 = 10^3\ \text{liters}$$

$$20.4\ \text{liters} = 1\ \text{mole}$$

Number of moles in $1\ m^3 = 10^3/20.4 = 4.9$ moles

Number of molecules in $1\ m^3 = 4.9 \times 6.02 \times 10^{23}$

$$= 2.95 \times 10^{24}$$

SUMMARY   Important definitions

$$\text{Mass density } d = M/V;$$

if mass M is in kg and V is in m^3, mass density d is in kg/m^3.

$$\text{Weight density } D = Mg/V;$$

often expressed in N/m^3 or lb/ft^3.

$$\text{Pressure } P = \frac{F}{A}\ ;$$

The units are lb/ft^2 and N/m^2. Pascal (Pa) is the basic unit. Pressure inside a liquid at a depth h is equal to hdg.

$$\text{Young's modulus } Y = \frac{\text{Stress}}{\text{Strain}} = \frac{F/A}{\Delta L/L}\ .$$

$$\text{Shear modulus } S = \frac{F/A}{\phi}\ .$$

$$\text{Bulk modulus } B = \frac{F/A}{\Delta V/V}\ .$$

Important Principles and Laws

Archimedes' Principle. Buoyant force acting on an object is equal to the weight of the displaced liquid.

Pascal's Law. Pressure applied at a point in a confined liquid is transmitted equally to all other points.

Boyle's Law. Volume of a gas is inversely proportional to pressure, provided the temperature remains constant.

Important Instruments

Hydrometers measure density.

Manometers measure pressure.

Barometer measures atmospheric pressure.

Hydraulic press is a machine based on Pascal's law.

QUESTIONS AND PROBLEMS

9.1 Twisting is related to shear. To what is bending related?

9.2 Ships made of steel float even though the density of steel is 8 times higher than that of water. Explain.

9.3 An ice cube floats in water with $\frac{9}{10}$ of its volume immersed in water. Calculate the specific gravity of ice. What will happen to the water level when the ice completely melts?

9.4 A hollow copper sphere of 6 in. radius floats with half of its volume outside the water (Fig. 9.27). Calculate the inner radius of the sphere.

9.5 The specific gravity of gold is 18.9. Calculate its mass density and weight density in English and SI units.

9.6 A log 4 m × 10 cm × 10 cm is floating in water with $\frac{7}{10}$ of its volume in water. Calculate its specific gravity.

9.7 How much more weight should be added to the log of Problem 9.6 so that it will be completely immersed in water?

9.8 A metallic ball weighs 975 g in air, 850 g in water, and 876 g in ethyl alcohol. Calculate the specific gravity of the object and the alcohol. Can you determine the metal of which the object is made?

Figure 9.27. Problem 9.4

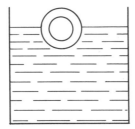

9.9   Calculate the pressure at a depth of 3 km under the sea. The specific gravity of sea water may be assumed constant and is equal to 1.03.

9.10   A gold crown is made with copper inside and having equal amounts of both copper and gold by weight. Calculate its density.

9.11   If a barometer is made with water as its liquid, what will be the height of the water column at sea level?

9.12   Water is not used in barometers. One reason is obvious from Problem 9.11. Can you think of another reason?

9.13   A barge loaded with coal appproaches a fallen tree lying over the canal. The coal is piled too high to go under the tree. What should you do, remove some coal or add some more?

9.14   Hare's apparatus is sometimes used to determine the specific gravity of a liquid. The liquid and water are allowed to enter the tubes by reducing the pressure inside. The stopcock is then closed. If the water level is 15 cm above the outside level and the liquid level is 9.8 cm, calculate the specific gravity of the liquid (Fig. 9.28).

9.15   A hydraulic press has pistons of 3 in. and 15 in. radius, respectively. If a force of 40 lb is applied on the smaller piston, calculate the force exerted by the larger piston. Also calculate the ideal mechanical advantage of this machine.

9.16   A hydraulic press used for cotton baling has an output force of 10,000 lb. If the radii of the pistons are 1.5 in. and 15 in., respectively, and the efficiency of the machine is 80%, calculate the necessary input force.

Figure 9.28 Problem 9.14. Hare's apparatus comparing the densities of liquids.

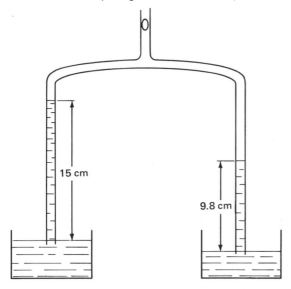

9.17   A 0.4 mm diameter wire supports a 5 kg mass. Calculate the stress on the wire.

9.18   A 3 m wire is stretched 0.3 mm. What is the strain?

9.19   A wire of 3 m long and 0.4 mm in diameter is stretched 0.3 mm when a mass of 5 kg is hung on it. Calculate Young's modulus.

9.20   A copper wire 5 m long and 0.5 mm radius is subjected to a tension of 980 N. Calculate the amount of elongation of the wire.

9.21   A crane lifts an object with a 30 ft long steel cable. Its diameter is 0.5 in. Tension on the cable is 10 tons. How much does the wire elongate? Calculate the maximum tension the cable can withstand. A factor of safety of 5 is recommended for this cable. What is the allowed maximum tension of the wire?

10

Temperature, Thermometry, and Expansion

An object moving with a velocity, or rotating with an angular velocity, has associated with it kinetic energy. This kinetic energy can be converted to other forms of energy and it can be used to do work. The reverse is also true most of the time. An object with a velocity, moving on a surface with friction, comes to a stop apparently not converting its kinetic energy to any other form of mechanical energy. What happened to the kinetic energy of the object? What do you observe during this process? We observe that the "temperature" of the object has increased during this process. Since conservation of energy is universally true it is easy to come to the conclusion that the increase in temperature is somehow related to the lost mechanical energy. What has actually happened, as we will see later, is that the mechanical energy is transformed into internal energy, that is, energy of the constituent particles, molecules and atoms, of the material. The increase in temperature is a measure of the increase in internal energy of the matter. In the example we discussed here transformation of mechanical energy produced a temperature change. Heat, another form of energy which we will discuss in the next chapter, also produces temperature change when added or removed from a material.

Everyone has some idea about temperature when he uses words such as "hot", "warm" and "cold." This classification is often made using the capacity of the human body to sense temperature differences. However, temperature as determined by using the human sense is completely arbitrary. The temperature sensation we get when we touch a warm object after dipping our hand in a bucket of hot water is different from the sensation we get if we touch the same object after dipping our hand in cold water. In the second case, the object appears warmer than in the first case. Another familiar example is that upon touching a piece of wood and a piece of copper, both kept at the same temperature, the piece of wood feels warmer than copper. This happens because copper which is a good conductor removes heat faster from the fingers than wood. Fortunately some physical parameters of materials such as length, volume, pressure, resistance, etc., change with temperature and anyone of these changes of a properly selected material can be used for accurate temperature measurement. Therefore, in this chapter we discuss these changes after a discussion of temperature scales. In the last section we study thermometers which utilize the above mentioned changes with temperature.

10.1 TEMPERATURE SCALES

On a meter stick, the distance between two lines located at the ends defines the unit of length known as the meter. The position of these two lines are usually copied from a standard kept at an institution like the National Bureau of Standards. Similarly we have two standard temperature points. They are the melting point of ice and the boiling point of water under normal atmospheric pressure. The temperature range between these two points are divided into 100 units on the Celsius scale and into 180 units on the Fahrenheit scale. These units are called a degree Celsius ($^{\circ}$C) and a degree Fahrenheit ($^{\circ}$F) on the respective scales.

A major difference between the two scales, in addition to the fact that the unit in the Fahrenheit scale is smaller than the unit in the Celsius scale, is illustrated in Fig. 10.1. The melting point of ice is taken as 0 degree on the Celsius scale while it is 32 degrees on the Fahrenheit scale. In converting temperatures from one scale to the other, both these differences should be taken into account.

To understand the conversion from the Celsius scale to the Fahrenheit scale or vice versa, let us refer to Fig. 10.2. In Fig. 10.2, a temperature difference t°C on the Celsius scale and t°F $-$ 32 on the Fahrenheit scale represent the same temperature

Figure 10.1. Relationship between Celsius and Fahrenheit scales.

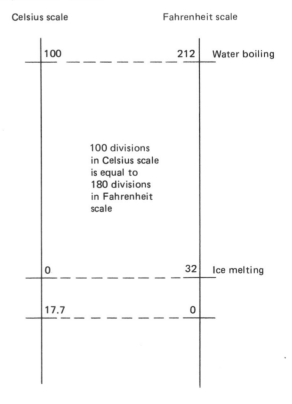

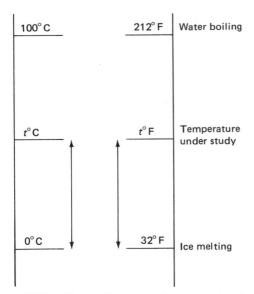

Figure 10.2. Figure illustrates the conversion from Celsius scale to Fahrenheit scale.

interval between the given temperature and the melting point of ice. Similarly, a difference of 100°C corresponds to a temperature difference of 180°F. Hence the following ratio holds:

$$\frac{t°C}{t°F - 32} = \frac{100}{180}$$

or

$$\frac{t°C}{t°F - 32} = \frac{5}{9}. \qquad (10.1)$$

Equation (10.1) can be used for conversion from one scale to the other as illustrated in the following example.

EXAMPLE 10.1

Aluminum melts at 658°C. Express this temperature on the Fahrenheit scale.

To convert the melting point from Celsius to Fahrenheit scale, we use Eq. (10.1):

$$\frac{t°F - 32}{t°C} = \frac{9}{5}$$

or

$$t°F - 32 = \frac{9}{5} \times t°C = \frac{9}{5} \times 658 = 1184.$$

Therefore, we get

$$t°F = 1216°F.$$

EXAMPLE 10.2

In regions of America the maximum temperature may go up to 100°F in the summer and the minimum temperature may go down to −25°F in the winter. Express these temperatures on the Celsius scale.

First we convert 100°F, using Eq. (10.1):

$$\frac{100 - 32}{t°C} = \frac{9}{5}$$

or $$t°C = (68) \times \frac{5}{9} = 38°C.$$

Similarly,

$$\frac{-25 - 32}{t°C} = \frac{9}{5} = -57 \times \frac{5}{9}$$

$$t°C = -57 \times \frac{5}{9} = -31.6°C.$$

Kelvin and Rankine scales, used by scientists and engineers, are related to the Celsius and Fahrenheit scales. As we have seen, the "zero" of the Celsius and the Fahrenheit scales are chosen arbitrarily. There is an "absolute zero" temperature, that is, the lowest temperature one can theoretically approach. In both the Kelvin scale and the Rankine scale, the zero is taken as this absolute zero. The relationship between the Celsius and Kelvin scales and Fahrenheit and Rankine scales are shown in Fig. 10.3.

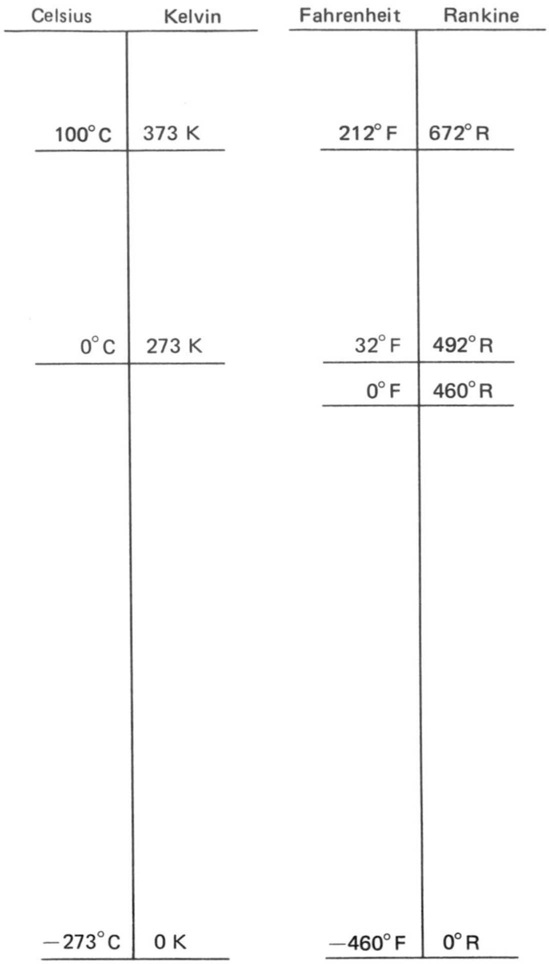

Celsius	Kelvin	Fahrenheit	Rankine
100°C	373 K	212°F	672°R
0°C	273 K	32°F	492°R
		0°F	460°R
−273°C	0 K	−460°F	0°R

Figure 10.3. Relationship between absolute, Celsius, and Fahrenheit scales.

10.2   LINEAR EXPANSION OF MATERIALS

Expansion of materials on heating and contraction on cooling are common phenomena. In our daily lives we observe effects of temperature variations such as the traffic signal hanging lower in the summer than in the winter (see Example 10.4) because of the expansion of the supporting cable. Even before the technological revolution our forefathers were aware of the effect of cooling and heating and used it to their advantage. For example, to obtain

a tight fit for the metallic rim over the wheel of his horse drawn wagon, the rim was heated and slipped over the wheel. Thus a tight "shrink-fit" was obtained when it cooled.

Let us first consider linear expansion, expansion along one of the dimensions, of a solid material. Figure 10.4 illustrates the expansion of the length of a bar, which increases from L at $t°C$ to L_1 at $t_1°C$. The increase in length $\Delta L \, (= L_1 - L)$ is proportional to the temperature increase $\Delta t \, (= t_1 - t)$. Hence we may write an equation relating ΔL and Δt:

$$\Delta L = \alpha L \Delta t. \qquad (10.2)$$

The proportionality constant α is called the coefficient of linear expansion, its value changes from material to material as shown in Table 10.1. The new length L_1 is given by

$$L_1 = L[1 + \alpha(t_1 - t)]. \qquad (10.3)$$

On rewriting Eq. (10.2), we obtain an equation for α in the form

$$\alpha = \frac{\Delta L}{L \Delta t}. \qquad (10.4)$$

The coefficient is, therefore, the fractional change in length for a temperature change of 1 degree. In Eq. (10.4) units of ΔL and L cancel each other and hence the unit of α is either $/°C$ or $/°F$.

Figure 10.4. Expansion of a rod with increase of temperature. Increase in length is proportional to the original length of the rod and to the increase in temperature.

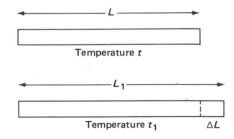

TABLE 10.1. Coefficient of Linear Expansion of Selected Materials.

MATERIAL	COEFFICIENT OF LINEAR EXPANSION	
	$10^{-5}/°C$	$10^{-5}/°F$
Aluminum	2.4	1.3
Brass	1.8	1.0
Copper	1.7	0.94
Lead	3.0	1.7
Silver	2.0	1.1
Steel	1.2	0.67

EXAMPLE 10.3

Two meter sticks made of aluminum and iron are calibrated at 20°C and are used at 40°C. What are the actual lengths at the time of their use? If the meter sticks are later used at 10°C, what are their lengths at this temperature?

The length of the meter stick at 20°C = 1m. The coefficient of linear expansion for aluminum is α_{Al} = 2.4 × 10^{-5}/°C; for steel, α_3 = 1.2 × 10^{-5}/°C.

Length of the aluminum meter stick at 40°C

$$= L[1 + \alpha_{Al}(T_1 - T)]$$

$$= (1 \text{ m})[1 + 2.4 \times 10^{-5}/°C \times 20°C]$$

$$= (1m)[1 + 4.8 \times 10^{-4}]$$

$$= 1000.5 \text{ mm}.$$

Length of steel meter stick at 40°C

$$= (1m) [1 + 1.2 \times 10^{-5}/°C \times 20°C]$$

$$= 1000.24 \text{ mm}.$$

Length of aluminum meter stick at 10°C

$$= (1m) [1 + 2.4 \times 10^{-5}/°C(10°C - 20°C)]$$

$$= 999.76 \text{ mm}.$$

Length of steel meter stick at 10°C

$$= (1m) [1 + 1.2 \times 10^{-5}/°C(10°C - 20°C)]$$

$$= 999.88 \text{ mm}.$$

This example shows that the accuracy of measurement sometimes depends on extraneous factors, this time the thermal property of the material of which the meter stick is made. It is also evident that errors can be minimized by selecting materials of desired properties; error introduced by a steel meter stick is less than introduced by an aluminum meter stick.

EXAMPLE 10.4

This example refers to Fig. 10.5. In the winter when the temperature is 32°F, the cable with the traffic light is nearly horizontal, while in the summer when the temperature is 100°F the signal sags. If the cable is made of aluminum and if its length is 50 ft, calculate how far it sags in the summer.

Figure 10.5. Sagging of signal light in the summer due to expansion of the cable. Example 10.4.

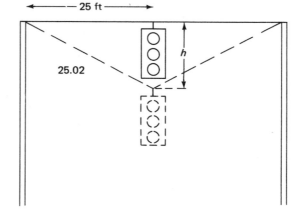

Coefficient of linear expansion of aluminum

$$= 1.3 \times 10^{-5}/°F.$$

Length of the cable in the summer

$$= (50 \text{ ft}) [1 + 1.3 \times 10^{-5}/°F(100 - 32)°F]$$

$$= 25.02 \text{ ft}.$$

The signal sags through a height h. h can be calculated from

$$(25 \text{ ft})^2 + h^2 = (25.02 \text{ ft})^2$$

or

$$h^2 = (25.02 \text{ ft})^2 - (25 \text{ ft})^2$$

$$h = 1 \text{ ft}.$$

The signal light sags through a height of 1 ft. One can list a large number of situations where thermal expansion of materials are of serious consequence, if precautions are not taken. One such situation is expansion of beams of bridges. If enough room is not provided, the expanding beams may produce huge strains on the supporting pillars. The following example illustrates such a situation.

EXAMPLE 10.5

A steel I beam is 60 ft long at 50°F and its cross-sectional area is 20 in.^{2} (Fig. 10.6). (a) Calculate

the increase in length when its temperature rises to 100°F; (b) what force would it exert on the supporting pillars, if it is assumed no room is left for expansion?

(a) The coefficient of linear expansion of steel is $0.67 \times 10^{-5}/°F$. Expansion of the I beam when the temperature increases from 50°F to 100°F

$$= (60 \text{ ft})(0.67 \times 10^{-5}/°F)(100°F - 50°F)$$

$$= 2.00 \times 10^{-2} \text{ ft} = 0.24 \text{ in}.$$

(b) The force acting on the pillars is equal to the force required to compress the steel by 0.24 in. Since

$$Y = \frac{F/A}{\Delta L/L},$$

and we are given

$$A = 20 \text{ in.}^2,$$

$$L = 60 \text{ ft},$$

$$\Delta L = 1.98 \times 10^{-2} \text{ ft},$$

$$Y(\text{steel}) = 30 \times 10^6 \text{ lb/in.}^2,$$

we have the relation

$$30 \times 10^6 \text{ lb/in.}^2 = \frac{F/(20 \text{ in.}^2)}{(1.98 \times 10^{-2} \text{ ft})/(60 \text{ ft})}$$

or

$$F = \frac{(30 \times 10^6 \text{ lb/in.}^2)(1.98 \times 10^{-2} \text{ ft})(20 \text{ in.}^2)}{60 \text{ ft}}$$

$$= 198,000 \text{ lb}.$$

The force acting on the pillars is huge; if enough room is not left for expansion, consequences are indeed disastrous.

Figure 10.6. Example 10.5.

60 ft

10.3 BIMETALLIC ELEMENTS

Bimetallic strips are made by welding together two dissimilar metal strips such as strips of iron and brass as shown in Fig. 10.7. At temperature t, room temperature, both sides have equal lengths. At t_2 ($t_2 > t$) the brass strip tends to expand more than the iron strip side ($\alpha_{brass} > \alpha_{iron}$). Since the metals have to stay together, the difference in lengths at t_2 produce a bending with the metal of larger coefficient of expansion on the outside. Bending in the opposite direction results when the temperature of the bimetallic elements t_1 is less than t.

Bimetallic elements are used as thermometers, thermally activated switches, etc. Operation of a bimetallic switch is illustrated in Fig. 10.8.

Figure 10.7. Bimetallic element. Heating the strip bends it to the side of the metal of low coefficient of expansion (iron in this example) and cooling bends it in the opposite direction.

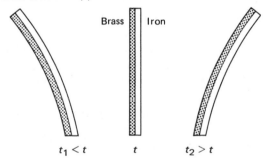

$t_1 < t$ t $t_2 > t$

Figure 10.8. A bimetallic switch. The switch completes the circuit when the temperature decreases below a set limit.

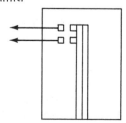

10.4 AREAL AND VOLUME EXPANSION

As the temperature changes, expansion or contraction takes place uniformly inside the material causing changes in the linear dimensions, area of the surfaces, as well as volume (Fig. 10.9). The changes in area and volume can be obtained from equations similar to the one used for linear expansion, namely,

$$A_1 = A\,[1 + \beta(t_1 - t)],\qquad(10.5)$$

$$V_1 = V\,[1 + \gamma(t_1 - t)],\qquad(10.6)$$

where A, V are the area and volume at temperature t and A_1, V_1 at temperature t_1. β is the coefficient of areal expansion and α, the coefficient of volume expansion. It can be shown that for a solid

$$\beta = 2\alpha;\qquad(10.7)$$

$$\gamma = 3\,\alpha.\qquad(10.8)$$

EXAMPLE 10.6

Prove that the coefficient of areal expansion for a solid, α_β, is 2α, where α is the coefficient of linear expansion.

Let ℓ represent the length and w the width of a

Figure 10.9. Length, area, and volume changes with temperature.

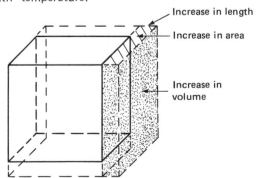

Increase in length

Increase in area

Increase in volume

flat rectangular sheet of area $A = \ell w$ (Fig. 10.10).
If the temperature increases by Δt, the new area is

$$A_1 = (\ell + \Delta \ell)(w + \Delta w)$$

$$= (\ell + \ell\alpha\Delta t)(w + w\alpha\Delta t).$$

Upon expansion we obtain

$$\Delta A = A_1 - A = \ell w + 2\ell w\alpha\Delta t + \ell w\alpha^2(\Delta t)^2 - \ell w.$$

Since $\ell w\alpha^2(\Delta t)^2$ is very small compared to the
second term, we have

$$A_1 - A = 2\alpha A \Delta t$$

or

$$\beta = 2\alpha. \qquad (10.7)$$

Note that by neglecting the term $\ell w\alpha^2(\Delta t)^2$ what
we have done is neglect the small shaded area $\Delta \ell$
$\times \Delta w$ in Fig. 10.10.

10.5   VOLUME EXPANSION OF LIQUIDS

Expansion of liquids, in general, can also be
represented by an equation similar to the one for
solids, that is,

$$V_1 = V[1 + \gamma(t_1 - t)], \qquad (10.6)$$

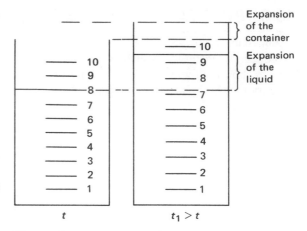

Figure 10.11. Increase in volume of liquid appears smaller because of the expansion of the container.

where γ is the coefficient of volume expansion, V
the volume at temperature t, and V_1 the new volume
at temperature t_1. In measuring the expansion of
liquids, we should be mindful of the fact that the
containers also expand. In Fig. 10.11, at temperature t, the liquid is filled to the 8th mark. As the
temperature increases, the capacity of the container
increases and therefore the positions of the marks
move up and consequently the increase in volume
of the liquid appears smaller. The apparent increase
in volume may be represented by an equation similar
to Eq. (10.6) with an apparent expansion coefficient γ_t. It can be shown that the difference between the true coefficient γ_t and the apparent coefficient γ_a is equal to the coefficient of the container
γ_c, that is,

$$\gamma_t = \gamma_a + \gamma_c. \qquad (10.9)$$

Therefore, to obtain the true coefficient (γ_t) we add
the apparent coefficient to the coefficient of the
container in which the measurements are made.

Volume expansion naturally affects the density,
density decreasing with increasing temperature. The
thermal behavior of water is the only known excep-
tion to this rule. The volume of water actually
decreases on heating from $0°C$ to $4°C$. For temperatures above $4°C$, the volume expands. Because of

Figure 10.10. Example 10.6. In the calculation of
coefficient of areal coefficient β, we usually neglect
the small shaded area.

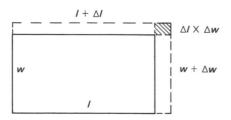

this anomalous behavior of water, at 4°C the density of water has a maximum value, and decreases on increasing or decreasing the temperature (Fig. 10.12).

10.6 EFFECTS OF TEMPERATURE ON GASES

Three quantities—temperature, pressure, and volume—specify the state of a given amount of gas. Boyle's law (Chapter 9) described the relationship between pressure and volume when the temperature was constant. Now we discuss the effect of temperature on volume when pressure is constant and on pressure when volume is constant. The volume of a gas, pressure remaining constant, increases linearly with temperature. The coefficient of volume expansion for all gases is approximately equal to 1/273 per °C. In comparison with the coefficient of expansion for solids (10^{-5} per °C) and liquids

(10^{-4} per °C), the coefficient for gases is high.

We may now write an equation similar to Eq. (10.6) for gases:

$$V_1 = V[1 + \gamma(t_1 - t)], \tag{10.10}$$

where V is the volume at t, V_1 at t_1, and γ the volume coefficient of expansion. If $t = 0°C$, for the volume at 0°C, designated by V_0, we have from Eq. (10.10)

$$V_1 = V_0(1 + \gamma t_1). \tag{10.11}$$

Substituting for γ, we get

$$V_1 = V_0[1 + (1/273)t_1]$$

$$= V_0 \frac{(273 + t_1)}{273}.$$

Figure 10.12. Variation of density of water (ice below 0°C) with temperature. Note that water has its highest density at 4°C.

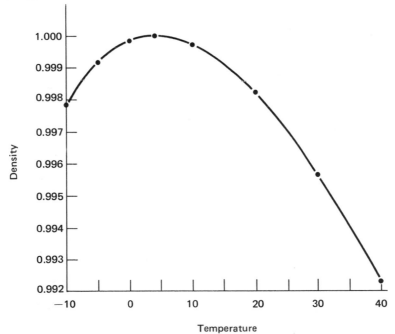

Since $273 + t_1 = T_1$, temperature on Kelvin scale

$$V_1 = (V_0/273)T_1, \tag{10.12}$$

that is, the volume of a gas is proportional to its temperature on the Kelvin scale,

$$V \propto T. \tag{10.13}$$

The statement that *the volume of an amount of gas is proportional to its temperature in the absolute scale when the pressure is held constant*, is known as Charles' law.

Figure 10.13 shows the variation of volume as a function of temperature. This plot as well as Eq. (10.13) tells us that the volume would reduce to zero at the absolute zero temperature if it remained as a gas. One cannot go below that point, that is, it is meaningless to talk about negative volume. Hence as mentioned before, 0 K is the theoretical minimum we can ever hope to achieve.

If V_1 and V_2 are volumes at T_1 and T_2, the proportionality may be written as

$$\frac{V_1}{V_2} = \frac{T_1}{T_2} \tag{10.14}$$

when pressure remains constant. This equation is helpful in solving problems, but remember that T_1 and T_2 are temperatures in absolute units.

EXAMPLE 10.7

A child's balloon filled at $27°C$ (room temperature) has a radius of 10 cm. If the balloon is taken outside on a winter day when the temperature is $-23°C$, what is its new radius?

$$27°C = 273 + 27°C = 300 \text{ K},$$

$$-23°C = 273 - 23°C = 250 \text{ K}.$$

Figure 10.13. Variation of volume of a gas as a function of temperature.

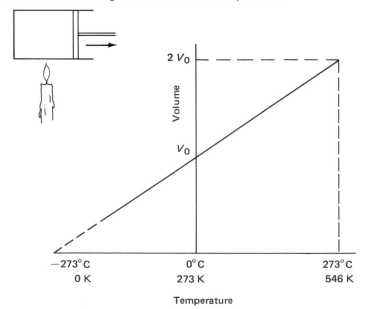

Volume of the balloon at $27°C = \frac{4}{3}\pi(10\text{ cm})^3$. Let the radius at $-23°C$ be equal to r_2 and $V_2 = \frac{4}{3}\pi r^3$. Using Eq. (10.14), we obtain

$$\frac{\frac{4}{3}(10)^3 \text{ cm}^3}{\frac{4}{3}r_2^3 \text{ cm}^3} = \frac{300\text{ K}}{250\text{ K}}$$

or

$$\frac{10^3}{r_2^3} = \frac{6}{5}, \quad r_2 = 9.41\text{ cm}$$

(see Fig. 10.14).

10.7 VARIATION OF PRESSURE WITH TEMPERATURE

When volume is kept constant, the absolute pressure of a gas increases linearly with the absolute temperature. The coefficient for increase in pressure is found to have the same value as the volume expansion coefficient, that is, 1/273 per °C. A simple experimental arrangement for the pressure variation at constant volume and the results of such an experiment are shown in Fig. 10.15. From this figure it is evident that the pressure varies directly proportional to the absolute temperature, that is,

$$P \propto T. \tag{10.15}$$

As in the case of volume, $P \to 0$ as $T \to 0$. If P_1 and P_2 are pressures at T_1 and T_2, from Eq. (10.1) we obtain the equation,

$$\frac{P_1}{T_1} = \frac{P_2}{T_2} \tag{10.16}$$

when the volume remains constant. This equation is in a form more suitable for solving problems.

EXAMPLE 10.8

A sealed tank contains oxygen at 27°C at a pressure of 2 atm. If the temperature increases to 100°C, what will be the pressure inside the tank.

To use Eq. (10.16) we convert the temperatures to absolute scale.

$$27°C = 273 + 27°C = 300\text{ K}$$

$$100°C = 273 + 100°C = 373\text{ K}$$

$$\frac{2\text{ atm}}{P_2} = \frac{300\text{ K}}{373\text{ K}}$$

$$P_2 = 2.49\text{ atm}.$$

The three equations, (9.40) (Boyle's law), (10.13) (Charles' law), and (10.15) can be combined into a single equation known as the general gas law or the ideal gas law,

$$\frac{P_1 V_1}{T_1} = \frac{P_2 V_2}{T_2}. \tag{10.17}$$

This equation relates the three quantities, absolute pressures, absolute temperatures, and volumes of a given mass of gas. If follows from that fact that P is directly proportional to T and proportional to V. When T is constant, Eq. (10.17) becomes

$$P_1 V_1 = P_2 V_2.$$

Figure 10.14. Example 10.7.

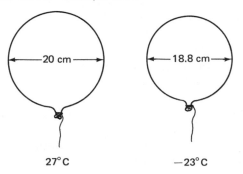

20 cm 18.8 cm

27°C −23°C

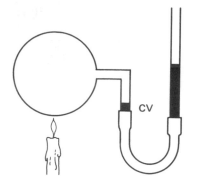

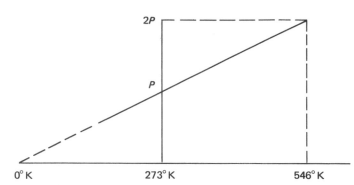

Figure 10.15. Variation of pressure as a function of temperature at constant volume. Before each pressure reading the mercury level is brought to the constant volume line (CV).

When P is constant,

$$\frac{V_1}{T_1} = \frac{V_2}{T_2} .$$

When V is constant

$$\frac{P_1}{T_1} = \frac{P_2}{T_2} .$$

EXAMPLE 10.9

A cylinder contains 2 m^3 of air at 27°C and a pressure of 1 atm. If the air is compressed by a piston to a volume of 0.5 m^3 and the temperature rises to 30°C calculate the new pressure. Since there is no change in the mass of the gas we use Eq. (10.19).

$$\frac{P_1 V_1}{T_1} = \frac{P_2 V_2}{T_2}$$

$$P_1 = 1 \text{ atm}, \ P_2 = \text{?};$$

$$V_1 = 2 \text{ m}^3, \ V_2 = 0.5 \text{ m}^3;$$

$$T_1 = 27°C = 303 \text{ K}, \ T_2 = 273 + 30°C = 300 \text{ K};$$

Substituting these values into the above equation, we have

$$P_2 = \frac{P_1 V_1}{T_1} \times \frac{T_2}{V_2}$$

$$= \frac{(1 \text{ atm})(2 \text{ m}^3)}{300 \text{ K}} \frac{303 \text{ K}}{0.5 \text{ m}^3}$$

$$= 4.04 \text{ atm.}$$

10.8 KINETIC THEORY OF GASES

The kinetic theory of gases presents a model in which it is thought that in gases the constituent particles are molecules that are in constant motion in random directions. These particles are assumed as a first approximation to be "point" particles occupying a negligibly small fraction of the volume of the container.

The forces between any two molecules is also assumed to be very small. However, perfectly elastic collisions (i.e., $\Delta KE = 0$) take place between these particles and with the walls of the container, resulting in a continuous change in the magnitude and direction of the velocities. The pressure exerted by the gas on the container can be considered as the net result of the change in momentum or impulse during these collisions.

Consider a ball making perfectly elastic collisions between the two surfaces shown in Fig. 10.16.

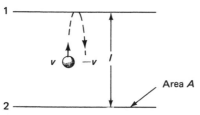

Figure 10.16. Elastic collision of a ball with two surfaces of area A.

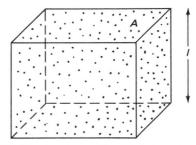

Figure 10.17 A volume V containing a large number of molecules. The molecules move around in random directions. The collisions between the molecules and the surfaces of the container are assumed elastic.

Each time a collision takes place at surface 1, the momentum of the ball changes by $-2mv$. An equal momentum change in the opposite direction ($= +2mv$) is experienced by the surface. The ball makes such collisions with surface 1, $v/2\ell$ times in 1 s. Hence, the force acting on the surface is given by

$$\text{Force} = \frac{\text{Change in momentum}}{\text{Time}}$$

$$= \begin{pmatrix} \text{Change in momentum} \\ \text{during one collision} \end{pmatrix}$$

$$\times \begin{pmatrix} \text{Number of collisions} \\ \text{in one second} \end{pmatrix}$$

Pressure on surface 1 due to 1 particle $= \dfrac{\text{Force}}{\text{Area}} = \dfrac{mv^2}{\ell A}$.   **(10.18)**

In a gas there are a large number, 10^{23}, of molecules; we can assume one third of them are colliding with the surface perpendicular to one of the axes. All gas molecules in a container do not have the same velocity, but they have a velocity distribution, hence the v^2 in Eq. (10.18) should replaced by an average of the square of the velocity, $\overline{v^2}$. Considering a volume (Fig. 10.17) containing N molecules, we have

$$P = \frac{1}{3} N \frac{\overline{mv^2}}{\ell A} = \frac{1}{3} N \frac{\overline{mv^2}}{V},$$ **(10.19)**

where P stands for pressure and V is the volume of the container. Rewriting Eq. (10.19), we obtain

$$PV \propto \frac{1}{2}\overline{mv^2},$$

since N is a constant for a given amount of gas. The product of pressure and volume is proportional to the absolute temperature, Eq. (10.17), hence

$$\frac{1}{2}\overline{mv^2} \propto T.$$ **(10.20)**

The average kinetic energy of a molecule is proportional to the absolute temperature. The increase in pressure with volume remaining constant is also easy to understand now. As T increases, velocity increases and consequently the forces exerted on the walls increase. The exact equation for the average kinetic energy is given by

$$(\text{KE})_{\text{avg}} = \frac{3}{2}kT.$$ **(10.21)**

The constant k has the value 1.38×10^{-23} J/K and is known as Boltzmann's constant.

Figure 10.18 shows the velocity distribution of molecules for two different temperatures. It is clear that the velocities range from zero to very large values and that the velocity distribution shifts to the higher velocity side as temperature increases. This again indicates that the average kinetic energy increases with temperature.

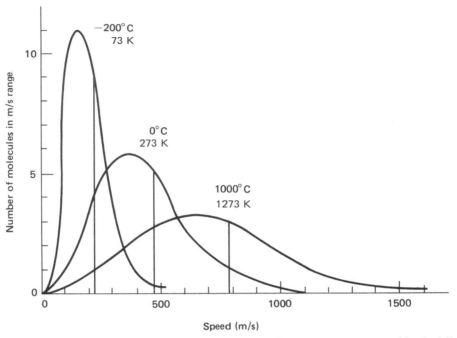

Figure 10.18. Velocity distribution of molecules at three different temperatures. Vertical lines indicate velocities that correspond to average kinetic energies for the given temperatures.

10.9 THERMOMETERS

As indicated earlier, many physical properties of matter (volume, pressure, resistance, etc.) change with temperature. Thermometers have been designed using many of these properties. Some of these thermometers are discussed below.

Liquid-In-Glass Thermometers

In the most commonly used thermometer, expansion of a mercury column with temperature is employed to read temperature. The construction of an ordinary thermometer is illustrated in Fig. 10.19. A thin-walled glass bulb is at the end of a glass stem containing a narrow capillary tube. When the bulb is in contact with a hotter object, the temperature of the bulb and the mercury in it rises. Since the coefficient of expansion of mercury is much higher than that of the glass, with increasing temperature,

the mercury column expands into the capillary. A thermometer contains a scale on its stem. The reading against the top of the mercury column gives us the temperature. Mercury thermometers can not be used below $-38°F$, which is the freezing point of mercury. For lower temperatures alcohol (freezes at about $-200°F$) is used. In clinical thermometers, there is a constriction between the bulb and the capillary tube. This constriction inhibits the flow of mercury back into the bulb, enabling us to read the temperature a few minutes after the thermometer is removed from the object (patient).

Bimetallic Thermometer

Bending of bimetallic elements are used in designing direct reading, rugged thermometers. As the temperature changes the bimetallic strip winds or unwinds (Fig. 10.20). This winding motion causes

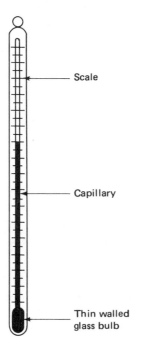

Figure 10.19. Mercury-in-glass thermometer.

Figure 10.20. Bimetallic thermometer.

the pointer to move over the scale indicating the temperature.

Resistance Thermometers

Every material offers resistance to the flow of current, in different degrees. Good conductors have low resistance, while insulators have high resistance. Resistance of conductors increases with increasing temperature. A coil of platinum wire is used to read

very low temperatures (-200°C) to high temperatures ($+1400^\circ$C). For very low temperatures carbon or germanium resistors are used.

The resistance of semiconducting materials lies in a range between those of conductors and insulators. Such materials, with impurity atoms (atoms that are different from the atoms of the material) added to it, show fast variations in resistance with temperature. A thermistor, made of a small bead of such a material placed between two conducting leads and covered with a thin shell of glass, is capable of detecting small changes in temperature of the order of 0.001°C.

Thermocouples

As shown in Fig. 10.21 two dissimilar metals, such as copper and Constantan, are fused together to form two junctions in a thermocouple. The voltage read by the meter is a function of the temperature difference between the two junctions; the reference junction is usually maintained at a constant temperature such as the melting point of ice. The useful temperature range of thermocouples varies with the materials used to make them. Some examples are as follows

Copper - Constantan, -190°C to 300°C

Iron - Constantan, -190°C to 700°C

Platinum - Platinum 10% rhodium, 0°C to 1700°C alloy

Chromel — Gold 0.07% iron, 0 K to 280 K

Thus by making the proper choice of a thermocouple, the temperature in the range of 0 K to 1900 K can be read.

Pyrometer

As the temperature of any object is increased, its color changes from dull red to red and finally to

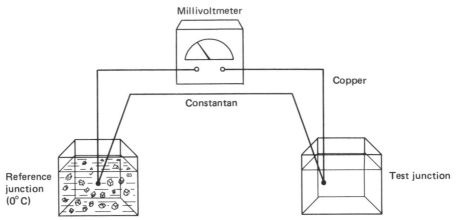

Figure 10.21. Thermocouples. The voltage read by the voltmeter is proportional to the temperature difference between the reference and the test junctions.

blue white. The color composition of the radiation thus depends on the temperature of the object. A pyrometer uses this fact to determine the temperature of objects at high temperatures. The radiation falling on the pyrometer and that produced by a hot filament are compared, and the current through the filament is adjusted till the colors match. From the current, the temperature can be determined. The instruments are calibrated to give a direct reading, that is, as the current is adjusted, the temperature scale automatically rotates so that the indicator reads the correct temperature.

SUMMARY   The equation to convert temperature from the Celsius to Fahrenheit scale is

$$t°C = (t°F - 32) \times \tfrac{5}{9}$$

and to convert from the Celsius to Kelvin scale is

$$t\,K = t°C + 273.$$

The dimensions of materials change with temperature changes. Such changes are given by

$$L_1 = L\,[1 + \alpha\,(t_1 - t)]\,,$$

$$A_1 = A\,[1 + \beta\,(t_1 - t)]\,,$$

$$V_1 = V\,[1 + \gamma\,(t_1 - t)]\,.$$

The volume expansion of liquids is also given by the above equation.

For gases we have the ideal gas equation,

$$\frac{P_1 V_1}{T_1} = \frac{P_2 V_2}{T_2}.$$

This equation can be used to determine the change in one or two parameters as a function of the others. When T is constant, we get

$$P_1 V_1 = P_2 V_2;$$

when P is constant, we have

$$\frac{V_1}{T_1} = \frac{V_2}{T_2};$$

and when V is constant, we have

$$\frac{P_1}{T_1} = \frac{P_2}{T_2}.$$

QUESTIONS AND PROBLEMS

10.1 Make a list of some of the properties of materials that can be used to measure temperature.

10.2 Water is not used as a liquid (instead of mercury) in thermometers. Why?

10.3 If a mercury-in-glass thermometer is immersed into a hot liquid, the mercury level first goes down and then goes up. Why?

10.4 Will temperature changes affect the following measurements? Why? (a) Length measurements by meter sticks, especially made of pure metals, (b) pressure measurements using a mercury column, (c) time measurement by a pendulum clock (period of oscillation of a pendulum is proportional to its length), and (d) specific gravity measurement by a specific gravity bottle.

10.5 To open tight metal caps, the caps are heated without heating the jars. Why?

10.6 The coefficient of linear expansion in $/°F$ unit is $\frac{5}{9}$ of the unit in $/°C$ unit. Explain.

10.7 A sealed Coca-Cola bottle explodes when it is frozen and when it is heated. Explain.

10.8 In winter water pipes break. Give the reason.

10.9 Show that $-40°C$ is equal to $-40°F$.

10.10 Normal body temperature is $98.6°F$. Express this in the Celsius scale.

10.11   Express $-100°F$, $-32°F$, $32°F$, $100°F$, and $500°F$ in Celsius and Kelvin scales.

10.12   Express $-100°C$, $-50°C$, $+50°C$, $+100°C$, and $1000°C$ in the Fahrenheit scale.

10.13   In an industrial plant 250 ft of iron pipe is laid at $70°F$. This pipe is to carry steam at $212°F$. How much will the pipe expand at $212°F$?

10.14   A horse drawn wagon has a 3 ft radius wooden wheel. The wheel is to be fitted with an iron rim (tire) of 2.99 ft at $20°C$. What should be the temperature of the rim so that it will slip over the wheel?

10.15   For the objects made of copper given in Fig. 10.22, calculate the changes in the indicated lengths, the sides of the rectangle, diameter of the hole, diameter of the disk, and the inner diameter of the spherical shell as the temperature increases by $60°C$. (Note that the diameter of the hole increases as the temperature increases. Thermal expansion is like photographic enlargement.)

10.16   Calculate the increase in area of the rectangle, the hole, and the disk in Fig. 10.22 as the temperature is increased by $60°C$.

10.17   Calculate the increase in the inner volume of the spherical shell in Fig. 10.22 (d) if the increase in temperature is $60°C$. (Did you use the coefficient of volume expansion of copper or air in this case. Why? Will the presence of air in any way affect the volume expansion?)

10.18   A person after walking outside tried to open his door. He found it difficult to insert the key in the key hole. (Because of the storm door, the door temperature is approximately the same as the inside temperature). Is the outside temperature higher or lower than the inside temperature?

10.19   A copper alloy rod of length 50 cm at $20°C$ expands to 50.1 cm at $100°C$. Calculate its coefficient of linear expansion. What will be its length at $0°C$? At what temperature will its length be equal to 49.97 cm?

10.20   Steel rail sections were laid with separation between them to allow for expansion. Each section was 60 ft long at $50°F$ and the temperature varied from $-30°F$ to $120°F$. What was the gap?

Figure 10.22.   Problems 10.15, 10.16, and 10.17. In (d) a spherical shell is shown.

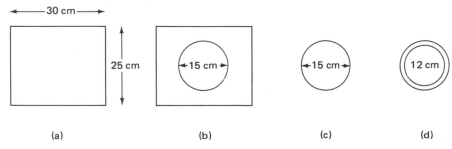

(a)   (b)   (c)   (d)

10.21 A Pyrex beaker is filled with mercury. The volume of the beaker at 20°C is 100 cm^3. How much mercury will overflow if the temperature is raised to 40°C. The coefficient of volume expansion for Pyrex is 9×10^{-6} /°C and for mercury it is 18.2×10^{-5}/°C.

10.22 A 2 m^3 steel tank is filled with water at 20°C. How much water will overflow if the temperature increases to 38°C. Assume the coefficient of expansion of water is 2×10^{-4}/°C in this temperature range.

10.23 A steel I-beam is 20 m long and its cross-sectional area is 12 cm^2. If its temperature is increased by 30°C, how much force should be applied to keep it from expanding?

10.24 The density of aluminum at 20°C is 2.9 g/cm^3. Calculate its density at 350°C.

10.25 A certain mass of gas occupies a volume of 0.5 m^3 at 10°C. Calculate its volume at 100°C, assuming the pressure remains constant.

10.26 An auto tire is inflated to 20 lb/in.^{2} gauge pressure at 45°F. After driving on an expressway for a while its temperature is found to increase to 125°F. Assuming there is no volume change, calculate its new pressure.

10.27 The pressure inside a bicycle tire is 24 lb/in.^{2} A boy is trying to pump air into it with a 2 ft long pump. How far should the piston be pushed in before the air can get in? Assume the temperature remains constant.

10.28 A steel tank contains 5 m^3 of gas at a pressure of 10 kPa at 20°C. If the temperature increases to 1000°C in a fire, calculate the pressure in the tank. Neglect the expansion of the tank.

10.29 Calculate the volume of 2 moles of helium at 50 lb/in.^{2} pressure and 60°C.

Heat: Its Effects and Measurement

In this chapter the meaning of heat and its measurement are first discussed. Following that we discuss the transfer of heat by conduction, convection, and radiation. The latter topic has become an extremely significant one these days, a clear understanding of how heat transfers from one medium to another (for example, from our house to the outside) is essential in conserving energy.

11.1 WHAT IS HEAT?

The word heat is an often misunderstood word. Originally heat was thought to be a fluid called "caloric" which was transmitted from a hot material to a cold material. However, no one could ever measure or detect this substance. To come to the conclusion that heat is a form of energy and not a fluid took years of work by several engineers and scientists. Foremost among them was a American, Count Benjamin T. Rumford who spent most of his lifetime in Europe. He supervised the boring of

cannon during his military service for the Bavarian monarch. He was puzzled by the fact that the temperature of the metal rose while it was being bored. He built a box around the part of the cannon where the boring occurred and filled the box with water. In 2½ hr the water boiled. He also found that continuing the boring resulted in a continuing rise in temperature. For him it was illogical to imagine a continuous generation of the "caloric" fluid during the boring process. From his observation Rumford concluded that the work of the drills on the metal of the cannon against friction must be responsible for the temperature rise. In other words, he concluded that heat was a form of energy just like mechanical energy.

Today, largely due to the work an Englishman, James Prescott Joule, heat is regarded as a form of energy. In the development of this concept, certain expressions are still retained from the past such as "heat contained in a body." When we supply heat to a material, its internal energy, both potential and kinetic energy of the constituent particles, increases. In an analogous situation when we do work such as lifting an object, its potential energy increases. We never say that the object contains work. Similarly, it is not correct to say that a body contains heat. The correct statement is that on supplying heat to an object without the performance of work its internal energy increases.

When two objects, one at a higher temperature than the other, are placed in contact, the temperature of the hotter object decreases and the temperature of the second object increases. Eventually the two objects will have the same temperature. When this happens, they are said to be in equilibrium. In reaching equilibrium, the internal energy of the first object decreased while that of the second object increased. The internal energy changes are caused by the transfer of heat from one object to the other.

Energy transferred from one object to another object because of their temperature difference is called heat. The net transfer of heat is always from a hot object to a cold object.

11.2 UNITS OF HEAT

Kilocalorie and Btu (British thermal unit) are used in the SI and English unit systems. In both unit systems heat is defined in terms of its effect—raising temperature—of a known quantity of water.

One kilocalorie is defined as the amount of heat required to raise the temperature of 1 kg of water through 1°C. Generally the temperature change is taken from 14.5°C to 15.5°C for the definition.

One Btu is defined as the amount of heat required to raise the temperature of 1 lb of water by 1°F. Here the difference of 1°F is the change from 63°F to 64°F. An amount of heat equal to 1 calorie, 1/1000 of a kilocalorie, raises the temperature of 1 g of water through 1°C.

The relationship between the different units are as follows:

$$1 \text{ kcal} = 1000 \text{ cal} = 3.968 \text{ Btu};$$

$$1 \text{ Btu} = 252 \text{ cal} = 0.252 \text{ kcal}.$$

In the definition of Btu, the weight of water instead of its mass is used mainly because of historic reason. However, the definition implies a quantity of water that weighs 1 lb, that is, 1 slug of water. Only in heat calculations, pound is used as a mass unit.

11.3 SPECIFIC HEAT CAPACITY

In the above definition we just saw that 1 kcal of heat raises the temperature of 1 kg of water 1°C. It is also natural to expect a temperature increase of 0.1°C if 10 kg of water receives the same amount of heat. By what magnitude does the temperature rise in other materials? Figure 11.1 illustrates what happens when heat is supplied to water and iron. The increase in temperature t is proportional to the heat input Q and inversely proportional to the mass m, that is,

$$\Delta t \propto \frac{Q}{m}.$$

We convert this proportionality to an equation by introducing a proportionality constant c:

$$Q = cm\Delta t. \tag{11.1}$$

The constant c is called the specific heat capacity of the material. The product of mass and specific heat capacity is called the heat capacity of the object. Rearranging Eq. (11.1), we have the definition of

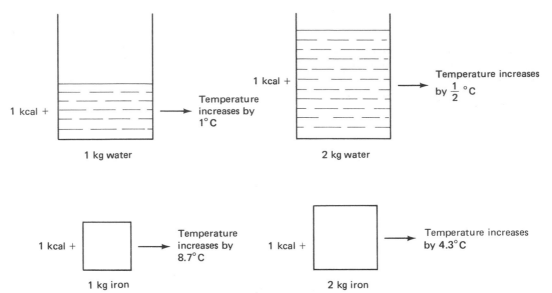

Figure 11.1. Effect of supplying 1 kcal of heat to 1 kg of water, 2 kg of water, 1 kg of iron, and 2 kg of iron.

specific heat capacity, namely,

$$c = \frac{Q}{m\Delta t}, \qquad (11.2)$$

that is, the specific heat capacity is the heat needed to raise the temperature of a unit mass of an object by 1°C (or 1°F). Units of c, therefore, are kcal/ kg · °C, cal/g · °C, or Btu/lb · °F. In the case of water 1 kcal produces a 1°C change in 1 kg of water; hence its specific heat capacity is equal to 1. This is true in all the units. Another consequence of the definition of kcal and other units is that specific heat capacity of a material is numerically equal in both unit systems. This can be readily proved by unit conversion:

$$\frac{1 \text{ Btu}}{\text{lb} \cdot °\text{F}} \times \frac{1 \text{ lb}}{0.454 \text{ kg}} \times \frac{9°\text{F}}{5°\text{C}} \times \frac{0.252 \text{ kcal}}{1 \text{ Btu}} = \frac{1 \text{ kcal}}{\text{kg} \cdot °\text{C}}$$

Table 11.1 lists the specific heat capacities of some common substances.

EXAMPLE 11.1

Calculate the amount of heat required to raise the temperature of 2 kg of iron from 27°C to 100°C.

The specific heat capacity of iron is 0.115 kcal/ kg · °C. Therefore the heat is

$$Q = cmt$$

$$= (0.115 \text{ kcal/kg} \cdot °\text{C}) \times (2 \text{ kg}) \times (100 - 27)°\text{C}$$

$$= 16.8 \text{ kcal}.$$

11.4 CALORIMETRY

When an object at higher temperature is mixed with another object of lower temperature, heat will transfer from the object at higher temperature to that at lower temperature. The heat transfer results in the lowering of temperature of the first object and the raising of the temperature of the second object. This process continues until both objects

TABLE 11.1 Specific Heat Capacities of Selected Materials

SUBSTANCE	SPECIFIC HEAT CAPACITY[a]
Air	0.24
Alcohol, ethyl	0.60
Aluminum	0.22
Brass	0.091
Copper	0.093
Glass	0.21
Gold	0.032
Ice	0.50
Iron (Steel)	0.115
Lead	0.03
Mercury	0.033
Silver	0.056
Steam	0.48
Water	1
Wood (avg)	0.42

[a]Values of specific heat capacity are the same in all units.

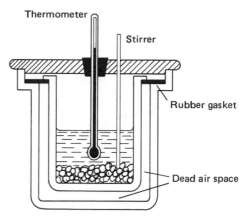

Figure 11.2. Details of laboratory calorimeter. The rubber gasket, the dead air space, and the polished surfaces of the inner and the outer vessels reduce the heat loss from the calorimeter (inner vessel).

reach the same temperature. How much heat is transferred in this process? If we assume there is no extraneous loss in energy, the decrease in the internal energy of one object has to be equal to the increase in the internal energy of the other object. This principle of energy balance is often expressed by the simple statement,

$$\text{heat lost} = \text{heat gained} \qquad (11.3)$$

which forms the basis of calorimetry, the science of measurement of heat.

In laboratories the mixing of a hot object and a cold object (usually water) is done in a "calorimeter." The calorimeter is designed (Fig. 11.2) so that proper mixing of the materials can be achieved and extraneous loss in energy is minimized. The inner vessel is usually made of aluminum and the outer vessel of aluminum. A stirrer inserted through the wooden cover is used to stir the water and thus achieve rapid temperature equilibrium. The rubber gasket supporting the inner vessel is nonconducting to reduce the heat loss from the calorimeter to the outside. Similarly the dead air space between the inner and the outer vessel prevents heat loss by air currents. Furthermore the surfaces of both metal vessels are polished to reduce loss or gain of heat by radiation. Let us designate the masses of the hot object, water, and the inner vessel be m_h, m_w, and m_c, respectively. Let t_h and t_c be the initial temperatures of the hot object and the calorimeter with water and t_f, the final equilibrium temperature. The heat balance equation in this case is

$$\underbrace{c_h m_h (t_h - t_f)}_{\text{Heat lost}} = \underbrace{m_w (t_f - t_c) + c_c m_c (t_f - t_c),}_{\text{Heat gained by water and calorimeter}}$$

$$(11.4)$$

where c_w $(= 1)$ is specific heat capacity of water, c_h that of the object, and c_c that of the calorimeter. The right-hand side can be written in the form

$$(m_w c_w + c_c m_c)(t_f - t_c).$$

The term $c_c m_c$ is the water equivalent of the calorimeter (including stirrer). Sometimes instead of the weight of the calorimeter its water equivalent is supplied by the manufacturers.

Use of Eq. (11.4) to determine the specific heat capacity of an object is illustrated in the example below.

EXAMPLE 11.2

In an experiment to determine the specific heat capacity of copper, a piece of copper weighing 50 g is first heated to 100°C in steam. It is then immersed into water at 27°C. The water in the calorimeter weighed 100 g and the inner aluminum cup weighed 50 g. If the final temperature is 30°C, calculate the specific heat capacity of copper.

We use Eq. (11.4) to calculate the specific heat capacity of copper, with the given data listed below:

$$m_h = 50 \text{ g}, \ m_w = 100 \text{ g}, \ m_c = 50 \text{ g},$$

$$t_h = 100°C, \ t_c = 27°C, \ t_f = 30°C.$$

The specific heat of aluminum is 0.22 cal/g · °C.

$$c_h m_h(t_h - t_f) = m_w c_w(t_f - t_c) + c_c m_c(t_f - t_c)$$

$$c_h \ (50 \text{ g})(100°C - 30°C)$$

$$= (100 \text{ g})(1 \text{ cal/g} \cdot °C)(30°C - 27°C)$$

$$+ (0.22 \text{ cal/g} \cdot °C)(50 \text{ g})(30°C - 27°C)$$

$$c_h(3500 \text{ g} \cdot °C) = 333 \text{ cal}$$

$$c_h = \frac{333 \text{ cal}}{3500 \text{ g} \cdot °C} = 0.095 \text{ cal/g} \cdot °C.$$

EXAMPLE 11.3

An aluminum pot weighing 5 lb was on a heater without anything in it. Its temperature increased to 190°F. How many pounds of water at 60°F should be poured into the pot to bring the temperature down to 90°F?

Heat lost by aluminum pot = Heat gained by water

$$(5 \text{ lb})(0.22 \text{ Btu/lb} \cdot °F)(190°F - 90°F)$$

$$= W_w c_w(90°F - 60°F).$$

Weight of water $W_w = \dfrac{5 \times 0.22 \times 100}{30} = 3.3 \text{ lb}.$

11.5 CHANGE OF PHASE

When a substance absorbs heat its temperature increases. Is this the only effect of heat? If we continuously supply heat, what else besides a rise in temperature can happen? To answer this question let us take a 1 kg piece of ice at −20°C and study the effects of absorption of heat by ice.

On supplying 1 kcal of heat to 1 kg of ice its temperature will rise by 2°C since its specific heat capacity is 0.5 kcal/kg · °C. The temperature rises at this rate for a heat input up to 5 kcal. Now we have ice at 0°C. At this point absorption of heat produces no increase in temperature, but the ice starts melting. For complete melting a heat input of 80 kcal is needed. Further heat input produces a rise in temperature, 1°C per 1 kcal till the temperature reaches 100°C. At this point again we note no increase in temperature for an additional heat input of 540 kcal. Water during this time is changing to steam. These changes are illustrated in Figs. 11.3 and 11.4.

The change of state of matter at constant temperature (solid to liquid or liquid to gas) on absorption of heat is called a phase change. The solid to liquid phase change is called fusion or melting and the temperature at which this happens is called the melting point. The amount of heat required to melt unit mass of solid to liquid at its melting point is called the latent heat of fusion.

For ordinary ice, the latent heat of fusion is 80 kcal/kg or 80 cal/g and in English units it is 144 Btu/lb.

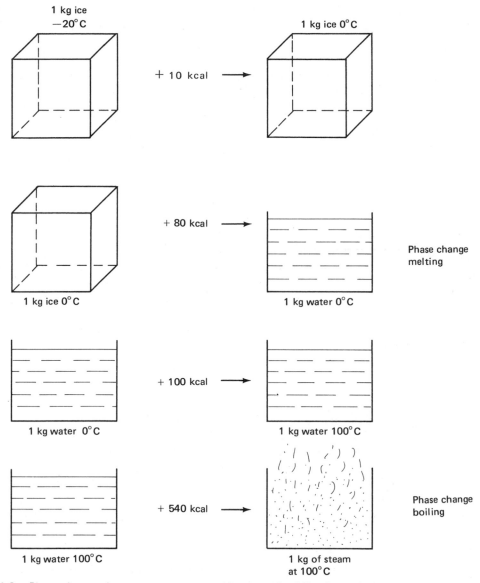

Figure 11.3.   Phase changes ice→water→steam and heat required for these changes.

If a heat input of Q kcal melts a mass m to liquid, its latent heat of fusion L_f is given by

$$L_f = Q/m. \qquad (11.5)$$

Similarly we have a phase change from liquid to gas called vaporization, which takes place at a tem-

perature called the boiling point. *The latent heat of vaporization is the heat required to completely vaporize a liquid at its boiling point and is given by*

$$L_v = Q/m. \qquad (11.6)$$

For water, the latent heat of vaporization is 540

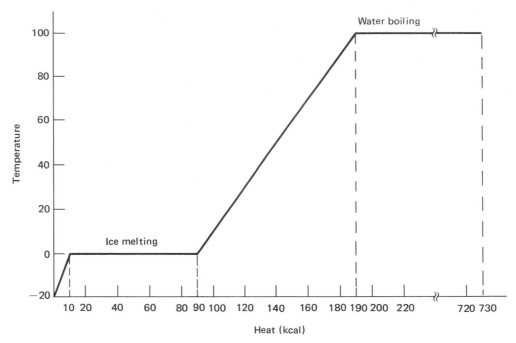

Figure 11.4. Figure illustrates the changes in 20 kg of ice as the heat absorbed increases. Temperature does not change during melting and boiling.

kcal/kg or 540 cal/g, or in English units 972 Btu/lb.

Heats of fusion and vaporization have the units of kcal/kg, cal/g, or Btu/lb. Heats of fusion and vaporization vary from material to material just as the boiling point and melting point.

On cooling, reversing the processes in Fig. 11.4, condensation and solidification takes place. Heat is removed during this process. Heat given up during condensation is called the heat of condensation which is equal to the heat of vaporization. Similarly the heat of solidification is equal to the heat of melting.

EXAMPLE 11.4

10 g of ice at $-10°C$ is put into 5 g of water at $27°C$. Obtain the final temperature after the ice has completely melted.

In this case the water loses heat and the ice gains heat. If t_f is the final temperature,

Heat lost by water $= mc_w(t_w - t_f)$

$$= (50\ \text{g})(1\ \text{cal/g} \cdot °C)\ (27°C - t_f).$$

Heat gained by the ice has three parts: (a) heat gained during the temperature rise from $-10°C$ to $0°C$:

$$mc_i\Delta t = (10\ \text{g})(0.5\ \text{cal/g} \cdot °C)(10°C) = 50\ \text{cal};$$

(b) heat gained during melting:

$$mL_f = (10\ \text{g})(80\ \text{cal/g}) = 800\ \text{cal};$$

(c) heat gained in raising the temperature of the

melted ice:

$$mc_w\Delta t = 10 \text{ g}(1 \text{ cal/g} \cdot {}^\circ\text{C})(t_f - 0) = (10 \text{ cal/}{}^\circ\text{C})t_f.$$

Since heat lost = heat gained

$$(50 \text{ cal/}{}^\circ\text{C})(27{}^\circ\text{C} - t_f) = 50 \text{ cal} + 800 \text{ cal}$$
$$+ (10 \text{ cal/}{}^\circ\text{C})t_f$$
$$t_f = 8.3{}^\circ\text{F}.$$

Note that the temperature difference is always taken as positive in heat balance equations.

EXAMPLE 11.5

Water weighing 60 lb at 60°F is heated to produce steam at 212°F. How much heat is supplied to the water?

Heat required to raise water to 212°F

$$= mc_w\Delta t = (60 \text{ lb})(1 \text{ Btu/lb} \cdot {}^\circ\text{F})(212{}^\circ\text{F} - 60{}^\circ\text{F})$$
$$= 9210 \text{ Btu.}$$

Heat required to boil 60 lb of water at 212°F

$$- mL_v - (60 \text{ lb})(972 \text{ Btu/lb}) - 58,320 \text{ Btu.}$$

Total heat = 9120 Btu + 53,820 Btu = 67,440 Btu.

11.6 EFFECT OF PRESSURE ON BOILING AND MELTING

In the above discussions, we assumed that the phase changes are taking place under conditions of standard (atmospheric) pressure. Pressure affects both the melting point and the boiling point. An increase in pressure raises the melting point of liquids which contract on freezing, and lowers the melting point of liquids which expand on freezing.

For water, which expands on freezing, an increase in pressure lowers the freezing point. This is demonstrated by a simple experiment, in which a small wire weighted at both ends is placed over a piece of ice (Fig. 11.5). The wire passes through the ice, the ice block still remaining as a single piece. This process known as regelation may be explained as follows. The wire with the weights increases the pressure on the ice below it and lowers its freezing point. Since the block as a whole is at 0°C, the temperature of the ice below the wire is slightly above its freezing point, it melts. Melting causes a small reduction in temperature. The wire settles down, squeezing the water above where pressure is the normal pressure and the temperature is slightly below zero. The water then freezes. The process of melting below the wire and freezing above the wire continues till the wire cuts through the block.

The same phenomenon occurs in ice skating. The weight of the skater acting over the thin skates results in a high pressure on the ice and, consequently, the ice melts. Thus one skates over a film of water. If the temperatures are extremely low as in Siberia, skating becomes more difficult, if not impossible.

The effect of pressure on the boiling point is more drastic. Increasing the pressure increases the boiling point and decreasing the pressure reduces it (Fig. 11.6). An example of the influence of pressure on boiling is provided by the pressure cooker.

Figure 11.5. An interesting demonstration of the effect of pressure on melting. The wire passes through the ice, the ice block still remaining as a single piece. This is known as regelation.

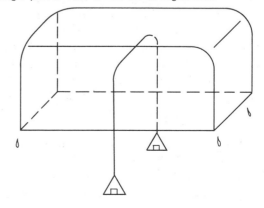

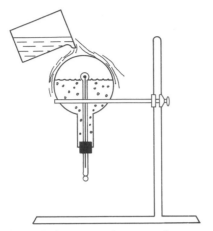

Figure 11.6. A demonstration experiment to show the effect of pressure on boiling. Boil some water in a flask. When the water is boiling remove it from the heater, close it with a stopper, and invert it. Pour cold water over the flask as shown in figure. As the flask cools, the pressure inside decreases and the water starts to boil again, now at a lower temperature.

The pressure inside makes the water boil above 100°C and thus the food is more thoroughly cooked. For a similar reason it is difficult to boil an egg at the top of a high mountain.

11.7 EVAPORATION

In liquids, the molecules move around in random directions like the molecules in a gas. Some of these molecules near or on the surface of the liquid have enough kinetic energy to escape the surface. A molecule located deep in the liquid, even if it has enough energy, will find it difficult to escape because of the high probability of colliding with another molecule on its way out. On the other hand, the escape of liquid molecules from the surface occurs all the time. This is called evaporation. As the temperature of the liquid increases, the number of molecules having sufficient energy to escape increases and hence the evaporation rate increases with temperature.

If the liquid is in a closed volume, some of the

escaped molecules (evaporated molecules) return (Fig. 11.7) to the liquid mainly as a result of collision with air molecules. The number thus returning to the liquid is proportional to the number of evaporated molecules in the air. Consequently, an equilibrium eventually will be reached when the number evaporating from the liquid is equal to the number returning to the liquid. The region above the liquid is now saturated with vapor and the pressure due to this saturated vapor is called the saturated vapor pressure. The saturated vapor pressure increases with temperature as more molecules are able to escape the liquid at the higher temperature (Fig. 11.8).

Molecules escape from the surfaces of solids too, but at a lower rate. Under certain conditions solid to gas transformation is favored. Such a transformation is called sublimation. Dry ice (solidified carbon dioxide) changes from the solid to the gas phase in this manner at room temperature.

The mechanism of boiling can now be explained. Boiling is a volume effect. The bubbles formed during boiling in the liquid, which are originally at the bottom but later rise to the surface, do not collapse. The pressure inside the bubble is the vapor pressure and the pressure outside is the atmospheric pressure plus the pressure due to the liquid column above. If the inside pressure is less the bubbles collapse; if the two pressures are equal the bubbles are stable. Hence boiling starts at a temperature when the vapor pressure is equal to the external pressure.

Figure 11.7. Liquid and vapor in equilibrium. The space above the liquid is saturated with vapor; the number of molecules leaving the liquid is equal to the number returning to it.

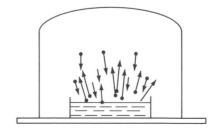

Figure 11.8 shows the variation of vapor pressure of water as a function of temperature. The curve also can be used to determine the boiling point at a given pressure.

The air is described sometimes as too dry or too humid; both conditions are uncomfortable to us. In one case the air contains very little water vapor and in the second case the water vapor content is very high. The weathermen, instead of telling us how much water vapor is in air in absolute units such as kg/m^3 or lb/ft^3 use a measurement called relative humidity,

Relative humidity

$$= \frac{\text{Amount of water vapor present}}{\text{Maximum amount of water vapor possible}}$$

$$= \frac{\text{Actual vapor pressure}}{\text{Saturated vapor pressure}}. \qquad (11.7)$$

Relative humidity is expressed in percentage; 50% humidity means the water vapor in air is only half of the maximum vapor the air can contain under saturation conditions.

Now we describe a method often used to determine relative humidity. Consider a situation where the temperature is 30°C and the pressure due to the water vapor is only 12.79 mm Hg. It is below the saturated vapor pressure of 31.82 mm Hg at this temperature. Relative humidity in this case is 12.79/31.82 = 40.2%. Now as we lower the temperature slowly, we will notice that at 15°C, where the vapor pressure of 12.79 mm Hg is equivalent to 100% relative humidity, condensation (water drops

Figure 11.8. Saturated vapor pressure of water as a function of temperature.

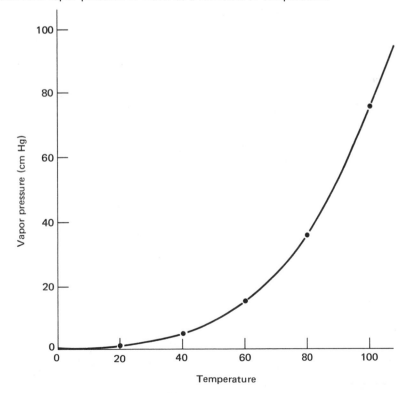

forming on the cooled surfaces) takes place. This temperature is known as the dew point. *Dew point is the temperature at which air becomes saturated with the water vapor it contains.* Knowing the dew point, we determine relative humidity as follows:

Relative humidity

$$= \frac{\text{Saturated vapor pressure at dew point}}{\text{Saturated vapor pressure at room temperature}}$$

$$(11.8)$$

Saturated vapor pressure at various temperatures is given in Table 11.2.

TABLE   11.2 Saturated Vapor Pressure of Water

TEMPERATURE (°C)	PRESSURE (mm Hg)
−10	2.15
−5	3.16
0	4.58
5	6.54
10	9.21
15	12.79
20	17.54
30	31.82
40	55.32
50	92.51
60	149.51
70	233.7
80	355.1
90	525.8
100	760.0
120	1489.1
140	2710.9
160	4636
180	7520
200	11,659

11.8 HEAT TRANSFER

It is our common experience that an object hotter than its surrounding loses its heat very fast. We often note that steaming hot food served to us loses its heat in a relatively short time. The principal methods by which an object loses energy are conduction, convection, and radiation.

Conduction

In conduction, heat energy is transmitted through a medium from the hot part to the cold part. This statement means that (a) a medium is necessary for conduction, (b) a temperature difference should exist, and (c) the heat flow is from a region of higher temperature to one of lower temperature.

In a metal poker, for example, heat flows from the end in contact with the fire to the end held by the hand. To protect the hand a handle of wood, a material of poor conductivity, is provided (Fig. 11.9). Note that in conduction there is no material flow, only energy transfer. Energy is transmitted through a mechanism of internal interactions. The atoms at the hot end are more energetic, they interact with

Figure 11.9. A metal poker. A handle made of wood (a material of poor conductivity) protects the hand.

the neighboring less energetic atoms through collisions and thus transfer energy.

We shall now derive an equation relating the heat flow per unit time, Q/T, to temperature difference and properties and dimensions of the medium. For this purpose we refer to Fig. 11.10. The two ends of the medium are at temperatures t_1 and t_2; $t_2 > t_1$. The rate of flow of heat energy, that is, the quantity of energy flowing per second, depends on several factors:

1 Area of cross-section of the medium. The energy flow rate is proportional to the area. (This is true in matter transfer as well, for example, water flow.)

2 The difference in the temperatures at the ends. The energy flow rate is proportional to $t_2 - t_1$.

3 Length of the medium. The energy flow rate is *inversely proportional* to the length. Combining 2 and 3, we see that the heat flow is proportional to $(t_2 - t_1)/L$, the temperature gradient. Temperature gradient can be considered as the slope of a temperature hill and analogous to the gradient of a hill. If the slope is steep, flow will be faster.

Combining the three factors and introducing a proportionality constant, we have the equation

$$\frac{Q}{T} = kA\,\frac{t_2 - t_1}{L}, \qquad (11.9)$$

where Q/T is the rate of heat flow. The proportionality constant k is called the thermal conductivity of

the material (Table 11.3). Units of thermal conductivity are

$$k = \frac{Q}{T}\,\frac{L}{A(t_2 - t_1)}$$

$$= \frac{kcal}{s} \times \frac{m}{m^2 \cdot {}^\circ C} = \frac{kcal}{s \cdot m \cdot {}^\circ C}$$

$$= \frac{Btu}{h} \times \frac{ft}{ft^2 \cdot {}^\circ F}$$

$$= \frac{Btu}{h \cdot ft^2 \cdot ({}^\circ F/in.)} \quad \text{(most commonly used)}.$$

EXAMPLE 11.6

Public buildings sometimes have glass windows with aluminum frames. The area of glass in such a window is 1 m^2 and the area of the aluminum part is 0.05 m^2. The thickness of the glass is 5 mm and the thickness of the aluminum frame is 1 cm. Calculate the amount of heat loss to the outside in 1 h if the inside temperature is 20°C and the outside temperature is −10°C.

The thermal conductivity of glass and aluminum are, respectively,

$$k_g = 1.6 \times 10^{-4} \text{ kcal/s} \cdot m \cdot {}^\circ C$$

$$k_{Al} = 5 \times 10^{-2} \text{ kcal/s} \cdot m \cdot {}^\circ C$$

The heat flow rate can be calculated using Eq. (11.9).

Figure 11.10. Heat conduction through a material of cross-sectional area A and length L.

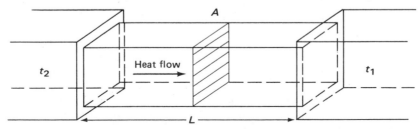

TABLE 11.3 Thermal Conductivities

SUBSTANCE	Btu $\cdot$ (in./ft)$^2 \cdot$ h $\cdot$ °F	kcal/m $\cdot$ s $\cdot$ °C
Aluminum	1451	5.0×10^{-2}
Brass	750	2.6×10^{-2}
Copper	2660	9.2×10^{-2}
Silver	2870	9.9×10^{-2}
Steel	320	1.1×10^{-2}
Asbestos	4.0	1.4×10^{-4}
Brick	5.0	1.7×10^{-4}
Concrete	12.0	4.1×10^{-4}
Corkboard	0.3	1.0×10^{-5}
Glass	5.8	2.0×10^{-4}
Air	0.14	5.3×10^{-6}
Water	4.15	1.4×10^{-4}

For the glass part, we get

$$\frac{Q}{T} = 1.6 \times 10^{-4} \; \frac{\text{kcal}}{\text{s} \cdot \text{m} \cdot \text{°C}} \frac{(1 \, \text{m}^2)[20-(-10)\text{°C}]}{5 \, \text{mm}(1 \, \text{m})/(1000 \, \text{mm})}$$

$$= 0.96 \, \text{kcal/s}.$$

The heat flow in 1 h through the glass window is

$$0.96 \frac{\text{kcal}}{\text{s}} \times \frac{3600 \, \text{s}}{\text{h}} = 3456 \, \text{kcal}.$$

The heat flow through the window frame is

$$\frac{Q}{T} = 5 \times 10^{-2} \; \frac{\text{kcal}}{\text{s} \cdot \text{m} \cdot \text{°C}} \times 0.05 \, \text{m}^2 \times \frac{30\text{°C}}{0.01 \, \text{m}}$$

$$= 7.5 \, \text{kcal/s}.$$

The energy flow (heat) through the window frame in 1 h is

$$7.5 \frac{\text{kcal}}{\text{s}} \times \frac{3600 \, \text{s}}{\text{h}} = 27,000 \, \text{kcal}.$$

The aluminum frame conducts 8 times more heat than the glass window. In any case it is easy to see that aluminum is not a good choice for a window frame.

R-Value

The thermal conductivity k is the rate of heat flow per unit area per unit thickness of a material for a unit temperature difference. Heat flow is obtained by Eq. (11.9a)

$$\frac{dQ}{dT} = kA \, \frac{\Delta t}{L}, \tag{11.9a}$$

where A is the area of the material, L the thickness, and Δt the temperature difference between the two sides of the material. Combining thermal conductivity k and thickness L into a single term $U (= k/L)$, we obtain

$$\frac{dQ}{dT} = \left(\frac{k}{L}\right) A \Delta t = UA \Delta t. \tag{11.10}$$

The term U is called the U-value which is equal to the rate of heat flow per unit area of the material per unit temperature difference. The unit of U-value in English units is Btu/h $\cdot$ ft$^2 \cdot$ °F. The inverse of the U-value (in English units) is called the R-value. Hence

$$R = \frac{1}{U}. \qquad (11.11)$$

When the U-value increases, the R-value decreases, that is, when heat flow per unit area is bigger, the R-value is smaller. In other words, the R-value is the resistance offered by the unit area of the material for heat flow. The unit of the R-value is h · ft^2 · °F/Btu, but engineers usually do not mention the unit. An advantage of using the R-value is that different materials placed together have a total resistance given by the sum of the individual resistance values, that is, the sum of the R-values. Note that the U- and R-values are not constants but depend on the thickness of the material.

EXAMPLE 11.7

A wall consists of dry wall ($R = 0.2$), insulating material ($R = 8$), insulation board ($R = 1.2$), and siding ($R = 0.2$). Calculate the total R-value of the wall and the heat flow per hour if the inside temperature is 70°F and the outside temperature is 20°F. The area of the wall is 200 ft^2.

$$\text{Total } R\text{-value} = 0.2 + 8 + 1.2 + 0.2;$$

$$U\text{-value} = \frac{1}{9.6} = 0.104 \; \frac{\text{Btu}}{\text{h} \cdot \text{ft}^2 \cdot {}^\circ\text{F}}$$

$$\text{Heat flow rate} = UA\,\Delta t$$

$$= (0.0104 \; \frac{\text{Btu}}{\text{h} \cdot \text{ft}^2 \cdot {}^\circ\text{F}})(200 \text{ ft}^2)(50^\circ\text{F})$$

$$= 104 \text{ Btu/h}.$$

Convection

In convection, masses of fluids, after gaining energy from a hot object, physically move carrying the energy with them. Consider a thin layer of air (defined by the dashed lines) immediately above a hot plate Fig. 11.11. The air in this region is hotter than the air above. Because of thermal expansion, the density of air closer to the hot plate is lower than that of the air above. Archimedes' principle now comes into play. Lighter air moves up, and denser colder air takes its place. The movement of air, hot air moving up and cold air coming down, sets up a circulation of air called convection currents, which lasts until the temperature of the hot plate and the surroundings equalize.

The rate of heat loss, Q/T, is rather difficult to calculate precisely because of its dependence on the property of the fluid, exposed area, and geometry of the hot object, as well as its nonlinear dependence on the temperature difference between the object and the surroundings. However, some general experimental observations that will lead us to an approximate equation are: (a) heat loss is proportional to the area, and (b) heat loss is proportional to the temperature difference, if it is not too high. This statement is known as Newton's law of cooling. Combining the two proportionalities and introducing a proportionality constant, we get

$$\frac{Q}{T} = hA(t_\text{h} - t_\text{s}), \qquad (11.12)$$

where h, the proportionality constant, is called the convection coefficient. The temperatures t_h and t_s

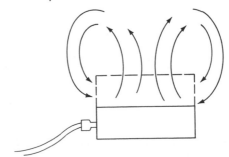

Figure 11.11. Air flow produces heat removal from the hot plate. This is called convection.

refer to hot object and the cold surroundings, respectively. Convection coefficients for generally used geometries are given in Table 11.4. Examples of convection are shown in Fig. 11.12.

In the discussion above, convection currents were produced by a natural process, buoyant forces acting on lighter material. There are also several examples of heat transfer produced by forced convection, such as blowing air through a radiator by a fan and heating a building by circulating hot air.

As an example of a situation in which the convection currents of air are reduced, we refer to storm windows. These windows confine the air to a limited space, resulting in a severe restriction of air circulation. Thus heat loss through convection is reduced. In addition, since air is a poor conductor of heat, the enclosed layer of air reduces the heat loss by conduction. Another example is the greenhouse. The sunlight entering through the glass heats up the interior of the greenhouse. Some of this energy is reradiated, but since energy is reradiated at a longer wavelength it is blocked by the glass enclosure. The property of some materials to transmit radiations of lower wavelength such as sunlight much more than radiation of higher wavelength is known as the "greenhouse effect." Recently it has been pointed out that the major reason for the effectiveness of a greenhouse is its ability to reduce the heat loss by convection. The greenhouse reduces convection currents and, as such, retains most of the heat trapped inside.

Radiation

By radiation we mean the release of energy in the form of electromagnetic waves by objects. In contrast to conduction and convection, radiation energy is transmitted between bodies without an intervening medium. The radiation from the sun reaches us through millions of miles of empty space. In the sun's radiation, energy is spread into all colors of the spectrum. The color content of the radiation is a function of the temperature. A detailed discussion of this may be found in Chapter 18. A few important things to remember are the following:

1 Energy transfer by radiation does not require any intermediate medium. Electromagnetic waves travel in vacuum too.

2 The energy radiated is proportional to the fourth power of the absolute temperature.

3 A good radiator is a good absorber of radiation.

4 An absorber that absorbs all radiation is an ideal absorber. Usually a black object comes close to being an ideal absorber.

5 Emissivity of an object is defined as the ratio of the energy absorbed by the object to the energy absorbed by a blackbody. Emissivity is close to 1 for a blackbody and 0 for a shiny silver surface (silver reflects all the radiation back).

Every object radiates heat whatever its temperature be. An object radiates energy to its surroundings as well as absorbs energy from the surroundings. If the object is hotter than the surroundings the energy radiated is more than the energy absorbed. The net energy lost is proportional to the difference between the fourth power of the temperatures. If the temperature difference is small, Newton's law of cooling applies to radiation too, that is, energy radiated is proportional to the temperature difference between the object and the surroundings.

A thermos flask (Fig. 11.13) cleverly uses our understanding of conduction, convection, and radiation to reduce the heat flow. To reduce conduction

TABLE 11.4 Convection Coefficients

GEOMETRY	h, kcal/m$^2 \cdot$ s $\cdot$ °C
Vertical	$(4.24 \times 10^{-4}) \sqrt[4]{\Delta t}$
Horizontal plate,	
facing up	$(5.95 \times 10^{-4}) \sqrt[4]{\Delta t}$
facing down	$(3.14 \times 10^{-4}) \sqrt[4]{\Delta t}$
Pipe of diameter D	$(1.0 \times 10^{-3}) \sqrt[4]{\Delta t/D}$

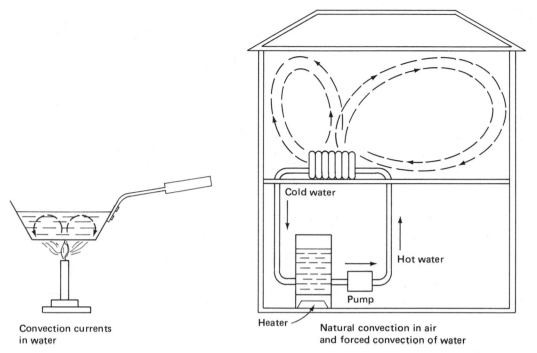

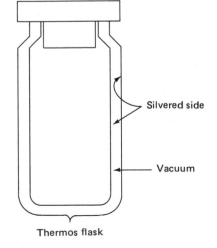

Convection currents
in water

Cold water

Hot water

Pump

Heater

Natural convection in air
and forced convection of water

Figure 11.12. Examples of heat loss by convection.

Figure 11.13. Thermos flask.

Silvered side

Vacuum

Thermos flask

it is made of glass—a poor conductor—and it is evacuated. Conduction through the remaining air is
very small. Convection is a minimum, the dead air
space with a very small amount of air cannot set up
convection currents. The inside of the outer bottle
and the outside of inner bottle are silvered to reduce
radiation.

SUMMARY   The units of heat were defined as follows: 1 kilocalorie is the heat needed to raise the temperature of 1 kg of water by 1°C. 1 Btu is the heat needed to raise the temperature of water weighing 1 lb by 1°F. The specific heat capacity is the heat required to raise the temperature of a unit mass of a material by a unit temperature difference. The specific heat capacity is numerically equal in all unit systems.

Calorimetric measurements are based on the principle of heat balance, namely, heat lost is equal to heat gained.

A change of state of matter at constant temperature is called a phase change such as melting and boiling.

The latent heat of fusion is the heat absorbed during melting at constant temperature per unit mass. The latent heat of vaporization is the heat absorbed during vaporization at constant temperature per unit mass.

$$\text{Relative humidity} = \frac{\text{Actual vapor pressure}}{\text{Saturated vapor pressure}}.$$

Heat is transferred from one object to another by conduction, convection, and radiation. Heat loss per unit time by conduction is given by

$$\frac{Q}{T} = kA\frac{t_2 - t_1}{L},$$

where k is the conductivity of the material. Heat loss by convection is proportional to temperature difference, area, and a factor that depends on the area of the hot object. For convection, the fluid should be present around the hot object, and the movement of molecules are responsible for energy transfer. Energy loss by radiation is proportional to the fourth power of the temperature. For heat transfer by this method no medium is needed.

QUESTIONS AND PROBLEMS

11.1   What is the difference between heat and temperature?

11.2   Is is true that heat input always increases temperature?

11.3   What are the effects of heat input on solids, liquids, and gases?

11.4   What is the material with the highest specific heat capacity?

11.5   The material with the highest specific heat capacity is good for storing heat and transferring heat. Give reasons.

11.6   The statement "Coffee with a little cream in it keeps warmer for a longer time" is often heard. Do you agree with this statement? If so why?

11.7   Wind blows from the sea to the land in the daytime and from the land to the sea in the evening. What are the reasons for it?

11.8   A fan cools you. How?

11.9   Steam at 100°C burns you more than water at 100°C. Why?

11.10 Air is a poor conductor, but you still use blankets to keep yourself warm. Why?

11.11 Two light blankets are better than a heavy blanket of a thickness equal to that of the two light blankets. Explain.

11.12 Two kilocalories of heat are given separately to 0.5 kg of water, 0.5 kg of iron, and 0.5 kg of aluminum. Calculate the increase in temperature in each case.

11.13 Fifty Btu of heat are supplied to 2 lb of water, 2 lb of mercury, and 2 lb of silver. Calculate the increase in temperature in each case.

11.14 What is the result of supplying 100 kcal of heat to 1 kg of ice at −10°C?

11.15 A heat exchanger in industrial plants uses 100 lb of steam every minute. The steam enters the exchanger at 212°F and leaves it at 52°F. How much heat is removed from the steam every minute?

11.16 An object weighing 0.2 kg was heated to 100°C and dropped into a hole in a block of ice. A mass of 0.07 kg of ice was melted. Calculate the specific heat capacity of the object.

11.17 The inner cup of a calorimeter is made of copper. It weighs 50 g. It contains 100 g of water at 20°C. If 50 g of silver heated to 100°C is dropped into the calorimeter, calculate the equilibrium temperature.

11.18 An aluminum calorimeter cup and stirrer weighs 25 g. It contains 50 g of water at 20°C. When a 50 g object is heated to 100°C and dropped into the water its equilibrium temperature is found to be 35°C. Calculate the specific heat of the object.

11.19 If two 2 g ice blocks at −10°C are added to a drink (mainly water) at 20°C weighing 20 g calculate the final temperature.

11.20 Latent heat of vaporization is experimentally determined by mixing known amounts of vapor and water in a calorimeter and noting the temperature changes. In such an experiment the following data were collected. The mass of aluminum calorimeter and stirrer was 40 g, mass with water 100 g, and the mass with water and condensed steam 108 g. Initial temperature of water was 20°C, temperature of steam 100°C, and the final temperature 85°C. Using these data calculate the latent heat of vaporization of water.

11.21 An electric hot plate made of steel is ¼ in. thick and has 1 ft^2 area. The inside of the hot plate is maintained at 600°F by an electric heater and the outside is at 200°F. Calculate the amount of heat flowing through the hot plate.

11.22 Water is boiling in an aluminum pan. Heat flows into the pan through its bottom which is 1/8 in. thick and has an area of 1.5 ft^2. If the temperature at the bottom surface is 400°F calculate the amount of water boiling per second.

11.23 A pond is covered with 2 in. thick ice. If the temperature below the ice is 0°C and above the ice is −10°C calculate the increase in thickness of ice in 1 h. The density of ice at 0°C is 999.87 kg/m^3.

11.24 Walls of a cold storage room consist of 6 in. of cork and 6 in. of concrete (Fig. 11.14). The temperature inside is −15°C and outside is 25°C. Calculate the temperature at the cork-concrete interface and calculate the heat flow through unit area of the wall in 1 s. (Hint: Heat flowing through the concrete has to be equal to the heat flowing through the cork).

11.25 A glass window has an area of 1.5 m^2 and a thickness of 8 mm. If the indoor temperature is 20°C and the outdoor temperature is 0°C calculate how much heat flows through the window.

11.26 Instead of a single glass, two glasses each 4 mm thick are used in the window of Problem 11.25. The space between the glasses is 4 mm thick and it is filled with air. Calculate the heat flow if the outside temperature is 0°C and inside temperature is 20°C.

11.27 A flat vertical wall (12 ft × 10 ft) is maintained at 120°F. The surrounding air temperature is 70°F. How much heat is lost by convection from both sides of the wall in 1 h.

11.28 A horizontal pipe has an outside diameter of 10 cm and its exposed length in a room is 2 m. If the outside temperature of the pipe is 90°C and surrounding air temperature is 10°C how much heat is lost per second by convection.

Figure 11.14. Problem 11.24.

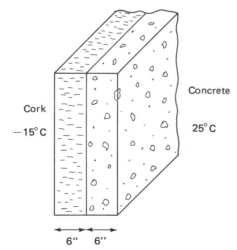

Cork

−15°C

Concrete

25°C

6″   6″

11.29 Temperature of a filament is increased from 600°C to 1000°C. Calculate the fractional increase in the amount of radiation.

11.30 In medicine "hot spots" in the body are detected by thermography (a photograph in which brightness is proportional to temperature). Two patches of skin of the same area are found to emit radiation in the ratio of 1:1.01. Calculate the ratio of temperatures at these two regions. Assume the properties of the skin with respect to emission are the same at both places.

12

Heat and Work

The discovery that heat is a form of energy was a milestone in the history of science. As we saw in Chapter 11, Count Rumford and several other scientists through their experiments established the equivalence of heat and mechanical energy. A quantitative relationship between the two was later established by James Prescott Joule. Since heat is energy, can it be converted to mechanical energy? This question seems superfluous at the present time when a large variety of heat engines are in use. The first successful heat engine was made in 1705 by Thomas Newcomen. The heat engines of today bear no resemblance in shape or performance to Newcomen's crude and inefficient steam engine. Advances in the last 300 years have not only perfected the steam engines, but have given us a variety of engines such the internal combustion engine, jet engine, and rocket engine.

Conversion of heat to work is the sole function of heat engines. Conversions, heat to work or vice versa, are subject to certain laws. These laws, the basis of the science of thermodynamics, and their implications are the main topics of this chapter. We begin with a discussion of Joule's experiment.

12.1 MECHANICAL EQUIVALENT OF HEAT

How much heat is equivalent to a unit of work? The answer to this question, as determined by

Joule, laid the foundation for quantitative thermodynamics. Joule, through a series of independent experiments, also showed that a unit of mechanical energy is equivalent to the same amount of heat. This result above all proved convincingly that heat is a form of energy.

One of the experiments Joule performed uses the simple equipment illustrated in Fig. 12.1. The paddle wheel inside the liquid is free to rotate. As the weight attached to the paddle wheel falls, the wheel turns, churning the water and thus increasing the temperature (read by the thermometer). Knowing the water equivalent of the calorimeter, the amount of water, and the increase in temperature, we can determine the amount of heat Q that will produce the same temperature change. The work done by the falling weight is given by

$$W = (PE + KE)_i - (PE + KE)_f.$$

Since $PE_i = mgh$, $KE_i = 0$, $PE_f = 0$, and $KE_f = \frac{1}{2}mv^2$, we obtain.

$$W = mgh - \frac{1}{2}mv^2,$$

Figure 12.1. Joule's experiment to determine mechanical equivalent of heat. As the mass m falls through a height h part of the potential energy converts to kinetic energy of the falling mass and the rest is spent in rotating the paddle wheel which increases the internal energy of the water.

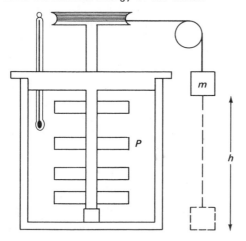

where h is the original height of the mass and v is its velocity as it reaches the ground, both of which can be experimentally determined.

The result of such experiments led Joule to the following relationship between heat Q and mechanical energy W, namely,

$$W = JQ. \qquad (12.1)$$

Equation (12.1) tells us that when $Q = 1$, $W = J$ or the constant J is the mechanical energy equivalent of a unit quantity of heat energy. Hence it is called the mechanical equivalent of heat sometimes also called Joule's constant.

The mechanical equivalent of heat is equal to the work done per unit quantity of heat.

From the units of W and Q, we obtain the following units for J:

Units of W	Units of Q	Units of J ($= W/Q$)
joule	cal	joule/cal
joule	kcal	joule/kcal
ft·lb	Btu	ft·lb/Btu

The accepted values of J in these unit systems are

$$J = 4.186 \text{ joule/cal,}$$
$$J = 4186 \text{ joule/kcal,}$$
$$J = 778 \text{ ft·lb/Btu.}$$

EXAMPLE 12.1

A piece of ice at $0°C$ is thrown at a wall. If 50% of the mechanical energy melts the ice and the temperature remains at $0°C$, determine the velocity of the ice just before the moment of impact.

Let M be the mass of ice in kg and v be the velocity of the ice in m/s. The heat required to melt M kg of ice is (M kg)(80 kcal/kg).

The mechanical energy equivalent of the heat absorbed by the ice is

$$= (M \text{ kg}) (80 \text{ kcal/kg}) (4186 \text{ J/kcal})$$
$$= (M \text{ kg}) (3.35 \times 10^5 \text{ J/kg}).$$

The kinetic energy of ice $= 2 \times$ mechanical energy equivalent of heat. Hence

$$\frac{1}{2} M v^2 = 2(M \text{ kg}) (3.35 \times 10^5 \text{ J/kg})$$

$$v = \sqrt{4 \times 3.35 \times 10^5} = 1156 \text{ m/s.}$$

12.2   HEAT-TO-MECHANICAL ENERGY

The conversion of heat to mechanical energy was achieved sometime around 200 BC by a Greek scientist of Alexandria named Hero. His heat-to-mechanical energy engine is shown in Fig. 12.2. As the steam shoots out, the reaction force rotates the drum; the energy of the steam of course is coming from the heat source.

Most of the modern engines' heat-to-mechanical energy transformation is accomplished in the following way: (a) Heat energy is supplied to a gas and it is allowed to expand. (b) As the gas expands, it causes a piston capable of doing work, to move. Therefore, we start our discussion by considering an amount of gas enclosed in a container shown in Fig. 12.3. The piston on the top is free to

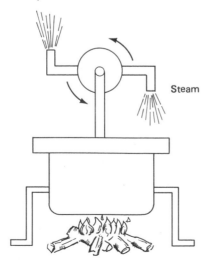

Figure 12.2. Hero's steam engine. Oldest steam engine developed around 200 BC.

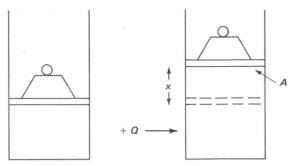

Figure 12.3. Gas in the cylinder expands on absorption of heat, lifting the piston and weight through a distance x.

move. On supplying heat Q, the gas expands moving the piston up without acceleration and thus lifting the weight with it. The work done in this case is equal to Fx, which can be written in the form

$$W = \frac{F}{A} Ax. \qquad (12.2)$$

In Eq. (12.2) F/A is the pressure acting on the gas and Ax is the change in volume ΔV. Hence

$$W = P\Delta V. \qquad (12.3)$$

Since the weight on the piston remains the same, the pressure in this example remains constant. Such processes are called isobaric.

Let us designate the initial temperature and volume by T_1 and V_1, respectively and the final temperature and pressure by T_2 and V_2, respectively. Since pressure is constant and since $V \propto T$, T_2 is greater than T_1. This means the internal energy of the molecules has also increased. If we denote the initial internal energy by U_1 and the final internal energy by U_2, the change in internal energy is given by

$$\Delta U = U_2 - U_1. \qquad (12.4)$$

On adding heat Q to the system, part of it transforms into mechanical energy (work) and the rest increases the internal energy of the gas, ΔU. Therefore, we get the following equation relating the heat input, work output, and increase in internal energy:

$$Q = W + U_2 - U_1 \qquad (12.5)$$

Equation (12.5) means *the heat added to the system is equal to the work done by the system plus the increase in the internal energy of the system.* This statement is known as the first law of thermodynamics. Equation (12.5) can be considered as another form of conservation of energy. Two points to remember are (a) in Eq. (12.5) heat input and work output are taken as positive and (b) to balance the units on both sides of the equation if heat is given in calories it should be converted to joules by multiplying by the mechanical equivalent of heat.

Special Cases of the First Law

Isochoric process, $W = 0$, $Q = \Delta U$: An isochoric process is a constant volume process ($\Delta V = 0$). Hence no work is done in this case. An increase in the internal energy is the only result of adding heat. Examples are heating a solid (volume change is negligible), fusion at melting point, etc.

Isothermal process, $\Delta U = 0$, $Q = W$: An isothermal process is a constant temperature process. For an ideal gas, the temperature remains constant, there is no change in internal energy and hence all the heat input goes into work. An example of an isothermal process is compressing a gas at a constant temperature. An amount of heat Q equal to the work input is removed from the system in this process (Fig. 12.4).

Adiabatic process, $Q = 0$, $\Delta U = -W$: An adiabatic process is one in which no energy transfer due to temperature difference between the system and the surroundings takes place. Assume the gas in the cylinder (Fig. 12.5) is suddenly expanded. If the expansion is very fast, the exchange of heat is very little; effectively $Q = 0$. Also if the system is completely isolated from its surroundings so that no heat can enter or leave, this process is adiabatic and

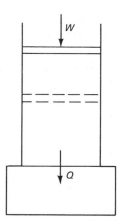

Figure 12.4. Isothermal compression; amount of heat Q released is equal to work done W on the gas.

again $Q = 0$. In both these cases the amount of work done comes out of the internal energy, and hence the latter decreases. A decrease in temperature follows the decrease in internal energy. Hence we can conclude the following: An adiabatic expansion decreases the temperature of an ideal gas. The reverse, adiabatic compression increases the temperature of an ideal gas, is also true.

We could find several examples to illustrate the effect of adiabatic expansion, such as escaping air from a pressurized tire causes the valve to become cooler. Another example is found in refrigerators and other cooling systems. In all cooling systems a fluid is allowed to undergo adiabatic expansion

Figure 12.5. Adiabatic expansion; internal energy of the gas decreases but no heat exchange takes place with external environment.

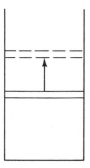

through a throttle valve made of a porous plug leading into a low pressure region. The region around the valve is properly insulated to ensure adiabatic ($Q = 0$) conditions. The temperature of the fluid on expansion is much lower than the temperature of the fluid before expansion on the high pressure side (Fig. 12.6).

EXAMPLE 12.2

One mole of an ideal gas occupies a volume of 22.4×10^{-3} m^3 (22.4 liters) at STP (standard atmospheric pressure and 0°C). It expands at constant pressure to 30×10^{-3} m^3 on absorbing 300 cal of heat. Calculate (a) the change in the internal energy of the gas and (b) the temperature of the gas after expansion.

(a) To solve the problem we use the first law of thermodynamics:

$$Q = W + \Delta U$$
$$Q = 300 \text{ cal} = 300 \text{ cal} \times 4.19 \text{ J/cal} = 1250 \text{ J}$$
$$W = P\Delta V = (1 \times 10^5) \text{ N/m}^2 \times (716 \times 10^{-3}) \text{ m}^3$$
$$= 7.6 \times 10^2 \text{ J}$$

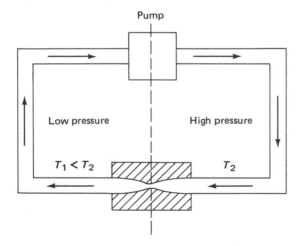

Figure 12.6. The free expansion of the gas into a low pressure region is called throttling. In the present case the temperature decreases in an adiabatic expansion of the gas.

(Note atmospheric pressure $= 1 \times 10^5$ N/m^2.) Therefore,

$$\Delta U = Q - W = 5 \times 10^2 \text{ J}.$$

(b) To calculate the new temperature we use the equation

$$\frac{P_1 V_1}{T_1} = \frac{P_2 V_2}{T_2}.$$

$$P_1 = P_2 = 1 \times 10^5 \text{ N/m}^2,$$
$$V_1 = 22.4 \times 10^{-3} \text{ m}^3,$$
$$V_2 = 30 \times 10^{-3} \text{ m}^3,$$
$$T_1 = 273 \text{ K}.$$

On substitution of these values into the above equation

$$\frac{P_1(22.4 \times 10^{-3} \text{ m}^3)}{273 \text{ K}} = \frac{P_1(30 \times 10^{-3} \text{ m}^3)}{T_2}$$

$$T_2 = 365.6 \text{ K} = 92.6°\text{C}.$$

12.3   WORK AND *PV* DIAGRAMS

In the example of gas expanding at a constant pressure (see p. 198) work was found to equal $P(V_2 - V_1)$. The pressure as a function of volume is now shown in Fig. 12.7. Work in this case is equal to the area under the curve between V_1 and V_2 (the shaded area).

Let us now consider the case where P is not constant during expansion as shown in Fig. 12.8. For the small volume interval ΔV, indicated in Fig. 12.8, the pressure can be considered as essentially constant and is equal to P_A. Since the portion of the area under the curve is approximately equal to the area of the rectangle (shaded area) the work done during the interval of expansion is essentially $P_A \Delta V$. Adding up incremental works from V_1 to V_2, we get the total work. The total work is equal to the total area under the curve from

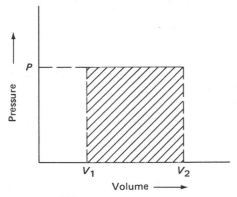

Figure 12.7. *PV* diagram. Work done is equal to area under the curve.

V_1 to V_2. Thus in all cases, whatever the shape of the *PV* curve, the area under the *PV* curve between two points V_1 and V_2 is equal to the work done during an expansion from V_1 to V_2.

12.4 CYCLIC ENGINE

It is not possible to obtain work continuously from an engine in which the piston is moving in one direction only, because of the physical limitations. A cyclic engine, on the other hand, is able to do work continuously. Since the piston and the rest

Figure 12.8. *PV* diagram in which pressure is not a constant. In this case also, work done is equal to area under the curve.

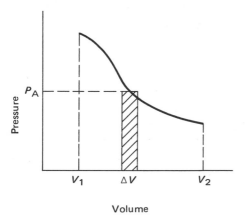

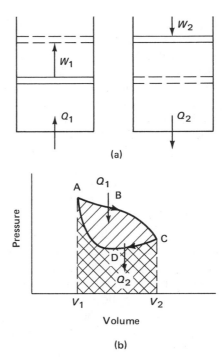

Figure 12.9. *PV* diagram for a cyclic engine. The net work done by the engine is the area enclosed by the closed curve ABCDA. In the first half of the cycle heat Q_1 is absorbed, gas expands and work W_1 is done. In the second half of the cycle (CDB), work W_2 is done on the gas. It releases heat Q_2. The net work done $W_1 - W_2$ is equal to net heat input $Q_1 - Q_2$.

of the system returns to the original state at the end of each cycle, the engine need not be too bulky.

Let us now consider an engine similar to the one shown in Fig. 12.9(a) which we discussed before, but now operating as a cyclic engine. The working substance in the engine, the substance that absorbs heat and does the work, is ordinarily a gas. For convenience, let us divide the cycle into two halves: the first half is where the working substance absorbs heat, expands, and does work on the piston. The return half is where work is done on the gas as it is compressed and it gives up some heat. On completion of the cycle, the gas returns to its original state. This implies its temperature, pressure, and volume are the same at the start of the cycle and at the end of the cycle. The curve ABC in the PV graph represents the first half of the cycle [Fig. 12.9(b)]. Work W_1 done by the gas is given by the area under the PV curve. During the return half of the cycle, if we take the same path in reverse, CBA, the work we have to do on the gas is exactly equal to the work done by the gas in the first half of the cycle. Hence there is no net gain in work if the same path is followed in each half of the cycle.

Consider a different path like CDA for the half of the cycle. The work done on the gas, W_2, is equal to the area under the curve CDA, which is less than W_1. The net work done by the gas, $W_1 - W_2$, is the difference between the two areas, that is, the area enclosed by the curve ABCDA. The system has returned to the original point in this cycle. The internal energy of the gas is the same as at the starting point. Now where did the work come from? The energy released during the return cycle Q_2 is less than Q_1. The net work done is equal to $Q_1 - Q_2$.

The efficiency of a cyclic engine is given by

$$\text{Efficiency } \eta = \frac{\text{Work}}{\text{Heat input}} = \frac{Q_1 - Q_2}{Q_1} . \quad \textbf{(12.6)}$$

Can the efficiency be made 100%? From Eq. (12.6) it is clear that 100% efficiency can be achieved if Q_1 is infinitely large, an impossibility, or $Q_2 = 0$. The latter implies that the work done on the engine during the reverse half of the cycle is zero, that is, the engine returns through a path BV_2V_1A in the PV diagram. This requires the pressure to drop to zero during part of the cycle. This would require the temperature to fall to absolute zero which cannot be achieved. Therefore, a cyclic machine with 100% efficiency cannot exist. In a practical engine like an automobile engine, the efficiency runs about 30%.

12.5   SECOND LAW OF THERMODYNAMICS

The first law restates the principle of conservation of energy, but it does not restrict the conversion of all the input heat into work and thus obtaining a 100% efficient engine. From our experience we know that such an ideal machine is one we all hope for but cannot achieve. The discussion given above supports the view. The second law of thermodynamics, in one of its numerous forms, is a statement regarding the fact that it is impossible to make a cyclic engine whose function is to transform heat completely to work (Fig. 12.10).

Second Law

It is impossible to construct an engine, whose sole function is to transform into work heat extracted continuously from a source.

In another form, the second law states that it is not possible for a cyclic engine to extract heat continuously from a source and transport it to another body at a high temperature without input work to the machine. (See Fig. 12.8.)

The second form of the second law prohibits us from having an ideal refrigerator which has only one function, that is, to remove heat continuously from a colder object to a hotter object without work being done.

Figure 12.10. Second law prohibits the complete transformation of heat to work and the continuous removal of heat from a low temperature source to a region of high temperature without doing work.

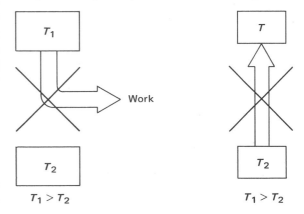

12.6 CARNOT'S CYCLE

Even without the consideration of energy losses (such as that due to friction) according to the second law, a perfect engine with 100% efficiency does not exist. A question now arises. What is the maximum efficiency of an engine which operates between two temperatures? An ideal engine with the maximum efficiency was described by Sadi Carnot (1776–1832). The engine is called the Carnot engine and its cycle is called the Carnot cycle.

The ideal Carnot engine is described here with the aid of Figs. 12.11 and 12.12. The working material of the engine is an ideal gas that follows the ideal gas law described in Chapter 11. The gas is enclosed in a cylinder, the cylindrical side of which is made of a perfect insulator and its bottom of a perfect conductor. Heat is absorbed into the gas from a source maintained at a constant temperature T_1 and the heat is given out to a sink maintained at constant temperature T_2 ($T_2 < T_1$) (see Figs. 12.12 and 12.13). It is assumed that the removal of heat from the source does not alter its temperature. Similarly the sink temperature does not change during the part of the cycle when heat is exhausted into the sink.

The Carnot cycle consists of four parts:

In Step 1 the state of the gas inside the cylinder is initially described by the quantities P_1, V_1, and T_1 and is indicated by the point A in the PV diagram (Fig. 12.12). The cylinder containing the gas is then placed on the heat source where it absorbs heat at a constant temperature T_1. The gas expands pushing the piston out and thus performing work. The pressure decreases as the volume increases, the product of PV remaining a constant. The heat Q_1 absorbed during the expansion appears as work in this process. The work done is the area under the curve AB in the PV diagram. (We note that since T is constant, ΔU for an ideal gas = 0. Hence $Q = W$.)

Step 2 is an adiabatic expansion. To obtain this adiabatic expansion, the system is isolated by keeping it on a block of insulator. As in any adiabatic expansion work output results from the change in internal energy. The expansion is allowed to take place till the temperature decreases to the tempera-

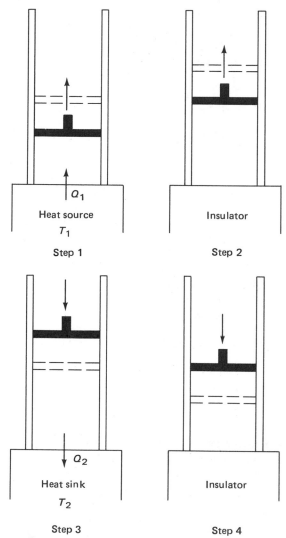

Figure 12.11. Four steps of the Carnot cycle. Step 1, the cylinder is kept on a heat source at T_1. Gas absorbs heat and expands isothermally. Step 2, gas expands adiabatically; temperature falls to T_2. Step 3, the piston is compressed at T_2 and the gas releases Q_2 to the sink. Step 4, gas undergoes adiabatic compression; temperature increases to T_1, and the gas attains its original pressure, volume, and temperature.

ture of the sink, that is, during this expansion the temperature falls from T_1 to T_2. Curve BC represents this change.

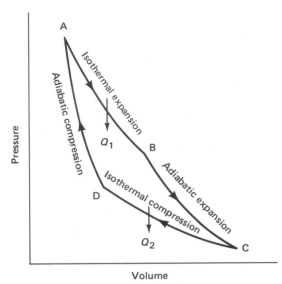

Figure 12.12. A Carnot cycle consists of isothermal (AB) and adiabatic (BC) expansions and isothermal (CD) and adiabatic (DA) compressions. The net work done is equal to the area enclosed by the curve ABCDA.

In Step 3 the engine is kept on the sink maintained at T_2. The gas is compressed isothermally.

Figure 12.13. Relationship between work, heat intake, and heat output in a heat engine.

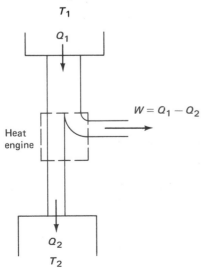

The engine gives up heat Q_2 to the sink. CD represents this change.

Step 4 is an adiabatic compression, which enables the gas to return to its original state. The curves AB and CD are called isotherms and the curves BC and AD are called adiabats.

The net work output of the Carnot engine is $Q_1 - Q_2$; and the efficiency of the engine η is given by

$$\eta = \frac{Q_1 - Q_2}{Q_1}. \tag{12.7}$$

The cycle can be repeated indefinitely and is a reversible process. It can also be shown that the efficiency is a function of the absolute temperatures T_1 and T_2. Hence we have

$$\eta = \frac{T_1 - T_2}{T_1}. \tag{12.8}$$

This equation again shows that for η to be 100% the sink temperature T_2 has to be 0, that is, part of the cycle should take place at absolute zero temperature. However, according to the second law, 100% efficiency can never be achieved. Moreover, based on the third law beyond the scope of this text, a temperature of absolute zero can never be achieved.

EXAMPLE 12.3

An engine is made of aluminum and the exhaust temperature is atmospheric temperature (20°C). What is the highest efficiency we can obtain with this engine?

Even though the highest temperature of the working gas should be well below the melting point of aluminum (658°C), for the calculation of the highest efficiency we use the melting point temperature as the source temperature. Hence we have

$$T_1 = (273 + 658°C) = 931 \text{ K}$$
$$T_2 = (373 + 20°C) = 293 \text{ K}$$
$$\eta = \frac{931 \text{ K} - 293 \text{ K}}{931 \text{ K}} \times 100 = 68.5\%.$$

The above example illustrates how material properties place limitations on their use. In addition to temperature, can you think of other physical quantities that should be considered in designing engines? Also in this example we assumed that the temperature of the engine walls is the same as the temperature of the working substance. As you will see in the next section this is not always true. In internal combustion engines the temperature of the working substance can be much higher than the engine walls.

12.7 INTERNAL COMBUSTION ENGINES

An ideal Carnot engine receives heat from an outside source at high temperature. In a similar fashion steam engines receive all the required heat from an outside source. The source produces heat by combustion and from there it is transferred to the working medium, water or steam, in a boiler. The temperature of the steam cannot be raised too high, the limiting temperature being dictated by the properties of the material of the boiler. This limiting temperature of the steam sets a limit to the efficiency of the steam engine.

Internal combustion engines differ from steam engines in the following significant ways: (a) Heat is produced inside the engine by chemical reaction between fuel and air. (b) The temperature of the combustion products, the working substance of the engine, may reach a value of 4000°F or higher. (c) The amount of heat transferred from the combustion products to the engine walls is very small and therefore its temperature remains comparatively low (400°F). (d) Because the temperature of the working substance remains high in combustion engines, higher efficiencies than in steam engines can be achieved. (e) Above all, the internal combustion engine is extremely compact.

Automobile engines are an important class of internal combustion engines. The Otto and Diesel engines named after Nikolaus Otto and Rudolf Diesel are two of the common types of automobile engines. In the Otto engine, a chemical reaction is initiated by an electric spark; hence it is called a spark ignition engine. In the Diesel engine the rise in temperature during an adiabatic compression initiates the chemical reaction and it is called a compression ignition engine. In both these engines, combustion takes place inside the engine. Once the reaction is completed, combustion products are not capable of producing more heat. Therefore, at the start of each cycle, a fresh quantity of fuel and air have to be taken in and, at the end, the reaction products have to be exhausted.

The general features of an internal combustion engine are shown in Fig. 12.14. The engine contains a cylinder in which a close-fitting piston moves. In the cylinder head, there are two values, the inlet valve for admitting the fuel and the exhaust valve for discharging the combustion products. The normally closed valves are opened at the appropriate times during each cycle by cams connected to the engine crankshaft. In an Otto engine, there is a spark plug in the center of the cylinder head and, in a Diesel engine, there is a fuel injection nozzle in its place. The movement of the piston from one end to the other is called a stroke.

All automobile engines are four-stroke engines. We first discuss the four strokes in each cycle of an Otto engine. Its idealized PV diagram is shown in Fig. 12.15. The four strokes are as follows:

1 Intake stroke. The intake valve admits a mixture of air and fuel vapor into the engine from the carburetor. Fuel and air in proper proportions is mixed in the carburetor.

2 Compression stroke. The mixture is compressed adiabatically during this stroke, with both valves closed. The line ab in the PV diagram represents this compression. The temperature and pressure increases during this compression.

3 Ignition and power stroke. At the end of the compression stroke ignition is started by the spark plug. The sudden combustion of the fuel supplies a quantity of heat Q to the system and the resulting temperature increase produces an instantaneous pressure rise (line bc in the PV diagram). Next the increased pressure causes

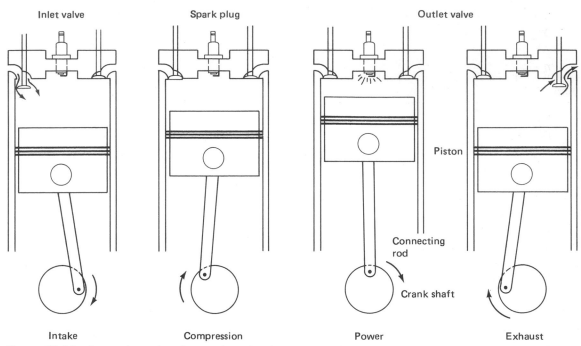

Figure 12.14. Internal combustion engine. In the case of a Diesel engine, instead of a spark plug there is a nozzle through which fuel under pressure is injected into the engine.

Figure 12.15. Idealized *PV* diagram of an Otto engine. The line a'a represents the intake stroke and aa' the exhaust stroke.

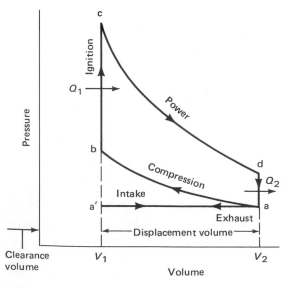

the piston to move out. The piston is performing the work during this stroke which, for this reason, is called the power stroke. The expansion of the gas in this case is adiabatic and is represented by the line cd. The gas pressure [75 psi (lb/in.^{2})] and temperature (2500°F) remain relatively high as this stroke is completed. The exhaust valve opens at this point. The combustion products escape the cylinder because of its high pressure. Heat Q_2 leaves the system at a constant volume as shown by line da.

4 Exhaust stroke. The remaining combustion products are now exhausted by the inward moving piston.

The net work done by the engine is equal to the area abcda in the *PV* diagram. The size of the area depends on the values of V_1 and V_2. If V_1 is larger compared to V_2, we get more output work. The efficiency of an Otto engine depends on the ratio

V_1/V_2, known as the compression ratio, and is given by

$$\eta = 1 - \frac{1}{(V_1/V_2)^{0.4}}. \qquad (12.9)$$

The larger the ratio V_1/V_2, the greater is the efficiency of the engine. Most Otto engines have compression ratios in the range of 7 to 12. The volume V_1 is the volume of the space above the piston when it is in the extreme top position and the volume V_2 is equal to V_1 plus the volume swept by the piston in a stroke.

In the Otto cycle consisting of four strokes, work is done by the engine only in one stroke, that is, only once in every two revolutions of the crank shaft. For smooth operation, multicylinder engines, at least two, are used; each cylinder supplying the power to the crank shaft at uniform intervals. A flywheel of high moment of inertia attached to the crank shaft also helps in the smooth running of the engine.

EXAMPLE 12.4

In ordinary automobile engines the compression ratio is about 8. Compute their theoretical efficiency.

$$V_1/V_2 = 8.$$

Substituting this value of the compression ratio into Eq. (12.9), we get

$$\eta = 1 - \frac{1}{8^{0.4}} = \frac{1}{1.23} = 57\%.$$

As mentioned earlier, the efficiency of an ordinary auto engine is about 30%, close to half of the ideal maximum we obtained in the above problem.

The four strokes of a Diesel engine are discussed below. Its idealized PV diagram is shown in Fig.

12.16. The four strokes in each cycle of the Diesel engine are as follows:

1 Intake stroke. The intake valve is open and air at atmospheric pressure is taken in during this stroke (a'a in Fig. 12.16). No fuel is taken in at this time in contrast to Otto engines where, in the first stroke, air and fuel are taken in.

2 Compression stroke. Air is compressed almost adiabatically to a fraction (1/16) of its original volume. The compression stroke is represented by line ab in the PV diagram (Fig. 12.16). The compression ratio (V_2/V_1) in Diesel engines ranges from 12 to 18.

3 Power stroke. In the first part of this stroke, fuel is injected at constant pressure. The temperature of the compressed air is high enough to burn the injected fuel. After the constant pressure burning of the fuel, an adiabatic expansion (cd) takes place. Work is performed during this stroke. At the end of this stroke, an exhaust valve opens releasing part of the

Figure 12.16. Idealized PV diagram of a Diesel engine. Ratio $V_2/V_1 \sim 12$ to 18.

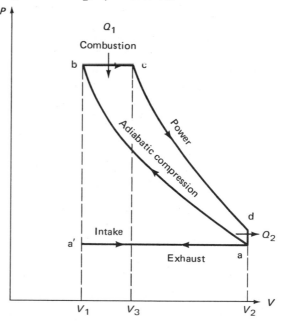

combustion products at constant volume (line da).

4 Exhaust stroke. Burnt fuel is forced out to make the engine ready for the next cycle.

The compression ratio of the Diesel engine is higher than that of the Otto engine. The efficiency of Diesel engines are higher (35%) than that of Otto engines because of the higher compression ratio. The theoretical value depends not only on the compression ratio V_2/V_1, but also on the expansion ratio V_3/V_1.

Another advantage of the Diesel engine is that the fuels used are cheaper than the fuels used in Otto engines. There are some disadvantages too. Diesel engines are made bulky to withstand the high peak pressures and the fast rise in pressure. With new advances in material technology, the size of the Diesel engines are decreasing. Even though the Diesel engines do not have spark plugs and associated circuits, the nozzle to inject fuel under pressure adds to the cost of the engine because the nozzle and the associated injection system are precision made. On cold days (temperature below 40 to 50°F), starting the Diesel engine becomes very difficult and often some starting aids such as electric glow plugs or electric resistance heaters are used.

Thus Diesel engines have the following drawbacks in comparison to Otto engines: (a) they are heavier for a given power; (b) they are more expensive (close to two times as costly per horse power); (c) Diesel engine exhausts contain more oxides of nitrogen than the stringent levels due to be enforced in 1980.

12.8 REFRIGERATION

By refrigeration we mean removal of heat. Heat engines absorb heat from a source at a higher temperature, using part of it to do work and discarding the rest to a sink at a lower temperature. This process is reversed in refrigerators. They use mechanical work to take heat energy from a colder region to a warmer region (Fig. 12.17). The efficiency of a

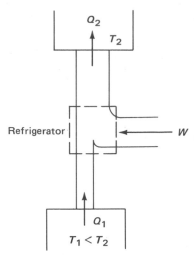

Figure 12.17. Heat Q_1 is taken from a region of low temperature T_1 and heat Q_2 is exhausted at a higher temperature T_2. The relation $Q_1 + W = Q_2$ holds in the case of an ideal refrigerator.

refrigerator is measured in terms of the heat it can remove for the amount of work done on it in each cycle. For this purpose, we define a term called coefficient of performance in the following way:

$$\text{COP} = \frac{Q_1}{W} = \frac{Q_1}{Q_2 - Q_1}. \tag{12.10}$$

An ideal refrigerator is a Carnot engine working in the reverse direction. The coefficient of performance for the ideal refrigerator can be written in terms of the absolute temperatures as

$$\text{COP} = \frac{T_1}{T_2 - T_1}. \tag{12.11}$$

COP becomes ∞ when $T_2 = T_1$, that is, the efficiency of a refrigerator is very high when the two temperatures are equal. As in the case of heat engines, theoretical efficiency is never achieved because of energy losses due to friction, etc.

A schematic diagram of a refrigerator is shown in Fig. 12.18. It consists of a closed system, which absorbs heat at one point—the evaporation chamber

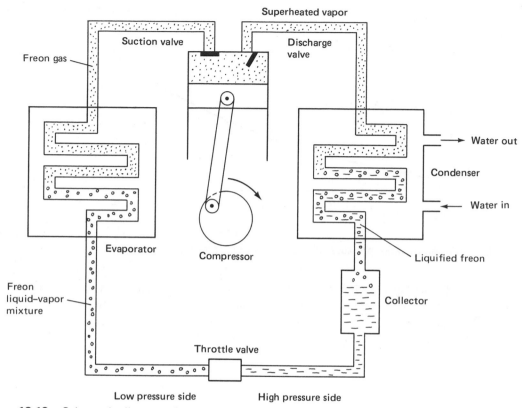

Figure 12.18. Schematic diagram of a refrigerator using freon-12 as its working substance.

in Fig. 12.18—and gives up heat at another point— the condenser. The working substance in the closed system, called the refrigerant, is a fluid with a low boiling point below the melting point of ice at standard pressure. Two such fluids are ammonia (bp, $-28°F$) and freon-12 (bp, $-21.6°F$).

Freon-12 is the refrigerant used in most of the modern refrigerators. The vaporization curve of freon is given in Fig. 12.19. At atmospheric pressure freon boils at $-22°F$, while, if the pressure is 110 psi (lb/in.^{2}) freon boils at about $100°F$ (Fig. 12.19). All cooling systems absorb heat during vaporization of the working substances and release heat to the outside during its liquefaction. Vaporization takes place in the vaporator and liquifaction in the condenser.

To analyze the operation of the refrigerator, let

us start at the compressor (Fig. 12.18). The compressor takes in vapor at slightly above atmospheric pressure and at room temperature and compresses the vapor to a high pressure (110 psi). This compression, since almost adiabatic, also increases the temperature of the vapor. Heat from the compressed vapor is removed by circulating air (in household refrigerators) or other fluids (water in commercial refrigerators) around the condenser. The temperature decreases below the boiling point (100°F) of freon-12 at high pressure. The freon, condensed into a liquid under the high pressure, is released into a low pressure region by a throttle valve. Some liquid vaporizes, cooling the rest. In addition, cooling produced by the expansion of the vapor into the low pressure region contributes to the cooling of the liquid-vapor mixture. As this

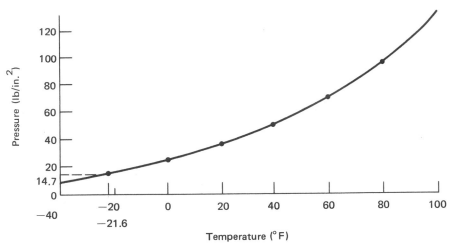

Figure 12.19. Variation of vapor pressure of freon as a function of temperature.

mixture goes through the evaporator it absorbs heat, and finally all of it converts to vapor at a temperature above its boiling point. This vapor then enters into the compressor and the cycle repeats.

not operating) expands, and further cools. This cooled liquid and vapor mixture absorbs heat from the outside at lower temperature. When the room needs to be cooled the flow reverses, and throttle valve II now operates instead of I.

12.9 HEAT PUMP

Suppose that we keep the evaporator of a cooling system outside, and the condenser inside a house. Even when the temperature outside is lower than the temperature inside we will be able to remove heat from outside and deliver it to the inside (Fig. 12.20). The heat delivered will be higher by the amount of the work W done by the compressor. This idea of the heat pump was suggested by Lord Kelvin in 1852, but only now such systems are becoming popular. This system, by reversing the flow of the fluid, can be made a cooling system in the summer when the house needs to be cooled.

A schematic diagram of a heat pump is shown in Fig. 12.21. In winter months the compressed fluid flows from the compressor to the house where the heat is removed from the fluid to heat the room. The liquid passing through throttle valve I (valve II is

Figure 12.20. Schematic representation of a heat pump. It takes heat from the cooler outside and transfers the heat to the inside at a higher temperature. Upon reversing the cycle, it can be made to work like an air conditioner in the summer.

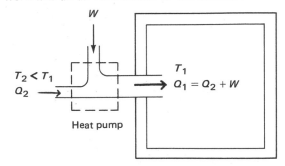

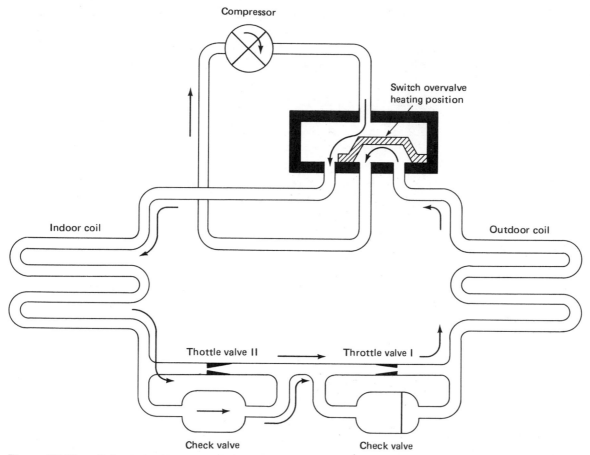

Figure 12.21. Schematic diagram of a heat pump. When it works as a heat pump (direction of flow indicated) throttle valve II is bypassed. To cool the room, fluid flow is reversed and expansion takes place at valve II instead of valve I.

SUMMARY

Heat is a form of energy. Equation (12.1) gives the relationship between heat and mechanical energy. J is known as the mechanical equivalent of heat.

The first law of thermodynamics expresses the interchangeability of work, heat, and internal energy, i.e., $Q - W = U_f - U_i = \triangle V$. The second law states that it is not possible to construct a heat engine with 100% efficiency.

The Carnot engine has the maximum possible efficiency of an engine working between two temperatures. Efficiency = $[(T_1 - T_2)/T_1] \times 100$ where T_1 is the temperature of the source and T_2 the temperature of the sink.

Internal combustion engines use heat produced by the combustion of fuel inside the engine to do work. The Otto engine uses a spark plug to ignite the fuel. The Diesel engine depends on a temperature increase during the compression stroke.

Refrigerators use work to remove heat at a lower temperature and exhaust it at a higher temperature. Heat pumps are reversible refrigerators.

QUESTIONS AND PROBLEMS

12.1 Is it correct to use the same units for both energy and heat? What will be the value of *J* in this case?

12.2 Is it possible to convert mechanical energy completely to heat?

12.3 Is it possible to convert heat completely to mechanical energy?

12.4 Explain the terms (a) mechanical equivalent of heat, (b) isothermal process, (c) adiabatic process, (d) isobaric process, (e) heat engine, (f) refrigerator, (g) efficiency of a heat engine, and (h) coefficient of performance of a refrigerator.

12.5 Is it possible to cool a room by opening the door of a refrigerator?

12.6 A liquid solidifies at −50°F. For a cooling machine designed to solidify this liquid can you use freon as a refrigerant?

12.7 Find some reasons why Otto engines are more commonly used than Diesel engines.

12.8 During adiabatic expansion, does a gas do any work? If so where is the energy coming from?

12.9 A loosely covered cooking pot opens and closes when the water boils in it. Is there any net gain in work?

12.10 A 50 lb block traveling with a velocity of 55 mi/h stops in 10 ft on a rough floor. How much heat is produced?

12.11 In an experiment to determine the mechanical equivalent of heat, a mass of 20 kg falls through a height of 2 m rotating the paddle wheel. The velocity of the object as it touches the ground is 1 m/s. The water equivalent of the paddle wheel and calorimeter including water is 20 g. The increase in temperature measured is 4.6°C. Obtain the mechanical equivalent of heat.

12.12 A gallon of gasoline, when burnt, produces 63.7 Btu of heat. If the engine is 30% efficient, how much mechanical energy will be produced by the engine for each gallon of gasoline?

12.13 The heat of combustion of methane gas (nature produces it as marsh gas by anaerobic decay of organic materials; farmers produce this gas from animal waste) is 24,000 Btu/lb. Calculate the mechanical energy equivalent of 200 lb of gas.

12.14 The heat of combustion of coal is 12,000 Btu/lb. A coal fired engine produces 1000 W of power with 10% efficiency. How much coal is burnt every hour?

12.15 The line ab (Fig. 12.22) represents a volume expansion from 20 m^3 to 100 m^3 at a constant pressure of 120 kPa (kilopascals). Calculate the amount of work done in this expansion.

12.16 The line bc (Fig. 12.22) represents a constant volume process. Is any work done by the gas in this process? Calculate the change in temperature during this process? Is there any change in the internal energy during this process?

12.17 How much work is done in compressing the gas from 100 m^3 to 20 m^3 at 60 kPa (line cd in Fig. 12.22)?

12.18 Calculate the net work done in the cyclic process described by the *PV* diagram in Fig. 12.22.

12.19 A Carnot engine operates between 20°C and 500°C. Calculate its efficiency.

12.20 A Carnot engine operating between 10°C and 400°C receives 800 cal of heat at the higher temperature. What is the efficiency of the engine? How much heat is exhausted at the lower temperature? What is the net work done by this engine?

12.21 If the Carnot engine of Problem 12.20 operates in reverse as a refrigerator and removes 1000 cal in each cycle, how many calories are exhausted at the higher temperature? How much mechanical work is done in each cycle?

12.22 The standard unit of refrigeration capacity is the ton of refrigeration. This is equivalent to freezing 1 ton of water at 32°F to ice at 32°F in a day. Obtain the equivalent of 1 ton of refrigeration in Btu per day.

12.23 The coefficient of performance of a refrigerator is 5. If the refrigerator removes 2500 cal of heat from a quantity of food, (a) how much work is

Figure 12.22. Problems 12.15, 12.16, 12.17, and 12.18.

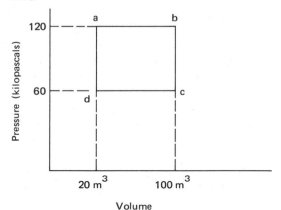

done by the electric motor and (b) how much heat is exhausted into the air.

12.24 The coefficient of performance of a heat pump is 4, when it works as a cooler. How much mechanical work should be done to deliver 1 Btu heat energy to a house? What fraction of the heat is taken from the outside?

12.25 The maximum (theoretical) efficiency of an Otto engine is 60%. Obtain its compression ratio.

13
Vibrations and Waves

The type of motion known as vibration or oscillation is different from both linear motion and circular motion which we have studied in previous chapters. In oscillatory or vibratory motion an object swings back and forth about an equilibrium position. In this type of motion, the magnitude of the force continuously changes from zero at the equilibrium position to its greatest value at the maximum displacement position. The force is, in fact, proportional to the displacement and directed towards the equilibrium position. Since the oscillating object swings to both sides of the equilibrium position, the direction of the force changes twice during each complete oscillation. In circular motion, on the other hand, the direction of the force is continuously changing and it is directed towards the center. The motion of a swing (Fig. 13.1), the motion of a stretched spring, and the motion of a pendulum in a clock are a few examples of vibration.

Oscillating objects are also of interest to us as sources of waves. Some of the properties of the waves, such as their frequency and the intensity (energy density), are determined by those of the source. Several fields of physics and engineering, optics, acoustics, seismology, ultrasonics, microwaves, communications engineering (audio and video) and a host of others, have as their foundations the topics on vibrations and waves we discuss in this chapter.

Figure 13.1. Oscillation of a swing.

13.1 VIBRATIONS

For the purpose of defining some of the basic terms associated with vibrations, let us consider a spring with an attached mass illustrated in Fig. 13.2. The mass m is resting on a frictionless surface.

The position of the mass for the unstretched spring marked E is the equilibrium position of the oscillating object. The maximum displacement of the object in either direction from the equilibrium position is called the amplitude of the oscillation (marked A in Fig. 13.2). The time taken for one complete oscillation is called the period of oscillation. The letter T stands for the period of oscillation.

What is one complete oscillation? In one complete oscillation the particle comes back to its original position with the original velocity in the same direction. For example, motion P_1 to E to P_2 to E to P_1 is a complete oscillation. Also $E - P_1 - E - P_2 - E$ is a complete oscillation.

The number of oscillations per unit time is called the frequency of oscillation. If the period is T, then the frequency f is given by

$$f = \frac{1}{T}. \tag{13.1}$$

The unit of frequency is oscillations per second or cycles per second which is now called hertz (Hz).

Figure 13.2. Oscillation of a mass attached to a spring. E is the equilibrium position and A its amplitude.

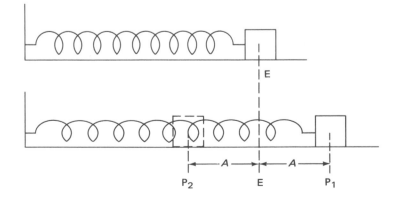

EXAMPLE 13.1

The period of oscillation of a simple pendulum is $\frac{1}{25}$ s. Calculate its frequency.

The frequency f is given by Eq. (13.1),

$$f = \frac{1}{T},$$

$$T = \frac{1}{25} \text{ s}$$

$$f = \frac{1}{(1/25) \text{ s}} = 25/\text{s} = 25 \text{ Hz}.$$

13.2 ELASTIC POTENTIAL ENERGY

A spring resists stretching. The force needed to overcome this resistance increases with the amount of stretching. The spring in Fig. 13.3 is stretched through a distance x by applying a force F_a, whose magnitude is proportional to x. This proportionality holds as long as we remain within the elastic limits (see Chapter 9) and it can be represented by the equation

$$F_a = kx, \tag{13.2}$$

where k is the force constant of the spring. The mass m remains in a fixed position as long as a specified F_a is acting on the mass. The latter implies that the spring exerts a force F_r in the opposite direction to F_a and of a magnitude equal to F_a in order to maintain equilibrium. The force F_r, known as the restoring force, is given by

$$F_r = -kx. \tag{13.3}$$

The minus sign indicates that the restoring force is in the opposite direction to the displacement. Note that the restoring force is always directed towards the equilibrium position.

The work done in stretching a spring through a distance can now be calculated in the following approximate manner. The force F_a is not a constant, but increases linearly with the displacement x from 0 to kx. The average force is, therefore, $\frac{1}{2}kx$ and the work done in stretching is equal to the average force multiplied by the distance:

$$W = \frac{1}{2}kx^2. \tag{13.4}$$

This amount of work is equal to the potential energy of the spring–mass system and changes gradually to kinetic energy on releasing the mass (or removal of F_a).

Let us now consider a spring–mass system that is oscillating with an amplitude A. In Fig. 13.4(a) the spring is in the fully compressed position, its potential energy is $\frac{1}{2}kA^2$, and a restoring force kA is acting on it. The restoring force accelerates the mass to the equilibrium position, and consequently its velocity increases. On the way to the equilibrium position, the displacement x is decreasing,

Figure 13.3. Forces acting on a mass displaced from the equilibrium position (E). To keep the mass in this position a force F_a, equal to the restoring force F_r, should be applied.

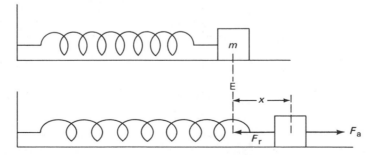

and hence its potential energy is decreasing. Thus the potential energy decreases and kinetic energy increases. At $x = 0$, potential energy is zero, kinetic energy has its maximum value. Once it has reached the equilibrium position with this kinetic energy, it continues to move toward the maximum displacement position on the other side. Now as x increases the restoring force in the opposite direction starts to act on the mass, resulting in a decrease in velocity, decrease in kinetic energy, and increase in potential energy. Finally at $x = A$ [Fig. 13.4(b)] kinetic energy is zero and potential energy has attained its maximum value ($= \frac{1}{2} kA^2$). The changes in v, F, and KE and PE during an oscillation are illustrated in Fig. 13.4. If there is no loss in energy due to friction, the spring would oscillate continuously with the changes indicated in Fig. 13.4. In reality, there is always some friction

resulting in an eventual decrease in the amplitude to zero.

EXAMPLE 13.2

A spring stretches 0.1 m when a force of 5 N acts on it. This spring with a 0.5 kg mass attached to it oscillates with an amplitude of 0.2 m. Calculate (a) the maximum velocity of the mass and (b) its velocity when its displacement is half its amplitude.

(a) First we obtain the force constant using Eq. (13.1).

$$F = kx \text{ or } k = F/x$$

$$k = \frac{5 \text{ N}}{0.1 \text{ m}} = 50 \text{ N/m}.$$

Figure 13.4. Changes in potential energy, kinetic energy, velocity, and force during oscillation of a mass attached to a spring.

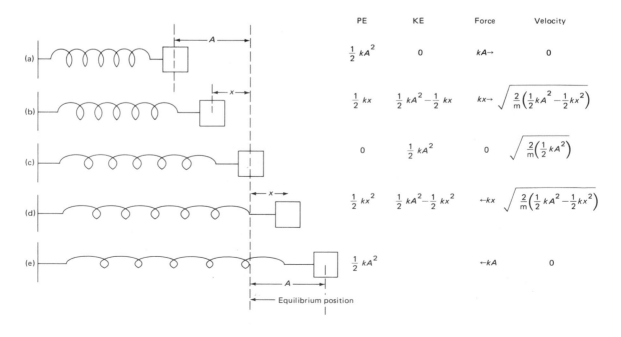

The maximum velocity of the oscillator can be obtained from its maximum kinetic energy which is equal to its maximum potential energy.

Maximum kinetic energy
= Maximum potential energy

$$\text{KE}_{max} = \frac{1}{2} kA^2 = \frac{1}{2} (50 \text{ N/m}) (0.2 \text{ m})^2 = 1 \text{ J}.$$

Since kinetic energy is given by

$$\text{KE} = \frac{1}{2} mv^2$$

$$v = \sqrt{2\text{KE}/m}$$

$$v_{max} = \sqrt{\frac{2 \times 1 \text{ J}}{0.5 \text{ kg}}} = 2 \text{ m/s}.$$

(b) The displacement at half amplitude is 0.1 m.

Potential energy for 0.1 m displacement
$= \frac{1}{2} (50 \text{ N/m}) (0.1 \text{ m})^2 = 0.25 \text{ J}.$
Kinetic energy at this displacement
$= \text{Total energy} - \text{Potential energy}$
$= 1 \text{ J} - 0.25 \text{ J} = 0.75 \text{ J}.$
Velocity at this displacement

$$= \sqrt{\frac{2\text{KE}}{m}} = \sqrt{\frac{2 \times 0.75 \text{ J}}{0.5 \text{ kg}}}$$

$$= 1.73 \text{ m/s}.$$

13.3 SIMPLE HARMONIC OSCILLATION

The displacement of an oscillator as a function of time can be traced out by an arrangement similar to the one shown in Fig. 13.5. A pointer attached to the mass records the motion on a moving paper. Both displacement and time can be measured directly from the trace. The lateral distance from the equilibrium position measures the displacement and the length of the paper from the starting point measures the time. The time at the starting point, when the mass is at its equilibrium position is gen-

erally taken as zero. Figure 13.5(a) shows the plot resulting from one complete oscillation. At $t = 0$, the displacement $x = 0$. Also at the times given by $t = T/2$ and $t = T$, the displacements are zero. At the times given by $t = T/4$ and $t = 3T/4$, the displacement x is a maximum at each time and it is equal in magnitude to the amplitude. At $t = T/4$, $x = +A$ and at $t = 3T/4, x = -A$. A mathematical function that can represent such a plot is a sine function. For the present case the equation representing the plot is

$$x = A \sin (2\pi/T)t. \tag{13.5}$$

This equation satisfies the above mentioned characteristics of the plot, that is, for $t = 0, T/2, T, \cdots$, $x = 0$ and for $t = T/4, 3T/4, \cdots, x = |A|$. The graph [Fig. 13.5(a)] is called a sinusoidal curve.

The term, $2\pi/T$ $(= 2\pi f)$ is called the angular frequency of the oscillator. The symbol often used for angular frequency is ω, but note that the angular frequency is not the same as the angular velocity which also has the same symbol. However, there is a correspondence between angular velocity and angular frequency as we will see below.

Let us consider a particle in a circular path of radius A with an angular velocity ω. The radius A and angular velocity ω are equal to the amplitude and angular frequency of the oscillator, respectively. The time taken by the particle for a complete revolution is $2\pi/\omega$ and it is equal to the period of the oscillator. Hence as the particle makes one revolution the oscillator completes one oscillation. The angular position of the particle ωt at time t is shown in Fig. 13.5. The projection of this position along the x axis is indicated by m' and the distance of m' from the center by x. As the mass m moves around, its horizontal projection moves back and forth or oscillates with its equilibrium position at the center of the circle. The displacement of the projection has an absolute maximum value at $T/4$, $3T/4$, $\cdots$, and minimum value at $T = 0, T/2$, $T, \cdots$. The motion of the projection is identical to that of the oscillator with amplitude A and angular frequency ω. The displacement of the

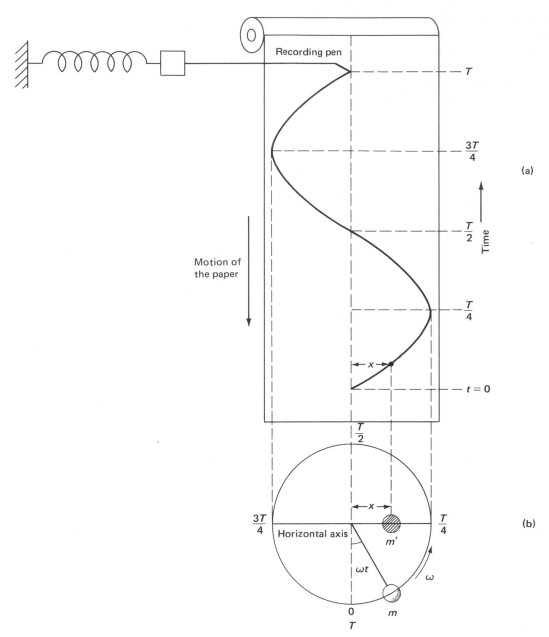

Figure 13.5. (a) An arrangement to trace the oscillation as a function of time. (b) The horizontal projection of an object in a circular path at a constant angular velocity corresponds to displacement of an oscillator.

projection is given by the trigonometric relation

$$x = A\sin\omega t = A\sin(2\pi/T)t, \qquad \textbf{(13.6)}$$

which is the same as Eq. (13.5).

The projection of the position of the particle

along the x axis is equivalent to the motion of an oscillator. Similarly, the projections of the velocity and acceleration gives us the velocity and acceleration of the oscillator. The equations for these projections are

$$v_x = A\omega\cos\omega t = A\omega\cos(2\pi/T)t, \qquad (13.7)$$

$$a_x = -A\omega^2\sin\omega t = -\omega^2 x. \qquad (13.8)$$

Equation (13.7) tells us, as is to be expected, that the velocity is a maximum at the equilibrium position, that is, for $t = 0, T/2, \cdots$, and zero at the maximum displacement, that is, at $t = T/4, 3T/4, \cdots$. From Eq. (13.8) it is evident that the acceleration is proportional to the negative displacement. Multiplying both sides of Eq. (13.8) by m, we get

$$ma_x = -m\omega^2 x.$$

From Newton's law, $F = ma$, we then have

$$F = -m\omega^2 x. \qquad (13.9)$$

This equation is similar to Eq. (13.3) which tells us that the restoring force is proportional to the negative displacement. The fact that Eq. (13.8) leads us to the correct force equation suggest that the acceleration given by Eq. (13.8) is true.

From Eqs. (13.3) and (13.9) we have

$$m\omega^2 = k.$$

Substituting for ω yields

$$m (2\pi/T)^2 = k$$

or

$$T = 2\pi \sqrt{m/k}. \qquad (13.10)$$

Equation (13.10) gives us the period of an oscillator in terms of its mass m and spring constant k. The period increases with m and decreases with k. The

unit consistency of Eq. (13.10) can now be readily checked.

$$\text{Units of } T = \sqrt{\frac{\text{Mass}}{\text{Force/Distance}}}$$

$$\text{Force} = \text{Mass} \times \text{Distance/Time}^2$$

$$\text{Unit of } T = \sqrt{\frac{\text{Mass}}{\dfrac{\text{Mass} \times \text{Distance}}{\text{Time}^2} \times \dfrac{1}{\text{Distance}}}}$$

$$= \sqrt{\text{Time}^2} = \text{Time}.$$

Now we may define a simple harmonic oscillator. Two equivalent definitions are given below.

If the displacement of an oscillator (like the spring) can be represented by a sinusoidal function, it is called a simple harmonic oscillator and the oscillations are called simple harmonic oscillations.

If the restoring force of an oscillator is proportional to the negative displacement, the oscillator is a simple harmonic oscillator and its oscillations are simple harmonic oscillations.

EXAMPLE 13.3

It is found that adding a 0.2 kg mass to a spring hanging vertically down elongates the spring through a distance of 10 cm. (a) Calculate the force constant of the spring. (b) If the mass is set in motion, what is the frequency of oscillation?

(a) The force constant is obtained by using the equation

$$F = kx$$

$$k = \frac{F}{x} = \frac{0.2 \text{ kg} \times 9.8 \text{ m/s}^2}{10 \text{ cm} \times (1 \text{ m})/(100 \text{ cm})} = 19.6 \text{ N/m}$$

(b) Period $T = 2\pi \sqrt{m/k}$

$$= 2\pi \sqrt{(0.2 \text{ kg})/(19.6 \text{ N/m})}$$

$$= 0.63 \text{ s}.$$

Frequency of oscillation $f = 1/T = 1/0.63$ s

$$= 1.6 \text{ Hz}.$$

EXAMPLE 13.4

If the amplitude of oscillation of the mass of Example 13.3 is 20 cm, calculate the maximum velocity of the mass.

The velocity of an oscillator is given by

$$v = A\omega\cos\omega t.$$

For the maximum value, we get $\cos\omega t = 1$. Hence we have

$$v_{max} = A\omega = A2\pi f = (\frac{20\ cm}{100\ cm/m})(2\pi)(1.6\ Hz)$$

$$= 2\ m/s.$$

The maximum velocity can also be obtained by first calculating the maximum potential energy $\frac{1}{2}kA^2$, and using the fact that when the oscillator has the maximum velocity, its kinetic energy is equal to the maximum potential energy.

Maximum potential energy $= \frac{1}{2}kA^2$

$$= \frac{1}{2}(19.6\ N/m)(0.2\ m)^2$$

$$= 0.392\ J.$$

Maximum kinetic energy $= \frac{1}{2}mv_{max}^2 = \frac{1}{2}kA^2$

$$= 0.392\ J.$$

Hence

$$v_{max} = \sqrt{\frac{2 \times 0.392\ J}{0.2\ kg}} = 2\ m/s.$$

13.4 THE SIMPLE PENDULUM

A simple pendulum (Fig. 13.6) consists of a concentrated mass at the end of a light cord usually assumed to have negligible mass. The mass executes its oscillations along an arc CED. From Fig. 13.6 it is clear that in this case also the system has a maximum potential energy for the maximum displace-

ment (positions C and D), and the potential energy (calculated with respect to the equilibrium position) decreases to zero and the kinetic energy increases to a maximum at the equilibrium position.

We will now show that the motion of a pendulum is simple harmonic provided that the angle θ is small. When the mass is at D, the only two forces acting are the tension T in the string and the weight mg. The mass has no radial motion, it has only tangential motion. This means that the net force along the radial direction, in the direction of the tension, is zero; that is, the component of mg along the direction of the string and T are equal and opposite in direction. Hence we get

$$T = mg\cos\theta. \tag{13.11}$$

The other component of mg in the tangential direction is $mg\sin\theta$. This component takes the mass to the equilibrium position and, therefore, is the restoring force in this case. Restoring force has magnitude equal to $mg\sin\theta$. For small angles $\sin\theta = x/l \simeq s/l$. So we have for the restoring force

$$F = -mg\ s/l = -(mg/l)s. \tag{13.12}$$

The minus sign is included to indicate restoring force is in the opposite direction to the displacement s. Equation (13.12) suggests that since restoring force is proportional to the negative displacement, the simple pendulum behaves like a simple harmonic oscillator.

A comparison of Eq. (13.12) with Eq. (13.3) shows that the present case has an equivalent force constant given by

$$k = mg/l.$$

Hence the period of oscillation

$$T = 2\pi\sqrt{m/k} = 2\pi\sqrt{\frac{m}{mg/l}}$$

$$T = 2\pi\sqrt{l/g}. \tag{13.13}$$

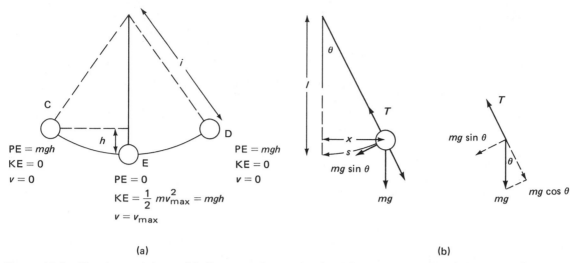

$$PE = mgh \quad \quad PE = mgh$$
$$KE = 0 \quad \quad KE = 0$$
$$v = 0 \quad \quad v = 0$$

$$PE = 0$$
$$KE = \frac{1}{2} mv^2_{max} = mgh$$
$$v = v_{max}$$

(a)   (b)

Figure 13.6. Simple pendulum. (b) Forces acting on the pendulum at an angular displacement θ. The un-balanced force $mg \sin\theta$ is the restoring force in this case.

For a simple pendulum the period of oscillation depends on (a) the length of the pendulum and (b) the gravitational acceleration at the location of the pendulum. It is independent of the mass of the pendulum. In the derivation of Eq. (13.13), we assumed the angle θ is small enough to make the approximation $x \simeq s$. Therefore, the error in the calculation of T increases with θ; it is about 0.2% for $\theta = 10°$ and about 1.7% for $\theta = 30°$.

Squaring the above equation

$$4\,s^2 = 4\pi^2 \times \frac{l}{9.8 \text{ m/s}^2}$$

or

$$l = \frac{(4 \text{ s}^2)(9.8 \text{ m/s}^2)}{4\pi^2} = 0.99 \text{ m}.$$

Figure 13.7. The oscillating plate O produces waves in the water. The frequency of the waves is equal to the frequency of the oscillator.

EXAMPLE 13.5

The period of oscillation of a simple pendulum is 2 s. Calculate the length of the pendulum, assuming $g = 9.8 \text{ m/s}^2$.

We are given $T = 2 \text{ s}, g = 9.8 \text{ m/s}^2$. Substitution of these values into Eq. (13.13) yields

$$2 \text{ s} = 2\pi \sqrt{\frac{l}{9.8 \text{ m/s}^2}}.$$

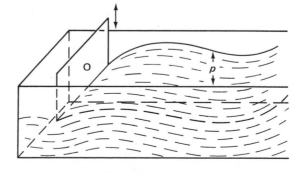

13.5 WAVES

As the chart paper in Fig. 13.5 is rolled out at a constant speed, it carries with it an "impression" of the oscillations which looks like a wave. In a somewhat similar fashion, if an oscillator is dipped in water, waves originating at the oscillator will be seen moving out to both sides of the oscillator (Fig. 13.7). In both cases frequency of the waves is equal to the frequency of the oscillator. However, there are some important differences. (a) When the real waves travel in a medium, particles of the medium do not travel with it. For example, a particle p in a water wave stays on the average in its place, executing oscillations about an equilibrium position, but it does not travel with the wave. (b) The velocity of the wave in a medium depends on the mechanical properties except for small variations with temperature and pressure. From the above discussion it is evident that only the wave profile moves carrying with it the energy but not the particles.

The waves produced in water (Fig. 13.7) are called periodic waves since they are produced one after another at constant time intervals. In contrast we may produce a "pulse" by dropping an object into water.

Refer to Fig. 13.8 to define some of the common terms associated with waves. In Fig. 13.8 the point experiencing the maximum displacement above the equilibrium level is called a crest and the point of maximum displacement below the equilibrium level is called a trough.

Figure 13.8. Figure defines some of the terms used in connection with waves. A is the amplitude of the waves and λ the wavelength.

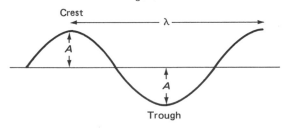

Crest

λ

A

A

Trough

Amplitude (A). The maximum displacement of each particle in the path of a wave is called the amplitude of the wave.

Wavelength (λ). The closest distance between two similar points in a wave is called a wavelength. For example, the distance between two adjacent crests or two troughs is equal to a wavelength.

Wave velocity (v). Wave velocity is the distance through which a wave travels in 1 second.

Frequency (f). The number of complete waves passing a point in 1 second is called the frequency. As mentioned before, the frequency of the waves is the same as that of the source.

A relationship among frequency, wavelength, and velocity can now be obtained. In 1 second f wave profiles pass a point. The distance through which the waves have traveled past that point is $f\lambda$ (Fig. 13.9). It is also equal to the velocity. Hence we have

$$f\lambda = v. \tag{13.14}$$

Wave velocity is equal to frequency multiplied by wavelength. Equation (13.14) can also be written in the form

$$vT = \lambda, \tag{13.15}$$

where T is the period which is equal to $1/f$.

EXAMPLE 13.6

The lowest frequency the human ear is sensitive to is 20 Hz. Calculate the wavelength of these waves in air and their period. (Assume the velocity of sound waves in air to be 340 m/s.)

We use Eq. (13.14) to calculate wavelength,

$$\lambda = \frac{v}{f} = \frac{340 \text{ m/s}}{20 \text{ Hz}} = 17 \text{ m}$$

$$T = \frac{1}{f} = \frac{1}{20 \text{ Hz}} = 0.05 \text{ s}.$$

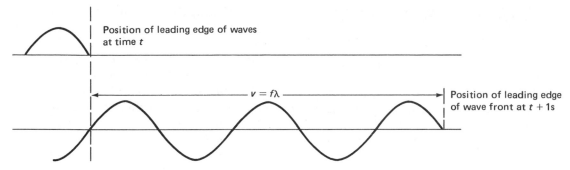

Figure 13.9. Positions of traveling waves at time t and $t + 1$ s. Frequency f is the number of waves passing a point in 1 s.

13.6 TYPES OF WAVES

When periodic waves, that is, a continuous train of waves with a constant frequency, travel through a medium, the particles in the medium execute simple harmonic oscillations. The direction of the oscillations of the particles is not necessarily the same as the direction of the waves.

For transverse waves the direction of oscillation is perpendicular to the direction of the waves. Waves in a rope or a stretched wire [Fig. 13.10(a)] are examples of transverse waves.

For longitudinal waves the direction of oscillation is the same as the direction of the waves. Sound waves and waves in a long helical spring shown in Fig. 13.10(b) are examples. As the longitudinal waves travel through the spring, each section of the spring is alternately compressed and extended. Distance between the centers of two nearest regions of compression is a wavelength in this case.

Waves on the surface of water (or any liquid) are interesting because they do not belong strictly to any of the two classifications above. Water molecules execute a circular motion [Fig. 13.10(c)] with the same period as the waves. At the crest, molecules are moving with the waves, while at the trough, against the waves and half way between these points perpendicular to the wave. The motion of the water molecules thus results from a combination of transverse and longitudinal motions.

13.7 VELOCITY OF WAVES IN A MEDIUM

As mentioned before, the velocity of waves depends only on the material properties. A qualitative discussion of this dependence is given here. Consider a series of balls (Fig. 13.11) attached to

Figure 13.10. Motion of particles in (a) transverse waves, (b) longitudinal waves, and (c) water waves. In all these cases, particles move about an equilibrium position and they never move with the wave profile.

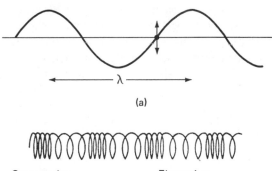

(a)

(b)

Compression   Elongation

(c)

Figure 13.11. Series of masses attached by springs.

each other by similar springs. (Atoms are bound to each other in materials in a somewhat similar manner.) If we disturb the first ball, by hitting it, the disturbance will travel from one ball to another like a wave pulse. To understand the dependence of the velocity on the stiffness (force) constant of the springs, let us consider two extreme cases: (a) The stiffness constant is infinite (very large). The balls move as a group, or in other words, the disturbance given to the first ball is instantaneously transmitted to the last ball; with an infinite velocity. (b) The stiffness constant is zero. In this case the balls act independently (except when they collide) and the disturbance does not travel from ball to ball. The velocity of the disturbance in this case is zero. In a real situation, where the stiffness constant lies between these two limits, the velocity also lies between these two limits of zero and infinity increasing with the stiffness constant. In addition to the stiffness constant, the inertia of the balls should also be considered. If the mass is heavier, it is harder to move it. Hence the velocity is inversely proportional to the mass.

In a solid material where we have a large array of atoms stringed together, an appropriate elasticity constant takes the place of the stiffness constant and mass density takes the place of the mass of the balls. The three elastic constants, Young's modulus, bulk modulus, and shear modulus, have the same units of force per unit area. The unit of the elastic constant divided by mass density is, therefore,

$$\frac{\text{Elastic constant}}{\text{Mass density}} = \frac{ML/T^2}{L^2} \bigg/ \frac{M}{L^3} = \frac{L^2}{T^2} = \text{Velocity}^2.$$

To obtain the velocity from this fraction the following easy step suggests itself to us:

$$v = \text{constant} \sqrt{\frac{\text{Elastic constant}}{\text{Mass density}}}. \qquad (13.16)$$

In the SI units the constant in Eq. (13.16) has the value of unity. Hence we have arrived at a simple equation for the velocity of waves from a qualitative study of the properties of the materials. Now let us consider specific cases.

Longitudinal waves in thin rods. As the waves travel through the rod, alternate sections of the rod experience compressions and elongations. Therefore the modulus involved here is Young's modulus. The velocity of longitudinal waves in thin rods is given by

$$v = \sqrt{Y/\rho}, \qquad (13.17)$$

where Y is the Young's modulus and ρ the mass density.

Transverse waves. As transverse waves travel through a medium, each section of the material experiences a shear stress (Fig. 13.12). Hence for the velocity of transverse waves we have

$$v = \sqrt{S/\rho}, \qquad (13.18)$$

where S is the shear modulus. Since fluids cannot support a shear stress in the interior, transverse waves are not transmitted through them. Fluids transmit longitudinal waves and their velocity is given by

$$v = \sqrt{B/\rho}, \qquad (13.19)$$

where B is the bulk modulus of the liquid. The water waves we discussed before are surface waves, the velocity of these waves being a function of g, surface tension, and the depth of the water.

Transverse waves are easily produced in a stretched wire or string. The stiffness in these cases

Figure 13.12. Each section experiences a shear stress as transverse waves pass through them.

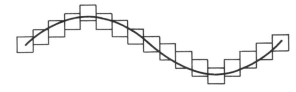

is proportional to the tension of the string. The velocity is given by

$$v = \sqrt{F/\mu}, \qquad (13.20)$$

where F is the tension and $\mu \, (= m/L)$ is the mass per unit length of the wire.

EXAMPLE 13.7

The Young's modulus of steel is $20 \times 10^{10} \, \text{N/m}^2$. Calculate the velocity of longitudinal waves in a thin steel rod. Its mass density is $7700 \, \text{kg/m}^3$.

$$v = \sqrt{\frac{Y}{\rho}} = \sqrt{\frac{20 \times 10^{10} \, \text{N/m}^2}{7700 \, \text{kg/m}^3}} = 5100 \, \text{m/s}.$$

EXAMPLE 13.8

The tension of a string is 10 N and its linear mass density is 1 g. Calculate the velocity of longitudinal waves in the string.

We use Eq. (13.20) to calculate the velocity

$$F = 10 \, \text{N}, \qquad \mu = \frac{1 \, \text{g}}{1 \, \text{m}} \times \frac{1}{1000 \, \text{g/kg}} = 10^{-3} \text{kg/m};$$

$$v = \sqrt{\frac{10 \text{N}}{10^{-3} \, \text{kg/m}}} = 100 \, \text{m/s}.$$

13.8 REFLECTION AND SUPER-POSITION OF WAVES

What happens when the waves reach the "end of the rope" (Fig. 13.13). If one end of the medium through which the wave is traveling is rigidly fixed, the crest is reflected as a trough. On the other hand if the end is free, the crest is reflected as a crest with no change. Figure 13.13 also shows that the wave is partially reflected and partially transmitted at the interface between two media. The transmitted wave is similar to the incident wave except for a change in amplitude. On the other hand, the reflected wave suffers a crest to trough reversal if the first medium

(a)

(b)

(c)

(d)

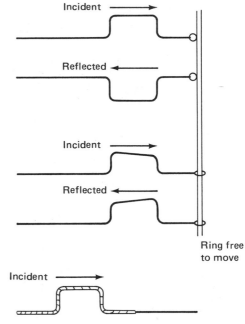
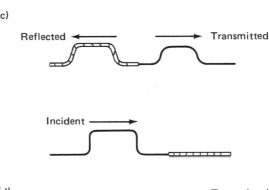
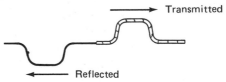

Figure 13.13. Reflection of waves. (a) Waves reflect out of phase at a rigidly fixed end of the rope. (b) Waves reflect in phase at loosely fixed end of the rope. (c) Partial reflection occurs at heavy–light interface. Both reflected and transmitted waves are in phase. (d) Reflected wave is out of phase at a light–heavy interface.

is lighter than the second medium and there is no reversal if the first medium is heavier than the second medium.

Let us now consider two waves traveling simultaneously through the same medium. The presence of the second wave does not affect the first wave and vice versa. Both travel through the medium unaffected by the other, producing a disturbance in the medium as though the other one is not present. The algebraic sum of the individual displacements gives us the result of the combined action of the waves at every point. This is known as the principle of superposition. Examples are shown in Fig. 13.14. In Fig. 13.14(a) the crests and troughs of both waves meet at the same place and at the same time. Hence the crests and troughs enhance each other. The waves are said to interfere constructively in this case. On the other hand in Fig. 13.14(b), the crest of one wave meets the trough of the other

wave cancelling each other. Here the waves are said to interfere destructively. The waves are further said to be in phase in the first case and out of phase in the second case.

An interesting pattern of waves we commonly observe is known as standing waves, which result from the interference between two periodic waves of equal amplitude and frequency traveling in opposite directions like the incident and reflected waves. To understand this phenomena in Fig. 13.15

Figure 13.15 Interference between two waves, traveling in opposite directions (a wave and its reflection), produce a standing wave pattern with points (N) remaining stationary and the points between them oscillating. The stationary points are called nodes and the central points between the nodes are called antinodes. The dashed lines represent the individual waves and the solid lines the resultant waves.

Figure 13.14. (a) Two waves of equal amplitude and in phase reinforce each other. This is called constructive interference. (b) Two waves of equal amplitude and out of phase cancel each other. This effect is called destructive interference.

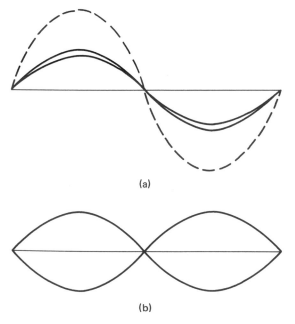

(a)

(b)

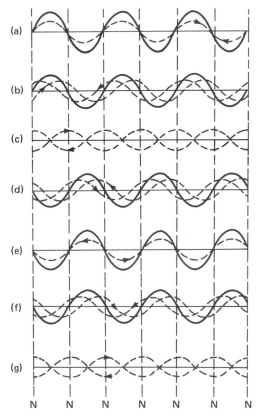

we consider two traveling waves. In Fig. 13.15(a) they are in phase and interfere constructively. In Fig. 13.15(c), after the waves have traveled a distance of $\lambda/4$ in their respective directions, or after a time of $T/4$, the waves cancel each other. The important feature of the resultant wave is that it appears standing with certain points in the medium (N) always remaining stationary and other points executing oscillations up and down. The points located at the center of the stationary points oscillate with maximum amplitude. The stationary points are called nodes and the points with maximum displacement are called antinodes.

It is also evident for this case from Fig. 13.15 that the distance between the nodal points is equal to $\lambda/2$. We note that in actual fact the nodes are not completely fixed in position for otherwise no energy would be transmitted along the medium.

A simple arrangement to demonstrate the standing waves phenomenon is shown in Fig. 13.16. The cord at one end is attached to an oscillator with a fixed frequency and the other end that passes over a pulley has a set of weights attached to it. For all practical purposes, we can regard the ends of the cord as essentially fixed and hence to be nodal points. Interference occurs between waves originating from the vibrator and waves reflected at the pulley. As the tension is increased slowly by adding weights, we start to see the formation of standing waves. First we see a large number of nodes and antinodes between the ends of the cord and even-

tually, fewer nodes and antinodes and down to one antinode as the tension increases and the corresponding wavelength increases.

The velocity of a wave in a cord is a function of the tension as given by the equation

$$v = \sqrt{F/\mu} = \lambda f, \qquad (13.20a)$$

where F is the tension and μ the mass per unit length. Since the frequency f is fixed, the wavelength λ $(= v/f)$, because of Eq. (13.20a), is a function of the tension. Only selected wavelengths can produce standing waves in a cord. Because the end points remain as nodal points, for standing waves, an integral multiple of $\lambda/2$ is equal to the length of the cord. If there are n vibrating sections in a cord, the corresponding wavelengths are given by

$$n\,\lambda/2 = l, \qquad (13.21)$$

where l is the length of the string between the two fixed end points. Combining Eqs. (13.20) and (13.21), we have

$$l = \frac{n}{2f}\sqrt{F/\mu}. \qquad (13.22)$$

This relationship gives the number of vibrating segments or antinodes one gets under the given

Figure 13.16. Experimental arrangement for producing standing waves in a string. String oscillates with six segments.

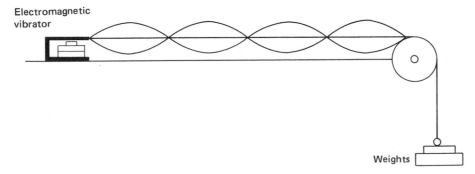

conditions of frequency of oscillation, tension, and mass per unit length. Since n is an integer this equation will be satisfied only for certain values of F. Also for a given F, we obtain standing waves only for certain characteristic frequencies.

EXAMPLE 13.9

A tuning fork is vibrating electrically at 250 Hz. A string 0.9 m long weighing 0.1 g, when attached to this tuning fork, is found to oscillate with 3 segments (3 antinodes). Calculate (a) the wavelength of the waves, (b) the velocity of the waves, and (c) the tension of the string.

(a) The length of each segment of 0.3 m.

Wavelength = 0.6 m.

(b) Velocity of the waves = $f\lambda$ = (250 Hz) $\times$ (0.6 m) = 150 m/s.

(c) The tension of the string can be calculated using Eq. (13.19):

$$v = \sqrt{F/\mu} \quad \text{or} \quad F = \mu v^2;$$

$$\mu = \left(\frac{0.1 \text{ g}}{1000 \text{ g/kg}}\right) \Big/ \left(\frac{1}{0.9 \text{ m}}\right) = 1.1 \times 10^{-4} \text{ kg/m};$$

Tension $F = (1.1 \times 10^{-4} \text{ kg/m}) (150 \text{ m/s})^2 = 2.5 \text{ N}$

SUMMARY

Several new terms were defined in this chapter:

Amplitude (A)—maximum displacement from the equilibrium position.

Frequency (f)—number of vibrations per second; also for waves, number of waves passing a point per second (or how frequently the waves come by).

Period (T)—time for one complete revolution: $f = 1/T$.

Wavelength (λ)—the distance between adjacent crests or troughs.

Velocity (v)—distance traveled by waves in 1 s: $f\lambda = v$.

Simple harmonic oscillation—oscillation of an object for which the restoring force is proportional to the negative displacement or one in which the displacement can be described by a sinusoidal function.

Energy of the oscillator is given by the maximum potential energy equal to $\frac{1}{2}kA^2$, where k is the force constant.

Period of oscillation—$T = 2\pi\sqrt{m/k}$ for a spring with an attached mass; $T = 2\pi\sqrt{l/g}$ for a pendulum.

Longitudinal waves—particles in the medium executing simple harmonic motion in the direction of the waves.

Transverse waves—particles vibrating perpendicular to the direction of motion.

Velocity of waves is given by $v = \sqrt{Y/\rho}$ for longitudinal waves in thin rods; $v = \sqrt{s/\rho}$ for transverse waves in solids; $v = \sqrt{B/\rho}$ for longitudinal waves in fluids; $v = \sqrt{F/\mu}$ for transverse waves in strings.

Constructive interference—two waves traveling through a medium enhancing each other.

Destructive interference—two waves cancelling each other.

Standing waves—interference between two waves with same frequency and amplitude traveling in opposite directions. Nodes are points of no motion in a standing wave. Antinodes are points of maximum displacement. The distance between two nodes is equal to $\lambda/2$.

QUESTIONS AND PROBLEMS

13.1 At what points during an oscillation is (a) the potential energy a maximum, (b) kinetic energy a maximum, (c) kinetic energy half of the maximum potential energy, and (d) potential energy half of the maximum kinetic energy?

13.2 In the derivation of the period of oscillation, the mass of the spring (and the cord) is neglected. How will the mass affect the period of oscillation?

13.3 For the position of maximum displacement, the tension of the cord of a simple pendulum is given by $mg\cos\theta$. Is this true for all angular positions?

13.4 How do temperature changes affect the period of oscillation of a pendulum? How would one correct for the temperature effect?

13.5 Compare the period of oscillation of a simple pendulum at sea level and on the top of the Himalayas.

13.6 Will the period of oscillation of a helical spring change with changes in gravitational constant.

13.7 Is the speed of waves the same as the speed of particles in it?

13.8 Can you think of an experiment to prove that energy is transmitted through waves?

13.9 How can you prove experimentally that the velocity of waves in air is independent of their frequency?

13.10 Can you think of some consequences of velocity changing with frequency?

13.11 What is the relationship between period and frequency? What is the relationship between period and amplitude?

13.12 Discuss qualitatively how the angle θ will affect the period of oscillation of a simple pendulum.

13.13 A spring stretches through 5 cm when a 20 g mass is attached to it. Calculate its force constant.

13.14 A 1 lb force stretches a spring by 6 in. (a) Calculate the force constant. (b) Calculate the force required to stretch it through 1 ft. (c) What will be the force constant of this spring if it is cut in half?

13.15 Calculate the period of an oscillator whose frequency is 1000 Hz.

13.16 A 0.5 kg mass is executing a simple harmonic oscillation with an amplitude of 0.1 m and with a period of 2 s. Calculate (a) the frequency of the oscillator, (b) the force constant, (c) the maximum potential energy, (d) the maximum kinetic energy, and (e) the magnitude of the force acting on the mass when the displacement is equal to 0.05 m.

13.17 A spring stretches through 5 cm when a mass of 5 g is attached to it. Calculate the period of oscillation of the spring with a 10 g mass attached to it.

13.18 An automobile can be considered as a heavy mass attached to a spring as far as its vertical motion is concerned. (a) If the period of oscillation is adjusted to be 4 s and the weight of the car is 4000 lb, calculate the force constant of the spring. (b) If four passengers, weighing 650 lb are in the automobile, calculate its new frequency.

13.19 Calculate the length of a simple pendulum whose period of oscillation is 1 s at a location where $g = 9.8 \text{ m/s}^2$.

13.20 Calculate the period of oscillation of the pendulum of Problem 13.19 on the moon.

13.21 A pendulum in a clock is 25 cm long at 27°C. If the thin rod (assumed massless) is made of steel, calculate the error in 1 h if the clock is in a room at 0°C.

13.22 A child's swing is 10 ft long. Calculate its period of oscillation.

13.23 Obtain the velocity of waves in a thin copper rod.

13.24 Calculate the velocity of longitudinal waves in water.

13.25 If 1000 Hz longitudinal waves are transmitted through water, calculate their wavelength.

13.26 A 500 Hz wave has a 2 m wavelength in a medium. Calculate the velocity of the waves in the medium.

13.27 The velocity of radio waves is 3×10^8 m/s. Calculate the wavelength of 5 kHz radio waves.

13.28 A metal wire of mass 200 g and length 2 m is under a tension of 210 N. Calculate the velocity of a transverse wave in this wire.

13.29 If the wire of Problem 13.28 is tied to a tuning fork that is vibrating at 360 Hz, calculate the wavelength of the waves in the wire.

13.30 A 1 m long cord oscillates in 20 segments when excited by a 200 Hz oscillator. The total mass of the cord is 5 g. Calculate (a) the wavelength of the waves, (b) the velocity of the waves, and (c) the tension of the cord.

13.31 A 20 Hz oscillator sets up a standing wave in a string. The length of each segment is 15 cm and the tension of the string is 4 N. Calculate the linear mass density of the string.

14

Sound

The word sound contains two aspects: (a) the physiological and psychological sensations it produces in our brain and (b) the physical nature of it, that is, longitudinal waves with frequencies lying in the audible range (about 20 Hz to 20,000 Hz). There are even two sets of terminology, each with emphasis on one of the above-mentioned aspects. The terms pitch, quality, and loudness refer to physiological and psychological aspects while their corresponding terms frequency, waveshape, and intensity refer to their physical properties.

14.1 SOURCES OF SOUND

Vibrating objects, with their frequencies in the audible range, are sources of sound; such as vibrating strings (violin, guitar, piano, and human vocal cords) and vibrating air columns (clarinet, trumpet, and pipe). Let us take the loudspeaker as an example (Fig. 14.1). The loudspeaker, when electrically energized, moves in and out (this motion is easy to observe, especially the motion of loudspeakers in low-frequency range), producing pressure changes in the air. Alternate compressions and rarefactions travel out in the air from the speaker. The pressure

variations, upon reaching the ear, set up vibrations in the eardrum, triggering a chain of physiological events which finally reach the brain carrying the information.

The pressure variations as a function of distance are plotted in Fig. 14.1(b), pressure being higher at the regions of compression and lower at the regions of rarefaction. The distance between the centers of adjacent compressions is equal to the wavelength. Figure 14.1(b) represents the pressure variations at a certain time (like a snapshot). If we measure the

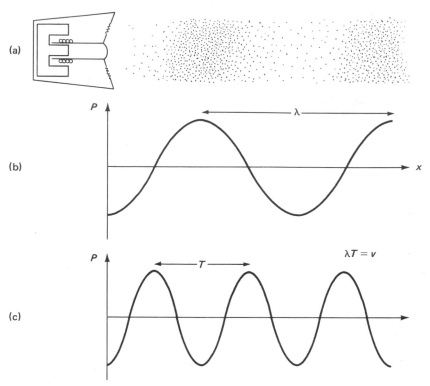

Figure 14.1. (a) Oscillations of a loudspeaker cone produce traveling pressure variations in air. (b) Pressure variations as a function of a distance for a single frequency wave. The curve of P vs distance is a sinusoidal curve. (c) Pressure variations at a point as a function of time. This curve is also a sinusoidal function.

pressure at a point in the path of the waves as a function of time, the results will give us another sinusoidal plot. The distance between consecutive peaks this time gives us the period of the waves.

A material medium is necessary for transmission of sound. If there is no material between the speaker and the listener, voice communication is impossible. For example, Astronauts Neil Armstrong and Edward E. Aldrin found themselves in such a situation on the moon and they had to rely on radiowaves for communication.

14.2 VELOCITY OF SOUND

The velocity of sound waves can be calculated using Eqs. (13.17), (13.18), and (13.19) in Chapter

13. For velocity in air, we had

$$v = \sqrt{B/\rho}, \tag{14.1}$$

where B is the bulk modulus which for air is shown to be $1.4\,P$, and $\rho = m/V$ (mass per unit volume). Therefore, we get

$$v = \sqrt{\frac{1.4P}{\rho}} = \sqrt{\frac{1.4PV}{m}}. \tag{14.2}$$

Since PV is proportional to absolute temperature, the velocity of sound in air is proportional to the square root of the absolute temperature. From Eq. (14.2) the velocity of sound in air at standard temperature and pressure (0°C and 1 atm) can be estimated (ρ may be obtained from tables). The velocity thus estimated and experimentally verified is 331.4 m/s or 1087 ft/s.

The velocity–temperature relationships are often expressed by simple approximate formulas, which enable us to calculate the velocity in a range of temperatures from the velocity at 0°C. They are

$$v_t = 331.4 \text{ m/s} + 0.61t°C, \qquad (14.3)$$

$$v_t = 1087 \text{ ft/s} + 1.98t°C, \qquad (14.4)$$

$$v_t = 1087 \text{ ft/s} + 1.1(t°F - 32). \qquad (14.5)$$

In fact, in the latter two cases, for each 1°C change in temperature, the velocity increases by approximately 2 ft/s and, for each 1°F change the velocity increases by 1.1 ft/s. Equations (14.3), (14.4), and (14.5) are approximate. The error in using one of these equations for estimating velocity at 50°C is about 0.2%.

14.3 PITCH AND QUALITY OF SOUND

The property of sound known as pitch is associated with its frequency. A high-pitched sound has a high frequency, while a low-pitched sound has a lower frequency. The ear is capable of distinguishing sounds differing in pitch.

A sound wave which can be represented by a single sinusoidal function is said to be a pure tone. A stretched string fixed at both ends vibrating as shown in Fig. 14.2(a) produces a pure tone. The tone corresponding to this frequency is called the fundamental tone or the first harmonic. The string may also vibrate at higher frequencies called overtones or higher harmonics. For example, the first overtone is the second harmonic.

The quality of the sound of a particular note produced by an instrument depends upon the number of overtones and their relative intensities. The quality of a note of a piano will be different from that of a violin mainly because the overtones and intensities are different. The human vocal tract differs from one person to the next for similar reasons. The presence of overtones gives a distinctive shape to the waveform as shown in Fig. 14.3. Hence the differences in the quality of sound show up as differences in the shapes of the waveforms.

14.4 INTENSITY AND SOUND LEVEL

How loud is the noise of a jet airplane? How soft is your whisper? How do we quantify loudness? A propagating sound wave carries energy with it. Also associated with it are pressure changes. Both of these quantities are used to define sound level.

Figure 14.2 Oscillations of a string fixed at both ends. String produces the first harmonic (a), second harmonic (b), and third harmonic (c) in vibrations shown in figure.

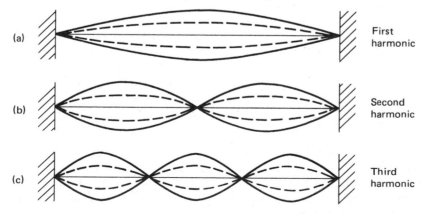

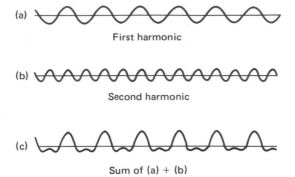

(a)

First harmonic

(b)

Second harmonic

(c)

Sum of (a) + (b)

Figure 14.3. Shape of the wave is determined by frequency and intensity of harmonics. The property of sound which depends on the frequency, the number, and the intensity of the harmonic is called the quality of sound.

The rate at which the energy is transmitted across a unit area perpendicular to the direction of flow in one second is called the intensity of the sound. Intensity has the units of watts/meter^2 (W/m^2).

The intensity of sound is proportional to the time average of the square of the excess pressure above the normal pressure in the absence of vibrations in air. Hence a significant quantity is the square root of the time average of the square of the excess pressure. This quantity is called rms (root mean square) pressure. What do you get if you calculate the average of the pressure for a sinusoidal wave instead of the square of the pressure?

Two similar scales for sound level, one in terms of intensities and the second in terms of the rms pressure P are given below.

$$\text{Sound intensity level} = 10\log (I/I_0), \quad (14.6)$$

where I is the intensity of sound and I_0 is a reference intensity taken as 10^{-12} W/m^2, which is the lowest intensity an average human ear can detect at about 2000 Hz. The unit of sound intensity level defined by Eq. (14.6) is decibels (dB). A bigger unit sometimes used is the bel which is 10 times bigger than the decibel. These units are named in honor of

Alexander Graham Bell, a pioneer in communications.

$$\text{Sound pressure level} = 20\log (P/P_0), \quad (14.7)$$

where P_0, the reference pressure level, is taken as 20 μPa (μN/m^2) which again is the threshold pressure level at 2000 Hz. The sound pressure level is also expressed in decibels. Except for sinusoidal motion, one uses Eq. (14.7) instead of (14.6). If there are several interacting sources, the intensity is not readily related to the pressure variations.

In both definitions above, we take the logarithm of the ratio of the numbers. This compresses the scale. For example, an average human ear can detect sound in the intensity range of 10^{-12} to 1 W/m^2. The enormous range is covered by a comparatively smaller range from 0 to 120 dB in sound intensity level. A tenfold increase in intensity or pressure produces an increase of only 1 dB in sound level.

EXAMPLE 14.1

Find the intensity level in dB of a sound intensity of 6×10^{-2} W/m^2 ($I_0 = 10^{-12}$ W/m^2).

$$\text{Sound intensity level} = 10\log (6 \times 10^{-2}/10^{-12})$$

$$= 10 \log(6 \times 10^{10})$$

$$= 107.8 \text{ dB}$$

The range of intensity for which the average ear is sensitive is shown in Fig. 14.4. The ear has its highest sensitivity for low-intensity sound in the 2000 to 4000 Hz range. The threshold of hearing in this range is below 10^{-12} W/m^2 while at 20 Hz it is 10^{-4} W/m^2 (approximately 100 million times). *The threshold of hearing is the minimum intensity of sound an average ear can hear.*

The threshold of feeling is the maximum intensity of sound an average ear can record without feeling discomfort. The threshold of feeling apparently has very little frequency dependence. We

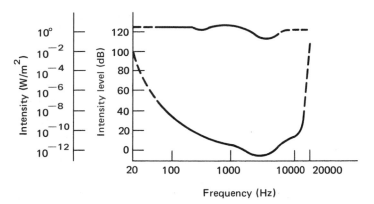

Figure 14.4. The figure shows frequency dependence, frequency range, and intensity level range of an average ear.

start feeling discomfort at about 120 dB and pain at about 140 dB.

For a long time it was thought that the decibel scale accurately represented the loudness sensation produced by sound. But this has been proved to be inaccurate. Just like the threshold of hearing, loudness is also found to be frequency dependent. However, the frequency dependence slowly decreases as the sound intensity level increases. Measuring loudness, a psychological effect which varies from person to person, is very complicated. Modified sound level scales have been recently adopted, which gives less weight to frequencies below 1000 Hz. A weighted scale which closely resembles the frequency dependence of the human ear at low sound levels is known as the A-weighted scale. The other scales known as B-weighted and C-weighted scales are used, respectively, for moderate and high sound pressure levels. The frequency dependence of the C scale is the least while the B scale lies between the A and C in its frequency dependence. In reporting results of measurements using these scales, it is customary to say the sound level is 40 dBA instead of saying A-weighted sound level is 40 dB.

Sound level meters read sound pressure levels, but not intensity levels. Most of these meters have microphones as detectors. They also contain amplifiers and display devices such as a galvanometer or a recorder. In addition to these components, most of the meters contain filter circuits located between the microphone and amplifier. The filter circuits make the meters frequency dependent. By switching in the proper filter circuit, the meters can be made to read in the A-, B-, or C-weighted scales. It is important to remember that the sound level measured by A-weighted meters are not identical to loudness response of the human ear even though their frequency response is close to that of the average human ear. However, their significance lies in the fact that they can be used for comparison purposes in ranking sound sources. The simplicity of sound level meters is one reason for their widespread acceptance.

Finally, we have seen the terms sound intensity level, sound pressure level, and sound level in this section. The term sound level is used for measurements in weighted scales.

14.5 NOISE POLLUTION

In the mechanical age in which we live, the noise pollution is increasing in certain areas to a dangerous level. In cities in the last 30 years it is estimated that the noise level has gone up by 30 dB. The different sources of noise pollution are shown in Table 14.1.

TABLE 14.1   Some Typical Sound Intensity Levels

SOUND LEVEL	EXAMPLES
0 (threshold of audibility)	
10	leaves rustling
30	audible whisper
60	conversation
65 (annoying)	
70	vacuum cleaner
85 (ear damage possible)	food blender
120 (threshold of feeling)	rock music, thunderclap
140 (threshold of pain)	jet engine
180	Apollo liftoff

Extended periods of exposure to high levels of noise cause damage to the ear. For example, 85 dB is generally considered to be the threshold level for hearing damage. Noise levels of 135 dB produce instantaneous damage to the ear. When noise levels are too great, the tiny sensory cells in the inner ear that transmit sound impulses to the brain may permanently be damaged producing permanent hearing loss. Complicating the situation further is the fact that there are no pain receptors in the inner ear to tell us when damage has been done. Often people are not aware that their inner ears are injured until considerable hearing loss has occurred.

Other factors in considering noise levels are the psychological effects of noise such as general annoyance, interference with work and sleep, and interference with speech. Background noise levels in excess of 50 dBA interferes with face-to-face conversation. Speech interference levels vary with distance, 55 dBA at 10 ft to 82 dBA at 0.5 ft. Some other health problems possibly created by high levels of noise are similar to stress-type diseases such as headaches and behavior problems such as irritability, aggressiveness, etc. Table 14.2 gives limits of noise exposures suggested in the Occupational Safety and Health Act (OSHA) of 1970.

A low-noise environment is, therefore, conducive to a healthy and peaceful life. How do we achieve it? We have to start with the sources. Dependence on noise-making appliances should be reduced to a

TABLE 14.2   Noise Limits Suggested by OSHA (1970)

NOISE LEVEL, dBA	EXPOSURE TIME LIMIT, H
90	8
95	4
100	2
105	1
110	1/2
115	1/4

Note that for an increase of 5 dBA in noise level exposure time should be reduced by half and at no time is the noise level is to exceed 115 dBA.

minimum. Of course, the person who lives near the expressway or an airport is almost helpless in this matter. But let us hope future actions on a community level will alleviate these intolerable situations. There are several ways to reduce noise reaching your house such as to plant trees, if possible evergreens, around your house, and to increase acoustical insulation of the house. Using good sound absorbing materials around the house is another approach we can take. Table 14.3 compares the absorption ability of different materials used in contruction.

14.6 DOPPLER EFFECT

From our common experience, we know that the pitch of an approaching automobile siren is higher than the pitch of a receding siren. The reason for the effect is illustrated in Fig. 14.5. The waves approach the listener on the right at a faster rate than the listener on the left. In Fig. 14.5(a), the distance between the wavefronts is λ.

TABLE 14.3 Sound Absorption

CONSTRUCTION MATERIALS	ABSORPTION COEFFICIENT β
Open window	1.00
Acousti–Celotex	0.82
Acoustic plaster	0.25
Carpet, heavy	0.40
Chalkboard	0.06
Draperies, cotton	0.50
Felt, heavy	0.70
Glass	0.025
Linoleum	0.03
Marble	0.01
Plaster wall, smooth	0.03
Wood wall, painted	0.04
Wood wall, floor, or paneling (unpainted)	0.08

It is also equal to the distance traveled by the waves in one period. Hence we have

$$\lambda = vT, \tag{14.8}$$

where v is the velocity of the waves and T its period. In Fig. 14.5(b), the distance between the wavefronts, λ_L, is less than λ by the distance the source has moved in one period. Hence

$$\lambda_L = \lambda - v_s T,$$

where v_s is the velocity of the source. Substitution for λ from Eq. (14.8) yields

$$\lambda_L = vT - v_s T.$$

The apparent frequency, as observed by the listener, is $f_L = v/\lambda_L$. Hence

$$f_L = \frac{v}{\lambda_L} = \frac{v}{vT - v_s T} = \frac{1}{T} \times \frac{v}{v - v_s}$$

Since $1/T$ is equal to the frequency f of the waves, we get

$$f_L = f \frac{v}{v - v_s} \tag{14.9}$$

for source and listener approaching each other. Since the denominator is smaller than the numerator, the frequency f_L is higher than the frequency of the source when the source and the listener are approaching each other. The apparent frequency, where the source and listener are separating from each other, is given by

$$f_L = f \frac{v}{v + v_s} \tag{14.10}$$

If the source is stationary and the listener is moving with a velocity v_L towards the source, waves reach the listener at a faster rate and the frequency in this case also appears higher. The relative velocity of the waves in this case is $v + v_L$. The apparent

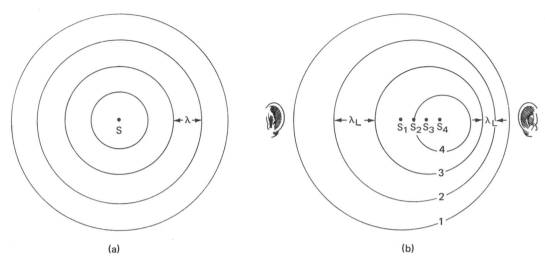

(a)   (b)

Figure 14.5. When the source is moving towards the listener, the waves reach him at a faster rate and, when the source is moving away, the waves reach him at a slower rate. Change in frequency due to the relative motion between the source and the observer is called the Doppler effect. (b) Waves 1, 2, 3, 4, are produced for the source positions S_1, S_2, S_3, S_4.

frequency, number of waves going past him in a second, is

$$f_L = \frac{v + v_L}{\lambda}.$$

Multiplying denominator and numerator by v, we obtain

$$f_L = \frac{v + v_L}{\lambda} \times \frac{v}{v} = \frac{v + v_L}{v} \times \frac{v}{\lambda}.$$

Since $v/\lambda = f$, it follows that

$$f_L = f\frac{v + v_L}{v}. \tag{14.11}$$

If the listener is moving away, we get a similar equation

$$f_L = f\frac{v - v_L}{v}. \tag{14.12}$$

Equations (14.9), (14.10), (14.11), and (14.12) can

be combined into one equation

$$f_L = f\frac{v \pm v_L}{v \mp v_s}, \tag{14.13}$$

where f_L is the apparent frequency, f the frequency of the waves, v the velocity of the waves, v_s the velocity of the source, and v_L the velocity of the listener. The $+$ in the numerator and $-$ in the denominator are used when the source and the listener are approaching each other and the $-$ in the numerator and the $+$ in the demoninator are used when they are separating from each other. You may have occasion (especially during a test) when you are not quite sure about the usage of these signs. Always remember that the frequency increases when the source and the listener are approaching each other and decreases when they are moving apart.

EXAMPLE 14.2

A train traveling at a speed of 100 ft/s is blowing a whistle at 600 Hz. Calculate the frequency of the whistle as heard by a listener when (a) the train is approaching him and (b) when the train is going away from him. Assume temperature to be 70°F.

(a) Velocity of the source of sound = 100 ft/s;
Velocity of the listener = 0;
Frequency of the sound = 600 Hz;
Velocity of sound in air = 1087 ft/s
$$+ 1.1(70°F - 32)$$
$$= 1128.8 \text{ ft/s.}$$

Frequency when train is approaching the listener is

$$f\frac{v}{v - v_s} = (600 \text{ Hz}) \left(\frac{1128.8 \text{ ft/s}}{1128.8 \text{ ft/s} - 100 \text{ ft/s}} \right)$$

$$= 658 \text{ Hz.}$$

(b) The frequency when the train is moving away from the listener is

$$f\frac{v}{v + v_s} = (600 \text{ Hz}) \left(\frac{1128.8 \text{ ft/s}}{1128.8 \text{ ft/s} + 100 \text{ ft/s}} \right)$$

$$= 551 \text{ Hz.}$$

EXAMPLE 14.3

Another train traveling at 100 ft/s is approaching the train of Example 14.2. What is the frequency of the whistle as heard by the conductor of the second train?

In this problem source and listener are moving towards each other. Hence

$$f_L = f\frac{v + v_L}{v - v_s} = (600 \text{ Hz}) \left(\frac{1128.8 \text{ ft/s} + 100 \text{ ft/s}}{1128.8 \text{ ft/s} - 100 \text{ ft/s}} \right)$$

$$= 715 \text{ Hz.}$$

14.7 BEATS

Alternately loud and soft sounds are heard when we listen to two sources of sound having slightly different frequencies. This results from interference between the waves produced by the two sources which oscillate at 49 Hz and 56 Hz the individual waves have equal amplitude. To understand this phenomena, let us consider two sources which oscillate at 49 Hz and 56 Hz, respectively. Assume they start oscillating at the same time. The wave trains starting from sources s_1 and s_2 are shown in Fig. 14.6. If the listener is at a distance equal to the velocity of sound, the initial waves A and B reach him in 1 s. These waves are in phase and they add up to produce a loud sound. The waves behind A and B progressively get out of step until we reach the fourth peak in the second wave train. (In the time the second source produces 4 waves, the first source produces only 3.5 waves). At this time, the wave from source s_1 is out of phase with that from s_2 and hence they add up to give zero amplitude. This resultant wave with zero amplitude reaches the listener $1/14$ (= 4/56) s after he hears the first loud sound. The waves behind the first minimum

Figure 14.6. Addition of waves of almost equal frequency produces beats.

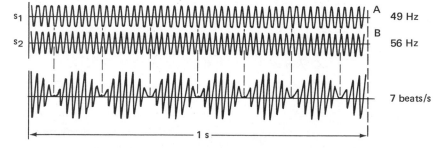

are slowly getting in phase. The 7th peak in the first wave train and the 8th peak in the second wave train are in phase. They add to give a loud sound which reaches the observer 2/14 s after the first loud sound. Thus the listener in this case hears loud and soft sounds 1/14 s apart. The time interval between consecutive maxima (or minima) is 2/14 s.

The sum of the two waves has a frequency equal to the average of their frequencies together with an intensity variation superimposed upon it. This variation in intensity has a frequency called the beat frequency and is given by the absolute difference of the two frequencies. In the example we discussed, the beat frequency is 7 Hz (= reciprocal of the time interval between maxima).

14.8 SOUND BARRIER AND SONIC BOOM

How can traveling sound waves produce a barrier? In Fig. 14.7 we see that as the speed of the source approaches the velocity of sound waves, the waves start to bunch together and finally when the velocities are equal, they pile up right at the source. When this happens, the pressure builds up as a barrier. If we attempt to increase the velocity at this point the object (supersonic planes do that very often) has to "crash" through this pressure barrier, called the sound barrier.

What happens if an object like an airplane travels at speeds faster than the speed of sound? The waves enhance each other on the surface of a cone with the airplane at its vertex. The high-pressure conical surface, with no pressure change in front of it, creates the well-known shock waves.

In Fig. 14.8(a) spherical waves at various positions of the moving object are shown. From any of these positions we obtain two distances, the distance to the surface of the cone which is equal to the radius of the spherical wavefront originating from that point, $v_s t$; and the distance to the present position of the object vt. The time t is the elapsed time since the source was at the position from which we calculated the above distances. We now obtain the angle θ, which the cone makes with the direction of the object, as

$$\sin\theta = \frac{v_s t}{vt} = \frac{v_s}{v}. \tag{14.14}$$

The velocity of a plane is sometimes expressed as the ratio of its velocity to the velocity of sound.

Figure 14.7. The waves in front of a moving source crowd together. When the velocity of the source is equal to the velocity of sound, the wavefronts bunch together forming a pressure barrier.

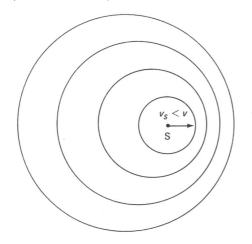

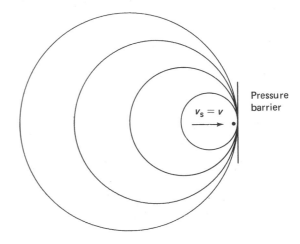

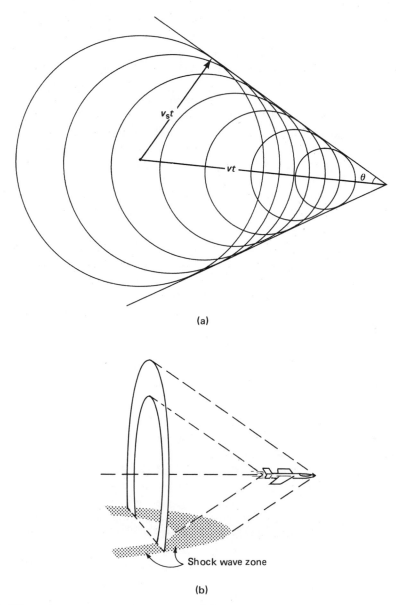

(a)

(b)

Figure 14.8 (a) When the source travels at a speed faster than the speed of sound, pressure builds up along the cone since the source of waves travels faster than the waves. (b) Figure shows the shock wave zone which travels with a supersonic airplane.

This ratio is known as the Mach number:

$$\text{Mach number} = \frac{\text{Velocity of the object}}{\text{Velocity of sound}}. \quad \textbf{(14.15)}$$

Obviously the Mach number of a supersonic plane is above 1.

EXAMPLE 14.4

A jet plane is traveling at a speed of 1.5 Mach number.   Calculate the angle made by the shock wave.

$$\text{Velocity of sound} = v_s;$$
$$\text{Velocity of the plane} = 1.5v_s.$$

We use Eq. (14.14) to obtain the angle

$$\sin\theta = \frac{v_s}{v}$$

$$= \frac{v_s}{1.5v_s} = 0.66.$$

$$\theta = \sin^{-1}0.66 = 42°.$$

14.9   APPLICATIONS OF SOUND WAVES

Sound waves are used in several areas of present day technology, such as prospecting, underwater ranging, ultrasonic cleaning, etc.   Some of those uses are discussed below.

The reflection of sound waves at the interface of two media of differing elastic properties is used in prospecting for petroleum and other minerals. Waves are produced by exploding small charges of dynamite or TNT underground.   The reflected waves are then picked up at appropriately located instruments to record their arrival time and intensity.   From these data experienced geologists are able to get some idea about the structure and nature of the subterranean levels.

Earthquakes involve the release of large amounts of energy, much larger than exploding charges in the above application, which trigger both transverse and longitudinal waves in the earth.   The velocities of these waves are different.   Therefore, the waves do not reach a distant point on the surface of the earth simultaneously.   The time difference in the arrival will enable us to locate the center of the disturbance. The instrument used to record these waves is called a seismograph. Earthquake detecting stations scattered around the world use seismographs to locate earthquakes.   Underground atomic blasts are also detected by sensitive seismographs essentially in the same way.

Ultrasound, waves above the human sensitivity limit of 20,000 Hz, finds application in several fields.   Recently Polaroid has released on the market a self-focusing camera.   This camera has a source of ultrasonic waves and detector.   From the time it takes for the waves to reflect from the object back to the camera, the latter estimates the distance of the object and automatically focuses the object on the film.   There are egg beaters (mixing the yolk and white) without breaking the egg!   Think about it! How is it done?   Ultrasonic cleaners are in wide use these days.   The object to be cleaned is immersed in water through which ultrasound is passing. The rapid agitation produced by the waves helps to shake off dirt particles.

Underwater ranging and the locating of sunken objects are done using ultrasound.   In both cases reflected waves are detected and the distances are estimated from the time it takes for the waves to reflect back to the surveying ship.   In medicine such techniques are used to map the body and locate abnormalities.   It is expected that someday ultrasound mapping techniques will replace the somewhat dangerous x-ray methods.

SUMMARY   Sound waves are longitudinal waves produced in air in a frequency range of 20 to 20,000 Hz called the audible range.

Sound does not travel through vacuum. It needs a medium to travel through. Sound velocity is given by

$$v = \sqrt{1.4P/\rho}$$

and the temperature variation of velocity is given by

$$v_t = 331.4 \text{ m/s} + 0.61t°C$$

or

$$v_t = 1087 \text{ ft/s} + 1.1(t°F - 32°C).$$

Pitch refers to the frequency and quality refers to the harmonic content of a sound.

Intensity is energy transmitted per unit area per second (W/m^2). Intensity level in dB is given by $l = 10 \log (I/I_0)$.

The Doppler effect is the frequency of sound change as the source and or the observer moves relative to the other, namely,

$$f_L = f_s \frac{(v \pm v_2)}{(v \mp v_s)}.$$

Beats are slow variations in intensity heard when two sources of approximately equal frequencies are energized simultaneously.

The sound barrier and sonic boom are results of the increase in pressure in certain regions due to the overlap of waves as the source travels at or above the velocity of sound.

QUESTIONS AND PROBLEMS

14.1 Is voice communication between two astronauts on the moon possible?

14.2 How do you tune a piano?

14.3 Calculate the velocity of sound at $-30°C$ and $50°C$.

14.4 Obtain the velocity of sound at $-70°F$ and $100°F$.

14.5 What are the frequencies of the fundamental and first three overtones of a copper wire 30 cm long? The velocity of the waves in the wire is 160 m/s.

14.6 A 1 m long metal wire weighs 0.5 g. This wire is to be tuned for a fundamental tone of 2200 Hz. What should be the tension of the wire?

14.7 A train whistle has a frequency of 1000 Hz. If the train is speeding at a velocity of 40 mi/h past a stationary observer, what is the apparent frequency (a) as the train approaches him and (b) as it moves away from him? Assume the velocity of sound in air to be 1100 ft/s.

14.8 A second train is approaching the first train in Problem 14.7 with a speed of 40 mi/h. Obtain the frequency of the whistle of the train of Problem 14.7 as heard by the driver of the second train.

14.9   The intensity level of a busy street is 60 dB. Calculate the energy entering through a window 0.6 m $\times$ 0.4 m of a house on this street.

14.10   The intensity of sound at a distance of 100 m from an airplane is 0.1 W/m^2. Calculate the intensity level of sound at that point.

14.11   Two tuning forks having frequencies 200 and 201 Hz produce a beat. What is the beat frequency?

14.12   Two sources of sound when struck together produce a slow variation of sound 5 times a second. The average frequency of the sources is 350 Hz. Calculate the individual frequencies.

14.13   Calculate the angle of shock waves produced by a jet plane traveling at a speed of twice the speed of sound. What is the Mach number?

14.14   A sonar range finder sends a pulse of ultrasonic waves. The waves return in 6 s. Assuming the velocity of sound in sea water is 4850 ft/s, calculate the depth at that point.

14.15   The self-focusing camera sends a pulse of ultrasonic waves which returns in 1/20 s after reflecting from the object. What is the distance between the camera and the object? The temperature is 70°F.

15
Light

Light is the most important physical means by which we learn about the world around us and the universe beyond. Just as in the case of sound, light has physical and physiological aspects to it. The physical aspect of light is that it is energy in the form of radiation and the physiological aspect is that it produces the sensation of vision upon impingement on the eye. As we will see below, light radiation has wave properties. While sound waves are mechanical, light waves are electromagnetic in nature. Electromagnetic waves are variations in electric and magnetic fields that travel through space. The ear is capable of distinguishing frequency differences or pitch of a sound. Somewhat in a similar fashion the eye distinguishes frequency differences of light as color differences. A difference of major consequence between sound and light waves is that for transmission of sound waves a medium is required, but light requires no medium. Before we further discuss the nature of light, a brief historical review of the theories of light is given below.

15.1 THEORIES OF LIGHT

An opaque object placed in front of a point source of light produces a shadow as shown in Fig. 15.1. The shadow on the screen (shaded area) is well defined. Light emitted within the cone defined by S1 and S2 is blocked by the object. The reason that no light falls on the shadow is that light does not bend, to any detectable extent, around an obstruction. Light emitted in directions outside the

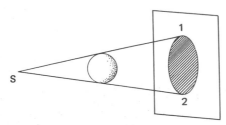

Figure 15.1. Formation of a shadow by an opaque object. The source S is a point source which produces a well-defined shadow.

cone 1S2 illuminates the region outside the shadow. A shadow illustrates in a simple and direct way the straight line motion of light. This phenomenon of straight line motion of light is called rectilinear propagation of light. (We ignore in this discussion diffraction effects.)

From similar observations on the rectilinear propagation of light, several investigators in the past supported the corpuscular (particle) theory of light, prominent among them being Sir Issac Newton. *According to this theory, light consisted of fast moving particles, which upon striking the retina of the eye, produced the sensation of vision.* At that time Christian Huygens, a contemporary of Newton, came forth with a wave theory of light which could also explain such phenomena as reflection and refraction of light. However, this latter theory was not taken seriously by the scientific community until after an English physician named Thomas Young explained the phenomenon of interference by this theory. Sir Issac Newton who first observed interference failed to explain it because he believed strongly in his corpuscular theory. Wave theory was further extended by Fresnel, Fraunhofer, and others.

As mentioned above, both wave theory and particle theory were used to explain reflection and refraction. According to the wave theory, the waves are supposed to slow down as they enter an optically denser medium, while the particle theory required a speeding up of the particles of light in the denser medium. An experimental measurement of velocity in an optically denser medium (e.g.,

water) by Foucault dealt the final blow to the corpuscular theory as known at that time.

Later a solid theoretical foundation to the wave theory was laid by James Clerk Maxwell by showing that the wave propagation of light and the rest of the electromagnetic waves can be represented by equations, known as Maxwell's equations. He derived these equations from known relationships between electric charge, electric field, electric current, and magnetic field. *According to Maxwell's theory, light waves are variations in electric and magnetic fields which travel through empty space (vacuum) as well as matter.* In other words light radiation is electromagnetic waves.

A modern theory of light, based on quantum mechanics, combines certain features of particle and waves theories. *According to this theory first postulated by Planck, light consists of wave packets, the wave character being predominant in phenomena like interference and diffraction and the particle character being predominant in light–matter interactions such as the photoelectric effect.* These wave packets are bundles of energy called photons. A photon contains the smallest amount of energy associated with that wave. The smallest energy of the light absorbed or emitted by a system is always equal to the energy of the photon.

15.2 LIGHT WAVES AND THEIR SPECTRUM

We further discuss wave aspects of light in this section since many important optical phenomena are explained in terms of the wave concepts.

When waves travel through a medium, we observe certain physical quantities changing with time at every point in the medium from the source to the wavefront. In water waves, the displacement of water molecules, and in sound, the pressure of the air, are the physical quantities that vary with time. To understand fully the nature of light waves, the following question should be answered first. What is the physical quantity that changes when light travels through space? Figure 15.2 is a pictorial

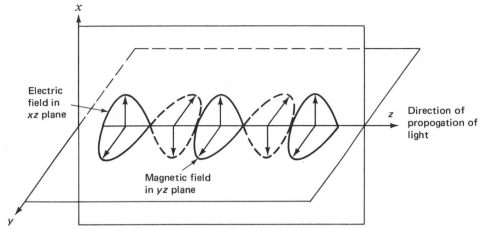

Figure 15.2. Pictorial representation of electromagnetic waves. The electric field lying in the *yz* plane and the magnetic field lying in the *xy* plane vary with time and distance from the source. Both fields are perpendicular to the direction of motion *z* of the waves.

representation of an electromagnetic wave. The quantities that change are the electric field and the magnetic field associated with the wave. The two fields are mutually perpendicular and they are both perpendicular to the direction along which the waves travel. Hence the electromagnetic waves are transverse waves. In Fig. 15.2 the waves travel in the *z* direction, while the electric field is in the *x* direction and the magnetic field is in the *y* direction. In our discussions from now on, we consider only the electric field associated with light, but keep in mind that the magnetic component is always present.

Visible light forms a very small part of the electromagnetic spectrum which contains gamma rays on the high frequency (short wavelength) side and radiowaves on the low frequency (long wavelength) side. (See Fig. 15.3.) Some sources such as the sun and incandescent bulbs radiate light in the wavelength range of 400 to 700 nm. Light from such sources produces an effect of "white light" in our eye. Some other light sources such as the neon discharge bulb radiate red light. White light can be separated into its component colors (Fig. 16.25) by refracting it in a glass prism or water droplets (rainbow). The rainbow colors in the order of their increasing wavelength are violet, indigo, blue, green,

yellow, orange, and red.

Why do certain objects appear colored? The reason for the color of an object is that it selectively reflects certain components of white light and absorbs (or sometimes transmits) the remaining components. A rose looks red because its petals reflect primarily red light and absorb primarily other colors. What will the color of a red rose be if illuminated by green light?

Wavelength of light is often expressed in units of nanometers (nm) and angstroms (Å)

$$1 \, \text{nm} = 10^{-9} \, \text{m},$$
$$1 \, \text{Å} = 10^{-10} \, \text{m}.$$

15.3 SPEED OF LIGHT

Electromagnetic waves (including light) travel in vacuum at a speed of 3×10^8 m/s. The speed of light is one of the important constants in physics. It is the maximum speed an object can attain. This postulate is basic to the special theory of relativity. Galileo was the first person to recognize that light might possibly have a finite velocity and he made attempts to measure the speed of light. Several successful experiments on the speed of light were

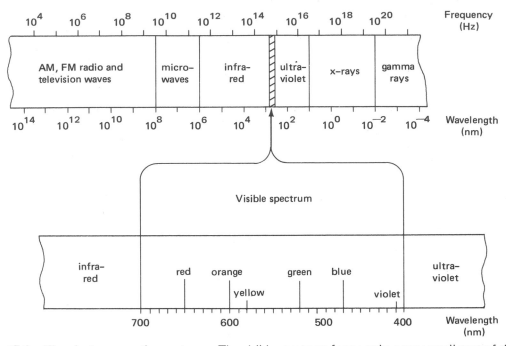

Figure 15.3. The electromagnetic spectrum. The visible spectrum forms only a very small part of the electromagnetic spectrum.

performed through the years since Galielo's experiment. An outstanding example of such an experiment performed by Michelson is described below.

Michelson's equipment consisted of a strong source of light S, a rotating octagonal mirror OM, a fixed mirror M, and a viewing telescope (Fig. 15.4). Light reflected from one of the sides of the octagonal mirror travels to the fixed mirror and reflects back to the rotating mirror. The distance between the two mirrors was 22 miles in Michelson's experiment. The reflected beam to the apparatus was viewed through a telescope. The beam will be observed only if (a) the rotating mirror remains still or (b) rotates through a multiple of 1/8 of a revolution by the time the light returns to the rotating mirror. In the latter case face 2 moves into the position of face 3 by the time the light reflected from face 1 returns to the octagonal mirror. If d is the distance between the rotating mirror and the fixed mirror, the time taken by the light to travel to

the fixed mirror and back is $2d/c$ where c is the velocity of light. In this period of time, the rotating mirror rotates through an angle $(2d/c)\omega$, where ω is its angular speed. The reflected beam stays in the telescope if

$$\frac{2d\omega}{c} = \frac{2\pi}{8}$$

or

$$c = \frac{16d\omega}{2\pi} \tag{15.1}$$

EXAMPLE 15.1

In a Michelson-type experiment the angular speed of the octagonal mirror is found to be 529 rev/s. If the distance between the mirrors is 22 mi, obtain the speed of light.

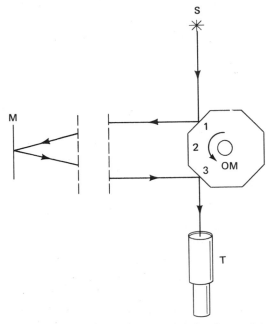

Figure 15.4. Michelson's apparatus for determining the velocity of light. S is a source, OM a rotating octagonal mirror, M a fixed mirror, and T a viewing telescope.

We use Eq. (15.1) to determine c

$$c = \frac{16 \times 22 \text{ mi} \times 529 \text{ rev/s} \times 2\pi \text{ rad/rev}}{2\pi \text{ rad}}$$

$$= 186,200 \text{ mi/s}.$$

The best determined value of the speed of light is 2.997×10^8 m/s or 1.86×10^5 mi/s.

The speed of light in other materials is less than that in vacuum. For example, in water the speed is 2.25×10^8 m/s and in common glass it is 2×10^8 m/s. In air the speed is about 7.0×10^4 m/s less than that in vacuo.

The speed of light is given by the equation

$$c = f\lambda, \tag{15.2}$$

where f is the frequency and λ is the wavelength. The unit for frequency is Hz (= vibrations/second).

EXAMPLE 15.2

Sources of light containing mercury vapor emit light having different characteristic wavelengths. One of these, the green light, has a wavelength of 518 nm. Calculate its frequency.

We use the equation $c = f\lambda$

$$c = 3 \times 10^8 \text{ m/s}; \lambda = 578 \text{ nm} = 578 \times 10^{-9} \text{ m}.$$

Therefore

$$f = c/\lambda = \frac{3 \times 10^8 \text{ m/s}}{578 \times 10^{-9} \text{ m}} = 5.19 \times 10^{14} \text{ Hz}.$$

15.4 POLARIZATION

In Fig. 15.2 the electric field vibration takes place in the xz plane, with the field directed along the x direction. If all waves in a beam of light behave like this, that is, their electric field and the field vibrations are confined to the xz plane, we have a polarized beam of light. In this case the xz plane is the plane of polarization and x is the direction of polarization. Light emitted by an ordinary source is unpolarized, that is, the electric field vibrations are not confined to one plane and the direction of the electric field could be any direction perpendicular to the beam direction as shown in Fig. 15.5.

Unpolarized light can be polarized by passing it through certain materials called polarizers. The light coming through the polarizer is plane polarized with the direction of polarization determined by the direction of the polarizer (Fig. 15.5).

If two polarizers are available one is used to polarize the light and the second one to "analyze" the light. As shown in Fig. 15.6(a) the polarized light is blocked when the direction of the analyzer is perpendicular to the direction of the polarizer. On the other hand, if the direction of polarization of the light is in the same direction as the direction of the analyzer, light will be transmitted through the analyzer. In the second case, the polarizing

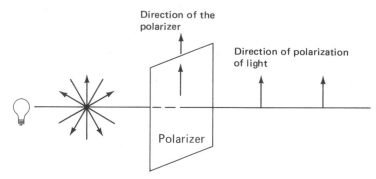

Figure 15.5. Light from an ordinary source is not polarized. When this light is passed through a polarizing material like polaroid, light gets polarized in a direction determined by the polarizing sheet.

directions of both polarizers are perpendicular to each other, often referred to as "crossed" and, in the first case, they are parallel to each other. One will also see that the intensity of light transmitted increases slowly from a minimum as the angle between the direction of the polarization of the polarizers decreases from 90° and the intensity reaches a maximum for 0°.

Figure 15.6 (a) Light is polarized by the sheet, and if the direction of the analyzer is in the same direction as the polarization direction of light coming through the first sheet, light will pass through the analyzer. If the polarizer and analyzer sheets are crossed (directions of polarization perpendicular to each other), no light will pass through the analyzer. (b) Mechanical polarizer and analyzer. If the directions of the polarizer and the analyzer are not the same, waves will not be transmitted.

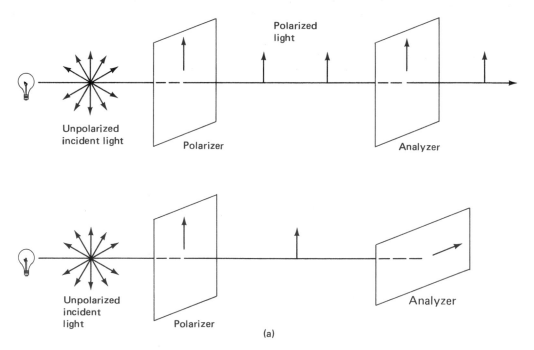

(a)

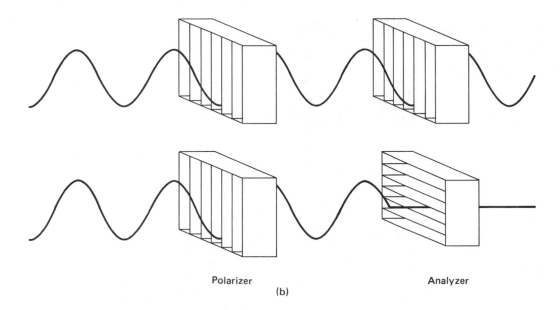

Polarizer

(b)

Analyzer

An analogous mechanical situation where a mechanical polarizer and a mechanical analyzer are used is shown in Fig. 15.6(b). When the polarization direction of both are the same, the wave in the rope goes through both. On the other hand, when the analyzer is 90° to the polarizer, the wave is stopped at the analyzer.

Polarizers find their most popular use in sunglasses. They reduce the glare of reflected light from the viewer's eyes. When light is reflected from a surface, partial polarization takes place. The direction of polarization of reflected light is parallel to the plane of reflection. Since most of the reflecting surfaces are horizontal, polaroid glasses with their polarization in the vertical direction can stop the glare from reaching the eye, thus providing a clearer view of objects ahead (Fig. 15.7). The same is evident in the polarized windshield of a car.

Industrial engineers and designers use polarized light to study stresses in machining parts. Since

Figure 15.7. Sunglasses reduce the glare of reflected light. (a) Photograph taken without the help of a polaroid. Interior of the car is not clearly visible. (b) Photograph taken by placing a polaroid sunglass in front of the lens.

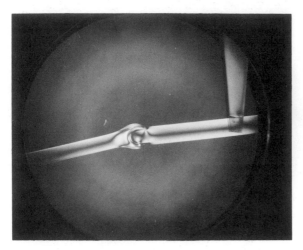

Figure 15.8. When the model of an object is studied in polarized light, bright and dark bands appear showing the strain distribution.

metals do not transmit light, plastic models of these parts are studied between polarizing sheets. The first sheet polarizes the light and the second sheet analyzes it. Before the plastic model is inserted the sheets appear uniformly bright or dark depending on the angle between the polarizers. When the model is introduced between the sheets, the strains become visible since they produce changes in the polarization of light passing through the plastic model. One will see (Fig. 15.8) bright and dark bands. In an area where the bands are crowded, stress is a maximum. Pictures of these bands (Fig. 15.8) can tell the engineer where machine parts need reinforcement and which parts are apt to fail under high stresses.

15.5 INTERFERENCE

Interference, as mentioned earlier, is a phenomenon that occurs because of the wave properties of light. To understand interference, first let us consider an analogous situation of interference between water waves. A floating cork will move up and down as the wave passes along the water surface shown in Fig. 15.9(a). In Fig. 15.9(b) there are two wave generators, connected together. They move

up and down in unison. Since the cork is half way between the sources, the crest from sources 1 and 2 reach the cork at the same time producing an enhancement in the amplitude of oscillation. If one of the sources is moved away from its position by a distance of $\lambda/2$, the crest of waves from source 1 will meet the trough of the other at the cork producing no motion [Fig. 15.9(c)]. In the first case, the superposition of the waves takes place constructively and, in the second case, destructively. These effects produce constructive interference and destructive interference, respectively. The waves reach the cork in phase in Fig. 15.9(b) while they reach the cork out of phase in Fig. 15.9(c). The phase difference in the latter case is produced because "path lengths" (distance from the source to the point under consideration) differ by an odd number of half wavelengths.

Interference of light can be demonstrated by using Young's double slit arrangement. Light waves from the slits S_1 and S_2 fall on a screen producing an interference pattern—alternate bright and dark lines. To understand the formation of this pattern, let us consider the arrangement in Fig. 15.10 where the two slits are kept behind a screen. The distance between the slits is d and the distance between the plane of slits and the screen is D. The light from the two slits comes from a single source symmetrically placed to the left of the two slits (the reason for this is explained later). If we assume that the light waves start in unison or in phase from both slits, the point P_1 will be bright or dark depending on whether the waves reach the point in phase or out of phase. In Fig. 15.10(b), the waves reach point P_1 in phase, that is, the path length difference is equal to one wavelength. The point P_1, therefore, is a bright point on the screen. Other bright points on the screen are P_0 (path length difference = 0), P_2 (path length difference = 2λ), etc. Between P_0 and P_1 there is a point where the path length difference is $\lambda/2$, and similarly between P_1 and P_2 where the path length difference is $3\lambda/2$. These points are dark.

Summarizing the above discussion, we can conclude that constructive interference occurs when the path length difference $L_2 - L_1 = n\lambda$, where n

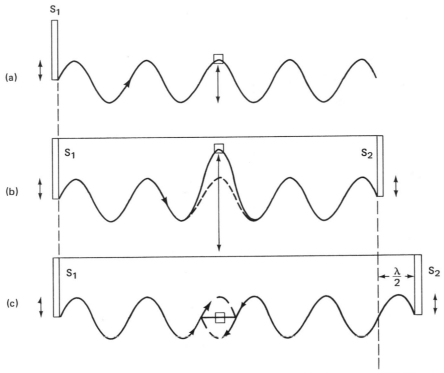

Figure 15.9. (a) A cork floating in water oscillates up and down with the water. (b) Waves from sources 1 and 2 reach the cork in phase. The two waves add and the cork oscillates with twice the amplitude. (c) Source 2 is moved through a distance of $\lambda/2$. The waves cancel each other at the cork because they are out of phase. In this case the cork is not moving. In all the above cases we are looking at the effect of waves only at one point in the vicinity of the cork.

$= 0, 1, 2, 3, \ldots$. Destructive interference occurs when the path length difference $L_2 - L_1 = n\lambda/2$, where n is an odd integer.

Now we derive an equation for the distance of the bright points, $P_1, P_2, \ldots$, from the center P_0. From Fig. 15.10(b), we get

$$L_2 - L_1 = S_2N.$$

In an experimental set up, generally, the distance S_1S_2 is very small, of the order of a few millimeters or centimeters, and the distance OP_0, a few meters. Therefore the line S_1N which is normal to S_2P is almost normal to OP_1 and the angle $S_2S_1N \cong \theta$. Comparing the triangles S_1S_2N and OP_0P_1, we have in this approximation that

angle $S_2N\,S_1 = 90°$, angle $OP_0P_1 = 90°$; angle $S_2S_1N \cong \theta$, angle $P_0OP_1 = \theta$.

The two triangles are thus similar. Hence we get

$$\frac{P_1P_0}{D} \cong \frac{S_2N}{d}.$$

Let X be the distance P_1P_0 from the central maximum to the first bright line. The path length S_2N in this case is λ. Therefore, as a first approximation, we have

$$X_1/D = \lambda/d \quad \text{or} \quad X_1 = \lambda D/d. \tag{15.3a}$$

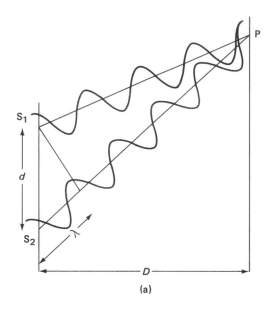

(a)

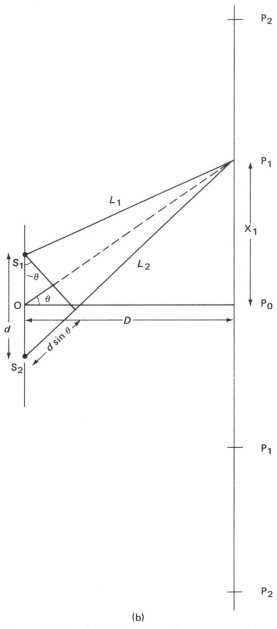

(b)

Similarly, the position of the second bright line is given by

$$X_2 = 2\lambda/Dd. \qquad (15.3b)$$

From Fig. 15.10(b) it is clear that the path length difference, $S_2N = d\sin\theta$. Hence the condition for maxima in terms of the angle θ is

$$d\sin\theta = n\lambda. \qquad (15.4)$$

EXAMPLE 15.3

Two slits separated by a distance of 1 mm are illuminated by green light ($\lambda = 550$ nm). Calculate the separation between the first two bright lines from the central maxima. The distance between the slits and the screen is 2 m.

We are given $d = 1$ mm $= 1 \times 10^{-3}$ m, $D = 2$ m, $\lambda = 550$ nm $= 550 \times 10^{-9}$ m. The distance of the first maxima x_1 is obtained by using Eq. (15.3a)

Figure 15.10. (a) Light waves from source S_1 and S_2 reach point P_1 in phase. The difference in path length $S_2P_1 - S_1P_1 = \lambda$. (b) Young's double slit arrangement. Bright lines appear at $P_0, P_1, P_2, \ldots$, because waves reach these point in phase. Distance between the sources is d and distance between the plane of the source and the screen is D.

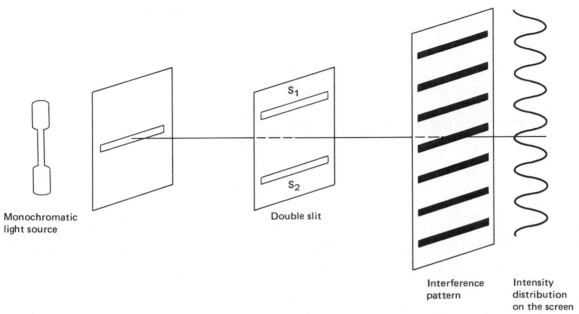

Figure 15.11. Young's double slit arrangement to demonstrate the interference of light waves from two slits. At the extreme right is a plot of the variation of intensity of the interference pattern on the screen.

$$X_1 = \frac{\lambda D}{d} = \frac{550 \times 10^{-9} \text{ m} \times 2 \text{ m}}{1 \times 10^{-3} \text{ m}}$$

$$= 1.1 \times 10^{-3} \text{ m} = 1.1 \text{ mm}.$$

Similarly, the distance of the second maxima is obtained using Eq. (15.3b)

$$X_2 = \frac{2\lambda D}{d} = \frac{2 \times 550 \times 10^{-9} \text{ m} \times 2 \text{ m}}{1 \times 10^{-3} \text{ m}}$$

$$= 2.2 \times 10^{-3} \text{ m} = 2.2 \text{ mm}.$$

In our discussion we assumed that the waves originating from S_1 and S_2 are in phase. Two independent light sources or even two slits kept in front of a broad source will not satisfy this condition. The waves from different parts of a light source have no correlation. Young overcame this difficulty by placing a slit between the light source and the double slit (Fig. 15.11). Waves through

this slit, which originate from a small part of the light source, reach the slits S_1 and S_2 in phase.

Why is the light from slits S_1 and S_2 in phase when another single slit is symmetrically placed between the source and the double slits? This

Figure 15.12. Wavefronts from a point source. Points 1, 2, and 3 in a wavefront are sources of new waves according to Huygens' principle. The envelope of new wavelets forms the new wavefront.

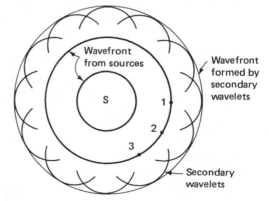

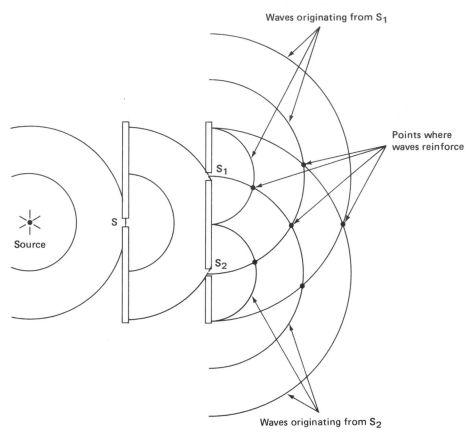

Figure 15.13. Huygens' principle is used to obtain wavefronts from slits S, S_1, and S_2.

question may be answered using *Huygens' principle which states that every point on a wavefront may be regarded as a new source of waves.* First let us understand Huygens' principle.

Consider a source S radiating light uniformly in all directions (Fig. 15.12). The wavefronts of the light waves are concentric spheres, like the concentric circular waves spreading in water when an object is thrown into water. According to Huygens' principle points 1, 2, 3, etc. in a wavefront are new sources of waves. Secondary wavelets starting from these points are spreading out simultaneously. The envelope of these wavelets is the new wavefront and these in turn send out secondary wavelets forming the next wavefront.

Figure 15.13 shows the construction of wavefronts at the slits of a Young's experimental arrangement. It is evident that the waves starting from S_1 and S_2 are in phase since they originate from two points on the same wavefront.

The wavefronts starting from S_1 and S_2 have the same frequency and they are in phase. They are examples of coherent waves. *In general if two sources with the same frequency radiate waves with a constant phase difference (in phase, out of phase, or fixed intermediate phase difference) the waves are coherent. Sources which produce coherent waves are called coherent sources. A steady interference pattern can be observed only with waves emanating from coherent sources.*

It is not very difficult to demonstrate Young's double slit experiment by using "homemade" equipment. Slits can be made by drawing lines on smoked glass slides. Light sources should be monochromatic. If more than one color is present, the interference patterns due to each color will overlap and may produce uniform illumination on the screen.

A large number of parallel slits, called a diffraction grating, produces an interference pattern consisting of sharp, narrow, bright lines. A common procedure to produce a grating is to rule grooves on glass with diamond; the clear regions between the grooves are the slits. A grating may contain 10,000 to 30,000 grooves per inch. Plastic impressions of glass gratings made by allowing liquid plastic to solidify on original gratings, called replica gratings, are widely used.

Figure 15.14 shows the experimental arrangement used to observe an interference pattern produced by a diffraction grating. The light rays at an angle θ with the horizontal focuses to the point P_1. Constructive interference will occur at P_1 if the path length difference between rays from neighboring slits is equal to an integral multiple of λ, that is, when

$$d \sin\theta = n\lambda, \quad n = 0, 1, 2, \ldots . \qquad (15.5)$$

Advantages of a grating over the double slit, as shown in Fig. 15.14, is that the interference pattern produced by the grating is much sharper than the broad lines produced by the double slit. Grating-produced lines are much brighter, as well as more sharply defined, since light energy distributes over a smaller area on the screen compared to a much larger area in the double slit case. Since the lines are sharper, the grating can be used for separating colors.

In this respect, the grating disperses colors like a prism. However, the intensity is distributed not into just one spectrum but into a number of spectra; also the spectra may overlap. Spectra occuring for $n = 1$ in Eq. (15.5) are called the first-order spectra, and for $n = 2$ are called the second-order spectra, etc. Because of the narrowness of the diffraction line, very accurate determinations of wavelength can be made using gratings.

EXAMPLE 15.4

A grating has 10,000 rulings per cm on it. Calculate the angular separation between the first-order maxima produced by the sodium yellow lights ($\lambda_1 = 589.0$ nm and $\lambda_2 = 589.6$ nm).

The grating has 10,000 lines per cm. Therefore, the distance between the slits, d, is given by

$$d = \frac{1}{10,000} \text{ cm} = 10^{-4} \text{ cm} = 10^{-6} \text{ m}.$$

The equation for maxima is

$$d \sin\theta = n\lambda.$$

In this example $n = 1$; hence

$$d \sin\theta = 1\lambda$$
$$\theta = \sin^{-1}(1\lambda/d).$$

Substituting the values of d and the two wavelengths, we get

$$\theta_1 = \sin^{-1} \frac{(1 \times 589 \times 10^{-9} \text{ m})}{10^{-6} \text{ m}} = 36° 5';$$

$$\theta_2 = \sin^{-1} \frac{(1 \times 589.6 \times 10^{-9} \text{ m})}{10^{-6} \text{m}} = 36° 8'.$$

The angular separation is

$$\theta_2 - \theta_1 = 0° 3'.$$

This angular separation is large enough to be observable in ordinary laboratory spectrometers.

15.6 THIN FILM INTERFERENCE

Interference occurring between light reflected from the top and bottom surfaces of a thin film of material has several practical applications. Multicolored patterns seen in soap bubbles and in thin oil films on water result from such interference effects.

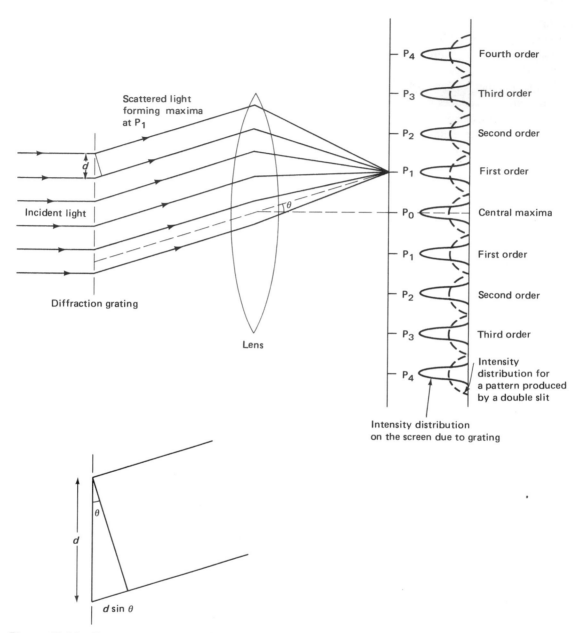

Figure 15.14. Experimental arrangement to study the interference pattern produced by a diffraction grating. A diffraction grating produces a sharp intensity pattern compared to the broad pattern produced by a double slit. Path length difference between waves from adjacent slits is equal to $d \sin\theta$.

In analyzing thin film interference, the following very important facts should be kept in mind:

1 When light travels through a medium of refractive index $\mu > 1$, its path length is given by the linear length in the medium multiplied by the refractive index.

2 When light is reflected at a less-dense to dense interface, a phase change of π occurs; in other words, when such reflection occurs, a path length decrease equivalent to $\lambda/2$ must be included.

The above two points are clarified in the following example.

EXAMPLE 15.5

Calculate the path length difference between rays 1 and 2 obtained by reflections as shown in Fig. 15.15. The wavelength of the incident light is 520 nm.

The path length difference between 1 and 2 due to the difference in the distances traveled assuming the angle of incidence is very small is equal to $2\mu t$ = $2 \times 1.5 \times 0.1$ mm = 0.3 mm. Ray 1 is obtained by reflection at a less-dense to dense interface, and, therefore, there is a phase change of π. On the other hand, for the reflection at the glass–air interface (dense to less-dense interface), there is no phase change.

Total path change = $2\mu t + \lambda/2$
$$= 3 \times 10^5 \text{ nm} + 260 \text{ nm}$$
$$= 3.0026 \times 10^5 \text{ nm}.$$

Let us consider an air wedge formed by two glass plates, which is illuminated by a monochromatic light (Fig. 15.16). Interference between light scattered from the top and bottom of the air wedge produces a pattern of bright and dark lines (Fig. 15.16). The bright fringes are equidistant, their separation being determined by the angle of the wedge. Total path length difference between rays 1 and 2 is equal to $2t + \lambda/2$; $\lambda/2$ is introduced because of the phase change in ray 2 during reflection. Note that we are assuming that the light is falling on the slides at normal incidence. When the total path length difference is equal to an odd multiple of $\lambda/2$, we get destructive interference. At the contact point, since $t = 0$, the total path length is $2 \times 0 + \lambda/2 = \lambda/2$; hence we get a dark fringe.

Figure 15.16. Air wedge. Interference between rays 1 and 2 occurs when they are brought together in the eye. The equidistant bright lines appear as shown in figure (bottom).

Figure 15.15. Example 15.5. Note the phase change at reflection at the top surface (less-dense to dense interface). Compare this phase change to phase changes in mechanical waves in Fig. 13.13.

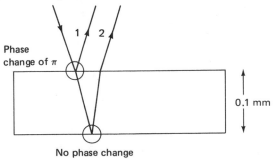

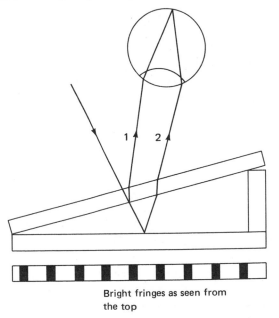

Bright fringes as seen from the top

The next dark fringe appears for $2t + \lambda/2 = 3\lambda/2$ or $t = \lambda/2$ and the next one for $t = \lambda$, etc. In other words, when one goes from one dark fringe to the next, the air gap thickness increases by $\lambda/2$ and, therefore, the fringes are equidistant. The bright fringes lie between the dark fringes. The measurement of a small thickness is a practical application of interference, as shown in the example below.

EXAMPLE 15.6

A wire of very small diameter (0.1 mm) is kept as shown in Fig. 15.17 between two glass plates to form a wedge. If the wedge is illuminated by sodium light of wavelength 589 nm, obtain the separation between bright interference fringes.

Angle θ of the wedge is given by

$$\tan\theta = D/L.$$

Let us assume that bright fringes form at distances l_1 and l_2 from the contact point; the thickness of the air gaps at these distances are $l_1 \tan\theta$ and $l_2 \tan\theta$, respectively. The difference in the two thicknesses should be equal to $\lambda/2$. Hence

$$l_2 \tan\theta - l_1 \tan\theta = \lambda/2$$
$$l_2 \, D/L - l_1 \, D/L = \lambda/2.$$

The distance between the fringes is given by $l_2 - l_1 = \lambda L/2D$. We are given $\lambda = 589 \times 10^{-9}$ m, $L = 10$

cm $= 10^{-1}$ m, and $D = 0.1$ mm $= 10^{-4}$ m. Therefore, we get

$$l_2 - l_1 = \frac{589 \times 10^{-9} \text{ m} \times 10^{-1} \text{ m}}{10^{-4} \text{ m}}$$
$$= 589 \times 10^{-6} \text{ m} = 0.58 \text{ mm}.$$

In this case, separation between the fringes is 0.58 mm, that is, there are about 2 fringes per mm and over the whole length of the glass slides there are about 172 fringes.

In an air wedge arrangement, the condition of destructive interference is satisfied when the thickness of the air gap is equal to a multiple of $\lambda/2$ including zero. If the two plates are perfectly plane, this condition is satisfied along straight lines parallel to the line of contact. Therefore, the fringes observed are straight line fringes. How do the fringes appear if one of the glasses is not perfectly flat? Instead of equally spaced straight line fringes, we may see a pattern similar to the one shown in Fig. 15.18. Looking at such a pattern, one gets an idea about the "flatness" of a piece of glass. In such a test one detects variations in thickness of the order of a wavelength ($\sim 5 \times 10^{-4}$ mm).

Thin coatings are used to block unwanted reflected or transmitted light. Nonreflecting coatings are found on good lenses and other glasses. Destructive interference of reflected light occurs (Fig. 15.19) if the path length difference, $2\mu t$, between the two reflected rays 1 and 2 is equal to $\lambda/2$.

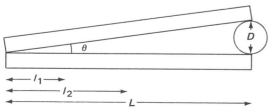

Figure 15.17. Example 15.6. l_1 and l_2 are distances from the contact point to two bright fringes.

Figure 15.18. Interference pattern produced when one of the glass pieces is not flat.

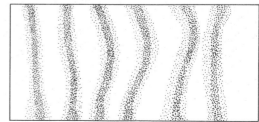

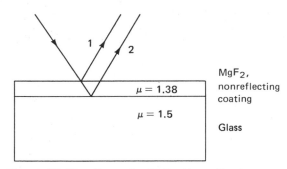

Figure 15.19. Example 15.7. Nonreflecting coating.

The condition for cancellation of reflected light is

$$2\mu t = n\lambda/2, \tag{15.6}$$

where n is an odd integer.

Thin film coatings are also used to stop unwanted transmitted radiation such as infrared (heat) radiation through glass sheets placed over intense light sources like the ones found in overhead projectors (e.g., score projector in bowling alleys).

EXAMPLE 15.7

Calculate the thickness of a MgF_2 coating over glass that will be sufficient to stop the reflection of green light (Fig. 15.19).

The condition for destructive interference between the reflected rays is

or
$$2\mu t = \lambda/2$$
$$t = \lambda/4\mu.$$

We calculate the thickness for the center of the visible spectrum by taking $\lambda = 550$ nm. Hence we get

$$t = \frac{550 \text{ nm}}{4 \times 1.38} = 100 \text{ nm}.$$

The thickness obtained is the smallest thickness. Why?

15.7 DIFFRACTION

Water waves and sound waves bend around obstacles in their path. Light waves also bend into a region not directly exposed to the light. One can

Figure 15.20. Diffraction of a parallel beam of light at a sharp edge. Intensity variations take place on both sides within a few wavelengths distance of the edge.

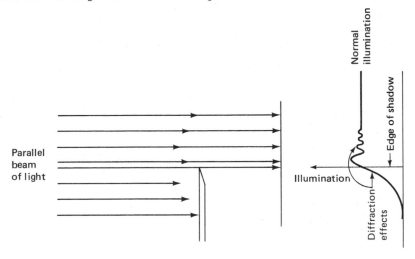

demonstrate this effect by shining monochromatic light on a screen with a sharp edge in its path.

Instead of a shadow with a sharp edge, a fuzzy shadow is observed. The illumination on the shadow side (Fig. 15.20) extends only for a few wavelengths from the geometrical edge of the shadow and it decreases very rapidly. Diffraction and interference effects cause rapid variations of illumination outside the shadow region. Another example is in pinhole photography. The hole is decreased to sharpen the image; but if it becomes too small, there are diffraction rings which obscure the image. The bending around corners is called diffraction.

In Fig. 15.13, light reached slits S_1 and S_2 from slit S because it was able to bend. Hence, in the double slit experiment and diffraction grating experiment, the bending of the light due to diffraction was necessary in order to have interference of the waves.

15.8 PHOTOELECTRIC EFFECT

While it has been possible to explain certain phenomena of light on the basis of a wave theory, other effects demanded a different explanation. One such effect is photoelectricity, which can be explained using the photon concept of light. First we explain the meaning of this new concept and later use it to explain the photoelectric effect.

A photon carries with it the smallest amount of energy associated with light of a given frequency. If the frequency is f, the energy of the photon is given by

$$E = hf,$$

where h is Planck's constant

$$h = 6.62 \times 10^{-34} \text{ J·s}.$$

Since frequency has a unit of Hz (= 1/s) the product of h and f is joules, the unit of energy.

EXAMPLE 15.8

Calculate the energy of a light photon of 555 nm wavelength.

Wavelength $\lambda = 555 \times 10^{-9}$ m;

Frequency $f = \dfrac{c}{\lambda} = \dfrac{3 \times 10^8 \text{ m/s}}{555 \times 10^{-9} \text{ m}} = 5.4 \times 10^{14}$ Hz;

Energy $= hf = 6.62 \times 10^{-34}$ J·s $\times 5.4 \times 10^{14}$ Hz

$= 3.6 \times 10^{-19}$ J.

What is the photoelectric effect? In the photoelectric effect, light having a frequency above a threshold value and falling on a metallic surface causes electrons to be released from the metal (Fig. 15.21).

Some of the experimental observations that required the photon concept are the following:

(a) For the electrons to be released, the incident light should have a minimum frequency. The kinetic energy of the electrons increases with increasing frequency of light. In terms of electromagnetic theory the energy would accumulate so that there would be no need for a minimum frequency contrary to experiment.

(b) Increasing the intensity does not increase the energy of the electrons. It increases the number of electrons.

Figure 15.21. Photoelectric effect.

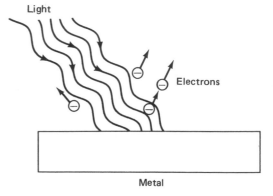

These results follow from the photon concept of light. Electrons gain energy from the photons; hence their energy should depend on the photon energy which is proportional to its frequency. Einstein expressed this relationship by the following equation

$$hf = KE + W,$$

where W is the minimum energy required to release the electron from the metal ($KE = 0$ in this case). The quantity W varies from metal to metal. It is called the work function of the metal. The photon theory also tells us that as the number of photons (intensity) increases, more electrons are released.

EXAMPLE 15.9

The work function of a metal is 4.0×10^{-19} J. Calculate the minimum (threshold) frequency of light that will produce photoelectrons from this metal. Calculate the kinetic energy of electrons released when illuminated by light of 8×10^{14} Hz frequency.

For the minimum frequency, the kinetic energy of the electrons released is zero. Hence

$$hf_{min} = W$$

or

$$f_{min} = W/h.$$

Substituting the values into the equation we obtain

$$f_{min} = \frac{4 \times 10^{-19} \text{ J}}{6.62 \times 10^{-34} \text{ J·s}} = 6.04 \times 10^{14} \text{ Hz}$$

To calculate the kinetic energy of the electrons released when 8×10^{14} Hz light shines on the metal we use the equation

$$hf = KE + W$$

or

$$KE = hf - W$$

$$= (6.62 \times 10^{-34} \text{ J·s})(8 \times 10^{14} \text{ Hz})$$
$$- 4.0 \times 10^{-19} \text{ J}$$
$$= 1.3 \times 10^{-19} \text{J}$$

EXAMPLE 15.10

Assuming 100% efficiency in photoelectric conversion, calculate the number of electrons released from the metal of Example 15.9 when 1 W of light shines on the metal for 10 s. The frequency of the light is 8×10^{14} Hz.

Total energy of light falling on the metal in 10 s is

$$1 \text{ W} \times 10 \text{ s} = 10 \text{ J}.$$

Energy of one photon is

$$hf = (6.62 \times 10^{-34} \text{ J·s})(8 \times 10^{14} \text{ Hz})$$
$$= 5.30 \times 10^{-19} \text{ J}.$$

Number of photons falling on the metal in 10 s is

$$\frac{10 \text{ J}}{5.30 \times 10^{-19} \text{ J}} = 1.9 \times 10^{19}.$$

The number of photoelectrons released = number of photons = 1.9×10^{19}. A photoelectric conversion efficiency of 100% is never achieved. An efficiency of $\sim$10% is obtained from carefully prepared metal surfaces. Photoelectric devices find several applications such as light controlled switches, video cameras, light-to-electricity conversion systems, etc.

SUMMARY   Theories of light

1   Corpuscular theory is based on rectilinear motion of light. According to this theory light consists of particles emitted by the source. This theory failed to explain interference and polarization.

2   Wave theory is successful in explaining several features of light such as polarization and interference. According to this theory, light consists of electromagnetic waves. The velocity of light is given by $c = f\lambda$.

3   Photon theory is successful in explaining photoelectric effect. According to this theory, light is emitted in small wavepackets containing the smallest amount of energy (hf) associated with the wavepacket.

Plane polarization of light means confining vibrations to one plane. Polarization has practical applications such as reducing reflected light, studying strains, etc.

Another phenomenon, interference, can be explained in terms of the wave properties of light. Double slit interference is a simple demonstration of this. Multiple slits (a grating) are used in resolving light into their components. Some applications of interference are (a) the measurement of small thicknesses using an air wedge (b) the determination of the flatness of glass plates, (c) non-reflecting coatings, and (d) accurate wavelength measurement by interference techniques. Constructive interference occurs when path length difference between light beams reaching a point is equal to a multiple of its wavelength.

**QUESTIONS
AND
PROBLEMS**

15.1   A light year is defined as the distance light travels in a year. How many meters is 1 light year? What is the distance of 10 light years away from us?

15.2   In a recent experiment, laser light was scattered from the surface of the moon and back to earth. How long does it take for the light to travel to the moon and back to earth? For comparison, is it possible to calculate the time sound waves will take for the same round trip?

15.3   Calculate the frequency and period of oscillation of green light of 550 nm and red light of 656 nm.

15.4   The period of oscillation of a light wave is 1.5×10^{-15} s. Calculate its frequency and wavelength.

15.5   The wavelength of light emitted by atoms in gaseous form is sufficiently stable, an accuracy of 1 part in 10^9 being achieved in the comparison of lengths using light waves. The wavelength of light emitted by krypton is now used as the international standard of length. One meter is now defined as 1,650,763.7 wavelengths of this light. Calculate the wavelength of this light.

15.6 A first maxima appears at a distance of 4 mm from the center on a screen placed 2 m from the sources (Fig. 15.22). Assuming light reaches S_1 and S_2 in phase, calculate (a) the path length difference between S_1P and S_2P, (b) angle θ, and (c) the distance between the two slits. The wavelength of the light is 486 nm.

15.7 In a double slit arrangement, the distance between the slits is 0.5 mm and the screen is 2 m away from the slits. What is the separation on the screen between the maximas due to light of 480 nm and 580 nm wavelengths?

15.8 A glass ($\mu = 1.50$) coated with a material ($\mu = 1.25$) is found to prevent reflection of light of wavelength 550 nm. Calculate the thickness of the coating.

15.9 Lenses with a nonreflective coating appear colored. Why?

15.10 Two glass plates 10 cm long touch at one end and are separated by a wire 0.05 mm in diameter at the other end. If light of wavelength 500 nm falls normally on the glass plates, how many fringes will be seen?

15.11 An oil film of refractive index 1.2 is floating on water. It is found that the film produces constructive interference for red light ($\lambda = 750$ nm) and destructive interference for green light ($\lambda = 500$ nm). What is the thickness of the oil film? The refractive index of water is 1.33.

15.12 A diffraction grating has 15,000 lines per inch. Calculate the angular separation between red light ($\lambda = 650$ nm) and yellow light ($\lambda = 520$ nm) in the second order. Will the dispersion increase with the order of the spectrum?

15.13 A diffraction grating has 12,000 lines per cm. Calculate the angle at which the 1st and 2nd peaks occur for a monochromatic light of wavelength 400 nm.

15.14 Calculate the energy of a photon of wavelength 200 nm.

15.15 How many photons cross an area of 1 m^2 in 1 s if the light energy flow is 100 W/m^2? The wavelength of the light is 200 nm.

15.16 Obtain the threshold frequency for a metal whose work function is 1.6×10^{-19} J. Calculate the energy of electrons released from this metal when 200 nm wavelength light shines on the metal.

15.17 Design a circuit to open and close a door containing a photoelectric tube. The door opens when a beam of light is interrupted. (Do this problem after studying the electricity chapters.)

Figure 15.22. Problem 15.6.

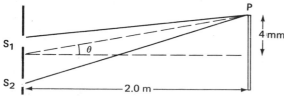

16

Reflection, Refraction, Mirrors, and Lenses

In this chapter light is considered as energy transmitted in straight lines and as such is called rays of light. During its travel it may experience reflection and/or refraction at an interface between two media. In both reflection and refraction the direction of a ray of light changes according to certain laws. In this chapter along with the discussion of reflection and refraction, basic principles of mirrors and lenses are also discussed.

16.1 WAVES AND RAYS

Spherical wavefronts [Fig. 16.1(a)] represent the light waves emitted by the source S. As explained in the last chapter (page 257), Huygens' principle can be used to obtain subsequent wavefronts from the preceding one. The lines S1, S2, etc. represent the direction in which energy is radiated from the source. These lines representing the energy flow are called rays of light. A bundle of rays in the same direction is called a beam of light [Fig. 16.1(b)].

16.2 TYPES OF REFLECTION

Consider a reflecting surface which is smooth like the surface of a plane mirror. All rays of light in a parallel beam proceed in the same direction after reflection from this mirror. Such a reflection is illustrated in Fig. 16.2(a) and is called specular reflection. However, if the reflecting surface is not smooth, reflected rays of light do not proceed in the same direction [Fig. 16.2(b)]. This case, most often found in our daily life, is known as diffuse

Wavefronts

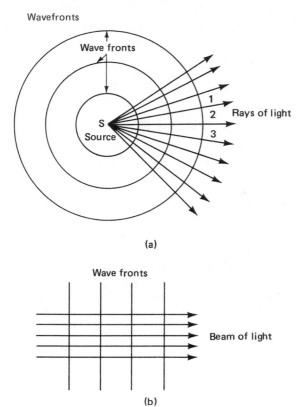

Figure 16.1. (a) Wavefronts and rays of light. (b) A beam of light consists of a bundle of light rays.

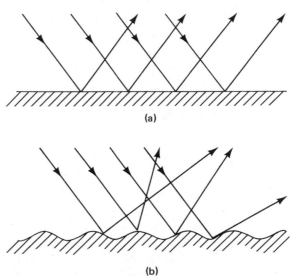

Figure 16.2. (a) Specular reflection; (b) diffuse reflection.

scattering. Surfaces capable of specular reflection can be produced by evaporation of silver or aluminum on a clean and smooth piece of glass or by polishing metallic surfaces.

With the help of the above definitions, we can state the laws of reflection. They are as follows:

(a) Angle of incidence is equal to the angle of reflection, that is $\theta_i = \theta_r$. (Since the two angles are always equal the subscripts i and r will not be used from now on).

(b) The incident ray of light, the normal, and the reflected ray of light all lie in the same plane. In other words, the lines IO, ON, and OR are all lying in the plane of the paper in Fig. 16.3.

Figure 16.3. Reflection of an incident ray of light IO from a mirror MM. OR is the reflected ray of light and ON the normal to the mirror. The angle of incidence θ_i and angle of reflection θ_r are equal.

16.3 LAWS OF REFLECTION

Figure 16.3 illustrates the reflection of a ray of light from a plane mirror. An incident ray of light IO falls on a plane mirror MM. The line OR represents the reflected ray of light. The line ON is the normal to the surface MM. Angle ION is the angle of incidence θ_i and angle NOR is the angle of reflection θ_r.

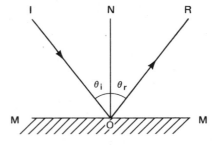

16.4 IMAGE FORMATION BY A PLANE MIRROR

When a person looks at a plane mirror, he sees his image behind the mirror. How is the image formed? We can answer this question with the help of Fig. 16.4. Two rays of light AO_1 and AO_2 start from point A of the object and after reflection, reach the observer's eye. For the eye these rays appear to come from point A'. Point A' therefore is the image of point A. Similarly light rays starting from other points of the object produce corresponding image points.

Since A' is the image of A, all rays of light originating from A and falling on the mirror will appear to come from A'. The ray of light AM perpendicular to the mirror after reflection will also appear to originate from A'. From the figure it is also clear that the two triangles AO_1M and $A'O_1M$ are congruent. Therefore the distances AM and A'M are equal; that is, the distance of the image from the mirror is equal to the distance of the object from the mirror.

In fully describing an image, in addition to the image distance, certain other characteristics of the image should also be given. They are briefly discussed below.

Images undergo inversion in plane mirrors. For example, the image formed by the plane mirror has

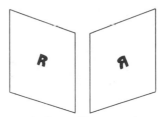

Figure 16.5 This figure illustrates lateral inversion of an image formed by a plane mirror.

lateral inversion as shown in Fig. 16.5. Sometimes one will see images with up-down inversions. (See the sections on lenses and spherical mirrors.)

Magnification, defined as the ratio of the height of the image to the height of the object, is another quantity usually of significance in describing an image. In the case of plane mirrors, the magnification is one, that is, the image size equals the object size.

As mentioned earlier light rays only "appear" to come from the image formed by the plane mirrors. Such an image is called a virtual image. On the contrary, if rays of light actually form the image, we have a real image. Since the real image is formed by rays of light, the image can be projected onto a screen by placing a screen at the image position.

Some interesting optical effects can be seen by using two or more mirrors. Figure 16.6(a) shows the four images formed by two mirrors kept at a 60° angle. In general the number of images is given by $360°/\theta - 1$, where θ is the angle between the mirrors. For example, if $\theta = 90°$ there are 3 images, if $\theta = 30°$, 11 images. Note that the given relationship is valid when 360° divided by θ gives an integer. Figure 16.6(b) shows the images obtained by reflections in two mirrors arranged so that the angle between them is small. Such an arrangement is called a kaleidoscope.

Figure 16.4. Formation of an image in a plane mirror.

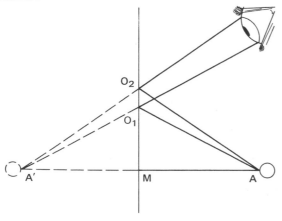

16.5 ROTATION OF A MIRROR

If a mirror is rotated through an angle ϕ, keeping the direction of the incident ray of light the same,

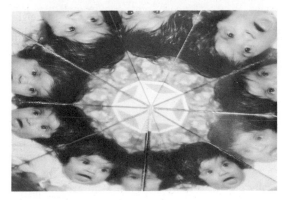

Figure 16.6 (a) Five images are obtained in two mirrors kept at 60°. (b) Symmetrical images are produced by a kaleidoscope.

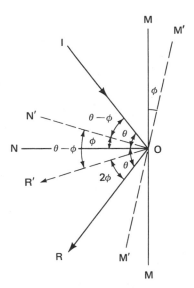

Figure 16.7. A reflected ray of light rotates through an angle twice as big as the angle of rotation of the mirror. M'M' indicates the new position of the mirror, ON' the new direction of the normal, and OR' the new direction of the reflected ray.

the reflected ray of light will change its direction. We are interested in finding the angle ROR' (Fig. 16.7) through which the reflected ray has rotated. Let MM represent the original position of the mirror and ON its normal. As the mirror is rotated to its new position, the normal rotates from its original position to a new position ON'. The angle NON' is ϕ. The new angle of incidence, the angle between the incident ray and the new normal N' is $\theta - \phi$, and therefore the angle between the incident ray of light and the new direction of the reflected ray of light is $2(\theta - \phi)$. Since 2θ is the angle between the incident and reflected rays in the original position, one can see that the reflected ray rotates through

2ϕ when the mirror rotates through ϕ.

An optical arrangement such as the one in Fig. 16.8, where a mirror is attached to a rotating system, is very often used to measure small rotations of the system. The rotations are measured by noting the changes in the position of the reflected beam on a transluscent scale. In a typical arrangement a circular scale of radius R is kept at a distance R from the mirror. When the mirror rotates through ϕ the position of the reflected beam changes by a distance of $2\phi R$. The sensitivity of this system depends on the mirror to scale distance. It is interesting to note that this arrangement can be considered as an amplifier of rotation with an amplification factor of 2.

16.6 CONCAVE AND CONVEX MIRRORS

In the preceding sections we discussed plane reflecting surfaces. In this section we are concerned with curved reflecting surfaces. Of special interest

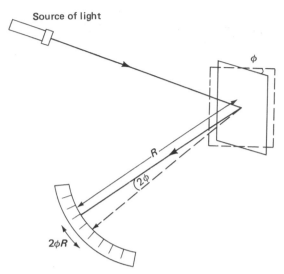

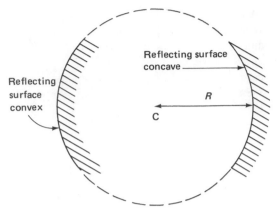

Figure 16.8. Optical arrangement to measure small rotations.

Figure 16.9. Concave and convex mirrors form sections of a mirror: for a concave mirror, the inside surface is reflecting, and for a convex mirror, the outside surface is reflecting. C is the center of the sphere and R the radius.

are concave and convex mirrors.

Reflecting surfaces of both concave and convex mirrors are sections of a spherical surface. If the inner surface of the section is reflecting, it will be a concave mirror; if the outer surface is reflecting, it will be a convex mirror (Fig. 16.9). Advantages of parabolic mirrors over mirrors formed by spherical

surfaces are discussed on page 289.

We now define a few terms in common usage in problems involving mirrors. The point V which is the mid-point of the mirror in Fig. 16.10 is called the vertex. The point C is the center of curvature of the mirror.

Figure 16.10. Figure indicates the axis, the center of curvature C, the focal point F, the vertex V, and the aperture AA' of spherical mirrors.

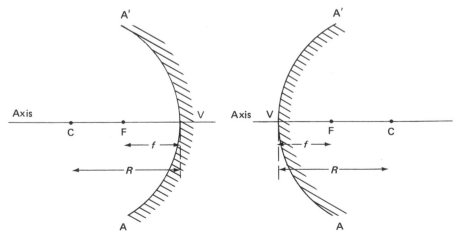

A line drawn through V and C is denoted as the axis or the principal axis and the distance VC is the radius of curvature, usually represented by the letter R. The point midway between V and C is the focal point (proof of this statement and further discussion is given below) and the distance VF is the focal length of the mirror. AA', the diameter of the circle defining the surface of the mirror is called the aperture of the mirror.

Figure 16.11. Reflection of a parallel beam of light from (a) concave and (b) convex mirrors.

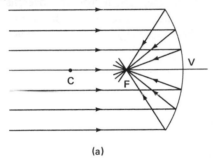

(a)

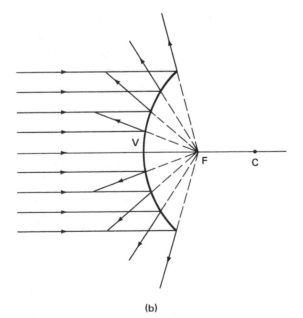

(b)

Concave surfaces focus a parallel beam of light to the focal point and hence are called converging mirrors (Fig. 16.11). However, convex surfaces cause a parallel beam of light to diverge such that the beam appears to the observer to originate from a virtual point, the focal point, behind the mirror and, therefore, are called diverging mirrors.

Now let us derive the relationship between the radius of curvature R and focal length f. In Fig. 16.12, a ray of light SP parallel to the axis CV is reflected at P. Since C is the center of curvature, CP is the normal to the reflecting surface at P. The reflected ray passes through a point F on the axis and it is by definition the focal point. Also the angles SPC and CPF are equal according to the laws of reflection. The line PV' is drawn perpendicular to the axis.

From Fig. 16.12 we get the trigonometrical relation, $PV' = CV' \tan \theta = FV' \tan 2\theta$. If $PV \ll R$, we can make the following approximations:

$$CV' \simeq CV = R, \quad FV' \simeq FV = f,$$

$$\tan \theta \simeq \theta \quad \text{and} \quad \tan 2\theta \simeq 2\theta.$$

Figure 16.12. The ray of light SP from the source S is reflected at P. Since SP is parallel to the axis, the reflected ray PF goes through the focal point.

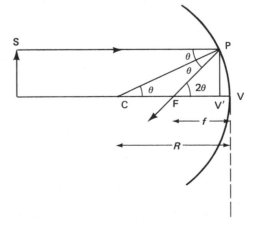

Using these approximations the above equation can be rewritten as

$$PV' = R\theta = f2\theta.$$

Consequently the relationship $R = 2f$ follows. The condition that $PV \ll R$ is equivalent to saying that the aperture of the mirror is very small. If this condition is not satisfied, all the rays of light in a parallel beam will not be focused to the same point, producing what is known as spherical aberration. Details of this are discussed later.

From the discussion above, we see that a parallel ray of light SP (Fig. 16.12) after reflection goes through the focal point (or appears to originate from the focal point in the case of convex mirrors). If we reverse this ray of light, we see that the ray of light passes through the focal point (or appears to go to the focal point) and after reflection goes in the direction parallel to the axis. There is no difficulty in seeing that the laws of reflection are satisfied in this case. Another ray of light, the path of which is easier to trace, is the one which goes through (or appears to go to) the center of curvature of the mirror. Since this ray of light is falling on the mirror normal to its surface, it will retrace its path in the opposite direction.

In geometrically locating the image of an object three rays of light, mentioned above, are usually used. They are (a) a ray of light parallel to the axis, (b) a ray of light passing or appearing to pass through the focal point, and (c) a ray of light that goes to or appears to go to the center of curvature. Examples of image formation for different object distances, for both types of mirrors, are given below. (Fig. 16.13). From these examples it will be clear that only two of the three rays are actually needed to locate the image. The third ray is used to confirm the position of the image.

EXAMPLE 16.1

Locate the image of object OS in Fig. 16.13(a).
The three rays of light selected to locate the image originate from point S of the object and they are, as mentioned earlier, a ray parallel to the axis (ray 1), a second ray passing through the focal point (ray 2), and a third ray passing through the center of curvature (ray 3). From the discussion above, it is easy to trace the paths of these rays of light after reflection. The first ray reflects through the focal point, the second ray reflects parallel to the axis, and the third reflects back along its own path.

The three rays of light after reflection meet at S'. If the aperture of the mirror is small, all rays of light originating from S and, falling on the mirror, essentially pass through S'. Point S' is, therefore, the image of S and O'S' is the image of the object.

The image is real since the light rays actually pass through the image. The image, moreover, has undergone an up–down inversion. Magnification, the ratio of the image height to the object height, is less than 1.

EXAMPLE 16.2

Locate the image of object OS in Fig. 16.13(b).
The difference between this example and Example 16.1 arises from the fact that since the mirror is convex in this example the focal point and the center of curvature are located behind the mirror. After reflection ray 1, the ray parallel to the axis, appears to start from the focal point. Ray 2 which goes in the direction of the focal point gets reflected parallel to the axis. Ray 3 which appears to go to the center of curvature reflects back at the surface of the mirror retracing its own path.

The three rays of light do not meet after reflection, but if they are extended behind the mirror, they meet at S'. For an observer in front of the mirror, the rays of light appear to come from S'. The point S' is the virtual image of S and O'S' is the virtual image of the object OS. The image is erect and the magnification is less than 1. This is true irrespective of the position of the object.

EXAMPLE 16.3

Locate the image of object OS in Fig. 16.13(c).
The procedure we are following here is the same

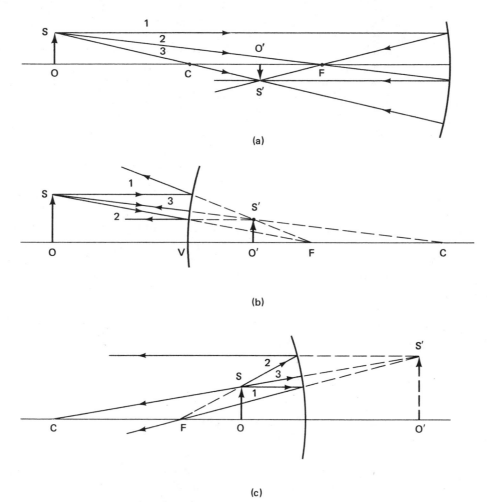

Figure 16.13. Figure illustrates the method of ray tracing to geometrically locate the image O'S' of object OS. (a) Example 16.1; (b) Example 16.2; and (c) Example 16.3.

as in Example 16.1. The object lies within the focal point, that is, between the focal point and the mirror. Here the image is virtual and erect. The magnification is greater than 1.

The method used in the above examples to locate the image is known as ray tracing.

Convex mirrors always form erect virtual images, with magnification less than 1, whereas concave mirrors form both types of images, a real image if the object is placed outside and a virtual image if the object is inside the focal point. Virtual images are generally erect.

16.7 MIRROR EQUATIONS

We have seen a useful equation relating the focal length f and radius R, namely,

$$f = R/2. \tag{16.1}$$

A second equation between magnification M, object distance D_o, and the image distance D_i is derived here. In Fig. 16.13, Example 16.1, consider the triangles OSC and O'S'C. Since the triangles are similar we have the relation

$$\frac{O'S'}{OS} = \frac{D_i}{D_o} \; .$$

But $O'S'/OS$ is the magnification. Therefore it follows that

$$M = \frac{O'S'}{OS} = -\frac{D_i}{D_o} \tag{16.2}$$

Note the minus sign in front of the ratio D_i/D_o, in Eq. (16.2). Its significance is explained later.

In addition to these two equations, if there is an equation relating D_i, D_o, and f, the position of the image can be calculated given the other two quantities D_o and f. Calculating the location of the image will be much easier than ray tracing. The equation we are looking for can be readily derived with the help of Fig. 16.14.

In Fig. 16.14 two rays of light SA and SA′ are used to locate the image O′S′. The rays SA and S′A′ are parallel to the axis OV. Lines AB and A′B′ are drawn perpendicular to the axis. It follows that, since the corresponding angles of the triangles are equal, the triangles OSF and A′B′F are similar and likewise ABF and O′S′F are similar. Therefore,

$$\frac{OS}{A'B'} = \frac{OF}{B'F} \quad \text{and} \quad \frac{AB}{O'S'} = \frac{BF}{FO'} \; .$$

As mentioned earlier SA and S′A′ are parallel to the axis. Hence $A'B' = O'S'$, $OS = AB$.

Figure 16.14. Rays of light SA parallel to the axis and SF passing through the focal point are used to obtain the position of the image O′S′. The figure is used to obtain the mirror equation.

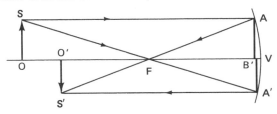

Therefore,

$$\frac{OF}{B'F} = \frac{BF}{FO'} \; . \tag{16.3}$$

If the curvature of the mirror is small, we have

$$B'F \simeq BF \simeq f.$$

Hence within this approximation,

$$FO' = D_i - f \quad \text{and} \quad FO = D_o - f.$$

Substitution of these relations into Eq. (16.3) yields

$$\frac{f}{D_i - f} = \frac{D_o - f}{f}$$

which upon simplification becomes

$$\frac{1}{D_o} + \frac{1}{D_i} = \frac{1}{f} \; . \tag{16.4}$$

Although Eq. (16.4) was derived for concave mirrors, this equation can also be used for convex mirrors. However, we have to follow the sign conventions:

1 All distances measured from the vertex to the points in front of the mirror are positive.

2 All distances measured from the vertex to points in back of the mirror are negative.

3 Heights measured above the axis are positive and below the axis negative.

It is appropriate to comment on the minus sign in Eq. (16.2) at this time. If we use the above sign conventions for the example in Fig. 16.12, magnification M, given by $OS/O'S'$, is negative. Distances D_i and D_o are both positive. Therefore, to get a consistent answer for M we need the negative sign in front of the ratio D_i/D_o.

EXAMPLE 16.4

The radius of a concave mirror is 40 cm. If an object is placed at 50 cm in front of the mirror where will the image be? Describe the image.

We are given that the focal length of the mirror = 40/2 cm = 20 cm and the object distance = 50 cm. Substituting these into the mirror equation, we get

$$\frac{1}{50} + \frac{1}{D_i} = \frac{1}{20},$$

$$D_i = +33.3 \text{ cm}.$$

The image is real. It is formed in front of the mirror. Its magnification is

$$M = -\frac{33.3}{40} = -0.83.$$

The image, moreover, is inverted. This problem is solved graphically in Fig. 16.15.

EXAMPLE 16.5

A concave mirror forms a virtual image at a distance of 10 cm for an object at a distance of 5 cm.

Calculate the radius of the mirror.

The image is virtual which means that the image is behind the mirror, that is, $D_i = -10$ cm. $D_o = 20$ cm. Using the mirror equation, we get

$$\frac{1}{5} + \frac{1}{10} = \frac{1}{f}$$

$$f = 10 \text{ cm}.$$

Radius of the mirror $= 2f = 20$ cm.

EXAMPLE 16.6

An object is kept 30 cm in front of a convex mirror of radius 50 cm. Calculate the position of the image and describe it.

$$f = R/2 = -25 \text{ cm}.$$

Radius and focal length are negative for convex mirrors. $D_o = 30$ cm. Substituting the values of D_o and f into the mirror equation we get

$$\frac{1}{30} + \frac{1}{D_i} = \frac{1}{-25}.$$

Figure 16.15. Example 16.4. First the given information is drawn to scale and the center of curvature and focal points are identified on the figures. After determining the position of the image, its distance is measured. Note that as in Examples 16.1, 16.2, and 16.3, we use three rays of light to locate the image.

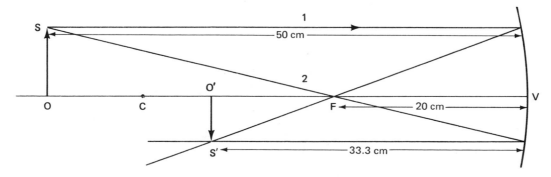

Solving the above equation for D_i yields

$$D_i = -13.6 \text{ cm},$$

$$M = -\frac{-13.6}{30} = +0.45.$$

The image is virtual and erect. This problem is solved graphically in Fig. 16.16.

16.8 REFRACTION

When light passes from one medium to another medium, its direction changes at the boundary. Figure 16.17 illustrates an example, where a ray of light changes its direction at the boundary between air and glass. Glass is considered optically denser than air and in such cases the ray of light bends towards the normal. In Fig. 16.17 θ_1 is the angle between the incident ray of light and the normal NN to the boundary, and θ_2 is the angle between the refracted ray and the normal. The relationship between these angles is given by

$$\mu_1 \sin \theta_1 = \mu_2 \sin \theta_2, \qquad (16.5)$$

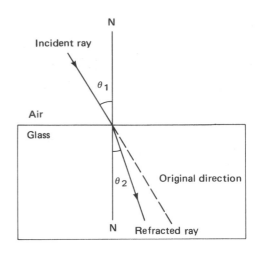

Figure 16.17. This figure illustrates the change in the direction of a ray of light at the boundary between air and glass. The change in direction is called refraction.

where μ_1 and μ_2 are called the refractive indices of the two media each relative to a vacuum (see Table 16.1 for examples). The refractive index is a property of the medium that varies with wavelength of the light and it is the ratio of the velocity of light in vacuum to the velocity of light in the medium. The refractive index of air is equal to 1.003 which is

Figure 16.16. Example 16.6.

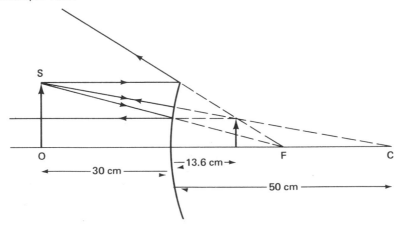

TABLE 16.1 Index of Refraction μ, for Various Substances, for Sodium Yellow Light (589 nm).

SUBSTANCE	INDEX
Diamond	2.417
Polycarbonates (Material used for eyeglasses)	1.60
Glass,	
Ordinary crown	1.517
Light flint	1.575
Dense flint	1.656
Ice at $-8°C$	1.31
Quartz, fused	1.458
Ethyl alcohol at $20°C$	1.360
Water	
at $0°C$	1.334
at $20°C$	1.333
at $40°C$	1.331
at $60°C$	1.327
Air	1.000293
Carbon dioxide	1.000450

almost equal to the refractive index of a vacuum with a value of 1. Hence in problems involving the refractive index of air, its refractive index may be taken as 1 without much error ($\sim 0.3\%$).

It is also true that the incident ray of light, the refracted ray of light, and the normal to the boundary lie in the same plane. This statement and Eq. (16.6) are known as Snell's laws of refraction. (Throughout this chapter we follow the convention that the quantities with subscripts 1 stand for the first medium and quantities with subscript 2 for the second medium.)

As mentioned earlier the speed of light changes when light travels from one medium to another medium. Using this fact and Huygens' principle, we construct the wavefronts on both sides of the interface as shown in Fig. 16.18. In a time t, the point B reaches C in the first medium while point A at the

interface reaches point D in the second medium. The distance $BC = V_1 t$ and the distance $AD = V_2 t$ where V_1 and V_2 are speeds of light in the two respective media. Hence from the triangles ABC and ACD, we obtain

$$\sin \theta_1 = V_1 t / AC$$

and

$$\sin \theta_2 = V_2 t / AC.$$

Hence

$$\frac{\sin \theta_2}{\sin \theta_1} = \frac{V_2 t}{V_1 t} = \frac{V_2}{V_1}$$

From (16.5) we get

$$\frac{\sin \theta_1}{\sin \theta_2} = \frac{\mu_2}{\mu_1} = \frac{V_1}{V_2}. \qquad (16.6)$$

If the first medium is a vacuum $V_1 = c$ and $\mu_1 = 1$. In this case the refractive index of the second medium with respect to the vacuum is equal to the ratio of the speed of light in vacuum to the speed of light in the medium. *The refractive index of a material is the ratio of the speed of light in vacuum to the speed of light in the material.*

EXAMPLE 16.7

If light falls on the surface of water and the angle of incidence is $60°$, how much does the light beam bend at the interface? (See Fig. 16.19.) The refractive index of water $= 1.33$.

We use Snell's equation, namely,

$$\mu_1 \sin \theta_1 = \mu_2 \sin \theta_2.$$

Since $\mu_1 = 1, \theta_1 = 60°$, and $\mu_2 = 1.33$, we have

$$\sin \theta_2 = \frac{\sin 60°}{1.33} = 0.65.$$

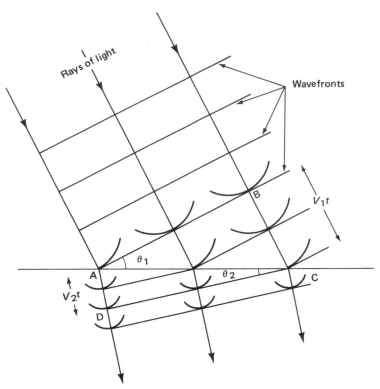

Figure 16.18. Change in the velocity of light in the second medium causes refraction. When the light travels a distance BC ($=V_1 t$) in the first medium it travels a distance of AD ($=V_2 t$) in the second medium.

Hence, $\theta_2 = 41°$. The angle through which the direction of the refracted ray changed from the original direction is $\theta_1 - \theta_2 = 60° - 41° = 19°$.

Figure 16.19. Refraction at air–water interface.

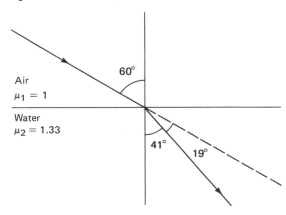

16.9 TOTAL INTERNAL REFLECTION

If the first medium is optically denser (i.e., of higher refractive index) than the second medium, the refracted light moves away from the normal (Fig. 16.20). Angle θ_2 increases as θ_1 increases and it is equal to $90°$ for a certain angle of incidence called the critical angle. The critical angle can be calculated for any medium using a relationship derived from Snell's law:

$$\mu_1 \sin \theta_1 = \mu_2 \sin \theta_2 .$$

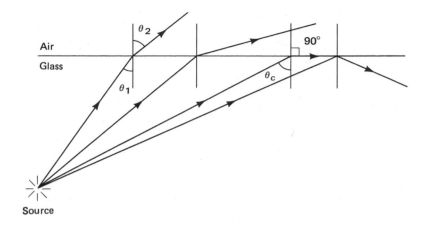

Figure 16.20. Figure shows that as light emerges from a dense medium to a less–dense medium, the rays bend away from the normal. For an angle θ_c, angle of refraction equals $90°$. For angles above θ_c, light does not emerge into the less–dense medium, instead total internal reflection takes place at the boundary.

Since $\mu_2 = 1$ and $\theta_2 = 90°$, we have

$$\mu_1 \sin \theta_c = 1,$$

or

$$\sin \theta_c = 1/\mu_1 .\qquad\text{(16.7)}$$

If the angle $\theta_1 \geqslant \theta_c$ light will not be transmitted to the second medium. Instead it will be totally internally reflected at the boundary. Total internal reflection obtained under this condition is called "total internal reflection." In the derivation of Eq. (16.7), the second medium is taken as air and therefore it is suitable only for calculating the critical angles for a denser-medium to air interface.

EXAMPLE 16.8

What should be the refractive index of the material of the prism (Fig. 16.21) to get total internal reflection for a ray incident perpendicular to the right face of the prism?

The angle of incidence θ_1 at the boundary where total reflection is expected is $45°$. To get total reflection this angle should be at least equal to

the critical angle. Therefore using Eq. (16.6), we obtain the refractive index of glass,

$$\mu_1 = \frac{1}{\sin 45°} = 1.41.$$

Note that there is no refraction of light where the light enters the prism and where it leaves the prism

Figure 16.21. Total internal reflection in a prism. Such prisms are used in binoculars and periscopes.

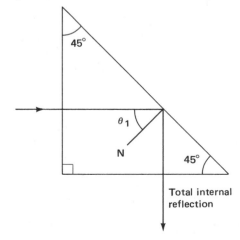

because in both instances the angle of incidence is zero. Such prisms are used whenever one needs to change the direction of light through 90° without loss of intensity. For example, periscopes and binoculars use totally reflecting prisms.

Optical Fibers

Optical fibers find wide uses in various fields of science and industry. Optical fibers, also known as light pipes, use total internal reflections to transmit light from one point to another (Fig. 16.22). The fibers, sometimes made of quarts, are so thin that they can be bent to "pipe" light around corners. Bundles of optical fibers are used in medical instruments for observation of internal organs.

16.10 PRISMS

Prisms are found in several optical instruments. The prism is made of transparent material, usually glass, and it has a triangular cross-section (Fig. 16.23). Light refracts at the first surface as it enters the prism and at the second surface as it leaves the prism. The total angle of deviation of a light ray, D, can be calculated by first obtaining the angles of deviation at both surfaces using Snell's laws.

EXAMPLE 16.9

Calculate the angle of deviation for red light (refractive index = 1.455) passing through a quartz prism (Fig. 16.24). The angle of the prism, A, is 30° and the angle of incidence is 20°.

Figure 16.22. Figure illustrates the transmission of light through optical fibers. Internal reflections keep the light inside the fiber and light emerges only through the end of the fiber.

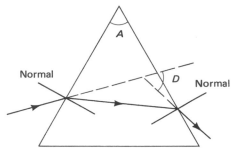

Figure 16.23. Figure shows the path of a ray of light through a prism. Angle D is the angle of deviation between the incident ray of light and refracted ray of light.

Angle θ_2 can be calculated by using Snell's equation:

$$\sin 20° = 1.455 \sin \theta_2$$

$$\sin \theta_2 = \frac{\sin 20°}{1.455} = 0.235$$

$$\theta_2 = 13°.$$

From Fig. 16.24 it can be seen that the angle of incidence at the second surface, θ_1', is equal to 17°. The angle θ_2' can be calculated by again using Snell's equation

$$1.455 \sin 17° = \sin \theta_2'$$

$$\theta_2' = 25°.$$

Angle of deviation at the first interface = 7°;

Angle of deviation at the second interface = 8°;

Total angle of deviation

= Sum of deviations at the two interfaces = 15°.

EXAMPLE 16.10

Calculate the angle of deviation for yellow light (refractive index = 1.470), everything else remain-

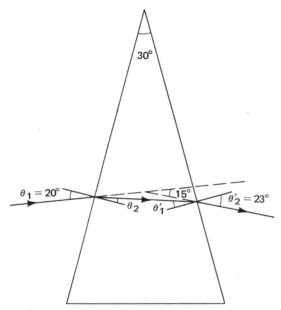

Figure 16.24. A quartz prism with an angle of 30°. Figure shows the refractions at both boundaries for an incident angle of 20°. Angle of deviation in this case is 15°.

ing the same as in Example 16.9.

Following the steps as in Example 16.9, one gets for the total angle of deviation, 1°.

Dispersion

From the above two examples, one can see that the prism is capable of separating the colors of light. Separation of the constituent colors of light is called dispersion. When a beam of white light, which in reality is composed of many colors, falls on a prism, the emerging beam is dispersed into a brilliant array of colors forming a spectrum. The major components are red, orange, yellow, green, blue, and violet in the order of increasing deviation as indicated in Fig. 16.25.

16.11 LENSES

Lenses, the most common components in optical instruments, are made of transparent materials and are usually defined by smooth spherical surfaces. There are two types of lenses—converging and diverging lenses. Typical shapes of commonly found lenses are shown in Fig. 16.26.

Figure 16.27 illustrates the behavior of a parallel beam of light passing through thin lenses. A converging lens [Fig. 16.27(a)] focuses a parallel beam of light, whereas in the case of a diverging lens the beam diverges out and appears to originate from a

Figure 16.25. Dispersion of light by a prism.

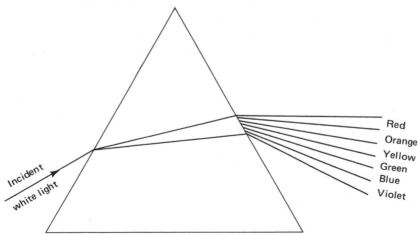

Figure 16.26. Typical lens configurations. The first group of converging lenses consists of (a) double-convex, (b) plano-convex, and (c) concave-convex lenses. The second group of diverging lenses consists of (d) double-concave, (e) plano-concave, and (f) convex-concave lenses. In each case, the names are derived from the shape of the refracting surfaces.

point. Figure 16.27 is also used to define the common terms associated with thin lenses. Point F_1, where the parallel beam focuses (or appears to focus) is the focal point and the distance VF_1 along the axis is the focal length. Since light can enter the lens from either side there are two focal points for a lens. The focal lengths VF_1 and VF_2 are always equal, even when the two surfaces of the lens are not symmetrical (e.g., a plano-convex lens). Also note that flipping the lens does not change focal lengths.

Ray tracing techniques can be used to locate the image produced by a lens. The procedure we follow is somewhat similar to that which we used for spherical mirrors. Three rays of light, usually selected for this purpose, are (a) a ray of light parallel to the axis, (b) a ray passing through the focal point F_2 (or appearing to go to F_2), and (c) a ray passing through the center of the lens. The ray passing through the center proceeds without any deviation. A few examples are given below.

The image formed by a converging lens when the object is beyond the focal point [Fig. 16.28(a)] is real and inverted. Its magnification is less than 1. The image formed by a diverging lens [Fig. 16.28(b)] is virtual, rays of light proceeding to the viewers side do not meet, but appear to start from point S'. The image formed by a converging lens when the object is within the focal point [Fig. 16.29(a)] is virtual, erect, and magnified. The image in Fig. 16.29(b) is virtual and erect and its magnification is less than 1.

Diverging lenses always produce virtual images, while converging lenses produce virtual images only if the object is within the focal point. In the latter case the image is also magnified.

16.12 LENS EQUATIONS

The focal length is a function of the refractive indices, radius r_1 of the first surface (surface nearer to the object), and radius r_2 of the second surface (Fig. 16.30). Note that C_1 and C_2 are the center of curvature of the two surfaces. The equation connecting these quantities, known as the lens maker's formula, is

$$\frac{1}{f} = \left(\frac{\mu_2 - \mu_1}{\mu_1}\right)\left(\frac{1}{r_1} - \frac{1}{r_2}\right), \qquad (16.8)$$

where μ_2 is the refractive index of the glass and μ_1 is that of the medium outside the lens. For air $\mu_1 \simeq 1$. The equation for magnification is

$$M = \frac{O'S'}{OS} = -\frac{D_i}{D_o} \qquad (16.9)$$

and the relationship between D_i, D_o, and f is

$$\frac{1}{D_o} + \frac{1}{D_i} = \frac{1}{f}. \qquad (16.10)$$

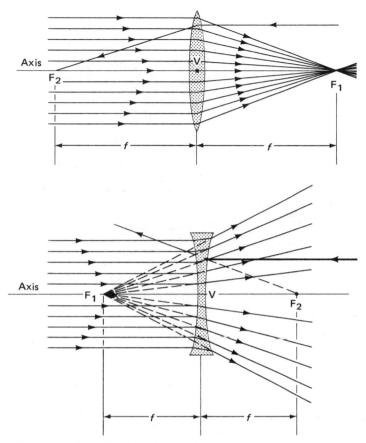

Figure 16.27. Figure illustrates the behavior of a parallel beam of light passing through a converging and a diverging lens.

Equations (16.9) and (16.10) are identical to the ones we used for mirrors and they can be derived in a somewhat similar manner. The sign assignments for the distances are a little more complicated in the case of lenses compared to mirrors. In the case of a mirror, for object and image, the side is the front of the mirror while the virtual side is the back of the mirror. Distances measured to points in front are considered positive and distances to points in back of the mirror are considered negative. For a lens, because of refraction, a real image forms on the opposite side of a real object. When combinations of lenses are used, the image formed by one lens becomes a real or a virtual object for the second

lens. (See Example 16.13.) The conventions we will adopt are the following:

1 Real object distances are always taken as positive.

2 Distances measured from the center of the lens to points lying on the real image side (side opposite to real object) are positive and distances measured to the virtual image side are negative. These sign assignments should be followed for radii of lens surfaces: r_1 in Fig. 16.30 is positive while r_2 is negative.

3 Consistent with the above two conventions, focal lengths are positive for a converging lens

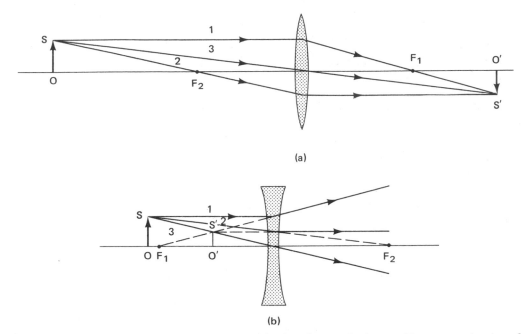

Figure 16.28. Figure illustrates the ray tracing method to locate the image. The converging lens forms a real inverted image and the diverging lens forms a virtual image.

and negative for a diverging lens. If the above conventions are followed, magnification is negative when the image is real and magnification is positive when the image is virtual.

4   Heights measured above the axis are positive while those measured below are negative.

EXAMPLE 16.11

A plano-convex lens (Fig. 16.31) is made of glass. The refractive index of glass is 1.5. The radius of the convex surface is 50 cm. If an object is placed 50 cm away from the lens, calculate the position of the image and describe it.

First we calculate the focal length using Eq. (16.6):

$$\frac{1}{f} = (\mu - 1)(\frac{1}{r_1} - \frac{1}{r_2}).$$

$\mu = 1.5, r_2 = -30$ cm, $r_1 = +\infty$.

$$\frac{1}{f} = (1.5 - 1)(\frac{1}{\infty} - \frac{1}{-30})$$

$$= 0.5\,(0 + \frac{1}{30})$$

$$f = +60 \text{ cm}.$$

We have calculated the position of the focal point F_1; it is 60 cm from the lens on its right-hand side. The focal point F_2 lies on the left-hand side of the lens and it is the point where a parallel beam of light coming from the right-hand side focuses. Hence to obtain the position of F_2 one has to assume the object is on the right-hand side and therefore, according to our conventions, distances measured to the right-hand side are negative and distances measured to the left-hand side are positive. Now we have $\mu = 1.5$, r_1 (radius of the surface close to the object) $= +30$ cm, and $r_2 = -\infty$. Substituting these

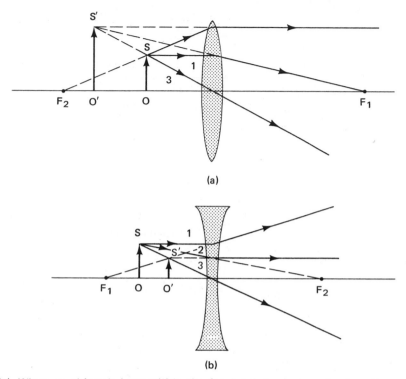

(a)

(b)

Figure 16.29. (a) When an object is kept within the focal point, a converging lens forms a virtual image; (b) diverging lens always forms a virtual image.

values into the lens equation, we obtain

$$\frac{1}{f} = 0.5(\frac{1}{30} - \frac{1}{-\infty})$$

$$f = +60 \text{ cm.}$$

The focal point F_2 is 60 cm from the lens on the left-hand side. This result supports the statement "the focal lengths are always equal even when the two surfaces of the lens are not symmetrical."

Figure 16.30. Figure defines the radius of curvature r_1 and r_2 of the first and second surfaces of the lens. C_1 and C_2 are the centers of curvature of the surfaces.

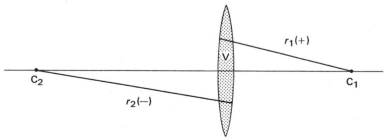

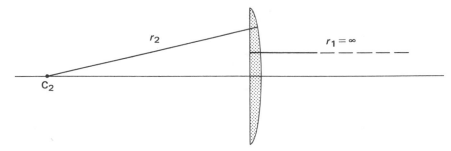

Figure 16.31. A plano-convex lens. Radius r_1 is infinite in this case.

To obtain the position of the image we use Eq. (16.10),

$$\frac{1}{D_o} + \frac{1}{D_i} = \frac{1}{f},$$

where $D_o = +50$ cm (given) and $f = 60$ cm. Therefore, it follows that

$$\frac{1}{50} + \frac{1}{D_i} = \frac{1}{60}$$

$$D_i = -30 \text{ cm}.$$

$$M = -\frac{-30}{+50} = +0.6.$$

The image is virtual and inverted.

EXAMPLE 16.12

A double concave lens with $\mu = 1.6$, $r_1 = 20$ cm, and $r_2 = 30$ cm is used to view an object kept 15 cm from the lens. Where is the image seen? Discuss the image.

We are given $\mu = 1.6$, $r_1 = -20$ cm, and $r_2 = +30$ cm. Substituting these into the lens maker's formula, we get

$$\frac{1}{f} = (1.6 - 1)(\frac{1}{-20} - \frac{1}{+30})$$

$$f = -20 \text{ cm}.$$

The lens is diverging.

Now we calculate the image distance using the lens equation and the given value of D_o.

$$\frac{1}{15} + \frac{1}{D_i} = \frac{1}{-20}$$

$$D_i = -8.6 \text{ cm}.$$

$$M = -\frac{-8.6}{15} = +0.57.$$

The image is virtual and erect.

In most of the sophisticated optical instruments one finds lens systems made up of combinations of two or more lenses. It will be illuminating to discuss a problem for such an arrangement. In these cases the image formed by the first lens acts as the object for the second lens.

EXAMPLE 16.13

Two lenses of focal lengths $+30$ cm and -60 cm are kept 20 cm apart. If the object is 60 cm from the first lens, where is the final image? Also calculate the total magnification and describe the image (Fig. 16.32).

For the first lens $f_1 = +30$ cm and $D_{o1} = +60$ cm. Substituting these values into the lens equation, we obtain

$$\frac{1}{60} + \frac{1}{D_{i1}} = \frac{1}{30}$$

$$D_{i1} = 60 \text{ cm}; M_1 = \frac{-60}{60} = -1.$$

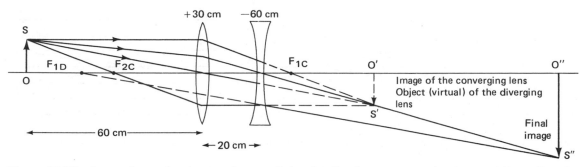

Figure 16.32. Image formation by two lenses. Subscript C refers to converging lens and D to diverging lens. The image formed by the first lens O′S′ acts as the object of the second lens.

This image formed by the first lens can be considered as the object for the second lens. It is a virtual object since the image, ignoring the refraction of the second lens, would be located on the right-hand side of the second lens, that is, D_{o2} is negative.

$$D_{o2} = -(60 - 20) = -40 \text{ cm}, \quad f_2 = -60 \text{ cm}.$$

Substituting the object distance and the focal length into the lens equation, we obtain

$$\frac{1}{-40} + \frac{1}{D_{i2}} = \frac{1}{-60}$$

$$D_{i2} = 120 \text{ cm}; \quad M_2 = -\frac{120}{-40} = 3.$$

The total magnification is given by

$$M = O_2'S_2'/OS = M_1 M_2 = -3.$$

The final image is real and inverted. Note in some problems the image produced by the first lens may fall between the two lenses. The object distance for the second lens D_{o2} is positive in such cases.

16.13 DEFECTS

Spherical mirrors and lenses share a common defect called spherical aberration. This occurs be-

cause the focal point for parallel rays changes as the distance from the axis of the parallel rays changes, the focal point moving closer to the lens for larger distances (Fig. 16.33). Spherical aberration is easily corrected in mirrors by changing their shape from spherical to parabolic. To minimize this aberration in photography it is good practice wherever feasible to reduce the size of the aperture and thus eliminate those rays farthest from the axis.

Another defect of a single lens which is absent in a mirror is chromatic aberration. A lens is close to a biprism in its shape (Fig. 16.34). When light is refracted through the lens, dispersion takes place. Consequently, the focal points for the different colors are different. The position of the image and its magnification depends on the focal length. The focal length is shorter for the blue light compared to the focal length for red light. Therefore the blue image will be less magnified and will be closer to the lens. A slightly exaggerated example illustrating these effects due to chromatic aberration is given below.

EXAMPLE 16.14

A lens has a focal length of 26 cm for red light and 25.5 cm for blue light. If an object is placed 30

Figure 16.33. Spherical aberration. Light farther from the axis focuses closer to the lens (mirror).

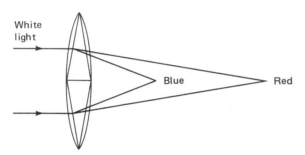

Figure 16.34. Chromatic aberration. Lens is close to a biprism in shape. The blue light is closer to the lens than the red light.

cm from the lens find the positions and magnifications of blue and red images.

For the red image, we have

$$\frac{1}{30} + \frac{1}{D_{ir}} = \frac{1}{26}$$

$$D_{ir} = 195 \text{ cm};$$

$$M_r = -\frac{195}{30} = -6.5.$$

For the blue image, we have

$$\frac{1}{30} + \frac{1}{D_{ib}} = \frac{1}{25.5}$$

$$D_{ib} = 170 \text{ cm};$$

$$M_b = -\frac{170}{30} = -5.6.$$

Note that the subscripts r and b stand for red and blue lights.

Chromatic aberration can be corrected by combining a converging lens and a diverging lens as shown in Fig. 16.35. The first lens is made of crown glass and the second lens of flint glass. The converging lens bends blue light inward more than red light whereas the diverging lens bends blue light outward more than red light. In this way the combination can reduce the chromatic aberration and hence it is called an achromatic combination.

16.14 BOUNCING LIGHT BEAM, MIRAGE, AND LOOMING

A bouncing light beam (Fig. 16.36) is an interesting demonstration of refraction and total internal reflection. To demonstrate this effect, a sugar solution with a concentration gradient, that is, concentration decreasing slowly from bottom to top, has to be prepared. This is accomplished by pouring sugar into the water. The sugar first settles down at the bottom. As the sugar dissolves in the water, the solution slowly diffuses to the top. After a few hours, one has a sugar solution of desired density gradient. The density variation produces variation in the refractive index which also decreases continuously from bottom to top.

To understand the effect shown in Fig. 16.36, let us consider the case where the density and hence the refractive index decreases in steps from layer to layer (Fig. 16.37). As the light goes from a more dense to a less dense layer (points 1 and 2), refraction takes place and bends the light beam away

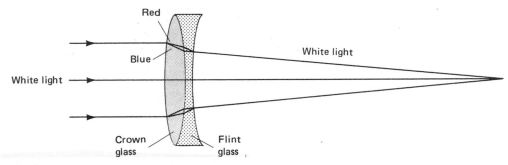

Figure 16.35. Achromatic combination of lenses.

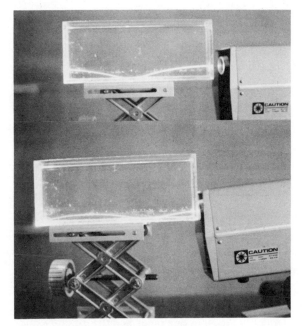

Figure 16.36. Bouncing light beam.

from the normal. At point 3 conditions are satisfied for a total internal reflection. From that point on, as the light goes from a less dense to a denser medium, refraction causes the beam to bend towards the normal. Since in the actual solution the density variation is continuous, the beam will form a continuous curve. As the beam hits the bottom, reflection takes place and the refraction process produces another arch, if conditions are favorable.

Natural phenomena such as mirage and looming occur because of the density variation in air as a result of temperature variations. On hot sunny days, the temperature near the ground is higher than the temperature above. Therefore, the density and the refractive index of air near the ground are lower than the density and refractive index of air farther above the ground. Density variation is continuous but in the opposite direction to what we found in the sugar solution. Therefore the light rays bend away from the ground, and to the eye it appears that the rays are coming from the inverted image

Figure 16.37. Density of the layers decreases from bottom to top. Refraction of light at points 1 and 2 bends the beam away from the normal. At point 3 total internal reflection takes place. Continuous density variation in a real solution will produce a continuously curving light beam as in Fig. 16.36.

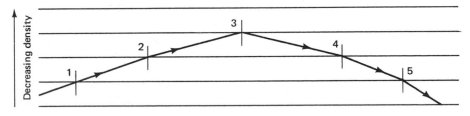

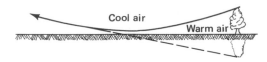

Figure 16.38. Mirage.

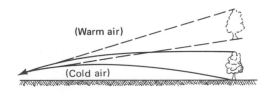

Figure 16.39. Looming.

below the ground (Fig. 16.38). The effect, known as mirage, is as if the distant objects are reflected on a quiet lake surface.

Looming sometimes occurs when the surface air is cooler than the air above. The bending of rays in this case is similar to that in the sugar solution. The result of looming is shown in Fig. 16.39.

SUMMARY   Reflection from a mirror obeys the following laws: (a) The incident ray, the normal, and the reflected ray all lie in one plane; (b) the angle of incidence equals the angle of reflection.

There are two types of curved mirrors: The concave mirror which focuses a parallel beam to a focal point, and the convex mirror, in which a parallel beam after reflection appears to diverge from a focal point behind the mirror.

In both cases, the magnitude of the focal length is one half of the radius of curvature. The equation between the image and object distances is

$$\frac{1}{D_o} + \frac{1}{D_i} = \frac{1}{f} = \frac{2}{R}.$$

At the boundary of two media of differing refractive indices, refraction or bending of the beam takes place. Snell's laws for refraction are (a) the incident ray, the normal, and the refracted ray are all in one plane and (b) $\mu_1 \sin \theta_1 = \mu_2 \sin \theta_2$ where μ_1 and μ_2 are the refractive indices and θ_1 and θ_2 are the angles the beam makes in the two media, respectively. When $\mu_1 < \mu_2$, $\theta_2 < \theta_1$, that is, the beam bends towards the normal in going from a less dense medium to a denser medium. On the other hand, for light going to a lighter medium, the beam bends away from the normal. In such a case, if the angle of incidence exceeds a certain value known as the critical angle, the light is totally reflected at the interface. The critical angle is given by the equation (when the second medium is air) $\sin \theta_c = 1/\mu_1$. Optical fibers that can carry light over a large distance work because of total reflection inside the fibers.

Separation of colors by a prism is known as dispersion.

Two types of lenses are the converging lens which focuses a parallel beam to a focal point, and the diverging lens, from which a parallel beam emerges diverging. In this case the beam appears to diverge from a focal point on the side of the lens where the light is incident. For both lenses, the relationship between object distance D_o, image distance D_i, and the focal length f is

$$\frac{1}{D_o} + \frac{1}{D_i} = \frac{1}{f}.$$

Magnification is given by $M = -D_i/D_o$.

After studying this chapter the student should be aware of the properties of images (real, virtual, inverted, etc.), the sign conventions associated with mirrors and lenses, and aberrations of mirrors and lenses.

QUESTIONS AND PROBLEMS

16.1   Is it possible to project a virtual image on a screen?

16.2   When you walk towards a plane mirror at a velocity V, (a) what is the velocity of the image relative to the mirror? (b) What is the velocity of the image relative to you?

16.3   When you raise your right hand, what hand does the image (in a plane mirror) raise?

16.4   Semitransparent mirrors are used as one way mirrors. When they are kept between a dimly lit and a brightly lit room a person in the dimly lit room can see people in the other room but not vice versa. Explain?

16.5   The radius of curvature of a plane mirror is infinite. Is it correct to use the spherical mirror equation for plane mirrors?

16.6   Which of the following statements are true:
(a) A plane mirror always forms a virtual image.
(b) A concave mirror always forms a real image.
(c) A convex mirror always forms a virtual image.
(d) A converging lens sometimes forms a virtual image.
(e) A diverging lens sometimes forms real images.

16.7   A lens has two focal points, while a mirror has only one focal point. Why this difference?

16.8   One of the most spectacular natural phenomena, the rainbow, is produced by combined effects of refraction, total reflection, and dispersion of sunlight by spherical water droplets. Explain how it forms. Why does it look circular?

16.9   Objects in water appear closer when viewed from above, while objects above water appear farther away when viewed from water. Discuss the reasons for this difference.

16.10 Images formed by mirrors do not have chromatic aberration. Why? What type of aberration is corrected by an achromatic lens system?

16.11 The high refractive index of diamond is responsible for its sparkle. Explain why.

16.12 A beam of light falls on a reflecting surface making an angle of 30° with the normal. Obtain the angle of reflection.

16.13 Three rays of light (1, 2, and 3) are falling on a curved reflective surface (Fig. 16.40). C is the center of curvature. Determine the angle of incidence for the three rays. Draw the reflected rays.

16.14 Four rays of light from a source in water falls on a water-air interface (Fig. 16.41). Calculate the angle of refraction for rays 1, 2, and 3. Calculate the critical angle for water. Will ray 4 be totally reflected? The refractive index of water is 1.33.

16.15 The refractive index of diamond is 2.42. Calculate the critical angle for diamond (See Problem 16.10).

16.16 Light falls perpendicular to one side (1) of a quartz prism of refractive index 1.46 (Fig. 16.42). Calculate the angle of incidence for the light at side 2. Is the angle of incidence larger than the critical angle? Do you get total reflection of light? Will the direction of light change at side 3?

Figure 16.40. Problem 16.13.

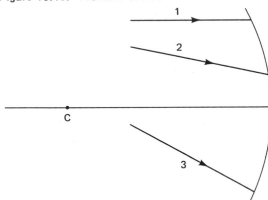

Figure 16.41. Problem 16.14

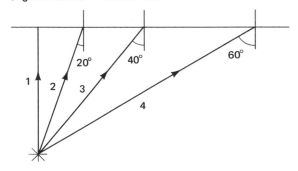

Figure 16.42. Problem 16.16.

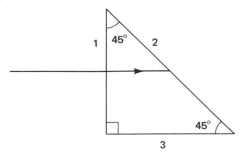

16.17 A prism similar to the one shown in Fig. 16.43 is usually used in binoculars. Determine the final direction of the rays of light from the object AB. On tracing rays 1 and 2 through the prism, one also observes that the reflected rays produce an up–down inversion.

16.18 A light beam falls on a glass slab as shown in Fig. 16.44. Trace the direction of the beam through the slab. Is the beam emerging from the slab parallel to the incident beam?

16.19 Trace the path of a beam of red light through the prism (Fig. 16.45). Calculate the angle of refraction at the first and second sides of the prism. Calculate the angle of deviation between the incident and the final beam. The refractive index of glass is 1.4.

16.20 A ray of light is totally reflected at the plane surface of a semicircular slab when the angle θ is equal to $50°$ or more (Fig. 16.46). Calculate the refractive index of the slab material.

Figure 16.43. Problem 16.17.

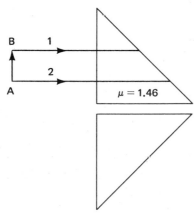

Figure 16.45. Problem 16.19.

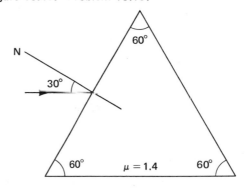

Figure 16.44. Problem 16.18.

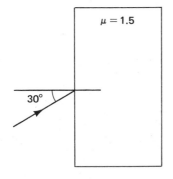

Figure 16.46. Problems 16.20 and 16.21.

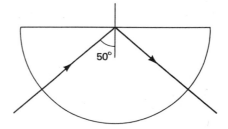

16.21 A layer of oil of refractive index 1.23 is placed on the plane surface of the slab of Problem 16.20. Will the light ($\theta = 50°$) be totally reflected now? Calculate the angle θ for which light will be totally reflected. This method can be used for an experimental determination of the refractive index of liquids.

16.22 The radius of curvature of a concave mirror is equal to 40 cm. What is its focal length?

16.23 A solar cooker uses a mirror with a 40 cm radius. What should be the distance between the mirror and the object to be heated?

16.24 An object is placed 50 cm in front of a concave mirror of radius 30 cm. Where is the image formed? Is the image real? Is it magnified? Locate the image by ray tracing.

16.25 An image is formed 30 cm behind a concave mirror. If the focal length is 40 cm, where is the object located?

16.26 An object 3 cm tall is placed 60 cm from the vertex of a concave spherical mirror which has a radius of 30 cm. (a) What is the focal length of the mirror? (b) How far from the mirrors is the image formed? (c) How tall is the image? (d) Discuss the image. (e) Will the image appear brighter than the object?

16.27 When an object is placed at the center of curvature of a concave mirror, a real inverted image forms at the same distance. Prove this statement.

16.28 A concave mirror produces a virtual image 20 cm behind it for an object 15 cm in front of it. Calculate the focal length of the mirror.

16.29 A convex mirror has a radius of 50 cm. What is its focal length.

16.30 Calculate the position of an image produced by the mirror of Problem 16.29 if an object is placed at (a) 100 cm, (b) 40 cm, and (c) 10 cm from the mirror. Discuss the image in each case. Locate the images by ray tracing.

16.31 Based on the results of Problem 16.30, can you make a statement "A convex mirror always produces a virtual image"?

16.32 A light bulb is 1 m under water. An observer outside sees a circle of light on the surface of the water. What is the radius of this circle?

16.33 A double-convex lens is made of glass with a refractive index $\mu = 1.5$. The radii of curvature of both surfaces are each 20 cm. Calculate the focal length of the lens.

16.34 Calculate the focal lengths of lenses shown in Fig. 16.47. The focal lengths for lenses (e) and (f) are equal. This result is expected. Why?

16.35 If the lens in Fig. 16.34(a) is immersed in water, will its focusing power diminish? Calculate its new focal length.

16.36 A converging lens has a focal length of 0.5 m. Find the positions and magnifications of the image for an object placed at 2 m, 1 m, 0.75 m, 0.3 m, and 0.1 m from the lens. Indicate whether the images are real or virtual.

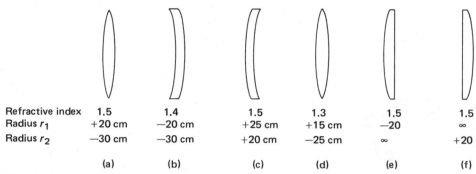

Refractive index	1.5	1.4	1.5	1.3	1.5	1.5
Radius r_1	+20 cm	−20 cm	+25 cm	+15 cm	−20	∞
Radius r_2	−30 cm	−30 cm	+20 cm	−25 cm	∞	+20
	(a)	(b)	(c)	(d)	(e)	(f)

Figure 16.47. Problem 16.34.

16.37 A diverging lens has a focal length of 0.2 m. Find the positions and magnifications of the image for an object placed at distances of 1m, 0.4 m, 0.2 m, and 0.1 m from the lens. Indicate whether the images are real or virtual.

16.38 A lens produces a real image with magnification 10 of an object placed at 30 cm from the lens. Calculate the focal length of the lens.

16.39 A virtual image, on the same side of the lens as the object, is formed at a distance of 16 cm from a thin lens. The magnification is $\frac{1}{3}$. Calculate the focal length of the lens. Is it a diverging lens or a converging lens?

17

Optical Instruments

The optical instruments and devices we discuss in this chapter have lenses and/or mirrors as their components. These instruments range from simple ones containing a single mirror or lens to complex ones containing several such components. Often these instruments help the eye in special ways. For example, in order to observe cell structure, microscopes comprised of several lenses are used and in order to observe distant stars, telescopes comprised of several lenses and mirrors are used. Since these instruments have different purposes, their systems of lenses and mirrors will have special features. These will be discussed later in this chapter.

17.1 SIMPLE OPTICAL INSTRUMENTS

Presently we consider two examples of simple systems each involving one mirror. Arrangements similar to those shown in Fig. 17.1(a) or 17.1(b) are used to obtain a parallel beam of light (for example, car headlights). The source of light is kept at the focal point of the parabolic mirror. In Fig. 17.1(b) a small mirror (designated secondary mirror) is kept in front of the source to reflect the diverging rays of light coming out of the source back to the mirror.

Figure 17.2 shows a solar radiation collector used for cooking purposes. The parabolic mirror made of polished aluminum focuses the energy from the sun to heat the pot kept at the focal point.

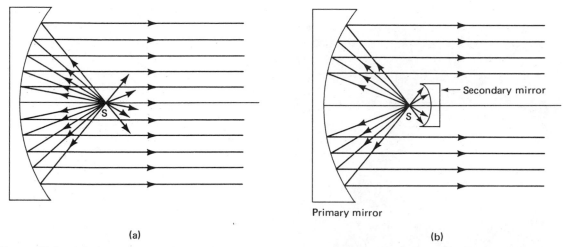

Figure 17.1. (a) Arrangement for obtaining a parallel beam of light. (b) A secondary mirror redirects the diverging light back to the main mirror.

EXAMPLE 17.1

In Fig. 17.1(b), the secondary mirror reflects the light back to the source. In other words, the image distance is equal to the object distance. Obtain the distance from source to secondary mirror in terms of the focal length of the mirror.

We use the equation

$$\frac{1}{D_o} + \frac{1}{D_i} = \frac{1}{f}.$$

In this problem $D_o = D_i$; therefore, we have

$$\frac{2}{D_o} = \frac{1}{f} \quad \text{or} \quad D_o = 2f = R.$$

The secondary mirror should be placed at a distance equal to its radius of curvature from the source of light.

EXAMPLE 17.2

A solar cooker has a diameter of 0.5 m. An aluminum pot weighing 50 g is kept at the focal point. If the pot contains 200 g of water how long will it take the water to heat to 60°C starting at 10°C.

This is not completely an optics problem, but because of the significance of this problem and because we have just discussed the solar cooker, it is included here.

First we have to calculate the total energy falling on the solar cooker (Fig. 17.3). The radiant energy falling on a unit area normal to the direction of the sun's rays on the surface of the earth is known as the solar constant. The value of the solar constant is 3.3×10^2 cal/m^2·s. If we assume that the reflector is directed towards the sun, the energy falling on the cooker is given by the solar constant times the area (area = $\pi \times$ radius^2).

Energy falling on the mirror per second

$$= (3.3 \times 10^2 \text{ cal/m}^2 \cdot \text{s})\pi(0.5/2 \text{ m})^2$$

$$= 65 \text{ cal/s} = 0.65 \times 10^2 \text{ cal/s}.$$

The energy input raises the temperature of the water. The increase in temperature can be calculated by using the equation (see Chapter 11):

$$(c_{Al}M_{Al} + c_w M_w)\Delta T = \text{Energy input}$$

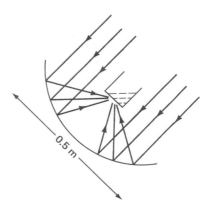

Figure 17.3. Example 17.2.

Figure 17.2. Solar cooker.   The spherical mirror focuses heat radiation from the sun onto the pot. Courtesy of Edmund Scientific Co.

where c_{A1} is the specific heat capacity of aluminum, M_{A1} the mass of the aluminum pot, c_w the specific heat capacity of water, and M_w the mass of water. Hence we get

$$(0.21 \text{ cal/g} \cdot {}^\circ\text{C} \times 50 \text{ g} + 1 \text{ cal/g} \cdot {}^\circ\text{C} \times 200 \text{ g}) \, 50^\circ\text{C}$$

$$= (0.65 \times 10^2 \text{ cal/s})t$$

$$t = 162 \text{ s} = 2 \text{ min } 42 \text{ s.}$$

Note that the solar constant is the energy reaching a unit area of the surface of the earth per unit time, if there is no absorption or scattering of energy in the atmosphere.

17.2 SEXTANT

The sextant used to be the only instrument which provided the navigator with information about his position on the high seas. He determined his position by measuring the angle between two directions, the direction to the horizon and the direction to the sun or a star.

The sextant contains a mirror M_1 which rotates about a pivot P, a fixed mirror M_2, and a telescope T (Fig. 17.4). The mirror M_1 is known as the index glass and the mirror M_2 is called the horizon glass. The horizon glass is fixed to the frame of the sextant which also contains a circular scale. The vernier scale that moves along the main scale rotates with the mirror M_1 enabling one to read the angular position or the orientation of M_1. The lower half of the horizon glass is clear and the other half is reflecting on the viewer's side.

Through the clear half of M_2, the telescope is first focused to see the horizon. Then M_1 is rotated so that the light H_2P from the horizon, after reflection from both mirrors, reaches the telescope. The "zero" position of the mirror is somewhat easy to determine. For this position the two views of the horizon, one viewed directly and the other through the mirrors, blend into each other. After obtaining the zero reading for this position of the mirror, it is rotated until the star appears in the center of the telescope. The new position of the mirror is then read on the main scale. The difference between the zero reading and the new reading is half of the angle between the horizon and the star.

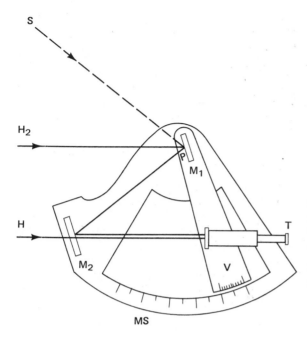

Figure 17.4. A pictorial representation of a sextant. M_1 is the rotating mirror, M the fixed mirror, P the pivot about which the mirror M_1 and vernier scale V rotates, MS the main scale, T the telescope, SP light from a star or any other celestial source, and H_2P light from the horizon.

and hence the amount of light falling on the film (Fig. 17.5).

To get a well-focused and correctly exposed picture, the following independent adjustments may be made in a modern camera: (a) the lens–film distance, (b) aperture size, and (c) exposure time. The effects of these adjustments are discussed below.

The distance between the lens and the film can be varied from a minimum distance equal to the focal length (focuses infinite objects) and a maximum distance equal to the focal length plus a few millimeters (focuses objects closer to the camera). The lens travel is limited between these two positions. The lens travel distance determines the minimum object distance for which a well-focused image is obtained. This is illustrated in Example 17.3.

EXAMPLE 17.3

The focal length of a camera lens is 50 mm and its travel distance is 5 mm. Obtain the minimum object distance.

When the object is at infinity $D_i = f = 50$ mm.

17.3 CAMERA

The camera is a simple optical device that contains a lens (usually a lens system). The system of lenses focuses the image on the film which is kept inside a light-tight box. Light enters the box through the lens system only when a shutter behind the lens is released. Modern cameras have devices inside to reduce the chance of double exposure. After exposure of a frame, the shutter release mechanism will work again only after the exposed frame is rolled into a light-tight container and a new frame of the film is brought into the correct position. In addition to the shutter, most cameras have a "diaphragm" controlling the aperture of the lens

Figure 17.5. Camera.

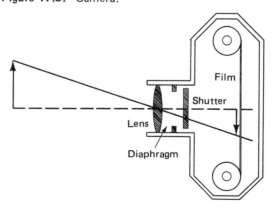

When the lens is at a position corresponding to the maximum lens travel (Fig. 17.6),

$$D_i = f + 5 \text{ mm.}$$

Minimum object distance in this case can be calculated using Eq. (16.4)

$$\frac{1}{D_o} + \frac{1}{D_i} = \frac{1}{f}$$

or

$$\frac{1}{D_{o,min}} + \frac{1}{f+5} = \frac{1}{f}$$

$$D_{o,min} = f\frac{(f+5)}{5}.$$

Since $f = 50$ mm

$$D_{o,min} = \frac{50 \times 55}{5} = 550 \text{ mm} = 55 \text{ cm.}$$

In the above example the minimum object to film distance is 60.5 cm (2 ft).

The diaphragm behind the lens helps the photographer to change the size of the opening (aperture) through which light falls on the film. The aperture size is adjusted externally by changing the position of an indicator to different numbers on an f - number scale. The f-number is defined by the equation

$$f\text{-number} = f/d, \qquad (17.1)$$

where d is the diameter of the aperture. From the definition, it is clear that as the f-number increases, the diameter of the aperture decreases and vice versa (Fig. 17.7). The amount of light entering the camera and, therefore, the brightness of the image is proportional to the area of the aperture and hence proportional to d^2. A simple calculation will show (example below) that the area decreases almost by half as the f-number indicator is moved from one number to the adjacent higher number.

EXAMPLE 17.4

The focal length of a camera lens is 50 mm. Obtain the diameters and areas of the diaphragm

Figure 17.6. Example 17.3.

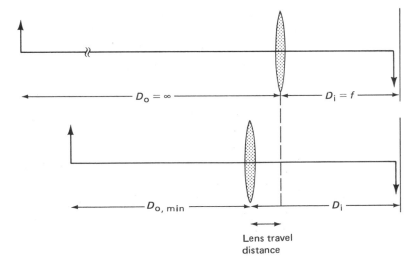

$D_o = \infty$ $D_i = f$

$D_{o, min}$ D_i

Lens travel distance

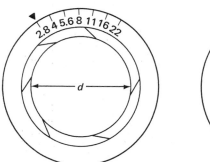 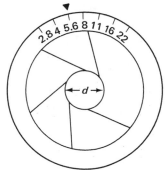

Figure 17.7 Figure illustrates the relationship between *f*-number and diameter of the diaphragm. As the *f*-number increases, the diameter decreases.

for the *f*-number settings 5.6 and 8.

The *f*-number is given by the equation

$$f\text{-number} = f/d.$$

Therefore, we have

$$d = f/f\text{-number}.$$

For *f*-number = 5.6, since we are given $f = 50$ mm,

$$d = \frac{50 \text{ mm}}{5.6} = 8.9 \text{ mm};$$
$$\text{Area} = \pi(d/2)^2 = 62.2 \text{ mm}^2.$$

For *f*-number 8

$$d = \frac{50}{8} = 6.25 \text{ mm};$$
$$\text{Area} = \pi (d/2)^2 = 30.69 \text{ mm}^2;$$

Ratio of the areas = 2.02.

The magnification of the image is just as important as the aperture size in the discussion of the brightness of the image. In Fig. 17.8 the height of the image is H_i and the area is H_i^2. The light from the object is distributed over this area. If the image height H_i doubles, the area increases 4 times, reducing the light received per unit area by ¼. H_i is approximately proportional to the focal length.

Therefore, the brightness is inversely proportional to the square of the focal length. Combining the above two factors that control the brightness of the image we have

$$\text{Brightness of the image} \simeq \frac{d^2}{f^2} = \frac{1}{(f\text{-number})^2}.$$

$$(17.2)$$

EXAMPLE 17.5

A picture of an object was taken with a 50 mm lens and 150 mm (telephoto) lens. Assuming the aperture diameter did not change, compare the brightness of the picture taken by the two lenses.

Figure 17.8. Relationship between sizes of the object and image.

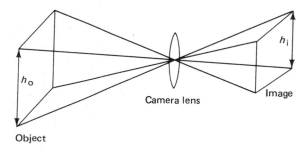

The image size increases as the focal length is increased. The image size is proportional to the focal length and hence the brightness is inversely proportional to the square of the focal length.

$$\text{Brightness of the first picture} \propto \frac{1}{(50 \text{ mm})^2};$$

$$\text{Brightness of the second picture} \propto \frac{1}{(150 \text{ mm})^2};$$

Ratio of the brightness

$$= \frac{\text{Brightness of the second picture}}{\text{Brightness of the first picture}}$$

$$= \frac{(50 \text{ mm})^2}{(150 \text{ mm})^2} = \frac{1}{9}.$$

In the above calculation we assumed that the change in the brightness is due to change in the image size only, i.e., other factors like the aperture diameter did not change.

When the object is not properly illuminated, the photographer has a choice of either decreasing the f-number or increasing the exposure time by adjusting the shutter speed. However, if the object is moving, the second option is undesirable. Another important factor to remember is that the depth of field, the range of distances for which images are reasonably focused, becomes smaller as the aperture diameter increases. Figure 17.9 illustrates this effect. In addition to the factors discussed above, film sensitivity should also be considered in adjusting the f-number and shutter speed.

Figure 17.9. Reducing the area of the aperture improves the depth of field of a photograph. (a) The camera is focused for object AB. Object A_1B_1 which is closer to the lens produces a fuzzy image on the film. (b) The fuzziness decreases as the diaphragm is reduced.

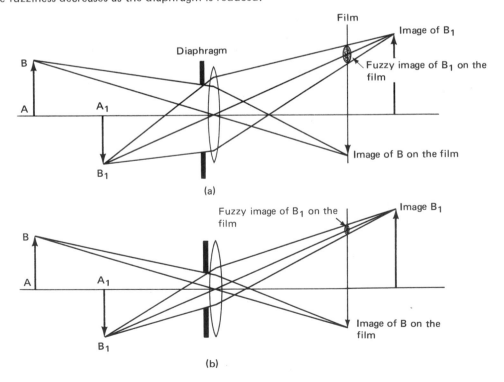

17.4 THE EYE

The camera, discussed in the last section, is modeled after the eye. The operation of the eye is somewhat like that of the camera, but the superiority of the eye lies in the fact that the focal length of its lens adjusts to produce sharp images on the retina for almost any object distance, provided it is not too close to the eye.

Figure 17.10 is a schematic representation of the eye. The crystalline lens, made of overlapping layers of plastic, transparent material, is attached to the ciliary body. The shape of the lens and hence the focal length can be changed by the action of the muscle of the ciliary body. The ability of one's eye to change its focal length and produce images of objects at different distances on the retina is called "accommodation."

The retina contains thousands of closely packed light-sensitive cells. These cells are in fact ends of optical nerves that carry information in the form of electrical impulses to the brain. In a section of the retina called the fovea, the cells are closely packed. The cells in this region, called the cones, are color sensitive. The fovea is the most sensitive section of the retina. Outside this region cells are not as closely packed as in the fovea. Among these cells are the rod cells which are not sensitive to color,

but they respond to low levels of light such as that encountered during night-time vision.

A normal eye, also called an emmetropic eye, without straining too much can accommodate objects at distances from 25 cm (10 in.) to infinity. The minimum distance is called the near point or the distance of distinct vision. The power of accommodation decreases as the muscles of the ciliary body lose their flexibility to produce the necessary shape changes in the lens due to old age or some physical disability. Two of the most common defects that may result from this are illustrated in Fig. 17.11. The nearsighted or the myopic eye focuses only objects closer to the eye and the far-sighted or the hypermetropic eye focuses only distant objects. As shown in Fig. 17.11, a diverging lens corrects short-sightedness and a converging lens corrects far-sightedness.

EXAMPLE 17.6

A person suffering from far-sightedness has his/her nearpoint at 60 cm. Calculate the focal length of the required correcting eyeglass.

A lens placed in front of the eye should be able to bring the near point from 60 cm to the normal nearpoint, 25 cm, that is, the lens should produce an image at 60 cm for an object placed at 25 cm (Fig. 17.12). Using the lens equation, we calculate the focal length

$$\frac{1}{-60 \text{ cm}} + \frac{1}{25 \text{ cm}} = \frac{1}{f}$$

("minus" in front of 60, because image is virtual)

$$f = 43 \text{ cm}.$$

See Problem 17.7 on the correction of myopic eyes.

17.5 MAGNIFYING GLASS

To the eye the apparent size of an object depends on how big an image it produces on the

Figure 17.10. Cross-section of an eye showing its inner structure. The numbers are the refractive indices of the different media.

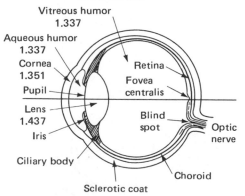

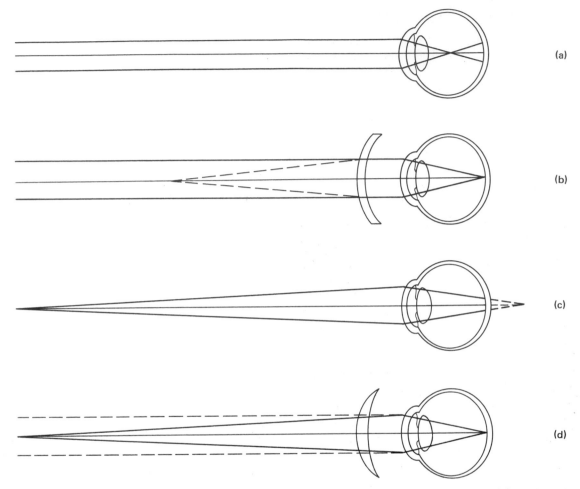

Figure 17.11. (a) Nearsighted eye. Objects at infinity are not focused on the retina. (b) A diverging lens corrects for near-sightedness. (c) Farsighted eye. Objects closer to the eye are not focused on the retina. (d) A converging lens corrects for far–sightedness.

Figure 17.12. Example 17.6.

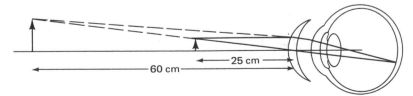

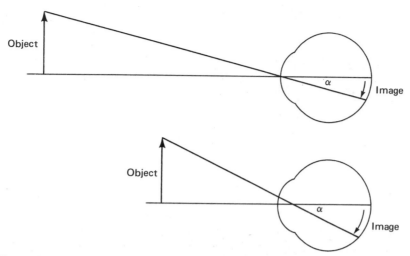

Figure 17.13. Figure illustrates the relationship between the size of the image formed on the retina and the object distance.

retina. Figure 17.13 shows the relationship between the size of the object and the image on the retina; image size increases as the object approaches closer to the eye. Also it is clear that the image size is proportional to the angle α. To increase the image size, one can bring an object closer to the eye, but the object cannot be brought closer than the near-point. To get around this limit a simple converging lens is often used. The lens, in this case, produces a magnified image at 25 cm for an object kept closer to the eye [Fig. 17.14(a)].

In Fig. 17.14(b) we note that α is the angle subtended at the eye when the object is at 25 cm from the eye. In Fig. 17.14(c) the object is placed closer to the eye so that the image is at 25 cm from the eye. The image now located at 25 cm from the eye subtends an angle β at the eye. Angle β is greater than α. The ratio of β to α is called the magnifying power, MP:

$$MP = \frac{\beta}{\alpha} \cong \frac{h_i}{h_o}. \qquad (17.3)$$

From Fig. 17.14(c)

$$\frac{h_i}{h_o} = \frac{25}{D_o}.$$

Using the lens equation we get

$$\frac{1}{D_o} + \frac{1}{-25} = \frac{1}{f}$$

or

$$D_o = \frac{25 f}{f + 25}.$$

Therefore, the magnifying power is given by

$$MP = \frac{h_i}{h_o} = \frac{f + 25}{f} = \frac{25}{f} + 1. \qquad (17.4)$$

The approximation $MP \simeq 25/f$ is sometimes used in this case. In the case of a magnifying glass, the magnifying power MP is approximately equal to the linear magnification h_i/h_o. [See Eq. (17.3).]

EXAMPLE 17.7

A newspaper is viewed through a magnifier of focal length 10 cm. The image forms at the near

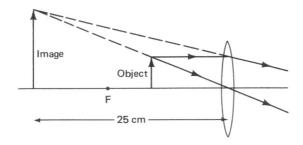

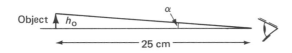

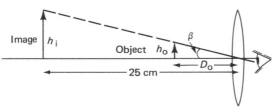

Figure 17.14. (a) Formation of a magnified virtual image by a lens. (b) Object at the near-point subtends an angle α. (c) Object is kept closer to the eye. Image at the nearpoint subtends an angle β.

point (25 cm). Calculate the magnifying power (angular magnification) and linear magnification produced by the lens.

We use Eq. (17.4) for the magnifying power. The magnifying power is given by

$$MP = \frac{25}{f} + 1 = \frac{25 \text{ cm}}{10 \text{ cm}} + 1 = 3.5.$$

The linear magnification is given by

$$M = \frac{D_i}{D_o}.$$

The image distance D_i is 25 cm. To obtain the

object distance D_o, we use the lens equation:

$$\frac{1}{D_o} + \frac{1}{D_i} = \frac{1}{f}$$

$$\frac{1}{D_o} + \frac{1}{-25} = \frac{1}{10}$$

$$D_o = \frac{250}{35}.$$

Hence we get

$$M = \frac{D_i}{D_o} = \frac{25/250}{35} = 3.5.$$

17.6 MICROSCOPE

A microscope uses two lenses (or lens systems) to magnify small objects. The lens closer to the object, called the objective lens, first produces a magnified real image (Fig. 17.15). This real image is viewed through a second lens, called the eyepiece, which acts as a magnifying glass.

The relative positions of the objects and the images with respect to the focal points of the lenses are the following: (a) The object O is just outside the focal point F_o of the objective. A real image with maximum magnification results from this arrangement (magnification decreases as the object to objective distance increases). (b) The eyepiece is located so that the real image of the objective falls within the focal length and the final image is 25 cm from the eye.

The total magnification M produced by the microscope is the ratio of the final image size to the initial object size (h_2/h_o). Its relationship to the magnification of the eyepiece, M_e ($= h_2/h_1$), and the magnification of the objective, M_o ($= h_1/h_2$), can be obtained in the following way:

$$M_e = \frac{h_2}{h_1}, \quad M_o = \frac{h_1}{h_o}; \tag{17.5}$$

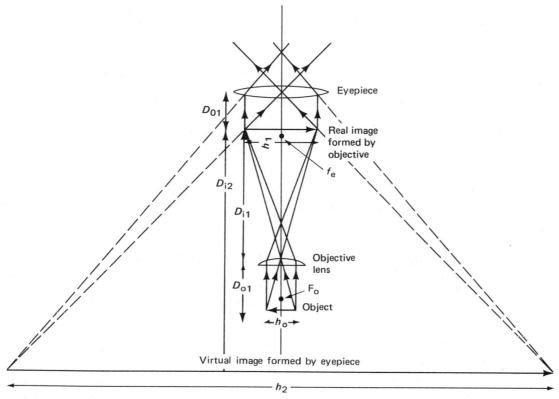

Figure 17.15. Microscope. The objective lens forms a real image behind the focal point of the eyepiece. The eyepiece forms a virtual image 25 cm from the eye.

$$M = \frac{h_2}{h_o} = \frac{h_2}{h_1} \times \frac{h_1}{h_o} = M_e M_o. \qquad (17.6)$$

To derive the equations for M_e and M_o let us designate by D_{o1} and D_{i1} the object and image distances from the first lens (objective lens) and by f_o its focal length. Similarly D_{o2} and D_{i2} stand for the object and image distances from the second lens (the eyepiece) and f_e stands for the focal length of the eyepiece. The lens equation for the objective is

$$\frac{1}{D_{o1}} + \frac{1}{D_{i1}} = \frac{1}{f_o}.$$

From this equation we get

$$M_o = \frac{D_{i1}}{D_{o1}} = \frac{D_{i1}}{f_o} - 1. \qquad (17.7)$$

For the magnification of the eyepiece we use Eq. (17.4), namely,

$$M_e = \frac{25}{f_e} + 1. \qquad (17.8)$$

Therefore, the total magnification is given by

$$M = M_o M_e = \left(\frac{D_{i1}}{f_o} - 1 \right)\left(\frac{25}{f_e} + 1 \right). \qquad (17.9)$$

Most often the following approximate equation is used in place of Eq. (17.9):

$$M = \frac{D_{il}}{f_o} \times \frac{25}{f_e}. \qquad (17.10)$$

The distance D_{il} is sometimes referred to as the tube length, since it is almost equal to the distance between the lenses, and the symbol L is used instead of D_{il}.

In Eq. (17.10) both focal lengths are in the denominator. From this it is evident that shorter focal lengths produce a larger magnification. However, shorter focal length lenses are thick at the center with highly curved surfaces. Such lenses always produce a significant amount of common aberration. Therefore, properly corrected lens systems are used in microscopes.

EXAMPLE 17.8

Focal lengths of an eyepiece and an objective in a microscope are 2 cm and 1 cm, respectively. Calculate the magnifications of both lenses and that of the microscope. The object distance from the objective lens is 1.05 cm.

To obtain the magnification of the objective, we first obtain the image distance D_{il} by using the lens equation:

$$\frac{1}{D_{o1}} + \frac{1}{D_{il}} = \frac{1}{f}$$

$$\frac{1}{1.05\text{cm}} + \frac{1}{D_{il}} = \frac{1}{1 \text{ cm}}$$

$$D_{il} = 21 \text{ cm.}$$

$$\text{Magnification of the objective} = \frac{D_{il}}{f_e} = \frac{21 \text{ cm}}{1 \text{ cm}} = 21;$$

$$\text{Magnification of the eyepiece} = \frac{25 \text{ cm}}{f_e} = \frac{25 \text{ cm}}{2 \text{ cm}}$$

$$= 12.5;$$

Total magnification $= M_o M_e = (21)(12.5) = 262.5.$

If we use the exact equations we get

$$M = \left(\frac{21}{1} - 1\right)\left(\frac{25}{2} + 1\right) = 270.$$

The difference in the two results is very small compared to the total magnification. Note: The magnifications written on the eyepieces and objectives of commercial microscopes are the values obtained by the approximate equations.

17.7   TELESCOPE

The telescope aids the eye in viewing objects located at large distances from the eye. An object, like a star or a moon, subtends a very small angle α at the eye [Fig. 17.16(a)]. The telescope produces a virtual image at 25 cm which subtends an angle β. Since β is much bigger than α, the object appears bigger than it would with the naked eye. The magnifying power of the telescope is defined by the equation.

$$\text{MP} = \beta/\alpha. \qquad (17.3a)$$

The telescope consists of an objective, which produces an inverted real image, and an eyepiece, which produces a magnified virtual image at the nearpoint [Fig. 17.16(b)]. The image produced by the objective is at or very close to its focal point and it falls within the focal point of the eyepiece. The magnifying power of a telescope, MP, can be shown to be given by the equation

$$\text{MP} = f_o/f_e, \qquad (17.11)$$

where f_o and f_e are the focal lengths of the objective and the eyepiece, respectively. Equation (17.11) tells us that to obtain a high magnifying power the focal length of the objective should be long and the focal length of the eyepiece should be short. It can also be shown that

$$\mathrm{MP} = \frac{f_o}{f_e} = \frac{D_o}{D_e},$$

where D_o and D_e are the diameters of the objective and the eyepiece lenses.

The telescope we have outlined above is an astronomical refracting telescope. It contains an objective of long focal length and eyepiece of very

short focal length. Also the objective is of larger diameter than the eyepiece. A larger diameter for the objective helps in collecting more light especially from weak sources, such as far away stars, thereby producing a brighter image. However, as the diameter increases the thickness of the lens increases. This will result in an increase in the absorption of light in the lens. Also it is very difficult to fabricate lenses of very large diameters.

A reflecting mirror telescope is shown in Fig. 17.17. The mirror takes the place of the objective lens of the refracting telescope. The mirror forms the image just within the focal length of the eyepiece like in refracting telescopes. Mirrors have the following advantages: (a) they are free of chromatic aberration (b) they are free of spherical aberration since usually parabolic mirrors are used, (c) large mirrors are easier to fabricate than large lenses, and (d) since the mirror is located at the bottom of the telescope, reflecting mirror telescopes are mechanically easier to support.

Figure 17.16. (a) A telescope produces an image that subtends an angle β which is much bigger than α, the angle subtended by the object. (b) A telescope. The objective forms a real image at its focal point. The real image acts as the object for the eyepiece. The final image is 25 cm from the eye.

EXAMPLE 17.9

A small telescope has an objective of 100 cm focal length and an eyepiece of 2 cm focal length. Find the magnifying power of the telescope. If this telescope is used to view the moon, what will be the

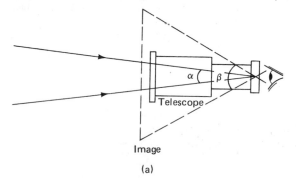

Image

(a)

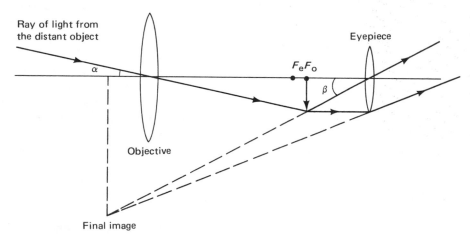

(b)

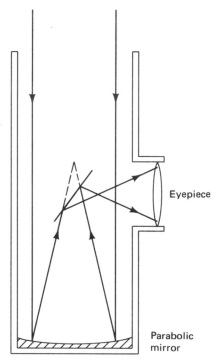

Figure 17.17. Reflecting mirror telescope.

Similarly, angle β is given by

$$\beta = \frac{\text{Diameter of the image}}{25 \text{ cm}}.$$

Angle β is also equal to 50α. Hence we get

$$\text{Diameter of the image}$$
$$= (25 \text{ cm})(50)(9.4 \times 10^{-3} \text{ rad})$$
$$= 11.75 \text{ cm}.$$

In astronomical telescopes the image is inverted. While studying astronomical objects this is not a serious problem, but in terrestrial telescopes one would like to obtain erect images. One method used is to introduce a third lens between the objective and eyepiece to erect the image. Another method is to use a convex lens as an eyepiece (Fig. 17.18). This eyepiece is placed so that it intercepts the converging rays from the objective before they converge to form a real image, diverging them so that a virtual image is formed. This type of telescope was developed by Gilileo.

size of the image of the moon?

$$\text{MP} = \frac{f_o}{f_e} = \frac{100 \text{ cm}}{2 \text{ cm}} = 50.$$

The magnifying power is the ratio of the angles α and β.

$$\frac{\beta}{\alpha} = 50.$$

Angle α subtended by the moon at the eye is

$$\frac{\text{Diameter of the moon}}{\text{Distance from the earth}} = \frac{3.5 \times 10^6 \text{ m}}{3.8 \times 10^8 \text{ m}}$$

$$= 9.4 \times 10^{-3} \text{ rad}.$$

17.8 BINOCULARS

Binoculars contain two identical telescopes, one for each eye, fixed to the same frame (Fig. 17.19). Both telescopes may be focused by a single thumb screw. In each telescope there are two totally reflecting prisms between the objective and the eyepiece. The reflections in the prisms make the final image erect. The path of the light is "folded" twice during these reflections. This allows for a shorter overall length for each telescope and results in easy-to-carry, compact binoculars. The prisms enable the objective lenses to be positioned farther apart than the eyepieces. The wider separation of the objective lenses provides a better stereoscopic effect, that is, a depth of perception at greater distances.

The quality of binoculars is rated in terms of its magnifying power and light-gathering ability. Manu-

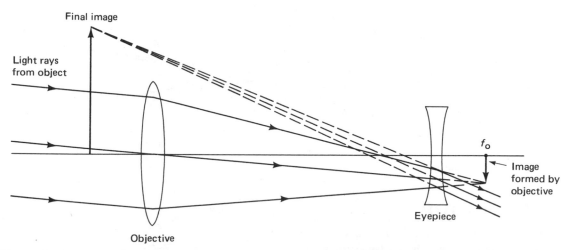

Figure 17.18. A terrestrial telescope. Note that the divergent eyepiece forms an erect image.

facturers designate their binoculars by two numbers, for example, you may find the numbers 6 × 35, 8 × 5, etc. written on them. The first number is the magnifying power and the second number is the diameter of the objective in mm. Light gathering power is proportional to the area of the objective and hence it is proportional to the square of the diameter.

Figure 17.19. Binoculars. Cut-away view of a binocular optical system. (Courtesy of Bushnell Optical Company.)

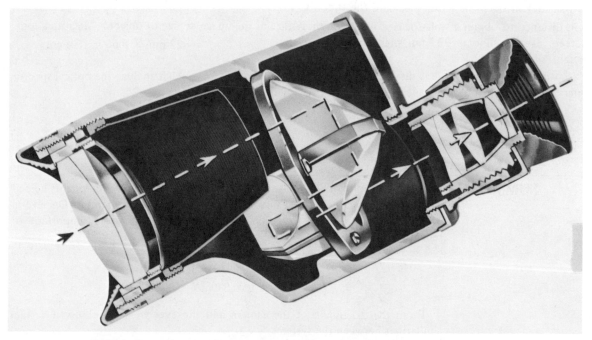

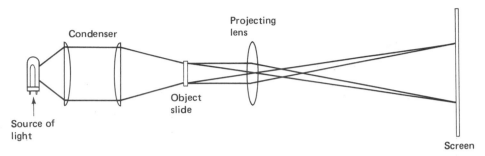

Figure 17.20. A projector.

17.9 PROJECTOR

The projector illustrated in Fig. 17.20 has two parts: (a) a condenser which is a lens system that focuses light from a high-intensity source onto a transparent film or a slide, and (b) a projector, which is a lens or a lens system that produces a magnified image on the screen.   The distance between the object and the lens is adjustable.

EXAMPLE 17.10

A slide projector has a lens of 15 cm. The screen is placed 1 m from the lens. Calculate the slide to lens distance to obtain a well–focused image on the screen. If the slide is 22 mm high, obtain the height of the image.

To obtain the slide to lens distance, we use the lens equation. The image distance is 1 m = 100 cm,

hence we have

$$\frac{1}{D_o} + \frac{1}{100 \text{ cm}} = \frac{1}{15 \text{ cm}}$$

or

$$D_o = 17.6 \text{ cm}.$$

Magnification is given by

$$M = \frac{D_i}{D_o} = \frac{100}{17.6} = 5.66.$$

Size of the image = Size of object × Magnification
$$= 22 \text{ mm} \times 5.66 = 12.4 \text{ cm}.$$

Spectrometers, which analyze the optical spectra, will be discussed in Chapter 24.

SUMMARY Our knowledge of the image forming properties of lenses and mirrors helped us to understand the optical instruments discussed in this chapter. On several occasions, the basic lens equation was used. It is, therefore, important to remember the equation

$$\frac{1}{D_o} + \frac{1}{D_i} = \frac{1}{f}.$$

From the discussion of the camera and the eye, we have observed certain similarities between the systems.

	Eye	Camera
Lens	Adjustable focal length to accommodate objects from 25 mm to ∞	Adjustable lens to film distance to accommodate objects at various distances
Diaphragm	Iris adjusts its diameter according to the intensity of light	Diaphragm diameter can be adjusted for varying intensity of light
Recording	Information about the image on the retina is transmitted to the brain.	Permanent record forms on the film due to chemical changes

We defined f-number by the equation f-number $= f/D$, where D is the diameter of the diaphragm. The brightness of the image is proportional to $(1/f\text{-number})^2$. In addition to f-number, one should be aware of lens travel distance, shutter speed, and film sensitivity.

A normal eye can see objects from the nearpoint to infinity. This ability is sometimes lost resulting in defects known as near-sightedness and far-sightedness. A diverging lens corrects the near-sightedness and a converging lens corrects the far-sightedness.

Magnifying power is given by

$$MP = \frac{25}{f} + 1,$$

where the image is formed at the nearpoint.

A microscope is used to magnify small objects kept close to the focal point of the objective. The real image formed by the objective is further magnified by an eyepiece lens which acts as a simple magnifying glass. The total magnification is given by the equation

$$M = \frac{D_i}{f_o} \times \frac{25}{f_e},$$

where f_o and f_e are the focal lengths of the objective and the eyepiece and D_i (sometimes referred to as the tube length) is the distance of the real image from the objective lens.

The telescope aids the eye in viewing distant objects by forming an image at the nearpoint which subtends an angle β at the eye much larger than the angle α subtended by the object. Magnifying power (not the linear magnification), defined as the ratio of these two angles, is given by

$$MP = f_o/f_e.$$

The objective should have a large area to collect light. Bigger telescopes are of the reflecting type because it is easier to fabricate large diameter mirrors than lenses.

Binoculars contain two telescopes. Reflections in prisms are used to correct the inversions in ordinary telescopes.

The projector is used to produce magnified real images of small slides.

QUESTIONS AND PROBLEMS

17.1   Ancient Greeks used concave mirrors to set fire to enemy ships at sea, an early example of the military application of science. Explain how it is done.

17.2   A telescope is sometimes referred to as a microscope held in the wrong way. It is true?

17.3   A swimmer cannot see properly in the water; however, goggles help very much in seeing underwater objects. Explain the reason for this.

17.4   The focal length of a camera lens is 25 mm. If the closest distance of the object that can be focused on the film is 1 m, what is the lens travel distance?

17.5   A camera has a focal length of 30 mm. Calculate the radius and area of the diaphragm for the f-number settings 2.8, 5.6, and 11.

17.6   A good picture of an object was taken with a camera for the following settings: f-number 11, shutter speed, 1/25 s. If the same object is to be photographed at a shutter speed of 1/50 s, what is the proper f-number setting? If the f-number is changed to 4, how will you change the shutter speed?

17.7   A myopic person has a nearpoint of 20 cm and a farpoint of 300 cm. Obtain the focal length of the glasses he should wear. What is his new nearpoint with this lens?

17.8   A person uses eyeglasses of focal length +55 cm. Calculate the nearpoint for the unaided eye.

17.9   Bifocal lenses of focal lengths +30 cm and −300 cm are worn by a person. Obtain the near and farpoints for the unaided eye.

17.10 A magnifying glass of 5 cm focal length is used to produce an image at 25 cm from the eye. Calculate the angular magnification and linear magnification produced by the magnifying glass.

17.11 A compound microscope has a magnification of 25 at the objective and 10 at the eyepiece. The real image produced by the objective is 15 cm from it and the final image is 25 cm from the eye. Calculate (a) the focal length of the objective, (b) the focal length of the eyepiece, and (c) the total magnification of the microscope.

17.12   The focal lengths of the objective and the eyepiece are 1.6 and 2.5 cm, respectively, in a compound microscope. The tube length is 15 cm and the final image forms at 25 cm from the eye. Calculate the total magnification of the lenses. Draw a diagram indicating the positions of the object, the lenses, and the images.

17.13   A telescope contains an objective of 200 cm focal length and an eyepiece of 15 cm focal length. Calculate the magnifying power of the telescope.

17.14   The telescope of Problem 17.13 is used to view the moon. What is the image size if the diameter of the moon is 3.5×10^6 m and the earth to moon distance is 3.8×10^8 m.

17.15   The lens of a slide projector has a focal length of 20 cm. The screen is set 10 m from the projector. If the height of the slide is 2 mm, what is the height of the image?

17.16   A 100 X magnified image is produced by a projector on a screen kept at 6 m from the projector. Calculate the focal length of the lens used in the projector.

18
Light Measurement and Light Sources

A source of light emits radiation and the radiation falling on a surface illuminates the surface. The surface illumination, the amount of radiation released, and the intensity of the source are all interrelated (Fig. 18.1). We shall define these quantities together with their units in this chapter and also derive relationships between them. We will also see that the illumination depends not only on the total energy of the radiation falling on the surface, but also on the response of our eye to different wavelengths or colors contained in the light. In the second part of this chapter we discuss common sources of light.

18.1 RADIANT FLUX AND LUMINOUS FLUX

Radiation emitted by sources such as an incandescent bulb contains radiation in the visible region (light) and radiation that we cannot see such as the infrared and ultraviolet rays. The total energy of the radiation emitted by the light source per second is called radiant flux. The radiant flux means the energy flow per second, hence it has a unit of watts (joules/second). The radiant flux Φ is always less than the power consumed in the bulb. For example, the radiant flux, energy released in the form of radiation per second, is less than 100 W from a 100 W bulb. The wattage rating of a bulb indicates

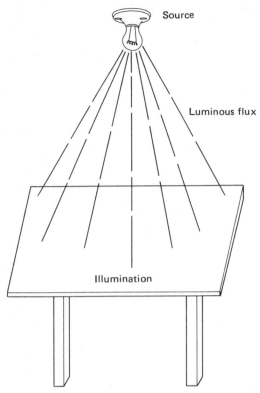

Figure 18.1. Intensity of the source, luminous flux emitted by the source, and illumination are all interrelated.

the electrical power consumed by the bulb, but not the radiant power.

It is possible to determine the fraction of power carried away by the radiation in the visible range of the spectrum out of the total radiant flux. The fraction of power thus calculated will not be a measure of the "brightness." Brightness is a physiological sensation which depends on the color composition of light as well as the radiant energy in each color.

To illustrate the above mentioned difference, let us consider two light sources: one radiates 1 W of green light and the other 1 W of blue light. For our eye the source of green light is brighter than the source of blue light. Thus the eye does not respond equally to the different colors of light. This variation in response as a function of wavelength is

shown in Fig. 18.2. The curve is obtained by comparing brightness sensation produced by light of various wavelengths with that produced by a 555 nm green light of equal power. *The brightness response of the eye is called relative luminosity or spectral luminous efficiency.*

It is evident from Fig. 18.2 that our eyes are most sensitive to green color of 555 nm wavelength. It is zero for radiation of wavelength less than 400 nm and larger than 700 nm; i.e., the eye does not respond to ultraviolet ($\lambda < 400$ nm) and infrared ($\lambda > 700$ nm) radiation. There had been claims of people seeing x-rays, but the relative luminosity represented in Fig. 18.2 is that for an average eye. In fact the data plotted in Fig. 18.2 are the relative

Figure 18.2. Graph shows the variation in the response of the eye to light in different wavelength regions of the same power.

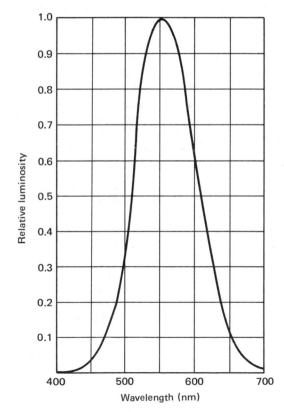

luminosity averages for a number of eyes to take care of individual variations.

As mentioned above, radiant flux (in watts) is the total energy flow per second for all radiation, but it is not a measure of the brightness of the radiation because brightness is a function of energy distribution in the spectrum. The term luminous flux is used for that. *Luminous flux is the energy flow weighted for the relative luminosity of the different wavelength components. The unit of luminous flux is the lumen. By definition 1 watt of radiant flux of green light of 555 nm is equivalent to 680 lumens of luminous flux. If the radiant flux is* Φ, *the luminous flux can be obtained by the equation*

$$F_{555} \text{ (lm)} = 680 \text{ (lm/W)} \times \Phi \text{ (W)}. \qquad (18.1)$$

F_{555} stands for luminous flux of 555 nm wavelength. In the following pages F_λ stands for flux calculated for light of wavelength λ and the symbol F without subscript stands for the total luminous flux.

For light of wavelengths other than 555 nm, the fact that the response of the eye is less than that for the green light should be taken into account. That is done by including the relative luminosity K_λ in the luminous flux calculations as follows:

$$F_\lambda \text{ (lm)} = 680 \text{ (lm/W)} \times \Phi \text{ (W)} \times K_\lambda. \quad (18.2)$$

For light containing more than one color, the total luminous flux is obtained by adding the contributions of each component.

EXAMPLE 18.1

A light source emits 50 W of green light (555 nm) and 40 W of yellow light (589 nm). Calculate the total luminous flux emitted by the source.

We have to calculate the luminous flux for both colors separately. For the yellow light we obtain its relative luminosity from Fig. 18.2 as 0.77:

$$F_{555} = 50 \text{ W} \times 680 \text{ lm/W} = 34,000 \text{ lm};$$

$$F_{589} = 40 \text{ W} \times 680 \text{ lm/W} \times 0.77 = 29,944 \text{ lm};$$

Total radiant flux $= F_{555} + F_{589} = 63,944$ lm.

18.2   ILLUMINATION

Illumination results from luminous flux (light) falling on a surface. *The magnitude of illumination known as illuminance is defined as the amount of luminous flux falling on a unit area. The unit lumen/meter^2 is called the lux (lx).* Illumination E in lux is given by the equation

$$E \text{ (lx)} = \frac{F \text{ (lm)}}{A \text{ (m}^2)}. \qquad (18.3)$$

When the source of light is very small, such as a bulb, light radiation is emitted uniformly in all directions. Such a source is called an isotropic source (see Fig. 18.3). An isotropic source illuminates a spherical surface equally at all points on the surface of the sphere when the source is at its center. If we denote the total luminous flux from the source by F, the illuminance on the spherical surface, whose total area is $4\pi r^2$, is given by

$$E = F/4\pi r^2. \qquad (18.4)$$

Figure 18.3. An isotropic source. Radiation is emitted in all directions with equal probability.

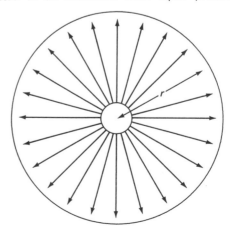

Equation (18.4) tells us that the illuminance decreases inversely as the square of the distance. For example, the illuminance at a distance of 3 m is smaller than the illuminance at a distance of 1 m by a factor of 1/9.

18.3 INTENSITY OF A SOURCE OF LIGHT

A light source produces luminous flux and luminous flux in turn produces illuminance. Now we define a unit for the intensity of a source in terms of the luminous flux emitted by the source. It is customary to define this unit in terms of light emitted in a small cone.

The intensity of the source is 1 candela if a flux of 1 lumen passes through a cone of unit solid angle. The intensity in candela (cd) can be defined by the following equation:

$$I\,(\text{cd}) = \frac{F\,(\text{1m})}{\Omega},\qquad (18.5)$$

where I stands for the intensity of the source, F the flux coming through the cone, and Ω the solid angle of the cone expressed in steradians.

What is the solid angle? Consider the sphere (Fig. 18.4) of radius r. A small cone with its apex at the center O of this sphere intersects an area on the surface of the sphere. If this area is A, the solid angle of the cone is defined by the ratio of A to the square of the radius, that is,

$$\Omega = A/r^2 \qquad (18.6)$$

or the surface area is given by the equation

$$A = r^2\Omega. \qquad (18.7)$$

If one considers the total surface area of the sphere ($=4\pi r^2$), the solid angle subtended by the sphere is 4π ($\Omega = 4\pi r^2/r^2$). The unit for solid angle defined by Eq. (18.6) is the steradian.

Equation (18.7), which gives us the area as equal to the solid angle in steradians multiplied by the square of the radius, is somewhat similar to the equation $S = r\theta$ which gives us the length of an arc on a circle as the product of the radius and the angle in radians (Fig. 18.5).

The relationship between intensity of the source, flux, and illuminance is shown in Fig. 18.6. The source is at the center of the sphere of radius 1 m. An area of 1 m^2 on the surface subtends a unit solid angle at the center. If the intensity of the source located at the center is 1 cd, a flux of 1 lm comes through the cone and, when this flux falls on the surface, it produces an illumination of 1 lx.

Figure 18.4. Figure defines solid angle. Area A is related to the solid angle by the relation $A = r^2$.

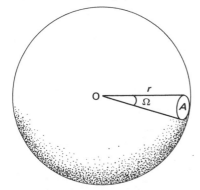

Figure 18.5. Figure shows the relationship between r and S.

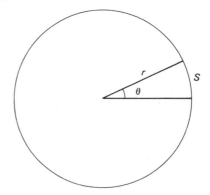

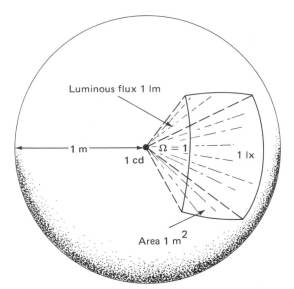

Figure 18.6. Figure illustrates the relationship between intensity of the source in candela, flux in lumen, and illumination in lux. The source intensity is 1 cd. The amount of flux going through the cone ($\Omega = 1$ sr) is 1 lm. This flux produces an illumination of 1 lm on the surface.

EXAMPLE 18.2

A bulb (assumed to be an isotropic source) is 4 m above the top of a table. The area of the table top is 0.5 m^2. (a) Calculate the solid angle subtended by the table at the bulb. (b) If the illuminance on the top of the table is 5 lx, calculate the intensity of the source.

(a) The solid angle is defined by Eq. (18.7) as

$$\Omega = A/r^2.$$

In this definition, area A is part of a sphere; but, in the example we are considering the area is a plane area. If the source to area distance is large, the error in using a plane area instead of a spherical area is very small and can be neglected. Hence the solid angle subtended by the table at the source can be calculated as follows

$$\Omega = \frac{0.5 \text{ m}^2}{(4 \text{ m})^2} = 0.03 \text{ steradians}.$$

(b) Illumination on the surface is 5 lx. Therefore the total flux falling on the table is

$$E \times A = (5 \text{ lx}) \times (0.5 \text{ m}^2)$$

$$= 2.5 \text{ lm}.$$

The intensity of the source is

$$\frac{F}{\Omega} = \frac{2.5 \text{ lm}}{0.03 \text{ sr}} = 80 \text{ cd}.$$

Illumination on a surface at a distance r from the source can be calculated from the intensity of the source. The equation for such a direct calculation is derived below.

Let the intensity of the source be I and the area at a distance r from the source be A (Fig. 18.7). The solid angle subtended by the area A at the source is A/r^2. From Eq. (18.5) we get the flux falling on the area A to be

$$F = I \frac{A}{r^2}.$$

Therefore, the illuminance of the surface is given by

$$E = F/A = I/r^2. \tag{18.8}$$

In this equation E is in lux, I in candela, and r in meters. In the derivation of the above equation the following assumptions were made: (a) the source is isotropic and (b) the area A is normal to the radius or the area forms part of a sphere with radius r.

Figure 18.7. Flux from a source of intensity I falls on area A. Area A subtends a solid angle Ω at the source.

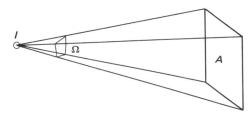

Equation (18.8) also can be considered as a state-ment of the inverse square law, that is, the illumina-tion by an isotropic source varies inversely as the square of the distance from the source.

A unit, named candle, was formerly the unit of intensity. The units candela and candle are approxi-mately equal.

The new unit candela was accepted by the International Committee on Weights and Measures in 1948. The unit candela is now defined as 1/60 the luminous intensity of a 1 cm^2 surface area of pure platinum at its melting temperature (ap-proximately 2043°K).

Illumination by a source of intensity equal to 1 candle at a distance of 1 ft from the source was known as 1 ft·candle. This unit is sometimes still used instead of lux. Illumination expressed in lm/ft^2 is sometimes loosely referred to as ft·candle.

EXAMPLE 18.3

An isotropic source has an intensity of 100 cd. Calculate the illumination produced by this source at a distance of 3 m in lux and ft·candle.

Illumination at a distance r from the source is given by the equation

$$E = I/r^2$$

$$E \text{ (lux)} = \frac{100 \text{ cd}}{(3 \text{ m})^2} = 11.1 \text{ lux}.$$

To obtain the illumination in ft·candle the distance has to be expressed in feet

$$d = 3 \text{ m} \times 3.281 \text{ ft/m} = 8.843 \text{ ft}.$$

$$E = \frac{100 \text{ cd}}{(9.843 \text{ ft})^2} = 1 \text{ ft·candle}.$$

Equation (18.8) was derived assuming that the light was falling normal to the surface. The modifi-cation of this equation to include situations where the surface is not normal is derived below. The sur-face A is at an angle θ to the normal surface A'

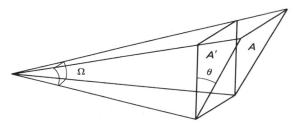

Figure 18.8. Solid angle subtended by the normal area A' and the area A are the same.

(Fig. 18.8). The solid angle in this case is

$$\frac{A'}{r^2} = \frac{A \cos\theta}{r^2}.$$

Hence the flux through the cone is given by

$$F = I\Omega = \frac{IA \cos\theta}{r^2},$$

where I is the intensity of the source. Hence for the illumination, flux per unit area, we get

$$E = \frac{F}{A} = \frac{I\cos\theta}{r^2}. \qquad (18.9)$$

EXAMPLE 18.4

A piece of paper of 0.1 m^2 area lies on a table 2 m away from a point directly below the bulb (Fig. 18.9). If the bulb of 100 cd is 4 m above the table, calculate the illumination on the paper.

Illumination in this case (Fig. 18.9) is given by

$$E = \frac{I\cos\theta}{r^2}.$$

Distance r from the source to the center of the paper is given by

$$r^2 = (4 \text{ m})^2 + (2 \text{ m})^2; \ r = 4.46 \text{ m}.$$

Since the adjacent side of the angle θ is equal to

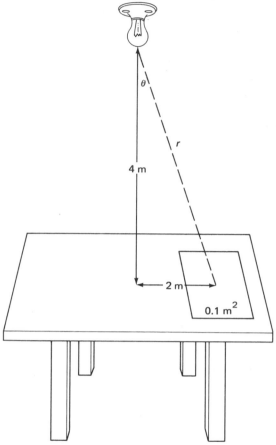

Figure 18.9. Example 18.4.

4 m, we get

$$\cos\theta = \frac{4\,m}{r} = \frac{4\,m}{4.47\,m} = 0.85;$$

$$E = \frac{I\cos\theta}{r^2} = \frac{100\,cd}{20\,m^2} \times 0.85 = 4.25\,\text{lx}.$$

The inverse square law applies only to a point source. How does the illumination change as a function of distance when the source is not a point source? The equations relating E, I, and r for a point source, a linear source, and an areal source are given below.

18.4   PHOTOMETER

Intensities of light sources can be compared by comparing the illumination produced by these sources. A photometer compares illumination due to two sources and determines the intensity of one in terms of the intensity of a second known source. A photometer developed originally by Robert Bunsen is illustrated in Fig. 18.10. The main part of the photometer is a paper with a translucent spot, i.e., a grease spot at the center of a white paper. Both sides of this paper are viewed from the top with the help of two mirrors. If the source on one

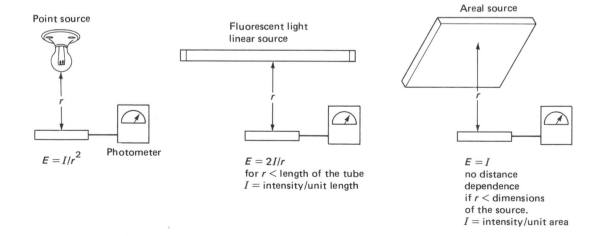

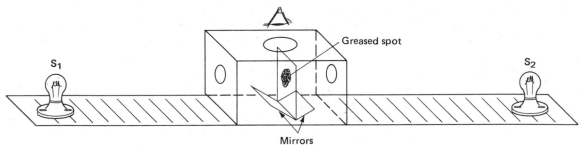

Figure 18.10. Photometer.

side is more intense, the grease spot on that side will look darker compared to the paper since more light is transmitted to the other side. When the illumination on both sides is equal, both sides of the grease spot will appear equally bright and almost indistinguishable from the rest of the paper.

To compare the intensities, one of the sources is kept fixed and the distance of the other source from the translucent paper is adjusted until both sides appear equally bright. If I_1 and I_2 are the intensities and r_1 and r_2 are the respective distances of the two sources, we have the following relationship between these quantities

$$\frac{I_1}{r_1^2} = \frac{I_2}{r_2^2}. \qquad (18.10)$$

This photometer, like others which employ visual comparison, are accurate if the two sources have nearly identical color composition.

Portable photoelectric photometers, which use cells that generate a voltage when radiant energy strikes the cells, are widely used these days. They are often fitted with color-correcting filters to match the spectral response of the human eye.

EXAMPLE 18.5

In an experiment to measure the intensity of a light source, a standard source of 100 cd is kept at a distance of 5 m from the translucent paper. Equal brightness on both sides is obtained when the

second source is placed at a distance of 3.3 m. Calculate the intensity of the second source.

We use Eq. (18.10):

$$\frac{I_1}{r_1^2} = \frac{I_2}{r_2^2}.$$

We are given $I_1 = 100$ cd, $r_1 = 5$ m, and $r_2 = 3.3$ m. Therefore, we get

$$\frac{100 \text{ cd}}{(5 \text{ m})^2} = \frac{I_2}{(3.3 \text{ m})^2}$$

or

$$I_2 = \frac{(100 \text{ cd}) (3.3 \text{ m})^2}{(5 \text{ m})^2} = 44 \text{ cd}.$$

18.5 LIGHT SOURCES

There are two types of commonly used light sources; the incandescent lamp and the electric discharge tube. The fluorescent lamps and the high-pressure discharge tubes (high-intensity street lights) belong to the latter group. The two groups differ in several respects such as the mechanism of light production, spectral content of the light, and luminous efficiency. The following discussion is a comparative study of these sources.

The incandescent lamp is the most common light source. It contains a metal filament in a glass envelope. When electric current passes through the filament, it gets hot and produces electromagnetic radiation including light. The color content of the radiation varies with the temperature of the filament. This can be demonstrated by a simple arrangement shown in Fig. 18.11. As the voltage applied to the filament is slowly increased its color changes from dull red to bright red and finally to blue white. One also observes that the brightness increases as the voltage increases. A detailed analysis of the light during such an experiment will show that (a) the radiation emitted by the filament has a continuous wavelength distribution (Fig. 18.12); (b) the frequency at which the radiant flux is maximum increases as the temperature of the filament increases; (c) the total radiant energy increases as the fourth power of the temperature. The above relationships are illustrated in Fig. 18.12. The second statement above can be written in mathematical form as follows

$$f_{max} \propto T \text{ or } \lambda_{max} \propto 1/T.$$

In fact, there is a relationship, called Wein's displacement law, which relates λ_{max} and T:

$$\lambda_{max} T = 2.9 \times 10^6, \qquad (18.11)$$

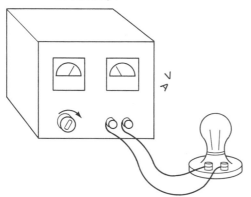

Figure 18.11. Experimental arrangement to study the intensity and color composition of light emitted by an incandescent lamp.

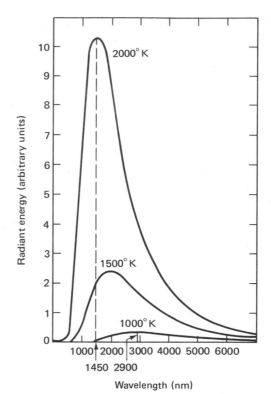

Figure 18.12. Figure shows the radiant energy distribution as a function of wavelength at three different temperatures. As the temperature increases the position of the peak shifts to the low-frequency side and the total energy (area under the peak) increases proportional to T^4.

where λ_{max} is in nm and T is in °K. Note that f_{max} is the frequency for which the radiant flux is a maximum and λ_{max} is the corresponding wavelength. Statement (c), which is known as the Stefan-Boltzmann law, can be expressed in mathematical form as

$$\Phi \propto T^4, \qquad (18.12)$$

where Φ is the total radiation flux and T the temperature on the Kelvin scale. According to Eqs. (18.11) and (18.12) doubling the temperature decreases λ_{max} by half (that is, the light output shifts to the low-wavelength and hence to the high-frequency side) and increases the total radiant flux sixteen times. This is illustrated in Fig. 18.13. For the curve for 1000°K, λ_{max} is 1450 nm and for 2000°K, λ_{max} is 2900 nm. Also the total radiated power (area under the curve) increases sixteen times as the temperature increases from 1000°K to 2000°K.

Since the radiant flux varies as T^4, can we increase the light flux output by increasing the temperature of the filament? The answer to this question is no. As we increase the filament temperature, the energy distribution shifts to the high-frequency side; consequently, the fraction of radiant flux in the visible region decreases. Figure 18.13 shows that the ratio of luminous flux to the total radiant flux, called the luminous efficacy, is a maximum at 6000°K.

From the above discussion, one can see that the material of the filament should be able to withstand high operating temperatures in addition to being an electric conductor. Thomas A. Edison in 1879 used a carbon filament for the first light bulb he developed. Even though the melting point of carbon is high (~ 3823°K), it evaporates very quickly at high temperatures blackening the bulb and weakening the filament. To reduce these effects, carbon filaments were operated at much lower temperatures than at the melting point of carbon and hence at very low luminous efficacy.

Around 1910, tungsten filaments were introduced. Tungsten has a high melting point (3683°K) and its evaporation rate is much lower than that of carbon. Both these properties allow operation of tungsten filaments at high temperatures with a high luminous efficacy. Filament temperature in a common bulb is in the range 2700°K to 3050°K while for a photoflash lamp it is in the range of 3400°K to 3450°K.

Most materials oxidize in air at high temperatures. To avoid oxidation, filaments used to be kept in vacuum, maintained in the glass envelope. Since 1913, however, the common practice is to fill the bulb with a mixture of inert gases, argon and nitrogen. The presence of these gases also reduces evaporation and hence increases the lifetime of the bulb.

In electric discharge lamps, a two-step process, excitation and deexcitation of atoms, is responsible for light production. The electrons in the atom (see Chapter 24 for details) are in orbits around the nucleus. The energy of electrons in each orbit is fixed. During a collision between an atom and a fast moving ion in a discharge tube, they exchange energy, part of the energy absorbed by the atom being used in exciting it, that is, raising one of the electrons in the atom to a higher state. Since the excited states are unstable, the electrons fall back

Figure 18.13. Luminous efficacy of incandescent filaments as a function of temperature. This curve is equally good for blackbody radiators.

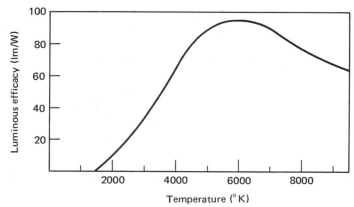

into the ground or the lowest energy state releasing the energy difference in the form of photons having a frequency given by $hf =$ energy difference. The spacings between the energy levels vary from atom to atom and hence the wavelengths of the light emitted are characteristic of the gas inside the discharge tube. For example, a light source containing mercury vapor produces light in different colors. The prominent radiation is yellow (578 nm), green (546 nm), violet (405 nm and 436 nm), and uv (280, 297, 302, 313, 334, and 365 nm). About 96% of the light is emitted in the ultraviolet (uv) region and the rest, about 4%, in the visible region.

The energy radiated in the uv range can be converted to energy in the visible range by taking advantage of a phenomenon known as fluorescence. Some materials absorb radiation in the uv range and reradiate in the visible range. Such materials are called fluorescent materials and the process is called fluorescence. A coating made of fluorescent material is applied over the inside surface of the

bulbs. This coating has uv to visible light conversion efficiency of about 33%. Light produced by the fluorescent material has a continuous spectrum from red to violet almost similar to sunlight. Therefore, light emitted in a fluorescent bulb has a continuous spectrum with superimposed peaks due to the characteristic radiation in the visible region mentioned earlier (Fig. 18.14).

The general features of a fluorescent lamp are shown in Fig. 18.15. The cylindrical glass envelope is coated with fluorescent material on the inside. The filaments at the ends of the tubes have two functions: they emit electrons when heated and they act as electrodes across which an electric potential can be applied. On applying a potential difference between the electrodes, the electrons accelerate and gain velocity in the electric field. These electrons collide with the atoms in the tube exciting and sometimes ionizing the mercury atoms, thereby causing the atoms to produce light.

The operating vapor pressure inside a fluorescent

Figure 18.14. Radiant energy distribution for fluorescent light sources. The peaks are light emitted by the mercury vapor and the continuous distribution by the fluorescent material. Energy distribution changes with the composition of the fluorescent material.

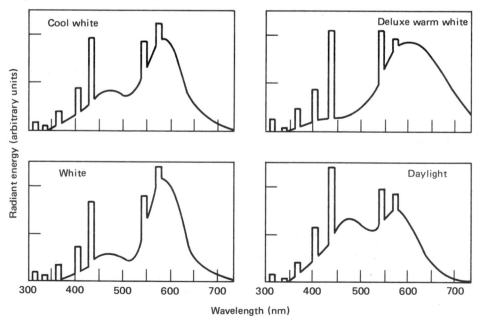

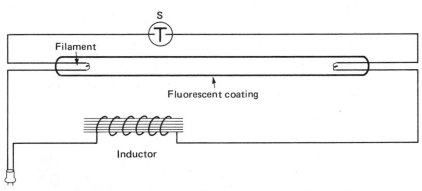

Figure 18.15. When the switch S is closed current flows through the filaments heating them. Hot filaments emit electrons. The switch is then opened and the resulting interruption of current induces sufficiently high voltage in the inductor to initiate an electric discharge by accelerating the free electrons. Automatic switching arrangements such as a neon tube with bimetallic element often replace the mechanical switch in the circuit.

bulb is 8×10^{-3} mm Hg which is close to the vapor pressure of mercury at 45°C. Light output and luminous efficacy reach optimum values around this temperature. It is interesting to note that the operating temperature of a fluorescent bulb, because of the basic difference in the mechanism involved in the emission of light, is much lower than that of an incandescent lamp.

There is a second class of discharge lamps which operate at a higher vapor pressure, about 1½ to 4 times atmospheric pressure. The main discharge tube is much shorter (4 to 12 in.) than the fluorescent lamp and is made of quartz, a glass which is capable of operating at high temperatures (1000°K). A small amount of argon enclosed in the tube plays an important role in initiating the main discharge. At room temperature the amount of mercury (or other metals like sodium) vapor is not sufficient to produce a discharge and the same is true with argon. However, an electric discharge in argon gas can start between the starting electrode and one of the main electrodes since they are closer (Fig. 18.16). This initial discharge increases the temperature near the electrodes causing mercury to evaporate. When the vapor pressure is sufficiently high, an electric discharge starts between the main electrodes. This discharge increases the temperature in the tube and,

consequently, the pressure of the metal vapor increases and finally a steady discharge through the vapor results. The lamp warmup process, from the starting discharge to normal operation, takes about 5 to 7 min.

The high-intensity discharge lamps are generally used for industrial or street lighting. Such bulbs, rating from 100 to 1500 W, are commercially available. The luminous efficiency of these bulbs, especially for the multivapor types, is higher than the other types of light sources (Fig. 18.17). The multivapor lamps contain iodides of sodium, thallium, and indine in addition to mercury. Since the vapor contains several elements, the light is distributed over a wider range of wavelengths than in single vapor lamps.

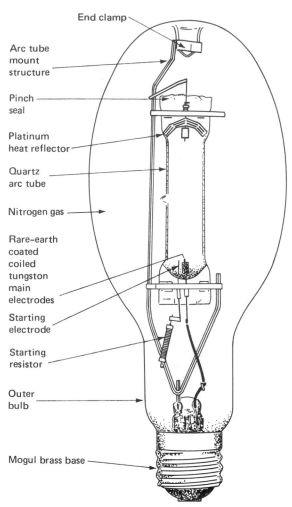

End clamp
Arc tube mount structure
Pinch seal
Platinum heat reflector
Quartz arc tube
Nitrogen gas
Rare-earth coated coiled tungston main electrodes
Starting electrode
Starting resistor
Outer bulb
Mogul brass base

Figure 18.16. Details of a high-intensity mercury discharge lamp.

lumens/watt
200
150
100
50
0
Carbon
Tungsten
Mercury
Fluorescent
Multivapor

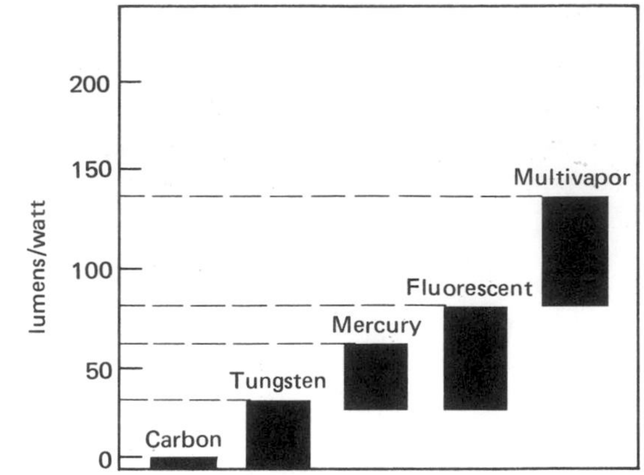

Figure 18.17. Efficacy of various light sources. Carbon and tungsten are incandescent sources; mercury, fluorescent, and multivapor are electric discharge sources.

SUMMARY

In this chapter we defined the following terms.

The intensity of a source measures the strength of the source. If a source emits 1 lumen per unit solid angle its intensity is 1 candela.

Luminous flux measures radiated power weighted for the relative sensitivity of our eye to different colors. Luminous flux in lumen equals 680 × radiated flux in watts × relative luminosity.

Illuminance measures the amount of luminous flux falling on a unit area surface. Illumination (lx) = lm/m^2. Also 1 lux (lx) of illuminance is produced by source of 1 candela (cd) at a distance of 1 m. Illumination produced by a source of 1 cd at a distance of 1 ft is close to the old unit, ft•candle.

The intensity of sources is compared using photometers. There are two types of light sources: (a) incandescent in which radiation is emitted as the temperature of the filament is increased; (b) discharge tubes in which kinetic energy gained by ions in the electric field is transformed into internal energy of the atoms (excitation of the atoms) during collisions, and during deexcitation of the atoms light is emitted. Luminous efficacy is very small (depends on temperature) for incandescent bulbs compared to discharge lamps. For example, fluorescent lamps are on the average two times more efficient in producing light than incandescent bulbs.

QUESTIONS AND PROBLEMS

18.1   An ice cream cone is 6 in. long and its cross-sectional area at 4 in. from the apex is 3 in^2. Calculate the solid angle of the ice cream cone.

18.2   A diverging light beam is produced by placing a cardboard with a 2 in. diameter circular hole at a distance of 10 in. from a 100 cd light source. Calculate (a) the solid angle of the beam and (b) the luminous flux flowing through the cone.

18.3   An isotropic bulb emits 200 W of light in the wavelength region of 555 nm. Calculate (a) the total luminous flux emitted by the bulb, (b) the intensity of the source in candelas, and (c) the resultant illuminance at a surface 1 m from the source.

18.4   A 4 ft long fluorescent light has a total intensity of 200 cd. Calculate the illuminance at a surface 12 ft below the surface.

18.5   The illumination due to a 100 cd source of light on a surface 2 m from the source is found to be equal to the illuminance by another source when kept 5 m away from the surface. Obtain the intensity of the second light source.

18.6   Intensity of a light source is 100 cd. Obtain (a) the illuminance at a surface 1 m away from the source and (b) the flux through a solid angle of 5 steradians.

18.7   Illuminance on a surface 1 m away from the source is 150 lx. Calculate the illuminance on a surface 2 m and 3 m away from the source.

18.8   A 50 cd lamp illuminates a table 2 m away from the lamp. The same illuminance is obtained from another source when it is kept 5 m away from the table. Calculate (a) the intensity of the second source and (b) the illuminance produced on the table.

18.9   Compute the illumination produced by a 40 cd source on a surface 30 cm from it. The normal to the surface makes a 30° angle with the flux.

Electric Charge and Electric Field

There are two types of electric charges: positive charges and negative charges. All electrically neutral matter (solids, liquids, and gases) is composed of atoms which contain equal numbers of positively charged and negatively charged particles. In an atom positively charged particles known as protons along with uncharged particles known as neutrons occupy a small volume in the center called the nucleus and the negatively charged particles called electrons form the outer shell of the atom. The magnitudes of the charges of the electron and the proton are equal (Table 19.1). Therefore, for neutrality of an atom, the number of electrons equals the number of protons. Figure 19.1 shows a schematic diagram of a carbon atom.

TABLE 19.1 Mass and Charge of the Constituent Particles of an Atom

PARTICLE	CHARGE	MASS
Electron	-1.6×10^{-19} C	9.1×10^{-31} kg
Proton	$+1.6 \times 10^{-19}$ C	1.6726×10^{-27} kg
Neutron	0	1.6749×10^{-27} kg

The force that keeps the electrons and protons together in an atom is the electrostatic force. This *electrostatic force is attractive between opposite charges and repulsive between similar charges.* Protons attract electrons, while they repel other protons. Similarly, electrons repel electrons.

19.1 COULOMB'S LAW

The force between two point charges q_1 and q_2 separated a distance r (Fig. 19.2) is given by the equation

$$F = k\frac{q_1 q_2}{r^2}. \qquad (19.1)$$

This equation is known as *Coulomb's law, which states that the electrostatic force is proportional to the charges and inversely proportional to the square of the distance of separation.* The force always acts along the line joining the two charges.

The basic unit for the charge is the coulomb (C). For example, the charge of an electron (e) is given

by

$$e = -1.6 \times 10^{-19} \, C.$$

Coming back to Eq. (19.1), if charge is expressed in coulombs, and r in meters, we obtain the force in newtons. Hence we have

$$F\,(N) = k\frac{q_1\,(C) \times q_2\,(C)}{[r\,(m)]^2}.$$

Solving for k, we note that its unit is $N \cdot m^2/C^2$. In SI units the value of k is $9 \times 10^9 \, N \cdot m^2/C^2$. Hence Eq. (19.1) becomes

$$F\,(N) = (9 \times 10^9\,\frac{N \cdot m^2}{C^2})\frac{q_1\,(C)\,q_2\,(C)}{[r\,(m)]^2}. \quad (19.1a)$$

EXAMPLE 19.1

Calculate the force acting on charge q_2 (2 C) due to q_1 (2 C) separated by a distance of 3.3 m (Fig. 19.2).

Figure 19.1. Pictorial representation of a carbon atom. The positive charge of the nucleus is balanced by the negative charge of the electrons.

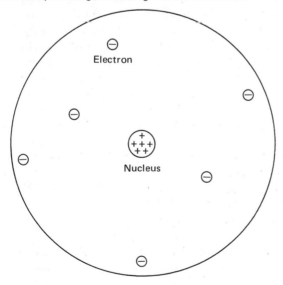

Electron

Nucleus

Figure 19.2. Electric forces exist between charges. Opposite charges attract each other and similar charges repel each other. Magnitude of the force is given by Coulomb's law. The force acting on charge q_1 is equal to the force acting on q_2, but they are in opposite directions.

Force F_2 on q_2 due to q_1 is

$$F_2 = (9 \times 10^9 \frac{\text{N} \cdot \text{m}^2}{\text{C}^2}) \frac{2\text{ C} \times 2\text{ C}}{(3.3\text{ m})^2}$$

$$= 3.3 \times 10^9 \text{ N}.$$

Force F_2 is a repulsive force and it is acting away from q_1 as indicated in Fig. 19.2. Also note that an equal force, in the opposite direction acts on q_1 due to repulsion of q_1 by q_2.

EXAMPLE 19.2

Calculate the total force acting on an electron placed midway between two charges $+ 1$ C each as shown in Fig. 19.3.

Force F_1 on the electron due to q_1 is

$$F_1 = (9 \times 10^9 \frac{\text{N} \cdot \text{m}^2}{\text{C}^2}) \frac{1\text{ C} \times 1\text{ C} \times 1.6 \times 10^{-19}\text{ C}}{(1\text{ m})^2}$$

$$= 14.4 \times 10^{-10} \text{ N to the left.}$$

Force F_2 on the electron due to q_2 is

$$F_2 = 14.4 \times 10^{-10} \text{ N to the right.}$$

The two forces F_1 and F_2 are of equal magnitude and in opposite directions and hence cancel each other. Therefore the net force on the electron equals 0.

Note that the total force acting on a charge due to a distribution of charges is not, in general, given by an algebraic addition of forces produced by the individual charges. Their direction should be taken into consideration as for all vectors. As a further illustration of these points, consider the following example which is a modification of Example 19.2.

EXAMPLE 19.3

Calculate the force on the electron for the arrangement shown in Fig. 19.4. This time forces F_1 and F_2 are acting in the same direction.

$$\text{Net force} = F_1 + F_2$$

$$= 2(9 \times 10^9 \frac{\text{N} \cdot \text{m}^2}{\text{C}^2}) \frac{1\text{ C} \times 1.6 \times 10^{-19}\text{ C}}{(1\text{ m})^2}$$

$$= 28.8 \times 10^{-10} \text{ N}.$$

The force on the electron is directed in the negative x direction.

19.2 ELECTRIC FIELD

When a small positive test charge q_t is brought near a charge Q (test charge is kept at point 1 in Fig. 19.5), it will experience a force given by Coulomb's law. There is no need for any medium or agent to exist to transmit this force between the two charges. The charge Q has associated with it a field, and the field, in turn, acts on the test charge placed at some point in this field. This is similar to the earth's gravitational field which produces an attractive force on any object brought into the field. Note that the field exists at every point around Q whether the test charge is present or not, but the existence of such a field can be verified by bringing

Figure 19.3. An electron is kept between two positive charges ($+ 1$ C each). Forces F_1 and F_2 due to charges q_1 and q_2, respectively, are in opposite directions. Hence the net force on the electron is zero.

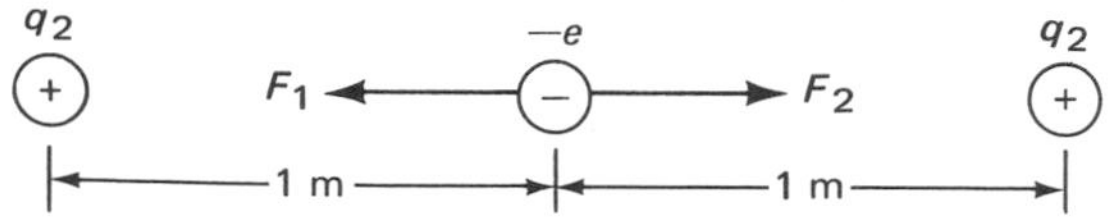

Figure 19.4. Forces acting on an electron placed between a positive and a negative charge are in the same direction.

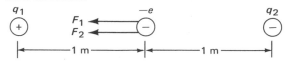

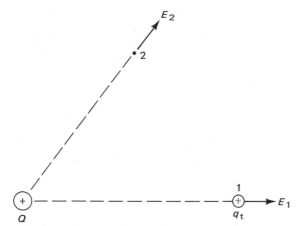

Figure 19.5. Directions of electric field due to charge Q at points 1 and 2 are indicated in figure. q_t is a small test charge.

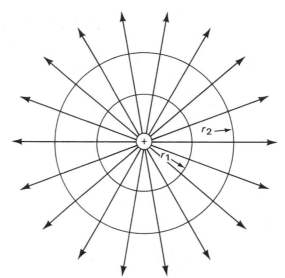

Figure 19.6. Electric field lines emanating from a positive charge.

a test charge into the field.

The electric field at any point is defined as the force acting on a unit positive charge. The direction of the field is the direction of the force on the positive test charge.

In Fig. 19.5, the electric field at point 1 is given by

$$E_1 = \frac{F_1}{q_t} = \frac{kQq_t}{r^2 q_t} = \frac{kQ}{r^2} \, . \tag{19.2}$$

The equation we just derived is for the electric field around a point charge. The direction of the field is along the line joining Q and point 1 and directed away from the charge as indicated in Fig. 19.5. Similarly, a field at point 2 is also along the line joining Q and point 2. In general. therefore, the electric field surrounding a point charge is radially directed away from the charge if it is positive and towards the charge if it is negative.

The unit for E is clearly newton/coulomb. It has another unit volt/meter; volt will be defined later. A representation of the electric field is shown in Fig. 19.6. The imaginary lines around the charge known as electric field lines are drawn such that (a) the line density, the number of lines per unit area, is proportional to the electric field, (b) the directon of the line at any point is the direction of the

field at that point, and (c) the field lines never cross each other. Consider the two spheres of radii r_1 and r_2 (Fig. 19.6). Even though the number of lines passing through both surfaces are the same, the line density is inversely proportional to r^2 just like the electric field.

Electric field lines between two charges are shown in Fig. 19.7 (a) and (b). In the case of two positive charges the field lines bend away from each other and thus avoid crossing.

Let us now consider the relationship between field and charge in a space defined by two metal plates, a system of some practical application. A total free charge $+Q$ is present on one of the plates and $-Q$ on the other one. Note that free charges always reside on the surface of a metal. Mutual attraction between the positive and negative charges brings them closer, and consequently, we get a uniform charge distribution on the inner surfaces of the plates (Fig. 19.8). The electric field lines by definition originate at the positive charges and terminate at the negative charges. Since the lines can not cross each other, the field lines are parallel in the center and are curved at the edges (Fig. 19.8). For the interior part in which the lines are parallel, the line

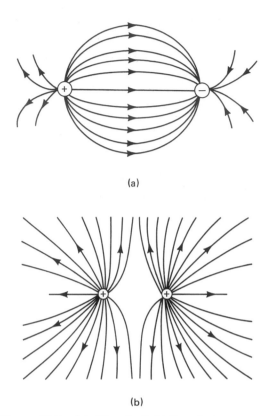

(a)

(b)

Figure 19.7. (a) Electric field lines between a positive and a negative charge. Note that field lines start on positive charges and end on negative charges. (b) Electric field lines between two positive charges. Note the field lines do not cross each other since, if they did, there would be a resultant field.

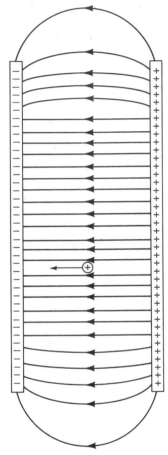

Figure 19.8. Electric field lines between two charged plates. The direction of the force acting on a positive charge and hence the direction of the field is from the positive plate to the negative plate.

density and hence the electric field is constant. The field in SI units in the interior is given by

$$E = Q/\epsilon_0 A, \qquad (19.3)$$

where A is the area of the plates and ϵ_0 is the permittivity of free space. $\epsilon_0 = 8.854 \times 10^{-12}$ $C^2/$ N $\cdot$ m^2 which is related to k in Coulomb's law by $k = 1/4\pi\epsilon_0$.

It is not difficult to see the proportionality, $E \propto Q/A$. As the charge density (Q/A) increases, one expects an increase in line density. The increased curvature of the lines at the edges indicates a de-

crease in the electric field and changes in its direction near the edges. This is called the edge effect. The edge effect increases as the distance between the plates increases. According to Eq. (19.3), the electric field is independent of d, the separation between the plates. However, one should be mindful of the increase in edge effect as d increases. A positive test charge placed between the plates will move in the direction from the positive plate to the negative plate. This direction, as indicated in Fig. 19.8, is the direction of the electric field.

EXAMPLE 19.4

Calculate the electric field at point P due to charges $q_1 = 4 \times 10^{-6}$ C and $q_2 = 4 \times 10^{-6}$ C, and determine the force on a proton placed at point P (see Fig. 19.9). The points P, q_1, and q_2 form an equilateral triangle with each side = 2m.

If a unit test charge is placed at P, it will experience a force F_1 due to q_1 and F_2 due to q_2. The two forces are acting in two different directions. To find the sum of these forces, we use the technique discussed in Chapter 1, that is, to resolve the forces along two mutually perpendicular directions. We choose x and y axes as shown in Fig. 19.9.

The sum of the x and y components of forces F_1 and F_2 are as follows:

<u>x Components</u>

$F_1 \sin (\theta/2)$

$-F_2 \sin (\theta/2)$

Sum: $F_1 \sin (\theta/2) - F_2 \sin (\theta/2)$

<u>y Components</u>

$F_1 \cos (\theta/2)$

$F_2 \cos (\theta/2)$

Sum: $F_1 \cos (\theta/2) + F_2 \cos (\theta/2)$

Since $q_1 = q_2$ and $r_1 = r_2$, $F_1 = F_2$. Therefore, the x components add up to zero.

Net force on the unit test charge

= Sum of the vertical components

$= 2(9 \times 10^9 \frac{\text{N} \cdot \text{m}^2}{\text{C}^2}) \frac{(4 \times 10^{-6}\text{ C})}{(2\text{ m})^2} (1\text{ C})\cos 30°$

$= 1.56 \times 10^4$ N.

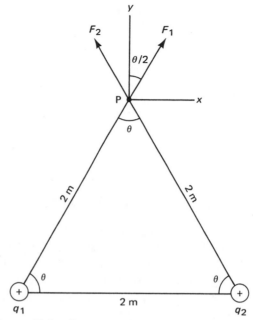

Figure 19.9. Example 19.4. F_1 and F_2 are, respectively, the forces acting on a test charge at P due to q_1 and q_2.

Electric field at P = Force per unit charge

$= 1.56 \times 10^4$ N/C.

Force on a proton kept at P = Eq.

$= (1.56 \times 10\text{ N/C}) \times 1.6 \times 10^{-19}$ C

$= 2.5 \times 10^{-15}$ N.

EXAMPLE 19.5

Assume the total charge on each plate in Fig. 19.8 is 5×10^{-10} C, the area of the plates is 10 cm^2, and the separation between the plates is 1 cm. Calculate the electric field between the plates.

Electric field $= \dfrac{Q}{\epsilon_0 A}$

$$= \frac{5 \times 10^{-10} \text{ C}}{(8.854 \times 10^{-12} \frac{\text{C}^2}{\text{N}^2 \cdot \text{m}^2})(10 \text{ cm}^2)(\frac{1 \text{ m}^2}{10^4 \text{ cm}^2})}$$

$$= 5.6 \times 10^4 \text{ N/C}.$$

19.3 ELECTRIC POTENTIAL

Let us consider the parallel plate system once again (Fig. 19.10). The total charge on each plate is Q and there exists an electric field between the plates. A positive charge q between the plates will experience a force Eq. To bring this charge from the negative plate to the positive plate a force at least equal to Eq should be applied in the opposite direction. Work done in moving the charge is

Figure 19.10. A force Eq acts on a charge placed in the electric field between the charged plates. To move the charge in the direction of the positive plate, a force at least equal to Eq in magnitude should be applied.

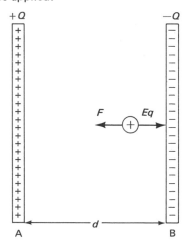

given by

$$W_{\text{B} \to \text{A}} = Eqd. \qquad (19.4)$$

This work will be stored as potential energy. *Work done per unit charge is called the electric potential.*

In the example above, the work done is from plate B to plate A. Therefore, the electric potential of plate A with respect to that of plate B, V_{AB}, is given by

$$V_{\text{AB}} = \frac{W_{\text{B} \to \text{A}}}{q} = Ed.$$

To define the units, Eq. (19.5) is written without the subscripts

$$V(\text{volts}) = \frac{W(\text{joules})}{q(\text{coulombs})}. \qquad (19.5a)$$

The potential V is commonly expressed in volts in SI units, W in joules, and q in coulombs. Also since $E = V/d$, we may now express E in volts/meter as well as newton/coulomb as mentioned earlier. It is important to remember electric potential energy, work done in bringing a charge to a point where the potential is V, is given by potential energy, $W = Vq$.

EXAMPLE 19.6

Calculate the potential difference between the plates for the given quantities in Example 19.5.

E, for the arrangement in Example 19.5, is 5.6 $\times$ 10^4 N/C; $d = 1$ cm $= 0.01$ m.

Potential difference between the plates

$$= (5.6 \times 10^4 \text{ N/C}) \times 0.01 \text{ m}$$

$$= 560 \text{ V}.$$

EXAMPLE 19.7

If a proton is released at the positive plate of Example 19.6, what will be its kinetic energy when it reaches the negative plate? (Mass of proton $= 1.67 \times 10^{-27}$ kg).

The potential energy of the proton at the positive plate equals the amount of work needed to bring it to the positive plate:

$$Vq = (560 \text{ V} \times 1.6 \times 10^{-19} \text{ C})$$

$$= 8.96 \times 10^{-17} \text{ J}.$$

As the proton reaches the negative plate its PE changes to KE.

$$KE = \tfrac{1}{2}mv^2 = 8.96 \times 10^{-17} \text{ J}.$$

Electric Potential Due to a Point Charge

The electric potential due to a point charge at a distance r from the charge is given by

$$V = k\frac{Q}{r} . \qquad (19.6)$$

The reference point is taken as infinity, that is, Vq is the amount of work needed to bring a positive charge q from infinity to that point.

EXAMPLE 19.8

A negative point charge of 5×10^{-6} C is located at point P. Calculate the electric field and potential at a point 0.5 m away from the charge. If a positive charge of 3×10^{-6} C is a distance of 0.5 m from the negative charge, how much work is done to separate the charges (Fig. 19.11)?

Figure 19.11. Example 19.8.

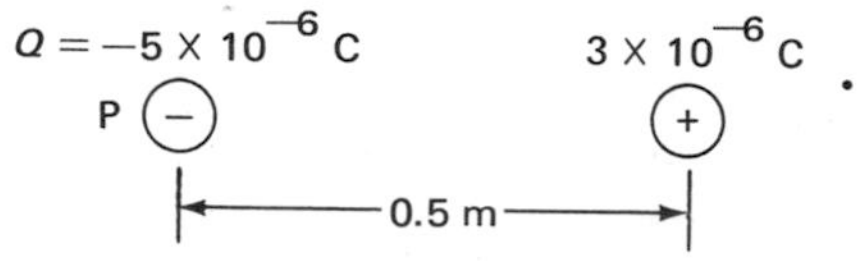

$$Q = -5 \times 10^{-6} \text{ C} \qquad 3 \times 10^{-6} \text{ C}$$

P (−)   (+)

|←——— 0.5 m ———→|

Electric field 0.5 m from the charge

$$= k\frac{Q}{r^2}$$

$$= (9 \times 10^9 \frac{\text{N} \cdot \text{m}^2}{\text{C}^2}) \frac{(5 \times 10^{-6} \text{ C})}{(0.5 \text{ m})^2}$$

$$= 1.8 \times 10^5 \text{ V/m}.$$

The direction of the field is towards the negative charge. The electric potential at 0.5 m is

$$k\frac{Q}{r} = (9 \times 10^9 \frac{\text{N} \cdot \text{m}^2}{\text{C}^2})(\frac{-5 \times 10^{-6} \text{ C}}{0.5 \text{ m}})$$

$$= -9 \times 10^4 \text{ V}.$$

Note that by keeping the negative sign for the charge, we obtained a negative potential. The potential around a negative charge is negative because to bring a positive charge from infinity to that point we do not have to do any work; the field does the work. On the other hand if we remove the positive charge by taking it away from the negative charge, we have to do work. Work to be done on a positive charge to remove it from the negative charge = Final potential energy − Initial potential energy = $V_\infty q - V_{0.5}q = 0 - (-9 \times 10^4 \text{ V}) (3 \times 10^{-6} \text{ C}) = 2.7 \times 10^{-1}$ J.

19.4 CAPACITOR

Two metal plates separated by a nonconducting material is a capacitor. Two parallel plates separated by air, which we discussed in the last two sections, is an example of a capacitor. If isolated, a capacitor can maintain its charge and, therefore, the potential difference. The equations we have relating these quantities are

$$E = \frac{Q}{\epsilon_0 A} \; ; \; V = Ed.$$

From these two we get an equation connecting Q and V,

$$Q = \frac{\epsilon_0 A}{d} V. \qquad (19.7)$$

The potential difference between the plates is proportional to charge. If the gap between the plates is filled by a material other than air, Eq. (19.7) has to be modified as follows

$$Q = (\frac{K\epsilon_0 A}{d}) V \qquad (19.8)$$

where K is the dielectric constant of the material; for air $K = 1.00054$ which is approximately equal to 1; Eq. (19.7) therefore is valid for an air-filled capacitor. Dielectric constants for a few selected materials are given in Table 19.2.

In Eq. (19.8) the terms inside the parentheses depend on the geometry of the arrangement and the material between the two conducting plates. By combining these terms together into one, represented by letter C, Eq. (19.8) can be rewritten in the form

$$Q = CV \text{ or } C = Q/V. \qquad (19.9)$$

In this case, C is called the capacitance of the system. According to Eq. (19.9) as charge increases voltage increases and vice versa, but there is a limit

on the charge one can place on a capacitor. It depends on the dielectric strength of the material between the plates, that is, the maximum field that can exist without the occurrence of electrical breakdown.

The unit of capacitance is the farad (F). If 1 V develops on a capacitor for 1 C of charge placed on it, the capacitor has a value of 1 F. One farad is usually too large and therefore, the microfarad (μF) and picrofarad (pF) are the units used more often.

EXAMPLE 19.9

Calculate the capacitance of an air-separated parallel plate capacitor. $A = 10$ cm^2; $d = 1$ cm. What is the maximum amount of charge one can place on the capacitor without causing electrical breakdown?

$$C = \frac{\epsilon_0 A}{d} = (8.85 \times 10^{-12} \text{ C}^2/\text{N} \cdot \text{m}^2)$$

$$\times (\frac{10 \text{ cm}^2}{10^4 \text{ cm}^2/\text{m}^2})(\frac{100 \text{ cm/m}}{1 \text{ cm}})$$

$$= 8.85 \times 10^{-13} \text{ F}$$

$$= 8.85 \times 10^{-7} \mu\text{F}.$$

If all the quantities are expressed in SI units, the answer will be in farads. The breakdown field in air is 0.8 kV/mm. Therefore, the maximum potential difference that can exist between the plates is

$$0.8 \text{ kV/mm} \times 1 \text{ cm} \times 10 \text{ mm/cm} = 8 \text{ kV}.$$

Maximum charge $= CV$

$$= (8.85 \times 10^{-7} \mu\text{F})(8 \times 10^3 \text{ V})$$

$$= (8.85 \times 10^{-13} \text{ F})(8 \times 10^3 \text{ V})$$

$$= 7.08 \times 10^{-9} \text{ C}.$$

TABLE 19.2 Properties of Selected Dielectrics

MATERIAL	DIELECTRIC CONSTANT	DIELECTRIC STRENGTH (kV/mm)
Vacuum	1.00000	∞
Air	1.00054	0.8
Water	78	—
Paper	3.5	14
Porcelain	6.5	4
Ruby Mica	5.4	160
Fused quartz	3.8	8
Teflon	2.1	60

EXAMPLE 19.10

A parallel plate capacitor with air has a capacitance of 2 μF. If the gap between the plates is filled with water, what is its new capacitance?

$$C_{air} = \frac{\epsilon_0 A}{d} = 2 \ \mu F$$

$$C_{water} = \frac{K \epsilon_0 A}{d} = 78 \times 2 = 156 \ \mu F.$$

19.5 CAPACITORS CONNECTED IN SERIES AND PARALLEL

If more than one capacitor is connected in series, what is its equivalent capacitance? A capacitor C is equivalent to capacitors C_1, C_2, and C_3 connected in series (Fig. 19.12) if it can maintain a charge equal in amount to that on each of the capacitors connected in series when the same voltage source is connected to it.

Note the symbols we are using for the capacitors: two vertical parallel lines of equal length and for the voltage source, two vertical parallel lines, the longer line representing the positive terminal and the shorter one the negative terminal. The positive terminal is at a higher potential than the negative terminal, the potential difference between the terminals being V. A voltage source (for example, a battery) maintains a constant voltage between its terminals. When such a source is connected to a system the potential difference between the two terminals of the system is equal to the source potential difference.

When the capacitors are connected in series, the charge on each capacitor, as stated above, will be the same. Consider the section of the circuit containing the negative plate of C_1, the connecting wire between C_1 and C_2 and the positive plate of C_2. When no voltage is applied, this section is neutral. When a voltage is applied charges separate, resulting in the accumulation of equal amounts of positive and negative charges on C_1 and C_2. Similarly the charges Q on C_2 and C_3 have to be equal.

The sum of the potential differences across the

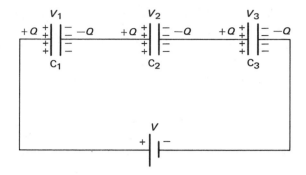

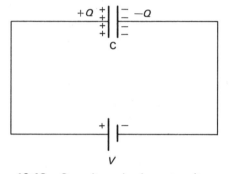

Figure 19.12. Capacitors having capacitances C_1, C_2, and C_3 are connected in series. The equivalent capacitor has a capacitance C.

capacitors should add up to the total potential difference at the terminals of the battery. Hence

$$V - V_1 + V_2 + V_3 = \frac{Q}{C_1} + \frac{Q}{C_2} + \frac{Q}{C_3}. \quad \textbf{(19.10)}$$

Similarly for the equivalent capacitor

$$V = \frac{Q}{C}. \quad \textbf{(19.11)}$$

From Eqs. (19.10) and (19.11), we get

$$\frac{1}{C} = \frac{1}{C_1} + \frac{1}{C_2} + \frac{1}{C_3}. \quad \textbf{(19.12)}$$

These equations give us the value of the equivalent capacitor in terms of the values of the individual capacitors connected in series.

Figure 19.13 shows three capacitors connected in parallel and the equivalent capacitor circuit. It is obvious that the charge on the equivalent capacitor should be equal to the sum of the charges on C_1, C_2, and C_3. Therefore, we get the following relationship:

$$Q = Q_1 + Q_2 + Q_3.$$

Since $Q = CV$, $Q_1 = C_1 V$, $Q_2 = C_2 V$, and $Q_3 = C_3 V$ we obtain

$$VC = VC_1 + VC_2 + VC_3.$$

Figure 19.13. Capacitors having capacitances C_1, C_2, and C_3 are connected in parallel. The equivalent capacitor has capacitance C.

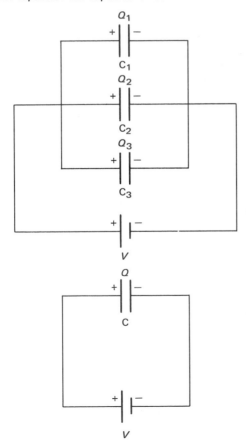

Dividing both sides of the equation by V yields

$$C = C_1 + C_2 + C_3, \qquad (19.13)$$

which gives the value of the equivalent capacitor as a sum of the values of the individual capacitors for a parallel connection.

EXAMPLE 19.11

Two capacitors of 10 μF and 20 μF are connected in series to a battery of 10 V (Fig. 19.14). Calculate (a) the equivalent capacitance, (b) the charge on each capacitor, and (c) the voltage across each capacitor.

The equivalent capacitance is given by

$$\frac{1}{C} = \frac{1}{C_1} + \frac{1}{C_2}$$

$$\frac{1}{C} = \frac{1}{10 \ \mu F} + \frac{1}{20 \ \mu F}$$

$$C = 6.66 \ \mu F.$$

Figure 19.14.   Example 19.11.

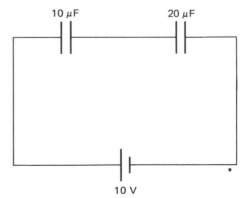

The charge on each capacitor equals the charge on the equivalent capacitor:

$$CV = (6.66 \times 10^{-6} \text{ F})(10 \text{ V})$$

$$= 6.66 \times 10^{-5} \text{ C}.$$

Voltage across $C_1 = \dfrac{q}{C_1}$

$$= \frac{6.66 \times 10^{-5} \text{ C}}{10 \times 10^{-6} \text{ F}} = 6.66 \text{ V}.$$

Voltage across $C_2 = \dfrac{q}{C_2}$

$$= \frac{6.66 \times 10^{-5} \text{ C}}{20 \times 10^{-6} \text{ F}} = 3.33 \text{ V}.$$

Note that the sum of the voltages across the capacitors is equal to the battery voltage.

EXAMPLE 19.12

Two capacitors 10 μF and 20 μF in parallel are connected to a battery of 10 V (Fig. 19.15). Calculate (a) the equivalent capacitance, (b) the voltage on each capacitor, and (c) the charge on each capacitor.

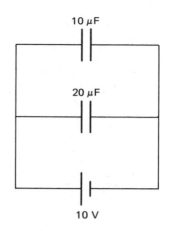

10 μF

20 μF

10 V

Figure 19.15. Example 19.12.

Equivalent capacitance $= C_1 + C_2 = 30$ F; the voltage across each capacitor $= 10$ V. The charge on each capacitor is

$$q_1 = C_1 V = (10 \times 10^{-6} \text{ F})(10 \text{ V})$$

$$= 1 \times 10^{-4} \text{ C};$$

$$q_2 = C_2 V = (20 \times 10^{-6} \text{ F})(10 \text{ V})$$

$$= 2 \times 10^{-4} \text{ C}.$$

SUMMARY   In this chapter, following the discussion of Coulomb's law, definitions of electric field, electric potential, and capacitance are given. The student should learn these definitions.

Coulomb's law relates the force between two point charges q_1 and q_2 at a distance r

$$F = k \frac{q_1 q_2}{r^2}.$$

The electric field at a point is defined as the force acting on a unit positive charge

$$E = F/q.$$

The field around a point charge is given by

$$E = k\frac{q}{r^2}.$$

The electric potential between two points A and B is the work needed to take a unit charge from A to B,

$$V_{AB} = \frac{W_{B \to A}}{q}.$$

The potential at a point located a distance r from a charge Q is

$$V = k\frac{Q}{r}.$$

The potential between two plates with a charge Q on them is given by

$$Q = \frac{\epsilon_0 A}{d} V.$$

The field between two plates is given by the following equations:

$$E = \frac{Q}{\epsilon_0 A} \quad \text{and} \quad E = \frac{V}{d}.$$

The capacitance C of two parallel plates with dielectric between them is given by

$$C = \frac{K \epsilon_0 A}{d}.$$

The equivalent capacitance is given by

$$\frac{1}{C} = \frac{1}{C_1} + \frac{1}{C_2} + \frac{1}{C_3}$$

For C_1, C_2, and C_3 connected in series, and

$$C = C_1 + C_2 + C_3$$

for C_1, C_2, and C_3 connected in parallel.

QUESTIONS AND PROBLEMS

19.1 A simple electroscope is shown in Fig. 19.16. The thin metal leaves at the end of the central electrode repel each other as the charge on it increases. Explain why.

19.2 The simple electroscopes played a very important role in the discovery of cosmic rays and the study of radioactivity (see Fig. 19.16). Radiation passing through the electroscope produces free charges. The amount of charges thus produced can be measured from the decrease in the deflection of the metal leaves. Explain how it works.

19.3 In winter, after walking on carpets, if one touches another object an electric shock results. Explain why.

19.4 Two small charged spheres are suspended by 10 cm long strings (Fig. 19.17). The charge on each sphere equals 10^{-7} C. If the separation between the charges is 5 cm, calculate the mass of each sphere.

19.5 In the hydrogen atom the distance between the electron and proton is 0.53×10^{-10} m. Calculate the electrostatic force between the two particles. You may now wonder why the two particles do not fall into each other. A satisfactory answer to this question was first suggested by Bohr.

19.6 The distance between protons in a nucleus is 10^{-15} m. Calculate the force between two protons in a nucleus.

19.7 Two positive charges, 5×10^{-6} C each, are separated by a distance of 10^{-6} m. Calculate the force acting on each of them.

19.8 Calculate the force on the central charge in Fig. 19.18(a). The charges at the corners are fixed $q = 1 \times 10^{-3}$ C, $Q = 2 \times 10^{-3}$ C, and $a = 10^{-3}$ m.

19.9 Repeat the calculation of Problem 19.8 for the case where two of the corner charges are negative as shown in Fig. 19.18(b).

19.10 Calculate the electric field at distances 10^{-2}, 10^{-1}, 10, and 10^2 m from a charge of 10^{-2} C.

Figure 19.16. Problems 19.1 and 19.2.

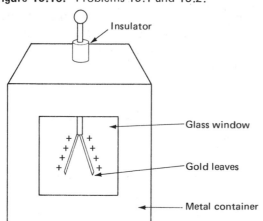

Insulator

Glass window

Gold leaves

Metal container

Figure 19.17. Problem 19.4.

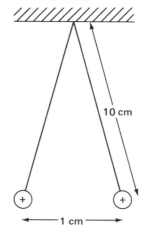

10 cm

1 cm

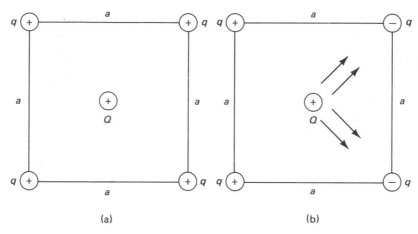

Figure 19.18. (a) Problem 19.8. (b) Problem 19.9.

19.11 Calculate the electric field between two parallel plates shown in Fig. 19.19.

19.12 Calculate the potential difference between the plates of Fig. 19.19.

19.13 Calculate the capacitance of the two plates in Fig. 19.19.

19.14 If an electron is placed between the two plates as in Fig. 19.19, what is the direction of the force acting on the electron? What is the magnitude of the force acting on the electron?

19.15 In a television tube electrons emitted by a hot filament are accelerated by applying an electric field, as shown in Fig. 19.20(a). If the accelerating potential is 10 kV calculate (a) the kinetic energy of the electrons and (b) their velocity as they emerge from the field.

19.16 The electrons of Problem 19.15 after acceleration go through two electric fields perpendicular to the velocity of the particle [Fig. 19.20(b)]. The strength of these fields determine the position on the TV screen where the electron will hit. The separation between the vertical deflection plate is 0.5 cm, the length of the plate is 1 cm, and the potential difference is 100 V. Obtain (a) the time spent by the electrons in the fields; (b) the force acting on the electron; (c) the total vertical deflection as the electrons emerge from the deflection plates.

Figure 19.19. Problems 19.11, 19.12, 19.13, and 19.14.

$Q = 1 \times 10^{-6}$ C

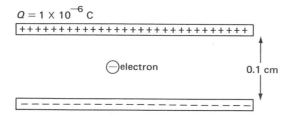

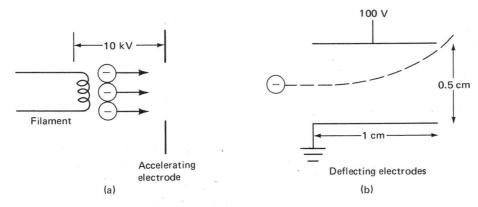

Figure 19.20. (a) Problem 19.15. (b) Problem 19.16.

19.17 How much work has to be done to take a charge of $+0.1$ C from a negative plate to a positive plate? The potential difference between the plates is 100 V (Fig. 19.21).

19.18 A charged oil drop is suspended between two plates by applying 700 V as shown in Fig. 19.22. The distance between the plates is 0.1 cm and the charge on the drop is 1.6×10^{-19} C. The oil drop remains stationary because the electric force balances the gravitational force. Calculate the mass of the oil drop. (Millikan used a similar arrangement to determine the charge of an electron in his historic oil drop experiment.)

19.19 Two metal plates of area 0.01 m^2 are separated by a distance of 0.01 cm (Fig. 19.23). (a) Calculate the capacitance. (b) If the gap between the metal plates is filled by mica ($K = 5.4$), calculate its new capacitance.

19.20 A 1 μF parallel plate capacitor is charged by a 10 V battery. Obtain the charge on each plate.

19.21 A 1 μF capacitor is first charged by a 10 V battery and then connected to an uncharged 1 μF capacitor (Fig. 19.24) in parallel. What is the voltage across the capacitor terminals?

19.22 Calculate the equivalent capacitance for the three capacitors shown in Fig. 19.25. $C_1 = 5$ μF, $C_2 = 10$ μF, and $C_3 = 20$ μF.

Figure 19.21. Problem 19.17.

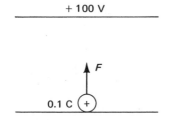

Figure 19.22. Problem 19.18.

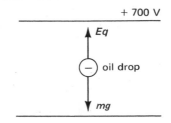

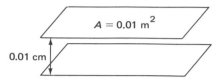

Figure 19.23. Problem 19.19.

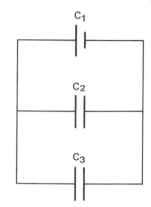

Figure 19.25. Problem 19.22.

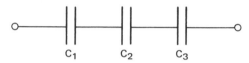

Figure 19.26. Problem 19.23.

19.23 Find the value of the equivalent capacitor for the three capacitors shown in Fig. 19.26. $C_1 = 2\,\mu F$, $C_2 = 10\,\mu F$, and $C_3 = 20\,\mu F$.

19.24 Find the equivalent capacitance of the system shown in Fig. 19.27. $C_1 = 10\,\mu F$, $C_2 = 7\,\mu F$, and $C_3 = 3\,\mu F$.

19.25 Find the potential difference between the terminals of each capacitor in Fig. 19.27.

Figure 19.24. Problem 19.21.

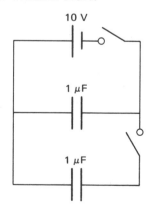

Figure 19.27. Problems 19.24 and 19.25.

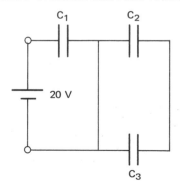

20
Electric Current and DC Circuits

Our daily life depends very much on the uninterrupted supply of electric power to our houses, offices, and factories. Imagine how uncomfortable life would be if electricity were turned off for long periods of time; all modern amenities of life like the air-conditioners, heaters, refrigerators, cooking ovens, radios, and televisions sets would then stop functioning. The flow of electric charge, or electric current, through these devices makes them work. Even though static electricity has some practical applications, current electricity is the most popular form of energy because of the convenience of transportation from the generator to the user and its variety of uses.

What is electric current? How is electric current maintained? What are some consequences of the flow of electric charge? How is electric current measured? These are some of the questions we attempt to answer in this chapter.

20.1 ELECTRIC CURRENT

Electric current is the flow of electric charges. Since there are both positive and negative charges, either or both may contribute to the current. Negatively charged electrons are responsible for the current through a metal wire, while electrons and positively charged ions, traveling in opposite directions, contribute to the current in an electric discharge tube (fluorescent bulb).

Current in amperes can be defined as the amount of charge in coulombs flowing through the cross-sectional area at a point in 1 second. This statement can be written in mathematical form:

$$I \text{ (amperes)} = \frac{Q \text{ (coulombs)}}{t \text{ (seconds)}}, \qquad (20.1)$$

where I stands for current and Q for charge flowing in a time interval t.

In the pictorial representation of the current flow in Fig. 20.1, a current of 1 A flows from left to right; that is, through the shaded cross-sectional area of the wire a charge of 1 C (sufficient number of charge carriers with a total charge 1 C) crosses every second. As mentioned earlier, electrons are the charge carriers in metals. In this case [Fig. 1(b)], therefore, a current flow from left to right is in fact produced by an electron flow in the opposite direction, right to left.

Figure 20.1. Pictorial representation of current flow through a wire. (a) Current is due to the flow of positive charges. Direction of current and charge flow are the same. (b) Current is due to the flow of negative charges. In this case direction of current flow is in the opposite direction to that of the charge flow.

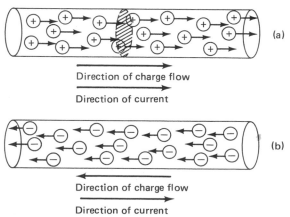

(a)

Direction of charge flow

Direction of current

(b)

Direction of charge flow

Direction of current

EXAMPLE 20.1

A current of 0.5 A flows through a wire for 5 min. Calculate the number of electrons that have crossed a point in the wire during this time.
We use

$$I = Q/t.$$

To obtain the charge flowing in a time t, we re-arrange the equation in the form $Q = It$. Q will be in coulombs if t is in seconds since I is defined as coulombs/second. $Q = I \, (\text{C/s}) \times t \, (\text{s})$. We are given $I = 0.5$ A, $t = 5$ min.

$$Q = (0.5 \text{ A}) (5 \text{ min} \times 60 \text{ s/min}) = 150 \text{ C}.$$

If n is the number of electrons carrying this charge, we have

$$ne = 150 \text{ C}.$$

Since e, the charge on each electron, is 1.6×10^{-19} C

$$n \, (1.6 \times 10^{-19} \text{ C}) = 150 \text{ C}$$

$$n = \frac{150 \text{ C}}{1.5 \times 10^{-19} \text{ C}} = 9.4 \times 10^{20}.$$

EXAMPLE 20.2

The flow of singly charged positive ions and electrons contributes equally to current flow in a fluorescent light bulb. The steady current through the tube is 1 A. How many positive ions flow through the cross-section of the tube per second?
The magnitude of charge of a singly charged positive ion is equal to that of the electron. Half of the total current is due to positive ions and the other half to electrons. If n_+ is the number of positive ions crossing a cross-section of the tube in 1 s, they carry with them a charge of ½ C, that is,

$$n_+ e = \text{½ C}$$

or

$$n_+ = \frac{1\ C}{2 \times 1.6 \times 10^{-19}\ C} = 3.1 \times 10^{18}.$$

20.2 OHMS' LAW

A metallic wire when connected to a battery provides a path for the current flow. The current flow through a conducting wire is proportional to the electric potential difference (battery voltage) between the ends of the wire and inversely proportional to the resistance of the wire. The word "resistance" has its literal meaning, that is, resistance to the flow of current. A wire of higher resistance allows a smaller current to flow compared to a wire of lower resistance. The dependence of current flow on potential difference and resistance is given by Ohm's law which states that, *the amount of current flowing in a metallic conductor is proportional to the potential difference and inversely proportional to the resistance of the conductor.* This statement can be written in mathematical form as

$$I = V/R, \tag{20.2}$$

where I is current in amperes, V the potential difference in volts, and R the resistance. The resistance R is measured in a unit called the ohm (Ω). The resistance is $1\ \Omega$ if a voltage difference of 1 V across a resistor produces a current of 1 A.

Ohm's law is illustrated in Fig. 20.2. In this simple case, the potential difference V is equal to the battery voltage (battery EMF). Also note the following symbols used in Fig. 20.2:

⎨⎧W⎫⎬, resistance;

⎢⎜⎟⎢$^+$, battery in which the longer line represents the higher potential or the + terminal;

——, connecting wire assumed to have zero resistance.

EXAMPLE 20.3

The filament of a flashlight battery has a resistance of $6\ \Omega$. Calculate the current through the filament when a 3 V battery is connected across it.

$$I = \frac{V}{R} = \frac{3\ V}{6\ \Omega} = 0.5\ A.$$

In Fig. 20.2 we have a completed electric circuit through which a current I flows. The current I that flows through the resistance outside a battery from b to a also flows through the battery. For the sake of simplicity, we assume a flow of positive charge in the direction of the current. An amount of charge equal to I coulombs flows through point a and reaches the negative side of the battery every second. This flow is somewhat like the motion of charges from the higher potential side to the lower potential side (as discussed in Section 3 of Chapter 19). However, in the flow through the battery from the negative to the positive terminal, the charges are moved to the higher potential side. Chemical reactions in the battery supply the necessary energy to do this work.

It is illuminating to compare the simple electric circuit of Fig. 20.2 with Fig. 20.3 where an analogous situation involving water flow is shown. The pump P maintains the water level in the tank by lifting water from a lower gravitational potential to a higher potential level. The pump P with the tank is similar to the battery in Fig. 20.2. The pipe AB provides the path for water flow. The rate of water flow through the pipe is determined by the physical properties of the pipe such as the diameter, length, etc. or the resistance of the pipe. It is also clear that

Figure 20.2. Ohm's law: current I is proportional to the voltage V, the proportionality constant being the reciprocal of the resistance R.

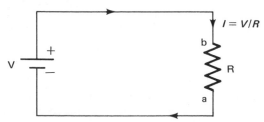

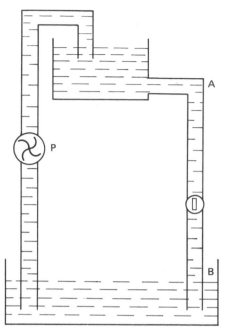

Figure 20.3. Water flow, analogous to current flow. The pump P maintains the water level in the tank like chemical reactions maintain the EMF of the battery.

the rate will be higher if the difference in water levels is higher. These two statements are analogous to Ohm's law. The chemical reaction in the battery "pumps" the charges from the lower potential side and maintains the potential difference between the terminals. This chemical energy, or to be exact, the amount of work done by the battery in moving a unit charge from one terminal to the other, is called the electromotive force, EMF for short. The unit for EMF is the volt. Since the EMF is work done per unit charge the unit of volt is consistent with the definition of volt in Chapter 19.

Verification of Ohm's Law

A circuit as shown in Fig. 20.4(a) is often used for verifying Ohm's law. The rheostat shown in Fig. 20.4(a) is a variable resistance which controls the current flowing through the circuit. The ammeter

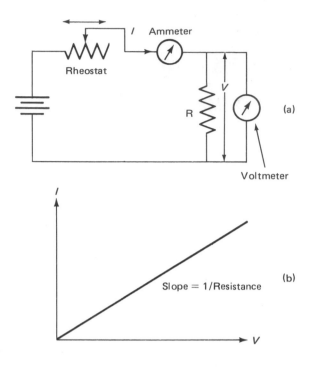

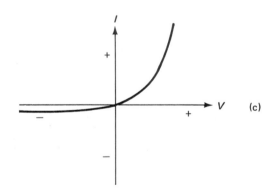

Figure 20.4. (a) Circuit to study Ohm's law. (b) Current versus voltage in a pure resistance. (c) Some materials do not obey Ohm's law. In a diode, the magnitude of the current depends on the polarity of the voltage. In this case, I is not proportional to V.

measures the current and the voltmeter measures the voltage across the resistance. Vary the current by adjusting the rheostat. Measure current for each rheostat setting and plot current versus voltage. The plot of the results of the experiment will be a straight line with slope $= 1/R$, verifying Ohm's law [Fig. 20.4(b)].

Ohm's law is not always obeyed. For example, if the resistance R in Fig. 20.4(a) is replaced by a diode, the results will be similar to the one shown in Fig. 20.4(c). The curve I versus V is not a straight line. This fact means that the resistance is not a constant but is a function of the applied voltage. Also note that the current flow depends very much on the direction of the battery.

20.3 INTERNAL RESISTANCE OF A BATTERY

The battery is capable of supplying a current at a constant EMF. Usually it is assumed that the battery has no internal resistance against the current flow through the battery. This is not true since the battery has a small internal resistance. Because of this small resistance, the potential difference across an external resistance connected to the battery is smaller than the EMF of the battery, V_B. In Fig. 20.5, the letter r stands for the internal resistance of the battery. Using Ohm's law we have

$$I = \frac{V_B}{R + r},$$

Figure 20.5. Effect of internal resistance r of a battery. Voltage V across the external resistance is less than the EMF of the battery, i.e., $V = V_B - Ir$.

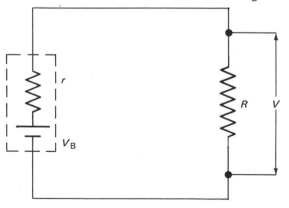

where I is the current flowing through the circuit. Since current I flows through R, the potential difference V across R is given by

$$V = IR = \frac{V_B}{R + r}R = V_B\left(\frac{1}{1 + r/R}\right). \quad (20.3)$$

From the equation we just derived, it is clear that $V < V_B$ and that $V \simeq V_B$ if $R \gg r$. The difference between V_B and V is insignificant if the external resistance is sufficiently high. However, there are situations where the external resistance is comparable to the internal resistance as shown in the example below.

EXAMPLE 20.4

A storage battery has an EMF of 12 V and internal resistance of 2 Ω. This battery is used to operate an ignition coil of resistance 1.5 Ω (Fig. 20.5). Calculate the maximum voltage that will appear across the ignition coil.

We use Eq. (20.3)

$$V = \frac{V_B}{R + r} \cdot R = \left(\frac{12\,\text{V}}{1.5\,\Omega + 2\,\Omega}\right)(1.5\,\Omega)$$

$$= 5.14\,\text{V}.$$

A capacitor when charged has a potential difference between its terminals. How is it different from a battery? A comparison between the two is given below (Fig. 20.6).

It is important to remember that shorting a battery (connecting a low resistance wire across the terminals) even for short periods of time may damage the battery due to excessive flow of current.

20.4 RESISTIVITY

Most materials at normal temperatures offer resistance to the flow of current. Resistance of a piece of material depends on (a) the material, specifically its capacity to conduct electricity, (b) length,

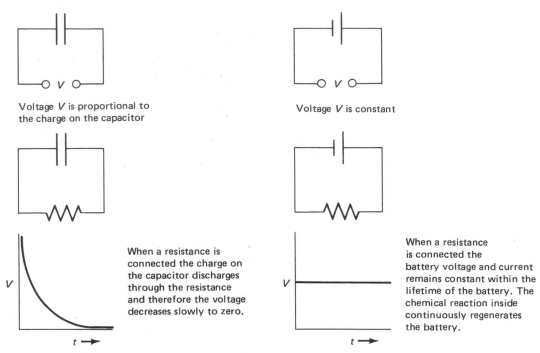

Voltage *V* is proportional to
the charge on the capacitor

Voltage *V* is constant

When a resistance is
connected the charge on
the capacitor discharges
through the resistance
and therefore the voltage
decreases slowly to zero.

When a resistance
is connected the
battery voltage and current
remains constant within the
lifetime of the battery. The
chemical reaction inside
continuously regenerates
the battery.

Figure 20.6. Comparison of a capacitor and a battery.

and (c) cross-sectional area. As in the case of water flow, resistance will be higher if the length is greater and cross-sectional area is less (see Fig. 20.7). Hence we can write

$$R \propto l/A \qquad (20.4)$$

which says that resistance is directly proportional to length and inversely proportional to area. By introducing a proportionality constant, Eq. (20.4) can be

written in the form

$$R = \rho \, l/A , \qquad (20.5)$$

where ρ (rho) is called the resistivity of the material. In Eq. (20.5), if $l = 1$ and $A = 1$, R is equal to ρ, that is, resistivity, is equal to the resistance of a piece of material of 1 m length and 1 m^2 area. The right-hand side of Eq. (20.5) contains three terms representing the three factors mentioned above; one of these terms, ρ, depends on the material and hence varies from material to material while the other two are dimensions, length and area, of the wire.

Rearranging Eq. (20.5), we get

$$\rho = \frac{R \, (\Omega) \times A \, (\text{m}^2)}{l (\text{m})} .$$

Therefore, the unit of resistivity is $\Omega \cdot$ m. The resistivities of a few common materials are given in

Figure 20.7. Resistance of a wire is proportional to the length and inversely proportional to the cross-sectional area of the wire.

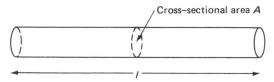

Cross–sectional area *A*

l

Table 20.1. These materials can be classified into three groups: insulators, semiconductors, and conductors in the order of decreasing resistivity. Note that silver is the best conductor of electricity and that copper is the next best.

When connecting wires are used to connect various components in an electric circuit, short wires made of low-resistivity material with large cross-sectional area are selected. A proper selection of such wires assures that no additional resistance is introduced into the circuit. In addition, low-resistance connecting wires minimize power loss.

EXAMPLE 20.5

Calculate the resistance of a 100 m long copper wire. The diameter of the wire is 1 mm. How much does the resistance decrease if the diameter is doubled?

The resistance of the wire is given by

$$R = \rho \, l/A.$$

Resistivity of copper $= 1.7 \times 10^{-8} \; \Omega \cdot m$.

Length of the wire $= 100 \, m$.

$$\text{Cross-sectional area } A = \pi (0.5 \text{ mm} \times \frac{1 \text{ m}}{1000 \text{ mm}})^2$$

$$= 7.8 \times 10^{-7} \, m^2.$$

Substituting these values, we get

$$R = \frac{(1.7 \times 10^{-8} \; \Omega \cdot m)(100 \, m)}{7.8 \times 10^{-7} \, m^2} = 2.2 \; \Omega.$$

When the diameter is doubled, the area increases by

TABLE 20.1 Resistivities of Some Selected Materials

SUBSTANCE	RESISTIVITY	
	$\Omega \cdot m$	$\Omega \cdot CM/ft$
Insulators		
Glass	$10^{10}-10^{14}$	$6 \times 10^{18}-6 \times 10^{22}$
Hard rubber	$10^{13}-10^{16}$	$6 \times 10^{21}-6 \times 10^{24}$
Mica	$10^{11}-10^{15}$	$6 \times 10^{19}-6 \times 10^{21}$
Semiconductors[a]		
Carbon	3.5×10^{-5}	2.1×10^4
Germanium	0.6	3.6×10^{18}
Silicon	2300	1.4×10^{12}
Conductors		
Aluminum	2.8×10^{-8}	16.9
Copper	1.7×10^{-8}	10.3
Gold	2.4×10^{-8}	14
Iron	12×10^{-8}	72
Silver	1.6×10^{-8}	97
Nichrome	110×10^{-8}	660

[a]Resistivity varies with impurity concentration.

four times. Hence the new resistance will be only $\frac{1}{4}$ of the original resistance ($2.2/4 = 0.55 \; \Omega$).

It is still common practice in industry to use a slightly complicated unit called ohm · circular mil/foot instead of ohm · foot (like ohm · meter). In this unit for area, instead of ft^2, circular mil is used. A mil is 1/1000 of an inch (or 0.001 in.). Area in circular mils is simply the square of the diameter in mils. (Note there is no multiplication by π and no conversion to radius by dividing the diameter by 2.) If d is the diameter in mils, area is not $\pi d^2/4$ but it is d^2 in circular mils (CM). The following example illustrates this difference.

EXAMPLE 20.6

A number 30 copper wire has a diameter of 10.03 mils. Calculate its area in CM and square inches.

In circular mils,

$$A = d^2 = 100.6 \; \text{CM}.$$

To calculate the area in square inches, the radius has to be obtained in inches.

$$\text{Radius} = \frac{\text{Diameter}}{2} = \frac{10.3}{2} \text{mil} \times 0.001 \; \text{in./mil}$$

$$= 0.00515 \; \text{in.}$$

$$A = \pi r^2 = 0.00019 \; \text{in.}^2$$

If area is expressed in CM, what is the unit for resistivity? To obtain the unit, let us start with Eq. (20.5):

$$R = \rho \frac{l}{A} \quad \text{or} \quad \rho = \frac{R \; (\Omega) A \; (\text{CM})}{l \; (\text{ft})}.$$

Since R is in ohms, A in CM, and l in feet, ρ has a unit of $\Omega \cdot \text{CM/ft}$.

EXAMPLE 20.7

Calculate the resistance of a 1000 ft long number 10 copper wire having a diameter of 101.9 mils.

Area of the wire is $(101.9)^2$ CM; resistivity (from Table 20.1) is $10.3 \; \Omega \cdot \text{CM/ft}$; length (given) is 1000 ft. Therefore, resistance is

$$\rho \frac{l}{A} = (10.3 \; \frac{\Omega \cdot \text{CM}}{\text{ft}})(\frac{1000 \; \text{ft}}{(101.9)^2 \; \text{CM}}) = 1 \; \Omega.$$

The numbers mentioned in Examples 20.6 and 20.7, namely, 30 and 10, refer to American wire gauge numbers indicating their sizes. As the number increases, the diameter of the wire decreases. The largest commercial size wire has a number 0000, the next four numbers are 000, 00, 0, and 1. From number 1, variations are denoted by increasing whole numbers up to a maximum 40. The smallest diameter wire commercially available is the number 40 wire.

20.5 ELECTRIC POWER

In the simple circuit shown in Fig. 20.8, a current $I \; (= Q/t)$ flows through the circuit. To maintain this current, during each time interval t, a charge Q is raised through a potential difference V. The work done by the battery is given by

$$W = VQ.$$

Figure 20.8. Current flow through a battery and a resistance.

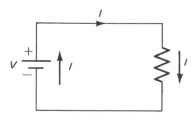

Therefore, the power output of the battery is

$$P = \frac{W}{t} = \frac{VQ}{t} = VI.$$

Using Ohm's law, Eq. (20.2), we can write P in the following forms

$$P = VI, \tag{20.6}$$

$$P = I^2 R, \tag{20.7}$$

$$P = V^2/R. \tag{20.8}$$

Where does this power output from the source of EMF appear? In the simple circuit illustrated in Fig. 20.8, it appears as the increase in internal energy of the resistance which produces a temperature increase. We are all familiar with toasters in which the electric power input into the resistance wire increases its temperature and finally the wire becomes red-hot. We are also familiar with other types of conversion of electrical energy to mechanical energy in electric motors and to light in incandescent bulbs. (See Examples 20.8 and 20.9).

Let us check the unit consistency of Eq. (20.6).

$$P \text{ (watts)} = V \text{ (volts)} \times I \text{ (amperes).}$$

Volt by definition is work per coulomb and ampere is coulomb per second. Hence we get

$$P = V \text{ (J/C)} \times I \text{ (C/s)}$$

$$= VI \text{ (J/s or W)}$$

The units are thus consistent.

EXAMPLE 20.8

A 40 W bulb is connected to a 120 V source. Calculate the current in the filament and the resistance of the filament.

The power input is 40 W. Since $P = VI$, we get

$$40 \text{ W} = (120 \text{ V})I \text{ or } I = \frac{40 \text{ W}}{120 \text{ V}} = \frac{1}{3} \text{ A.}$$

Resistance of the filament can be obtained by using Ohm's law:

$$I = \frac{V}{R} \text{ or } R = \frac{V}{I} = \frac{120 \text{ V}}{\frac{1}{3} \text{ A}} = 360 \ \Omega.$$

EXAMPLE 20.9

An electric motor is rated 1 hp and works on 120 V. Assuming it is 100% efficient, calculate the current in the power line.

First we have to obtain the power in watts using the relation

$$1 \text{ hp} = 746 \text{ W.}$$

To obtain current, we use the equation

$$P = VI \text{ or } I = P/V.$$

Substituting the given values, $P = 746$ W and $V = 120$ V, we get

$$I = \frac{746 \text{ W}}{120 \text{ V}} = 6.2 \text{ A.}$$

EXAMPLE 20.10

A 1 W electric immersion heater is used to heat 0.05 kg of water. Assuming 90% of the power input goes into heating water, calculate how much time it will take to raise the temperature from 20°C to 60°C.

The power input to heat the water is 0.9 W. The energy input in 1 s is

$$0.9 \text{ J} = \frac{0.9 \text{ J}}{4186 \text{ J/kcal}} = 2.15 \times 10^{-4} \text{ kcal.}$$

Let the increase in temperature of water in 1 s be Δt, then we have

$$M_w c_w \Delta t = \text{Heat input}$$

$$(0.05 \text{ kg})(1 \text{ kcal/kg} \cdot {}^\circ\text{C})\Delta t = 2.15 \times 10^{-4} \text{ kcal}.$$

The increase in temperature in 1 s is

$$\Delta t = 4.3 \times 10^{-3} \, {}^\circ\text{C}.$$

So the time needed for a 40°C increase is

$$\frac{40{}^\circ\text{C}}{4.3 \times 10^{-3} {}^\circ\text{C/s}} = 9302 \text{ s} = 2.6 \text{ h}.$$

20.6 RESISTORS IN SERIES AND PARALLEL

We often encounter circuits where resistors are connected in series (Fig. 20.9) or in parallel (Fig. 20.10). In both these cases, the total current flow from the battery and its power output can be calculated if the equivalent resistance is determined.

First let us look at a circuit containing a series of resistors R_1, R_2, and R_3 (Fig. 20.9). The same current I flows through resistors R_1, R_2, and R_3 from the positive terminal of the battery to the negative terminal. Can the current through R_2 be different from R_1? Think of two water pipes of different radii connected together. If the rate of flow in the second pipe is smaller than the rate of flow in the first pipe, what will happen? When resistors R_1, R_2, and R_3 are replaced by an equivalent resistance R_{eq} the same amount of current I should pass through it.

As a first step in obtaining the value of R_{eq}, let us isolate R_1 (Fig. 20.9). There is a current I in R_1 and, therefore, we can say there exists a potential difference across the resistance that maintains this current. The potential difference V_1 across R_1 is given by

$$V_1 = IR_1 \quad (\text{Ohm's law}).$$

Figure 20.9. (a) Resistors having resistances R_1, R_2, and R_3 are connected in series. (b) R_{eq} is the equivalent resistance of the combination R_1, R_2, and R_3 connected in series. (c) A current I passing through resistance R_1 develops a voltage V_1 between points a and b.

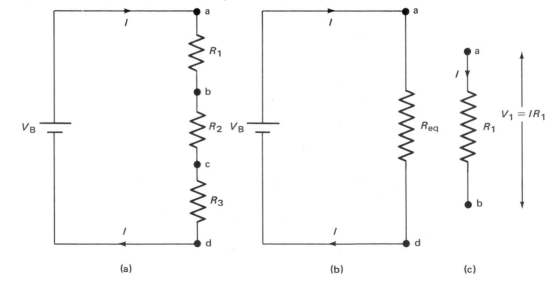

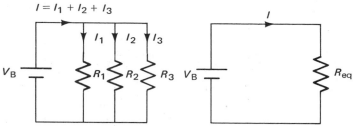

Figure 20.10. Resistors having resistances R_1, R_2, and R_3 are connected in parallel. Equivalent resistance R_{eq} has a current I passing through it which is equal to the sum of the currents through the resistors R_1, R_2, and R_3.

Similarly, potential differences across R_1 and R_2 are given by

$$V_2 = IR_2,$$

$$V_3 = IR_3.$$

The sum of these potential differences is equal to V_B, since the total potential difference between a and d is due to the battery EMF,

$$V_B = V_1 + V_2 + V_3$$

or $\quad V_B = IR_1 + IR_2 + IR_3.$ (20.9)

We have a similar equation for the equivalent resistance

$$V_B = IR_{eq}.$$ (20.10)

The right-hand sides of Eqs. (20.9) and (20.10) are equal. Hence we get

$$R_{eq} = R_1 + R_2 + R_3.$$ (20.11)

Figure 20.10 illustrates a circuit in which the resistors are connected in parallel. Since each resistor is connected to the battery, the potential difference across each resistor is V_B. The currents I_1, I_2, I_3 in resistors R_1, R_2, R_3 can, therefore, be obtained by Ohm's law:

$$I_1 = \frac{V_B}{R_1}, \quad I_2 = \frac{V_B}{R_2}, \quad I_3 = \frac{V_B}{R_3}.$$

The currents I_1, I_2, I_3 originate from the battery and hence the total current in the battery I is the sum of the individual currents. We get

$$I = I_1 + I_2 + I_3$$

or

$$I = \frac{V_B}{R_1} + \frac{V_B}{R_2} + \frac{V_B}{R_3}.$$ (20.12)

Since an equivalent resistance should draw the same amount of total current from the battery, we have

$$I = \frac{V_B}{R_{eq}}.$$ (20.13)

From Eqs. (20.12) and (20.13), we get

$$\frac{1}{R_{eq}} = \frac{1}{R_1} + \frac{1}{R_2} + \frac{1}{R_3}.$$ (20.14)

Equations (20.11) and (20.14) can be used to obtain the values of the equivalent resistance (or total resistance) when resistors are connected in series and parallel, respectively. The equation for series resistors is similar to the equation for capacitors in parallel and the equation for resistors in parallel is similar to the equation for capacitors in series. The following examples illustrate the use of these equations.

EXAMPLE 20.11

Calculate the current in each resistor and potential drops across each resistor for the circuit shown in Fig. 20.11.

Total resistance in the circuit is 35 Ω. The current from the battery is 7 V/35 Ω = 0.2 A. In a circuit where resistors are connected in series, the same current passes through each resistor. The potential drop across each resistor can now be obtained by using Ohm's law:

$$V_1 = 0.2 \text{ A} \times 5 \text{ Ω} = 1 \text{ V};$$

$$V_2 = 0.2 \text{ A} \times 10 \text{ Ω} = 2 \text{ V};$$

$$V_3 = 0.2 \text{ A} \times 20 \text{ Ω} = 4 \text{ V}.$$

The sum of the potential drops (1 V + 2 V + 4 V) is equal to the battery EMF, as it should be.

EXAMPLE 20.12

Three resistors 100 Ω, 200 Ω, and 300 Ω are connected in parallel to a 120 V source (Fig. 20.12). Calculate the equivalent resistance and the total current flow.

There are two ways of calculating the total current: (a) Calculate the equivalent resistance and then use Ohm's law; or (b) use Ohm's law and calculate the individual currents and add them up. The

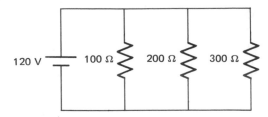

Figure 20.12. Example 20.12.

equivalent resistance is given by Eq. (20.14):

$$\frac{1}{R_{eq}} = \frac{1}{R_1} + \frac{1}{R_2} + \frac{1}{R_3}.$$

Substituting the given values, we get

$$\frac{1}{R_{eq}} = \frac{1}{100 \text{ Ω}} + \frac{1}{200 \text{ Ω}} + \frac{1}{300 \text{ Ω}} = \frac{6+3+2}{600 \text{ Ω}}$$

or

$$R_{eq} = \frac{600 \text{ Ω}}{11} = 54.5 \text{ Ω}.$$

$$\text{Total current} = \frac{120 \text{ V}}{54.5 \text{ Ω}} = 2.2 \text{ A}.$$

Individual current can be calculated as follows:

$$I_1 = \frac{120 \text{ V}}{100 \text{ Ω}} = 1.2 \text{ A};$$

$$I_2 = \frac{120 \text{ V}}{200 \text{ Ω}} = 0.6 \text{ A};$$

$$I_3 = \frac{120 \text{ V}}{300 \text{ Ω}} = 0.4 \text{ A}.$$

$$\text{Total current} = I_1 + I_2 + I_3 = 2.2 \text{ A}.$$

Very often we encounter circuits containing series and parallel resistors. The procedure to solve such problems is shown in the following example.

Figure 20.11. Example 20.11.

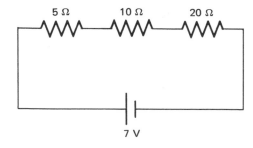

EXAMPLE 20.13

Calculate the current flow in each resistor shown in Fig. 20.13.

Since R_2 and R_3 are in parallel, first replace them by their equivalent resistance R.

$$\frac{1}{R} = \frac{1}{30} + \frac{1}{20} = \frac{2+3}{60}$$

or

$$R = 12\,\Omega.$$

The current I from the battery can be obtained by using Ohm's law:

$$I = \frac{110\text{ V}}{50\,\Omega + 12\,\Omega + 48\,\Omega} = 1\text{ A}.$$

Currents in R_1 and R_4 are each 1 A. The 1 A current divides at the junction of R_2 and R_3. In Fig. 20.13(b) resistances R_2 and R_3 are replaced by their equivalent resistance 12 Ω. Since 1 A flows through R_2 and R_3, the potential difference between points a and b is 12 V ($= 1$ A $\times$ 12 Ω).

Current I_2 through $R_2 = \dfrac{12\text{ V}}{20\,\Omega} = 0.6$ A;

Current I_3 through $R_3 = \dfrac{12\text{ V}}{30\,\Omega} = 0.4$ A.

20.7 KIRCHHOFF'S RULES

Kirchhoff's rules discussed below are useful in analyzing circuits more complicated than those we have discussed. The first rule concerns the division of current at a junction. A junction is a point in a circuit where more than two elements are connected. Point J in Fig. 20.14 is a junction. A current I flows towards the junction and currents I_1 and I_2 flow out of the junction. The analogy of the river forking at point J is worth considering. Water flow into the junction has to be equal to the water flow out of the junction. Similar considerations in the case of the electrical circuit give us the equation

$$I = I_1 + I_2. \tag{20.15}$$

Equation (20.15) is one of Kirchhoff's rules which states *the sum of the currents flowing into a junction is equal to the sum of the currents flowing out of the junction.*

To understand the second rule, let us consider the simple circuit in Fig. 20.15. Charge flow is from a→b→c→d→a. Since the battery is doing work on the charge, raising it from a lower potential to a higher potential, energy of the charge increases. The potential difference between a and b, potential energy per unit charge, is positive and equal to the EMF of the battery. The potential at any point in a good conductor ($R = 0$) is the same. Therefore, the potential at c is the same as the positive side of the battery and the potential at d is the same as the

Figure 20.13. (a) Example 20.13. (b) Resistors having resistances R_2 and R_3 are replaced by their equivalent resistance R.

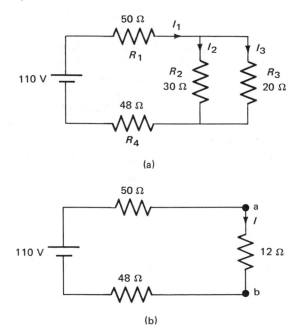

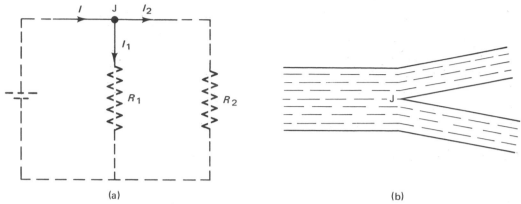

Figure 20.14. (a) Division of current at a junction; (b) division of water flow at a junction.

negative side of the battery. The potential energy of the charge flowing from c to d is therefore decreasing. The potential difference V_{cd} by Ohm's law is $-IR$. When the charge completes the path, a→b→c→d→a, it is coming back to the original point and the potential differences along the path should add up to zero because energy is conserved; that is,

$$V - IR = 0. \qquad (20.16)$$

Note that the above equation is a modified version of Ohm's law. Also it is important to remember that whenever there is a flow of current through a resistance there is a potential drop across the resistance equal to IR. Kirchhoff's second rule may be states as follows: *In a closed loop the sum of the potential difference is zero.*

What is a closed loop? A closed loop is a com-

Figure 20.15. abcd is a closed loop.

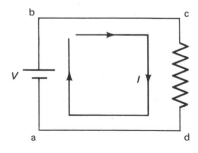

pleted path like a→b→c→d→a in Fig. 20.15. (Also see examples below.)

Correct assignment of signs to the potential differences is extremely important in using Kirchhoff's rules. The different possibilities we may encounter are summarized in Table 20.2.

EXAMPLE 20.14

Using Kirchhoff's rules, determine the currents in $R_1, R_2, R_3,$ and R_4 in Fig. 20.16.

The first step in solving this problem is to assign the assumed current through the resistors, namely, I_1 through R_1, I_2 through R_2, and I_3 through R_3. From J_1 to J_2 the electrical elements, R_3, V_{B2}, and R_4 are in series, and therefore the same current I_3 flows through all of them. Similarly, I_1 flows through V_{B1} and R_1 and I_2 flows through R_2. There are two junctions J_1 and J_2 in the circuit. Applying Kirchhoff's first rule, we get

$$I_1 + I_3 = I_2. \qquad (20.17)$$

The second rule can be applied to the two loops marked in Fig. 20.16. Do you see a third loop?

In loop 1, point 1 is the starting point to calculate the potential differences.

$$V_{B1} - I_1R_1 - I_2R_2 = 0. \qquad (20.18a)$$

TABLE 20.2. Signs of Potential Difference Across Circuit Elements

CIRCUIT ELEMENT: ARROW INDICATES ASSUMED CURRENT FLOW	SIGN OF V_{ab}	COMMENTS
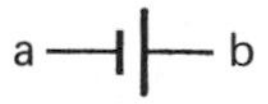	+	Potential increases in the direction of the negative terminal to the positive terminal.
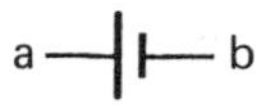	−	Potential decreases in the direction of + to −
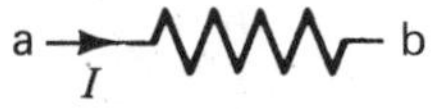	−	Potential decreases in the direction of the current in a resistance.
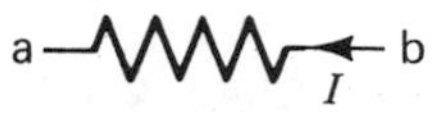	+	Potential increases in the opposite direction of the current in a resistance.

In loop 2, potential differences are calculated starting from point 2 and going around the loop in the counterclockwise direction as marked in Fig. 20.16.

$$V_{B2} - I_2R_2 - I_3R_4 - I_3R_3 = 0. \qquad \textbf{(20.18b)}$$

We have three equations [(20.17), (20.18), (20.19)] to solve for the three unknowns (I_1, I_2, and I_3). Substituting the values given, we get

$$I_1 + I_3 = I_2, \qquad \textbf{(20.17)}$$

Figure 20.16. Example 20.14.

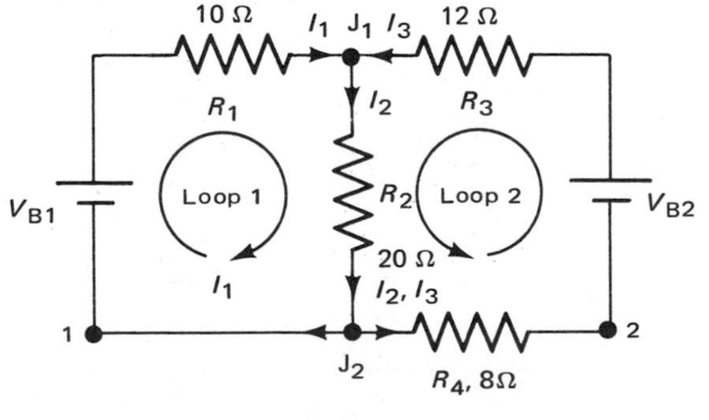

$$10\text{ V} - 10I_1 - 20I_2 = 0, \qquad \textbf{(20.18)}$$

$$15\text{ V} - 20I_2 - I_3(12 + 8) = 0. \qquad \textbf{(20.19)}$$

Equations (20.18) and (20.19) can be written as

$$10I_1 + 20I_2 = 10, \qquad \textbf{(20.20)}$$

$$20I_2 + 20I_3 = 15. \qquad \textbf{(20.21)}$$

Substitution for I_2 yields

$$10I_1 + 20(I_1 + I_3) = 10, \qquad \textbf{(20.22)}$$

$$20(I_1 + I_3) + 20I_3 = 15. \qquad \textbf{(20.23)}$$

Rearranging the terms, we have

$$30I_1 + 20I_3 = 10,$$

$$20I_1 + 40I_3 = 15.$$

Multiplication of Eq. (20.21) by 2 and Eq. (20.22) by 3 yields

$$60I_1 + 40I_3 = 20, \qquad \textbf{(20.24)}$$

$$60I_1 + 120I_3 = 30. \qquad (20.25)$$

Subtracting Eq. (20.24) from (20.25), we obtain

$$80I_3 = 10 \text{ or } I_3 = 1/8 \text{ A} = 0.125 \text{ A}.$$

Substituting the value of I_3 into Eq. (20.21), we obtain the value of I_2:

$$20I_2 = 15 - 20(1/8)$$

or

$$I_2 = 0.625 \text{ A},$$

and from Eq. (20.17), we obtain

$$I_1 = 0.625 - 0.125$$

$$= 0.5 \text{ A}.$$

EXAMPLE 20.15

Set up the necessary equations to solve for the currents I_1, I_2, I_3, I_4, I_5, and I_6 (Fig. 20.17). Complete solution is left as an exercise for the reader (see Problem 20.18).

Figure 20.17. Example 20.15.

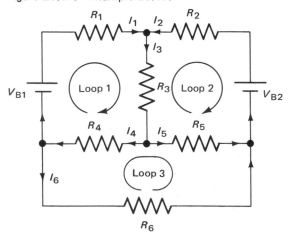

Using Kirchhoff's first rule, we obtain the equations

$$I_1 + I_2 = I_3,$$

$$I_4 + I_5 = I_3,$$

$$I_5 + I_6 = I_2.$$

From the three loops, we obtain the following additional equations:

$$V_{B1} - I_1R_1 - I_3R_3 - I_4R_4 = 0,$$

$$V_{B2} - I_2R_2 - I_3R_3 - I_5R_5 = 0,$$

$$I_4R_4 - I_5R_5 - I_6R_6 = 0.$$

We have six independent equations, sufficient to solve for the six unknown currents.

Sometimes it is possible to get a negative value for the current on solving Kirchhoff's equations. It means that the current is flowing in the opposite direction to the assumed direction. A very simple example is illustrated in Fig. 20.18. Assume the current flow is from d to c. Applying the second rule and starting from point a, we get

$$V + IR = 0$$

$$I = -V/R.$$

Figure 20.18. Since the current is taken as flowing from d to c, the solution of Kirchhoff's equation for this circuit will give a negative value of I.

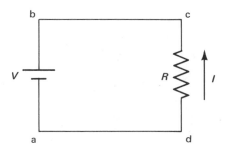

The negative sign means the current flow is actually
in the direction c to d.

SUMMARY   Several important laws, necessary for the analysis of electric circuits, were
discussed in this chapter. They are the following:
Ohm's law states

$$I = V/R.$$

Kirchhoff's laws state that (a) at a junction, the sum of the currents going in
equals the sum of the currents going out; and (b) the sum of the potential drops
in a closed loop is zero.
Power generated due to a current flow is given by

$$P = I^2 R.$$

Equivalent resistance for resistors connected in series is given by

$$R = R_1 + R_2 + R_3 + \cdots,$$

and for resistors connected in parallel is given by

$$\frac{1}{R} = \frac{1}{R_1} + \frac{1}{R_2} + \frac{1}{R_3} + \cdots.$$

QUESTIONS   *20.1*   Electrons accelerate in an electric field, increasing their velocity from
AND   zero to a maximum given by $\frac{1}{2} mv^2 = eV$, if the electron path is not
PROBLEMS   obstructed, as in vacuum tubes. However, when electrons flow through a
                        conducting wire, numerous collisions with the atoms take place. How
                        does it affect the velocity of the electrons?

20.2   As the temperature increases, atoms in a resistor vibrate about their
equilibrium positions more violently. How will the atomic vibrations
affect the current flow? Will the resistance increase with temperature?

20.3   Semiconductors are not good conductors like metals. The reason for this
is that the number of free electrons per unit volume in semiconductors is
less than in metals. The number of free electrons in such materials increases with temperature. Therefore, the resistance of semiconducting
materials vary considerably with temperature. Will the resistance increase with temperature for semiconducting materials in the neighbor-
hood of room temperature?

20.4 A current of 0.5 A flows through a wire. Calculate the number of electrons crossing an area of the wire in 1 s, 1 h, and 1 day.

20.5 To silverplate a copper ustensil, a 1 A current flows through the electrolyte solution for 1 h. (a) If the charge on each silver atom is 1.6×10^{-19} C, how many silver atoms reach the copper electrode? (b) Calculate the weight of a silver atom (weight of an atom = atomic weight/Avogadro's number). (c) Calculate the weight gain of the copper electrode. (Atomic weight of silver is 107.87.)

20.6 An accelerator produces a stream of particles at high energies. The output of the accelerator is often quoted in amperes. If an accelerator output is 500 mA, how many particles pass a cross-section of the beam per second? The charge of each particle is equal to 1.6×10^{-19} C.

20.7 In a vacuum electron tube, a current of 100 mA is maintained. How many electrons reach the anode (positive plate) every second?

20.8 The filament of an incandescent bulb has a resistance of 60 Ω. Calculate the current through the filament if it is connected to a 120 V outlet. What is the power expended by the filament?

20.9 A 100 Ω resistance is rated ½ W. What is the maximum value of the potential difference that can be applied across the resistor? Is it proper to connect this resistance to a 110 V outlet?

20.10 A 110 V outlet is connected to the main power lines through a 20 A fuse. What is the maximum current you can get from this outlet? What is the maximum value of the resistance you can connect at this outlet? What is the maximum power you get from this outlet?

20.11 A wire has a resistance of 10 Ω. If this wire is drawn to twice its length, what is its new resistance?

20.12 An aluminum wire is 100 ft long and its diameter is 0.01 in. The resistivity of aluminum is 16.9 Ω · CM/ft. Calculate (a) the area of the wire in circular mils and (b) the resistance of the wire.

20.13 Nichrome, used in electric heaters, has a high resisvitity in comparison to good conductors like copper. The resistivity of nichrome is 110×10^{-8}

Figure 20.19. (a), (b), and (c). Problems 20.14 and 20.15.

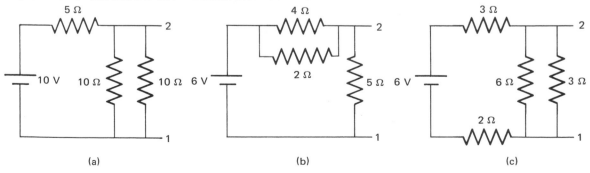

$\Omega \cdot$ m. Calculate the resistance of a nichrome wire 10 m long and 0.1 cm in diameter.

20.14 Calculate the current flowing through each resistor in Fig. 20.19(a), (b), and (c).

20.15 Calculate the potential difference between points 1 and 2 in Fig. 20.19(a), (b), and (c).

20.16 Using Kirchhoff's laws, solve for the current in each resistor shown in Fig. 20.20(a), (b), and (c).

Figure 20.20. (a), (b), (c). Problem 20.16.

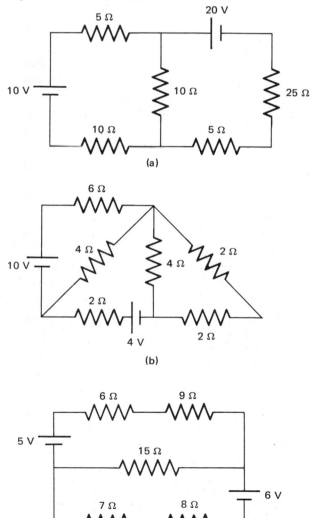

20.17 The arrangement in Fig. 20.21, known as a potentiometer, is used to measure the voltage of an unknown voltage source (V_2 in this case). To measure V_2, the contact on wire ab is slowly moved until there is no current through the ammeter. When there is no current, show that $V_2 = V_1 \, l/L$, assuming the wire is uniform.

20.18 Solve for currents I_1, I_2, I_3, I_4, I_5, and I_6 for the circuit shown in Fig. 20.17 where $V_{B1} = 100 \ V$, $V_{B2} = 50 \ V$, $R_1 = 50 \ \Omega$, $R_2 = 100 \ \Omega$, $R_3 = 200 \ \Omega$, $R_4 = 400 \ \Omega$, $R_5 = 600 \ \Omega$, and $R_6 = 300 \ \Omega$.

Figure 20.21.   Problem 20.17.

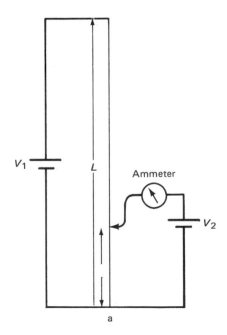

Magnetism

The magnetic properties of some minerals (compounds of iron) aroused the curiosity of men from time immemorial. Two properties of such materials were well known: their capacity to attract other magnets and pieces of iron—the Greeks attributed this property to an inner force or a "soul"—and their ability to orient themselves along the geographic north-south direction. The Chinese knew of this property and used it in designing compasses as early as 21 A.D. In spite of all this early knowledge, the revolution in the understanding and the uses of magnets came with the discovery by Hans Christian Oersted in 1820 that a magnetic field exists around a current-carrying wire and by Michael Faraday in 1831 that a change in a magnetic field induces an electromotive force in a loop of wire.

21.1 PERMANENT MAGNET

The magnets we are familiar with are the permanent bar magnets. The end of such a magnet which points to the geographic north when freely suspended is called a north pole and the other end is called a south pole (Fig. 21.1). Unlike poles attract each other and like poles repel each other (Fig. 21.2). The force of attraction or repulsion is proportional to the strength of the poles and inversely proportional to the square of the distance between them. In contrast to the electrical case where positive charges can be separated from negative charges, so far magnetic poles are found to be inseparable. A magnet, therefore, always contains both poles. If a magnet is divided we obtain two complete magnets,

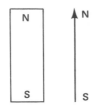

Figure 21.1. The end of a magnet directed to the geographic north is called a north pole (N). The other end is called a south pole (S).

each containing a north pole and a south pole (Fig. 21.3).

The magnetic field surrounding a magnet is represented by field lines as we have done in the case of the electric field. Just as in the latter case, these are imaginary lines with the following properties: the density of lines at a point is proportional to the field and the tangent to the line at a point gives the direction of the magnetic field at that point. Note

Figure 21.2. Unlike poles attract each other and like poles repel each other.

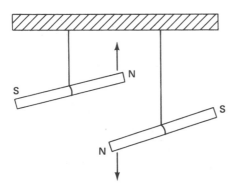

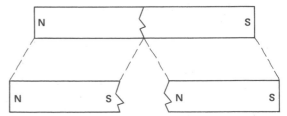

Figure 21.3. When a magnet is broken into two parts, each part is a magnet with a north and a south pole.

that the field lines are directed from the north pole to the south pole outside the magnet. These magnetic field lines form closed loops passing through the magnet. This is in contrast to electric field lines which originate and terminate on charges.

The magnetic field lines can be traced by simple methods illustrated in Fig. 21.4. In Fig. 21.4(b) a small compass is used to trace the field. In Fig. 21.4(c), iron filings align themselves along the field lines. The filings become small magnets in the field and these small magnets, like the compass in Fig. 21.4(a), trace out the field.

21.2 MAGNETIC INDUCTION AND FORCE ON A CHARGE AND CURRENT IN A MAGNETIC FIELD

The magnetic field strength, as mentioned earlier, is measured in terms of the number of lines crossing a unit area at a point. The magnetic field lines, often called the magnetic flux, is measured in units of weber (Wb) and lines per unit area; flux density is measured in weber/m^2 called tesla (T). This quantity, the flux density, is called the magnetic induction. Sometimes it is also called magnetic field strength. The symbol B is generally used to represent it.

$$\text{Magnetic induction } B = \frac{\text{Flux lines}}{\text{Area}}. \quad (21.1)$$

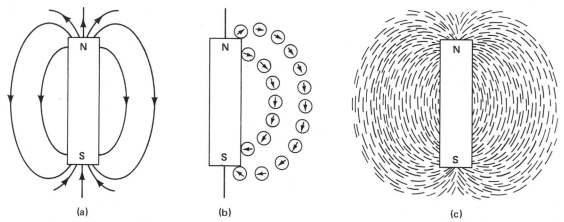

Figure 21.4. (a) Magnetic field lines are directed from the north pole to the south pole outside the magnet. (b) and (c) Small compasses and iron filings are used to trace the fields.

EXAMPLE 21.1

Calculate the magnetic induction at a point where 5 flux lines cross a 10 m² area.

The total number of flux lines is 5 Wb. Therefore magnetic induction is

$$B = \frac{\text{Number of lines}}{\text{Area}} = \frac{5 \text{ Wb}}{10 \text{ m}^2} = 0.5 \text{ T}.$$

In the above discussion we defined the magnetic field strength in terms of imaginary field lines. However, a more practical definition based on the force exerted by the field on a moving charge or on a wire carrying current is given below. We used a similar definition for the electric field. If a charge is placed in an electric field, a force acts on it and the force per unit charge is defined as the electric field at that point. To help us define the field, let us perform the following "thought" experiments on a positive charge in a magnetic field and determine the force acting on it. Four possible experimental conditions are shown in Fig. 21.5 and they are discussed along with their results.

1. A stationary charge is kept in a magnetic field. The force acting on it due to the magnetic field is zero.

2. A charge moving along the direction of the magnetic field also experiences no force.

3. A charge moving perpendicular to the magnetic field experiences a force. The direction of the force is perpendicular to the velocity of the charge and the direction of the field. The magnitude of the force is given by

$$F = qvB, \tag{21.2}$$

where q is the charge (coulombs), v its velocity (m/s), and B the magnetic induction (tesla). Force is maximum in this case when the charge is moving perpendicular to the field.

4. A moving charge making an angle θ ($0° < \theta < 90°$) with the direction of the magnetic field experiences a force. The magnitude of the force is less than that given by Eq. (21.2) and it has an angular dependence, such that

$$F = qvB \sin \theta. \tag{21.3}$$

Note that cases 2 and 3 are special cases of 4, for $\theta = 0°$ and $\theta = 90°$, respectively. The force is maximum ($F = qVB$) for $\theta = 90°$ and minimum for $\theta = 0°$.

Magnetic induction B (in tesla) can now be defined using Eq. (21.2) or (21.3). Using Eq. (21.2),

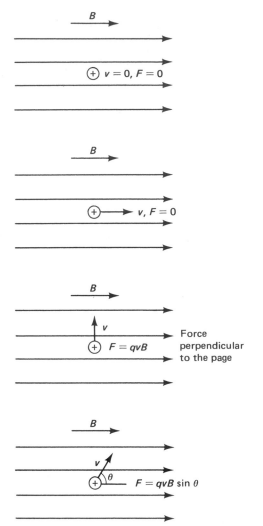

Figure 21.5. Forces acting on a charge in a magnetic field are a function of the magnitude and direction of the velocity of the charge in addition to their proportionality to the magnetic induction B and charge q.

we get the following definition:

$$\text{Magnetic induction} = \frac{\text{Maximum force on a charge}}{\text{Charge x Velocity}},$$

or in mathematical form

$$B = \frac{F_{\max}}{qv}. \tag{21.4}$$

The magnetic field is 1 tesla if a maximum force of 1 newton acts on a unit charge that moves with a velocity of 1 m/s.

As we have seen in Chapter 20, a current is equivalent to a flow of charges and, therefore, we expect the magnetic field to exert a force on a wire carrying current. Three orientations of the wire with respect to the field corresponding to three different directions of motion of charge (cases 2, 3, and 4 above) are discussed here (Fig. 21.6).

1. For current in the direction of the magnetic field (similar to case 2 above)

$$F = 0.$$

2. For current perpendicular to the field (similar to case 3 above), the force is maximum and is given by

$$F = IlB,$$

when I is the current in amperes and l the length in meters of the section of the wire in the magnetic field.

3. For the wire at an angle to the field (similar to case 4 above, charge moving at an angle to the field lines), the force in this case is

$$F = IlB\sin\theta. \tag{21.6}$$

Magnetic field can be defined with the help of Eq. (21.5) or (21.6). Using Eq. (21.5), we can write

$$B = \frac{F_{\max}}{Il}. \tag{21.7}$$

Magnetic induction is 1 tesla if a maximum force of 1 newton acts on a 1 meter long wire lying in the field that carries a current of 1 ampere.

The unit tesla (T) defined above is too big a unit

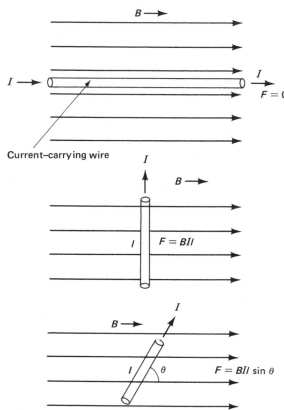

Figure 21.6. Force on a current-carrying wire in a magnetic field depends on the direction of the wire.

in most instances. Another unit often used is gauss (G)

$$1 \text{ T} = 10^4 \text{ G}. \tag{21.8}$$

The gauss does not belong to SI units. If the field is given in gauss, it should be converted to tesla before using it in any of the equations above.

What is the direction of the force acting on a moving charge and on the current-carrying wire? We did not pay much attention to the question. We mentioned that the force is perpendicular to both the velocity of the charge (or direction of the wire) and the magnetic field. In the case illustrated in Fig. 21.7, the magnetic field is directed out of the page and the velocity is directed up. There are two

directions perpendicular to B and v, the $+y$ and $-y$ directions. Which one of these is the direction of the force? To determine the direction, the right-hand rule is often used. The fingers of the right hand point in the direction of the magnetic field, and the thumb in the direction of the charge. Then the direction perpendicular to the palm gives the direction of the force.

It is interesting to note that a charge moving in a magnetic field experiences a force perpendicular to the magnetic field while in an electric field the force is along the field. Also the electric force is independent of the velocity of the charge.

EXAMPLE 21.2

Calculate the magnitude of the forces acting on each of the three charges shown in Fig. 21.8.

Charge 1 is at rest, charge 2 is moving along the direction of the magnetic field with a velocity of 100 m/s, and charge 3 moves with the same velocity (100 m/s) perpendicular to the field. Indicate the direction of the force in each case.

No force acts on charges 1 and 2 because charge 1 is not moving ($v = 0$) and charge 2 is moving along the field ($\sin\theta = 0$). The force acting on 3 is given by

$$F = qvB;$$

$$q = e = 1.6 \times 10^{-19} \text{ C};$$

$$v = 100 \text{ m/s};$$

$$B = 300 \text{ T}.$$

The force acting on charge 3 is

$$F = (1.6 \times 10^{-19} \text{ C})(100 \text{ m/s})(300 \text{ T})$$

$$= 4.8 \times 10^{-15} \text{ N}$$

The direction perpendicular to the palm oriented according to the right-hand rule (Fig. 21.8) is out of the paper. Hence the direction of force is out of the paper.

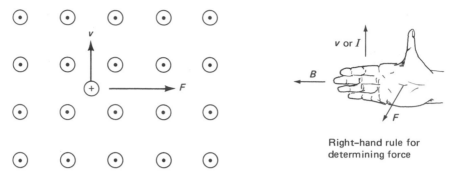

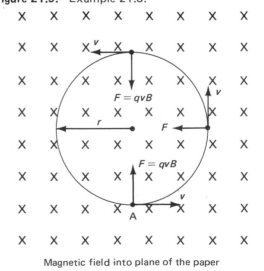

Right-hand rule for
determining force

Figure 21.7. Direction of the force acting on a charge in a magnetic field is given by the right-hand rule. If the thumb points in the direction of the charge and the fingers in the direction of the field, the direction of the force is away from and perpendicular to the palm of the hand. ⊙ stands for field lines directed out of the page and ✕ is used to indicate field lines directed into the page.

EXAMPLE 21.3

A proton with a velocity of 10^5 m/s is introduced into a magnetic field at point A. Calculate the radius of its circular orbit. The magnetic field is directed into the paper and has a magnitude of 0.01 T (Fig. 21.9).

Since the force acting on the charge is always perpendicular to its velocity, the charge continuously experiences an acceleration perpendicular to its velocity. In Chapter 6 we saw that such an acceleration produces a circular motion.

The radius of the circular orbit can be obtained by equating the centripetal force and the magnetic force:

$$\frac{mv^2}{r} = qvB \qquad (21.9)$$

or

$$r = \frac{mv}{qB}.$$

Figure 21.8. Example 21.2.

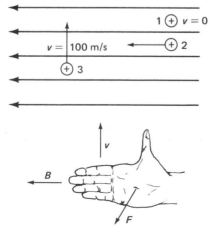

Figure 21.9. Example 21.3.

X   X   X   X   X   X   X   X

X   X   X   X   X   X   X   X

X   X   X   X   X   X   X   X

X   X   X   X   X   X   X   X

X   X   X   X   X   X   X   X

X   X   X   X   X   X   X   X

X   X   X   X   X   X   X   X

X   X   X   X   X   X   X   X

X   X   X   X   X   X   X   X

Magnetic field into plane of the paper

The mass of the proton is 1.67×10^{-27} kg, the charge of the proton is 1.6×10^{-19} C, the velocity of the proton is 10^5 m/s, and the magnetic field is 0.01 T.

Hence

$$r = \frac{(1.67 \times 10^{-27} \text{ kg})(10^5 \text{ m/s})}{(1.6 \times 10^{-19} \text{ C})(0.01 \text{ T})}$$

$$= 1.04 \times 10^{-1} \text{ m}.$$

EXAMPLE 21.4

A mass spectrograph (Fig. 21.10) is used to determine the masses of ions (positively charged atoms). Hydrogen and deuterium are two isotopes of hydrogen. The deuterium ion has twice the mass of the hydrogen ion (mass of deuterium ion, m_D = 3.34×10^{-24} kg; mass of hydrogen ion $m_H = 1.67 \times 10^{-24}$ kg). They are accelerated in an electric field and allowed to enter a magnetic field with a velocity of 10^4 m/s. The magnetic field has an induction of 0.8 T and the charge on the ions is 1.6×10^{-19} C. Calculate the distances, d_D and d_H, between the slit and the points where they hit the photographic plate. (Measuring this distance, one

Figure 21.10. Example 21.4. A mass spectrograph.

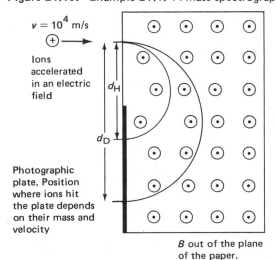

$v = 10^4$ m/s

$\oplus \rightarrow$

Ions accelerated in an electric field

d_H

d_D

Photographic plate. Position where ions hit the plate depends on their mass and velocity

B out of the plane of the paper.

can calculate the mass of unknown ions in the sample. This is a good analytical method for detecting chemicals in air samples, etc).

Both ions have circular orbits in the magnetic field, because the magnetic force is perpendicular to the velocity. The radius of the orbit is given by

$$\frac{mv^2}{r} = qvB \text{ or } r = \frac{mv}{qB}.$$

For deuterium $m_D = 3.34 \times 10^{-24}$ kg, $v = 10^4$ m/s, $q = 1.6 \times 10^{-19}$ C, and $B = 0.8$ T.

$$d_D = 2 \times \text{radius}$$

$$= 2 \times \frac{3.34 \times 10^{-24} \text{ kg} \times 10^4 \text{ m/s}}{1.6 \times 10^{-19} \text{ C} \times 0.8 \text{ T}}$$

$$= 5.2 \times 10^{-1} \text{ m} = 52 \text{ cm}.$$

For the proton $m_H = 1.67 \times 10^{-24}$ kg, $v - 10^4$ m/s, $q = 1.6 \times 10^{-19}$ C, and $B = 0.8$ T.

$$d_H = 2 \times \text{radius}$$

$$= 2 \times \frac{1.67 \times 10^{-24} \text{ kg} \times 10^4 \text{ m/s}}{1.6 \times 10^{-19} \text{ C} \times 0.8 \text{ T}}$$

$$= 2.6 \times 10^{-1} \text{ m} = 26 \text{ cm}.$$

21.3 COIL IN A MAGNETIC FIELD

In the last section we discussed the force acting on a current-carrying wire in a magnetic field. We now consider a loop or a coil placed in a magnetic field (Fig. 21.11). The current I in section ab is flowing in the $+x$ direction while in section cd is flowing in the $-x$ direction. On the other hand, the direction of the magnetic field is the same in sections ab and cd. Therefore, the direction of the force acting on ab is in the $+z$ direction and that on cd is in the $-z$ direction. The magnitude of these

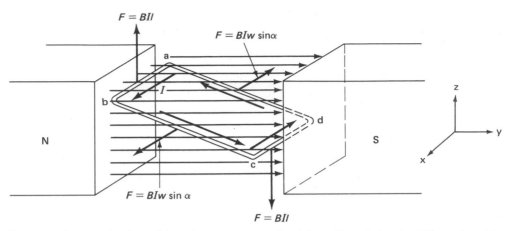

Figure 21.11. Forces acting on the four sides of a current-carrying loop of length l and width w placed in a magnetic field.

forces is the same and the forces are given by

$$F_{ab} = -F_{cd} = BIl, \qquad (21.10)$$

where B is the magnetic induction and l the length of section ab (and cd).

Similar reasoning shows that the forces acting on sections cb and da are equal in magnitude and opposite in direction. The forces are

$$F_{cb} = -F_{da} = BIw\sin\alpha,$$

where w is the width of the coil and α is the angle the coil makes with the field (Fig. 21.11). The sum of the forces acting is zero and hence the coil is in translational equilibrium. However the coil is not in rotational equilibrium since there is a net torque acting on the coil. The torque on section ab about an axis passing through the center of the coil (Fig. 21.12) is $BIl\,(w/2)\cos\alpha$ (= force × perpendicular distance) and similarly on section cd it is $BIl\,(w/2)\cos\alpha$. Both torques have the same sense of rotation (clockwise) and they add up to give a total torque acting on the coil:

$$T = BIl(w/2)\cos\alpha + BIl(w/2)\cos\alpha$$

$$= BIlw\cos\alpha. \qquad (21.11a)$$

Since the area of the coil is $A = lw$, Eq. (21.11a) becomes

$$T = BIA\cos\alpha. \qquad (21.1b)$$

The torque experienced by the coil is a maximum for $\alpha = 0°$, in which case the coil is along the magnetic field and a minimum for $\alpha = 90°$ in which case the coil is perpendicular to the field.

Figure 21.12. Torque acting on the coil is $2[(w/2)\cos\alpha]\,BIl$.

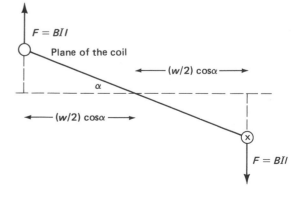

What is the effect of this torque on a coil placed in a magnetic field. Let us assume the coil starts from the $\alpha = 0°$ position. The angular velocity of the coil increases as it approaches the $\alpha = 90°$ position. Even though the torque is zero at the $90°$ position, its angular velocity will take the coil past this position. The coil, when it reaches the other side of the perpendicular position, starts experiencing a reverse torque. Consequently, the angular velocity first decreases to zero and then increases in the reverse direction. Thus the resulting motion is an oscillatory motion about the $\alpha = 90°$ position (i.e., perpendicular position).

In most of the practical arrangements, instead of the loop, several turns of wire are used to increase the torque. If there are N turns, the total torque is

$$T = NBIA \cos\alpha.$$

EXAMPLE 21.5

A rectangular coil has the following dimensions: $l = 10$ cm, $w = 5$ cm, number of turns $N = 200$, and $I = 10$ A. If this coil is placed in a magnetic field of flux density 5×10^{-3} T, calculate the torque acting on the coil at $45°$. Also calculate the maximum and minimum torques.

$$T = NBIA \cos\alpha$$

$$= (200)(5 \times 10^{-3} \text{ T})(10 \text{ A})(0.1 \text{ m} \times 0.05 \text{ m})(0.707)$$

$$= 0.035 \text{ N} \cdot \text{m}.$$

The maximum torque is obtained for $\alpha = 0°$ or $\cos\alpha = 1$:

$$T_{\max} = NBIA$$

$$= (200)(5 \times 10^{-3} \text{ T})(10 \text{ A})(0.1 \text{ m} \times 0.05 \text{ m})$$

$$= 0.05 \text{ N} \cdot \text{m}.$$

The minimum torque $= 0$, for $\alpha = 90°$.

21.4 GALVANOMETER

A meter used to measure small currents is called a galvanometer. The majority of these meters make use of the fact that a current passing through a coil produces a torque when placed in a magnetic field. In a common laboratory galvanometer (Fig. 21.13), a coil wound around a soft iron cylinder is kept between the poles of a magnet. The coil rotates on jewelled bearings about a vertical axis.

There are two spiral springs restraining the rotation of the coil. When the coil is rotated, there is a restoring torque due to the springs, T_R, proportional to the angle of rotation, which may be written

$$T_R = k\theta. \tag{21.12}$$

This equation is Hooke's law for twisting a wire, where k is called the torque constant. In Eq. (21.12), T_R is torque (N · m), θ the angle (radians), and k the torque constant (N · m/rad).

Figure 21.13. Galvanometer.

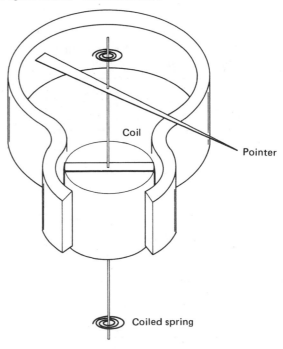

Coil

Pointer

Coiled spring

The poles of the magnet are shaped so that the magnet with the soft iron cylinder provides a radial magnetic field which is uniform within the range of the rotation of the coil (Fig. 21.14). Since the radial field is parallel to the coil for any angular position of the coil within this range, the magnetic torque has no dependence on the orientation of the coil.

The deflection of the coil is indicated by a needle which moves over a properly calibrated scale. The direction of the deflection depends on the direction of the current. For the direction of the current indicated in Fig. 21.14, the deflection is towards the left. The coil comes to rest at an angular position for which the restoring torque is equal to the electromagnetic torque, that is,

$$NBAI = k\theta \qquad (21.13)$$

or

$$\theta = \frac{NBA}{k} I. \qquad (21.14)$$

Figure 21.14. Top view of a galvanometer. Note that the field is radial in the center.

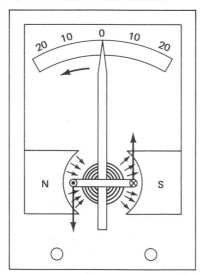

The deflection θ is thus proportional to the current. The sensitivity of a galvanometer, measured as θ/I, deflection per unit current, is proportional to $N, B,$ and A and is inversely proportional to the torque constant. In other words, increasing $N, B,$ and A gives us a larger deflection for a given I. Decreasing k also achieves the same purpose.

EXAMPLE 21.6

A galvanometer has the following design constants: $B = 0.4$ T, $N = 40$, length of the coil = 0.01 m, width of the coil = 0.01 m, and $k = 4 \times 10^{-6}$ N · m/rad. Maximum deflection is limited to 50°. Calculate the amount of current that will produce this deflection.

We use Eq. (21.13) to solve for I. We are given $B = 0.4$ T, $N = 40$, $l = 0.01$ m, $w = 0.01$ m, and $k = 4 \times 10^{-6}$ N · m/rad.

$$BANI = k\theta$$

$$0.4 \text{ T} \times 0.01 \text{ m} \times 0.01 \text{ m} \times 40 \times I$$

$$= 4 \times 10^{-6} \frac{\text{N} \cdot \text{m}}{\text{rad}} \times \frac{50°}{57.3°/\text{rad}}$$

$$I = 2.18 \times 10^{-3} \text{ A}.$$

As is made clear in this example, galvanometers are extremely sensitive instruments giving full deflection for current in the micro and milliampere ranges. They are also mechanically very delicate instruments. They should be properly handled, always making sure that the current does not exceed the limit for full scale deflection.

Voltmeter and Ammeter

A galvanometer is often converted to read potential differences in volts or currents in the ampere range. When used for these purposes, it is called a voltmeter or an ammeter, respectively.

The galvanometer is converted to a voltmeter by adding a high resistance R in series with the internal resistance R_G of the coil (Fig. 21.15). Addition of this resistance reduces the current flow through the galvanometer. If a current I_G gives the maximum deflection in the galvanometer, the maximum voltage that can be read with the meter is

$$V = I_G(R + R_G). \tag{21.15}$$

In a multirange voltmeter, a multipole switch enables us to introduce resistors of appropriate values into the circuit.

EXAMPLE 21.7

A current of 1 mA produces the maximum deflection in a galvanometer with an internal resistance of 50 Ω. The galvanometer is to be converted to a voltmeter in the ranges 0–1 V and 0–10 V. Obtain the values of the resistances that should be connected to make this conversion.

Resistance is selected so that only 1 mA current flows through the galvanometer for the applied maximum voltage. The equation relating these quantities is

$$V = I_G(R + R_G).$$

In the first case $V = 1$ V,

$$1\text{ V} = (1 \times 10^{-3}\text{ A})(R + 50\ \Omega)$$

$$R = 950\ \Omega.$$

In the second case $V = 10$ V

$$10\text{ V} = (1 \times 10^{-3}\text{ A})(R + 50\ \Omega)$$

$$R = 9950\ \Omega.$$

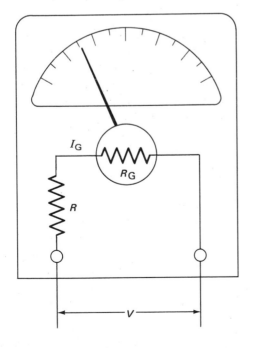

Figure 21.15. A galvanometer is converted to a voltmeter by connecting a resistance R in series.

From the above discussion we see that the resistance R connected in series to the galvanometer to convert it to a voltmeter performs the following functions: (a) It controls the amount of current flowing through the galvanometer and thus decides the range of the voltmeter. (b) The resistance allows only a very small fraction of the current to flow through the galvanometer and thus it reduces the effect of the measuring instrument on the rest of the circuit.

Remember voltmeters are connected in parallel to the electrical element across which the potential difference is to be measured. What does the voltmeter read when it is connected in series? How else will it affect the circuit?

Ammeters measure the current flow through a circuit; they are connected in series so that the current flows through them (Fig. 21.16). Since they are connected in series, they are designed to have an effective low resistance in order to introduce no appreciable resistance into the circuit. An ideal ammeter is supposed to have very little effect on the existing current flow.

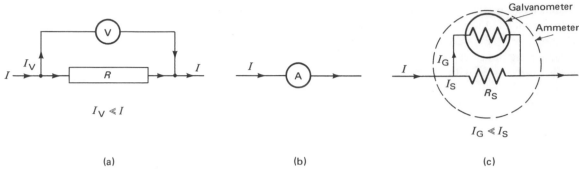

(a)   (b)   (c)

Figure 21.16. (a) A voltmeter is always connected parallel to the electrical element R, across which potential difference is to be determined. Ideally, a voltmeter will not alter the existing conditions if its resistance
is infinite (no current flow through voltmeter). (b) An ammeter is connected in the circuit in series with
other elements so that the current we want to measure flows through the ammeter. Ideally resistance of
the ammeter should be zero in this case. (c) The low shunt resistance connected to the galvanometer reduces the resistance of the ammeter. Also note that only a small fraction of the current passes through
the ammeter.

The changes we make in a galvanometer to convert it to an ammeter should achieve the following:
Reduce the effective resistance and channel a small
fraction of the current into the galvanometer.
These criteria are obtained by connecting a low
resistance, called a shunt, parallel to the coil of the
galvanometer (Fig. 21.17). If R_S is the shunt resistance, the effective resistance of the ammeter R_A is

Figure 21.17. A shunt resistance is connected in
parallel to the galvanometer coil to convert it to
an ammeter.

$$R_A = \frac{R_G R_S}{R_G + R_S} \qquad (21.16)$$

which is smaller than both R_G and R_S. Also since
the shunt is of low resistance, a large fraction of the
current flows throug the shunt instead of the galvanometer coil. I_G is the galvanometer current and I_S
is the shunt current, since I, the total current, divides
into the galvanometer current I_G and shunt current
I_S we have

$$I = I_G + I_S. \qquad (21.17)$$

Applying Kirchhoff's second rule to the loop
containing R_G and R_S, we get

$$I_G R_G = I_S R_S. \qquad (21.18)$$

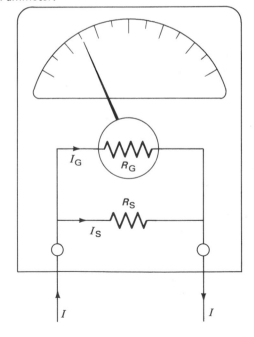

From Eqs. (21.17) and (21.18) we have

$$I_G = I \frac{R_S}{R_G + R_S} \quad . \qquad (21.19)$$

Equation (21.19) tells us that I_G is only a fraction of the total current I.

EXAMPLE 21.8

A current of 0.5 mA produces maximum deflection in a galvanometer. To convert this galvanometer to an ammeter in the range of 0–5 A, what should be the value of the shunt resistance? The resistance of the galvanometer is 50 Ω.

When 2 A flows through the ammeter, 0.5 mA should flow through the galvanometer coil and the rest through the shunt resistance. Using Eq. (21.18), we get

$$(0.5 \times 10^{-3} \text{ A})(50 \text{ } \Omega) = (1.9995 \text{ A})R_S.$$

$$R_S = \frac{0.5 \times 10^{-3} \text{ A}(50 \text{ } \Omega)}{1.9995 \text{ A}} = 0.012 \text{ } \Omega.$$

21.5 MOTORS

A current-carrying coil experiences torques in a magnetic field, but unless special arrangements are made the sense of the torque reverses as the coil rotates, as explained earlier (Fig. 21.18). What can

Figure 21.18. (a), (b) and (c) show direction of forces as the coil rotates in the magnetic field. In (a), the torque on the coil is clockwise, and in (c) it is counterclockwise. (d) A commutator helps to reverse the current flow as the coil rotates from position (a) to (c). The resulting torque will have the same sense throughout the rotation of the coil.

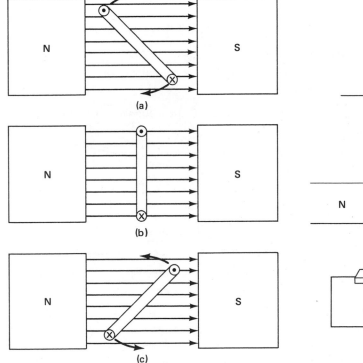

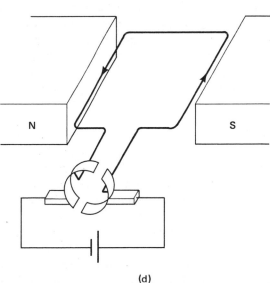

be done to achieve continuous rotation instead of oscillation? The obvious solution to this problem is to reverse the direction of the current flow as the coil rotates through the $\theta = 90°$ orientation of the coil. One method uses a commutator, which consists of two half rings connected to each end of the conducting coil and insulated from each other [Fig. 21.18(d)]. As the loop rotates, the brushes contact alternate half rings with which they make electrical contact in each $180°$ rotation and thus reverse the current flow. In an actual motor there are several sets of coils distributed around an iron core called an armature. The torque is increased and made uniform by this arrangement. The motor described above is a dc motor.

The second method uses a voltage source which changes its direction (called ac) and when connected to the coil, the current automatically reverses. The frequency of the voltage source should be synchronized with the frequency of the motor. This motor is an ac motor. An example is the motor used in electric clocks.

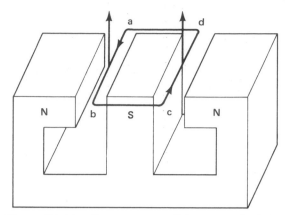

Figure 21.19. Force on a coil in magnetic fields produced by a three–pole magnet.

the form of a voltage that changes in amplitude and direction with time is fed into the coil, a current with similar characteristics is produced in the coil. The resulting force, that changes with time, produces oscillations of the loudspeaker cone (loudspeaker cone is firmly attached to the coil).

21.6 THE LOUDSPEAKER

In the cases we discussed above, a current-carrying coil experiences a torque in a magnetic field. In a loudspeaker, on the other hand, the coil experiences a net force but no torque. To produce a net force on a coil, a three-pole magnet that produces fields in opposite directions (the direction of the field between the first pole and the central pole and the direction of the field between the central pole and the third pole are in opposite directions) is used (Fig. 21.19). It is evident from Fig. 21.20 that the direction of the current through section cd is opposite to that of the current through ab and also the direction of the magnetic field at cd is opposite to that of the field at ab. Since both current and field simultaneously reverse their directions, the forces acting on both sides of the coil are in the same direction. In a speaker a circular coil and circular field are used, so that the force acts on all parts of the coil (Fig. 21.20). The figure shows the cross-sectional view of the loudspeaker. When a signal in

21.7 MAGNETIC FIELDS

Permanent magnets are not the only sources of magnetic fields. A current-carrying wire has a mag-

Figure 21.20. Cross-sectional view of a loudspeaker.

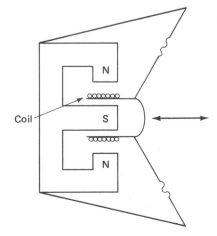

netic field around it. The existence of a magnetic field around a current-carrying wire was first discovered by Oersted in 1820. One of the obvious advantages of an electrically produced magnetic field is that it can be turned on and off at will. Also with the availability of superconducting materials (extremely low-resistance wires), it has become possible to produce very high magnetic fields unattainable with permanent magnets.

A simple experimental arrangement illustrated in Fig. 21.21 can be used to demonstrate the presence of a magnetic field around a current-carrying wire. A wire is passing at right angles through a cardboard with iron filings sprinkled on it. When current passes through the wire and the cardboard is lightly tapped, the filings form circles around the wire tracing the field lines. The field lines around the wire are concentric circles with the wire at its center. The direction of the field is obtained by using the rule given below.

Grasp the wire with the thumb of the right hand in the direction of the current. The direction of the fingers around the wire is the direction of the magnetic field. This right-hand rule gives us the direction of the field around a current. The first right-hand rule we saw in the beginning of this chapter was to determine the direction of force acting on a charge (current) in a magnetic field.

The general equation for the calculation of the magnetic field is not that simple, but there are some special cases where the field can be obtained from rather simple equations. These are discussed below.

Field Around A Long Wire

The magnitude of the field at a distance r from a long current-carrying wire (Fig. 21.22) is

$$B = \frac{\mu_0 I}{2\pi r}, \qquad (21.20)$$

where μ_0 is the permeability of free space and it is equal to $4\pi \times 10^{-7}$ T $\cdot$ m/A. To obtain the field in tesla, the current should be in amperes and r should be in meters.

EXAMPLE 21.9

Calculate the magnetic field at a point 10 cm away from a long wire carrying a current of 10 A.

We use Eq. (21.20) to calculate the field

$$B = \frac{\mu_0 I}{2\pi r} = \frac{(4\pi \times 10^{-7} \text{ T} \cdot \text{m/A})(10 \text{ A})}{2\pi \times 0.1 \text{ m}}$$

$$= 2 \times 10^{-5} \text{ T} = 0.2 \text{ G}.$$

Figure 21.21. Magnetic field around a wire carrying current. Figure also illustrates the right-hand rule to determine the direction of the field.

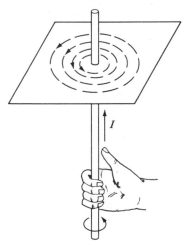

Figure 21.22. Field around a long wire.

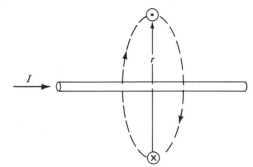

EXAMPLE 21.10

Two 50 cm long wires, kept 5 cm apart, carry a current of 15 A each. Calculate the force between the two wires (a) when the currents are in the same direction and (b) when they are in opposite directions.

(a) To solve this problem we consider the second wire to be in the magnetic field set up by the first wire (Fig. 21.23). The direction of this field is into the paper (verify using the rule given above). The force acting on the second wire, by applying the right-hand rule, is directed toward the first wire. The force is an attractive force. Remember there is an equal force acting on the first wire directed

toward the second wire. Magnetic field at the second wire due to the first wire is

$$B = \frac{\mu_0 I_1}{2\pi d} = \frac{(4\pi \times 10^{-7} \ \text{T} \cdot \text{m/A})(15 \ \text{A})}{2\pi \times 0.05}$$

$$= 6 \times 10^{-5} \ \text{T}.$$

The force acting on the second wire is

$$BI_2 l = (6 \times 10^{-5} \ \text{T})(15 \ \text{A})(0.5 \ \text{m})$$

$$= 4.5 \times 10^{-4} \ \text{N}.$$

(b) The magnitude of the force is the same as in part (a) but the force is repulsive.

Wires carrying current in the same direction attract each other, while wires carrying currents in opposite directions repel each other and the magnitude of the force per unit length of the wire in both cases is

$$F = \frac{\mu_0 I_1 I_2}{2\pi d}, \tag{21.21}$$

where I_1 and I_2 are the currents in the wires and d is the distance between the wires.

Figure 21.23. Force between two wires carrying current in the same direction is attractive. If the currents are in opposite directions, the force is repulsive.

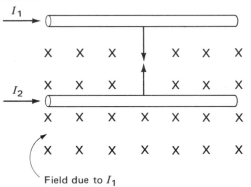

Field due to I_1

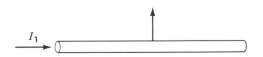

Field due to I_1

Figure 21.24. Bending a wire into a coil increases the flux inside the coil.

Field Due to a Coil

Bending a wire into a coil increases the field in the region within the coil since every section of the coil contributes flux lines to this region as shown in Fig. 21.24. The field at the center of the circular area defined by the coil is

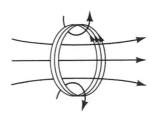

Figure 21.25. Magnetic field near a coil.

$$B = \frac{\mu_0 I}{2r}. \tag{21.22}$$

The field is about three times bigger than the field due to a straight wire for the same distance r. The field can further be increased by increasing the number of turns (Fig. 21.25). For a coil with N turns, the field is

$$B = \frac{\mu_0 N I}{2r}. \tag{21.23}$$

EXAMPLE 21.11

A coil, 5 cm in diameter, has a current of 10 A flowing through it. (a) Calculate the field at the center of the coil. (b) If this coil is stretched into a wire, obtain the field for the same current at a point 2.5 cm away from the wire.

(a) The magnetic field at the center of the coil is

$$\frac{\mu_0 I}{2r} = \frac{(4\pi \times 10^{-7} \text{ T} \cdot \text{m/A})(10 \text{ A})}{2 \times 2.5 \times 10^{-2} \text{ m}}.$$

$$= 2.5 \times 10^{-4} \text{ T}.$$

(b) The magnetic field 2.5 cm away from the wire is

$$\frac{\mu_0 I}{2r} = \frac{(4\pi \times 10^{-7} \text{ T} \cdot \text{m/A})(10 \text{ A})}{2\pi \times 2.5 \times 10^{-2} \text{ m}}$$

$$= 8 \times 10^{-5} \text{ T}.$$

As mentioned earlier, at the center of the coil we get a higher field.

Solenoid

A helical coil (like a slinky) with a large number of turns is called a solenoid (Fig. 21.26). The field

Figure 21.26. (a) A loosely wound solenoid. Magnetic field near the axis is uniform. (b) Magnetic field produced by a solenoid is similar to that produced by a permanent magnet.

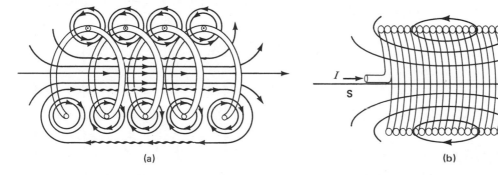

(a)   (b)

in the interior of the solenoid is

$$B = \frac{\mu_0 NI}{l} = \mu_0 \, nI, \qquad (21.24)$$

where N is the total number of turns, n is the number of turns per unit length, and l is the length of the solenoid. If the coil is closely packed, the field in the interior is uniform and is almost independent of the radial distance from the axis of the solenoid (except near the ends of the solenoid).

An electromagnet (a magnet energized by an electric current) consists of a solenoid filled with a magnetic material in its core. The presence of magnetic material inside the solenoid increases the field by several hundred times. The equation for the field in this case is

$$B = \frac{\mu IN}{l}, \qquad (21.25)$$

where μ is the permeability of the core material. The permeability of magnetic materials is much higher ($\sim$3 orders of magnitude) than the permeability of free space. The ratio of μ/μ_0 is called relative permeability.

EXAMPLE 21.12

A 10 cm long solenoid has 500 turns. A current of 1.5 A is passing through the solenoid. Calculate the field inside the solenoid (a) with no material inside the core and (b) with soft iron inside the core. The relative permeability of soft iron is 1300.

(a) The field inside the solenoid is

$$\frac{\mu_0 NI}{l} = \frac{(4\pi \times 10^{-7} \text{ T} \cdot \text{m/A})(500)(1.5 \text{ A})}{0.1 \text{ m}}$$

$$= 9.4 \times 10^{-3} \text{ T}.$$

(b) The field with iron in the core is

$$\frac{\mu NI}{l} = 1300 \times 9.4 \times 10^{-3} \text{ T}$$

$$= 12.2 \text{ T}.$$

21.8 MAGNETIC MATERIALS

As we have seen in the discussion above, an existing magnetic field is enhanced very much by introducing a magnetic material like iron. There are other materials such as copper and aluminum whose effect on an existing magnetic field is negligible. There is a classification of materials based on their effect on existing fields. Three major types of materials are the following:

1 Ferromagnetic materials. They increase the field and $\mu/\mu_0 \gg 1$.

2 Paramagnetic materials. They increase the field only slightly. For these materials, μ/μ_0 is very slightly greater than 1.

3 Diamagnetic materials. They reduce the field slightly. μ/μ_0 is numerically less than but close to 1.

Magnetism is a bulk property, but our model discussed below states that it results from certain properties of the atoms. Some atoms, like iron and nickel atoms, are like miniature magnets. In a ferromagnetic material these atoms align in one direction producing a net external magnetism (Fig. 21.27). The atomic forces favoring the alignment sometimes are not strong enough to overcome the thermal energy of the individual atoms. Under these conditions, the random motion associated with the thermal energy breaks up the orderly alignment of

Figure 21.27. In a ferromagnet, small atomic magnets align in the same direction.

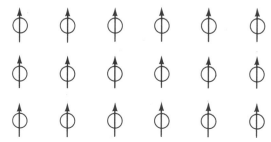

the atoms. The temperature at which this happens is called the Curie temperature. Iron, for example, is ferromagnetic below its Curie temperature (760°C) and it is paramagnetic above this temperature. In addition to temperature, chemical composition and crystal structure play important roles in deciding whether a material is ferromagnetic or paramagnetic. For example, certain stainless steels (alloys of iron) are paramagnetic and there are several iron compounds which are paramagnetic. Therefore, the presence of magnetic atoms in a material does not guarantee that the material is magnetic.

Let us now investigate the puzzling behavior of some magnetic materials. Take two bars of iron that do not show any external magnetism. They do not attract each other. Bring one of the bars near a permanent magnet. They attract each other, as if the bar has now become a magnet as shown in Fig. 21.28(b). If the bar is above the Curie temperature,

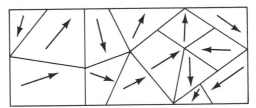

Figure 21.29. Domains in a magnetic material. Atomic magnets are aligned in the same direction in a domain. Direction of magnetization of the domains is random in an unmagnetized material.

the attractive force is much weaker. Evidently below the Curie temperature the bar becomes a stronger magnet under the influence of an external field, but by itself it shows no bulk magnetism.

The behavior of magnetic materials is explained in terms of magnetic domains. Magnetic domains are very small regions inside the material where atomic magnets aligned in the same direction, but the direction of alignment varies from domain to domain. The net result is that the bar has no bulk magnetism (Fig. 21.29). To understand why domains have random directions, let us consider three bar magnets. It is hard to keep them aligned in one direction (Fig. 21.30). On the other hand, a configuration like the one shown in Fig. 21.30 offers the least resistance and it is favored over the configuration where all of the magnets are aligned in the same direction. The triangular configuration also effectively produces no external magnetism. The behavior of the randomly oriented domains are somewhat like the behavior of these magnets.

In terms of the domain theory it is easy to understand the results of the above experiment (Fig. 21.28). Ordinarily the bars are not magnetized. In Fig. 21.28(b) the field of the U magnet forces the domains to orient in its direction and thus the field induces magnetization in the bar. Now the strong attractive force is between the permanent magnet and the induced magnet. If the bar is heated above the Curie temperature the bar becomes paramagnetic reducing the force between the bar and the magnet.

Figure 21.28. (a) There is no magnetic force between two bars of soft iron. (b) A permanent U magnet attracts the bar. The force between the U magnet and the bar is much weaker when the bar is heated above 760°C.

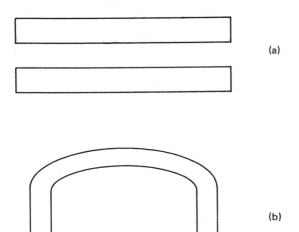

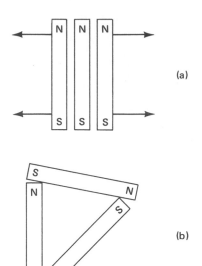

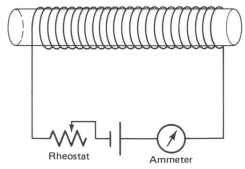

Figure 21.31. Experimental setup to study magnetic induction inside the cylindrical material as a function of the current through the coil.

Figure 21.30. (a) Repulsive forces between similar poles make it difficult to align three magnets as shown in this figure. (b) Attractive force between unlike poles makes the triangular arrangement stable. This arrangement produces no net external magnetic field.

domain occurs, while in other materials a few domains with atomic magnets in the direction of the external field grow in size at the expense of other domains. These processes become harder as the number of reoriented atomic magnets increases. This explains why the rate of increase of B is fast at first and slow later. Finally, we reach saturation magnetization where alignment is achieved throughout the material.

In the consideration above, magnetization of the sample is discussed in terms of the current passing through the solenoid. Another quantity used in similar discussions, in place of the current, is the

An external field induces magnetism in a magnetic material. A qualitative discussion of the phenomenon was given above. To further study this effect, we use a solenoid wound over a cylindrical rod of the material under study (Fig. 21.31). The current in the solenoid produces the field which causes the magnetization in the material. Therefore, we look at the variation of magnetic induction inside the material as a function of the current. As we increase the current, the magnetic induction B increases first at a fast rate and then at a slower rate (Fig. 21.32). The permeability $\mu = Bl/NI$, in this case, is not a constant.

Figure 21.32. Variation of magnetic induction as a function of current inside the coil of Figure 21.31.

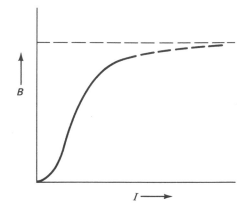

The behavior of B inside the rod can now be explained in terms of the magnetic domains. As the current and hence the inducing field—field due to the solenoid above—increases, more and more atomic magnets of the magnetic material align in the direction of the field of the coil. In some materials, simple reorientation of atomic magnets in each

magnetic intensity. Magnetic intensity H is defined by

$$H = \frac{B}{\mu} = \frac{NI}{l}. \qquad (21.26)$$

The magnetic intensity is independent of the material. The intensity inside a solenoid is the same whether the core is empty or filled with some material. The intensity H is proportional to the current I in a solenoid and hence the discussion of magnetization in terms of H or I is equivalent. Also note that the unit of H is ampere/meter.

Let us continue with our experiment on magnetization. After reaching point a on the magnetization curve (Fig. 21.33), the magnetic intensity is slowly decreased to zero (by decreasing the current). The magnetic induction does not go down to zero. Some magnetization (point b) is retained in the material. The value of B at point b is called the retentivity of the material or residual magnetism. Materials used for permanent magnets have high retentivity while "soft magnetic" materials have low retentivity. The magnetic intensity is slowly increased in the opposite direction till we reach a point where B is zero. The value of H at point c is known as the coercive force. The coercive force is a measure of the magnitude of the external field needed to reduce the residual magnetization to zero. If the coercive force is small, the magnetic field can be reduced to zero by a small external field.

The results of taking the material through a complete cycle of magnetization (returning to point a) is shown in Fig. 21.33. The BH curve does not trace the same path. This behavior of magnetic material resulting from B *lagging* H is called hysteresis. The area enclosed by a BH curve can be shown to be equal to the amount of work done in orienting and reorienting the domains in one cycle. This work appears as heat.

In transformers, magnetic materials are used in their cores. The rapid reversal of the direction of the current through the coil takes the core through successive hysteresis loops (60 times a second if ordinary electric power sources are used). If the

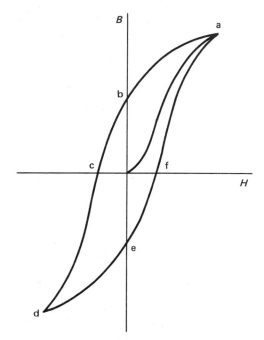

Figure 21.33. Hysteresis loop, B versus H, of a magnetic material. Induction at point b is the field remaining as H is reduced to zero and is called residual magnetism.

hysteresis loop is wide, its area will be large and, consequently, a large loss of energy as heat results. Hence for transformer cores, materials with narrow hysteresis loops are used to give low energy loss.

SUMMARY   The maximum force acting on a charge moving through a magnetic field is

$$F = qvB.$$

Maximum force on a current is

$$F = IlB$$

These equations are used to define magnetic field induction.

The torque on a coil is given by

$$T = IAB.$$

In a galvanometer, the deflection of the coil is proportional to current. A galvanometer is converted to a voltmeter by adding a resistance in series with the coil. A galvanometer is converted to an ammeter by adding a shunt of low resistance in parallel.

In a dc motor, the torque on the coil is maintained in the same sense by reversing the current by a split-ring commutator. In an ac motor, current reverses as the voltage changes polarity.

Current passing through a wire produces a magnetic field. Magnetic induction is given by

$$B = \frac{\mu_0 I}{2\pi r} \qquad \text{(straight wire)}$$

$$B = \frac{\mu_0 NI}{2r} \qquad \text{(coil)}$$

$$B = \frac{\mu NI}{l} \qquad \text{(Solenoid filled with a material of permeability } \mu\text{)}$$

Magnetic field intensity H is given by

$$H = B/\mu.$$

The area of a BH curve, or hysteresis curve, gives the heat lost during a cycle of magnetization.

The domain theory explains most of the features of the magnetization curve.

QUESTIONS AND PROBLEMS

21.1 How can you prove convincingly that a given metallic bar is a magnet?

21.2 A charge passing through a region is found to deflect. Does it prove there is a magnetic field in that region?

21.3 A charge deflects and its speed increases. Can a magnetic field produce both these effects? Can you think of fields which can cause both these effects?

21.4 Heating a magnetic material causes it to lose its magnetization. Explain.

21.5 Design an electric bell. Give complete details.

21.6 Draw diagrams showing the proper connections of an ammeter and a voltmeter in an electric circuit. Discuss the errors introduced by an ammeter and a voltmeter.

21.7 An ammeter should never be connected to a circuit without a resistance or load in series with it. Why?

21.8 One end of a coil is fixed and the other end is dipped slightly in a pool of mercury. A dc power supply is connected as shown in Fig. 21.34. What will happen to the coil?

21.9 For a lifting magnet what is preferable, a permanent magnet or an electromagnet. Why?

21.10 When superconducting magnets are made with wires through which high currents can pass, magnetic cores are not used. Why?

21.11 Find out why superconducting magnets are not ordinarily used. What are their limitations?

21.12 Three particles, a neutral atom, a negatively charged atom, and a positively charged atom, are shot into a magnetic field with their velocity perpendicular to the magnetic field. Describe their motion in the field. Draw a sketch illustrating their motion.

21.13 A uniform magnetic field B points horizontally from east to west; its magnitude is 1.5 T. If protons with velocity 3×10^{-7} m/s are directed vertically down, calculate the force acting on them.

Figure 21.34. Problem 21.8.

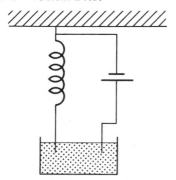

21.14 Electrons with velocity 6×10^7 m/s are shot into a magnetic field perpendicular to the field. The magnetic field B is 0.01 T. Calculate (a) the force acting on the electrons, (b) radius of the orbit of the electrons in the field, and (c) time taken by an electron to complete an orbit.

21.15 To compare the masses of two isotopes of iron (iron-56 and iron-57) singly ionized (net charge $+ 1.6 \times 10^{-19}$ C) atoms were first accelerated in an electric field of 100 kV. They were then introduced into a magnetic field where $B = 0.8$ T. Calculate the separation between the lines produced by the isotopes on the photographic paper (Fig. 21.10). The masses of the two isotopes are 56.953 u, and 57.951 u (1 u = 1.66 $\times 10^{-27}$ kg).

21.16 An alpha particle travels in a circular path of radius 0.25 m in a field with $B = 0.9$ T. Calculate (a) the speed of the alpha particle and (b) its period of revolution.

21.17   A wire, 0.5 m long carrying a current of 15 A, is kept in a magnetic field as shown in Fig. 21.35. If the field is given by $B = 0.9$ T, calculate the force acting on the wire for the three positions of the wire.

21.18 A 1 kg rod is 1 m long. The rod is suspended by two springs. To balance its weight, keeping the springs unstretched, it is kept in a magnetic field and a current of 10 A is passed through it (Fig. 21.36). Find the direction and magnitude of the field.

21.19 Find the direction and magnitude of a magnetic field that produces a vertical force of 4 N on a horizontal wire 0.5 m long carrying a current of 10 A.

21.20 A rectangular coil is 3 cm^2, has 100 turns, and carries a current of 100 μA. It is in a field (Fig. 21.13) of 0.5 T. Calculate the torque on the coil.

Figure 21.35. Problem 21.17.

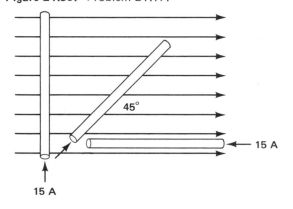

Figure 21.36. Problem 21.18.

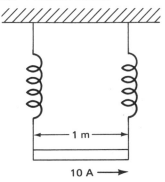

21.21 The coil of Problem 21.20 is pivoted by frictionless bearings and has a spring with a torque constant of 1.5×10^{-6} N · m/rad attached to it. Calculate the maximum deflection for 100 μA.

21.22 The coil of a galvanometer has a resistance of 50 Ω. A current of 1 mA produces full deflection. Calculate the resistance to be added to convert this to a voltmeter with a range of (a) 0–1, (b) 0–5, and (c) 0–10 V.

21.23 Calculate the resistance to be added to convert the galvanometer of Problem 21.22 to an ammeter in the range of (a) 0–1, (b) 0–5, and (c) 0–10 A.

21.24 A straight wire carries a current of 10 A. Calculate the magnetic field 3 cm, 5 cm, and 10 cm from the wire.

21.25 Two straight wires 1 m long carry 10 A each in the same direction. Calculate the force between the wires if they are kept 6 cm apart. Show that the force is attractive.

21.26 A coil of radius 10 cm and 50 turns carries a current of 5 A. Calculate the magnetic field at the center of the coil.

21.27 A 20 cm long solenoid has 200 turns. It carries a current of 1 A. Calculate the magnetic field (B) inside the solenoid.

21.28 The interior of the above solenoid is filled with iron whose relative permeability is 1500. Calculate the field inside the coil at point 1 and just outside the coil at point 2 (see Fig. 21.37).

21.29 Calculate the field intensity at points 1 and 2 for Problem 21.23.

Figure 21.37. Problem 21.28.

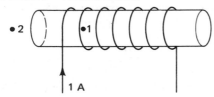

Induced EMF

Changes in magnetic flux in a closed loop of a conducting wire induce an electromotive force in it. This discovery, made by Faraday and simultaneously by Joseph Henry in 1831, paved the way for the development of modern electric generation and distribution systems.

22.1 INDUCED EMF

A simple experiment (Fig. 22.1) can be used to demonstrate several aspects of induced EMF. The galvanometer connected to the coil ordinarily shows no deflection since there is no source of EMF in the circuit. When a magnet is introduced into the coil, we observe a deflection. Reverse deflection occurs when the magnet is withdrawn. Similarly moving the coil with the magnet held fixed produces deflection. Keeping the magnet stationary in the coil produces no effect. From such observations Faraday concluded that an electromotive force is induced by the relative motion between the coil and the magnet. The galvanometer deflection also depends on the speed of the relative motion between the magnet and the coil; deflection is directly proportional to the relative motion. The greater the speed, the greater the change in the magnetic flux threading the coil. Similar simple experiments were performed by Joseph Henry, an American scientist. From the results of such experiments Faraday concluded that *the induced EMF in a coil is proportional to the rate at which the magnetic flux is changing in the coil.* This statement, known as Faraday's law, may be represented by the equation

$$\mathcal{E} = -N\frac{\Phi_2 - \Phi_1}{t}, \tag{22.1}$$

where $\mathcal{E}$ is the EMF, N the total number of turns in

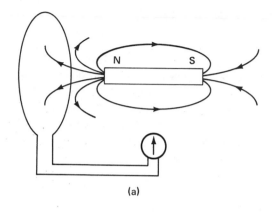

(a)

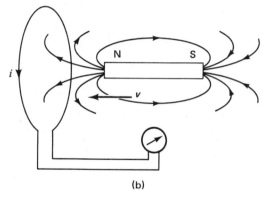

(b)

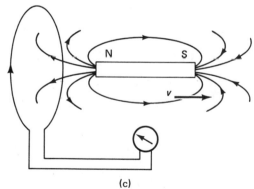

(c)

Figure 22.1. Relative motion of a magnet with respect to a coil produces induced EMF in the coil. (a) No relative motion, no induced EMF. (b) Magnet is moved into the coil with a velocity v. The deflection of the needle of the voltmeter indicates the magnitude of the induced EMF. (c) The magnet is removed from the coil with a velocity v. The needle deflects in the opposite direction.

the coil, and $\Phi_2 - \Phi_1$ the change in the flux enclosed in the coil in a small time t. The change in the flux in a short time interval is often written as $\Delta\Phi/\Delta t$ so that Eq. (22.1) may be written in the form

$$\mathscr{E} = -N\frac{\Delta\Phi}{\Delta t}. \qquad (22.1a)$$

As mentioned before Δ stands for change in the quantity following it. The flux Φ at any time is

$$\Phi - BA, \qquad (22.2)$$

where B is the magnetic induction and A the cross-sectional area of the coil. Therefore, a change in the strength of the field or a change in the area of the coil (B remaining constant) results in a flux change inside the coil and hence an induced EMF. Both these cases are illustrated below.

Changing Field

The magnetic field in Fig. 22.2 is directed into the plane of the paper. The flux enclosed in the coil of area A is

$$\Phi = BA.$$

Figure 22.2. Changing magnetic induction induces an EMF in a coil lying in the field.

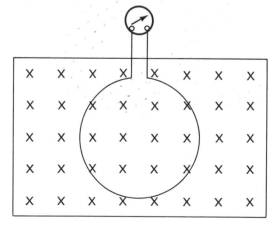

Let us suppose the magnetic induction reduces from B to zero in a time t. The induced EMF can be obtained by applying Faraday's law. In this case $N = 1$, $\Phi_2 - \Phi_1 = 0 - BA$,

$$\mathscr{E} = -N\frac{\Phi_2 - \Phi_1}{t} = -1\left(\frac{0 - BA}{t}\right)$$

$$= \frac{NB}{t}. \qquad (22.3)$$

Changing Area

The wire AB lying on two wires AD and BC is moving as indicated in Fig. 22.3, so that the area enclosed is continuously increasing and hence the flux enclosed by the coil is increasing. If the wire AB moved to a position A′B′ in a time t the increase in the flux is given by $BL\Delta x$; $L\Delta x$ is the change in the area. The magnitude of the EMF induced in this case is

$$\mathscr{E} = \frac{BL\Delta x}{\Delta t}.$$

Since $\Delta x/\Delta t = v$, we have

$$\mathscr{E} = BLv. \qquad (22.4)$$

In this case, $BL\Delta x$ is the number of flux lines cut by the moving wire in time Δt. Therefore, the EMF is the number of flux lines cut by the moving wire per unit time. A wire placed perpendicular to a field (Fig. 22.4), when moved with a velocity v, will induce an EMF given by

$$\mathscr{E} = BLv,$$

where L is the length of the wire lying in the field. The arrangement in Fig. 22.4 can be used as a good demonstration of induced EMF.

Lenz's Law

In both cases discussed above, we did not consider the polarity of the induced voltage and the direction of the resulting current. Lenz's law which is often used to find the direction of the current

Figure 22.3. When wire AB moves in the magnetic field, the flux enclosed by the coil and the wire changes, inducing an EMF in the coil.

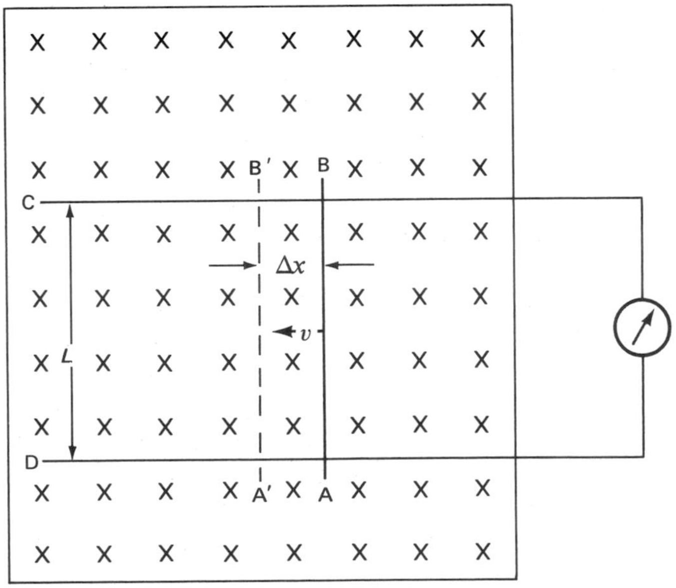

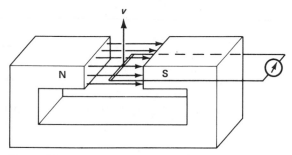

Figure 22.4. A wire cutting the flux lines produces an EMF.

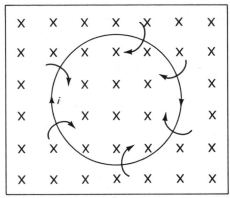

Figure 22.5. When flux inside the coil decreases, current due to the induced EMF in the coil tries to maintain the flux density by producing field lines in the same direction.

states: *The current induced in a coil is in such a direction as to oppose the change that causes it. This opposition to the change is indicated by the minus sign in Faraday's law.*

Two ways of interpreting Lenz's law are given here: (a) "Opposition to change" is taken literally, which means that if the change involves a decreasing flux, the induced current tries to increase the flux and vice versa. (b) The direction of the induced current is in such a direction that the force exerted on this current by the field opposes the change, like the motion of the wire in the second example. Both these approaches give the correct direction of the current.

Let us now look at the first example to determine the direction of the current. The magnetic field inside the coil is decreasing. To oppose this change the induced current produces a field in the same direction as the existing field inside the coil. The direction of the current that results in such a flux increase is indicated in Fig. 22.5.

The second example of the wire moving to the left with a velocity v will be analyzed in terms of both interpretations. To oppose the motion of the wire, the magnetic force on the induced current must be directed to the right (Fig. 22.6). To obtain this direction the induced current must be from B to A (verify this using the right-hand rule).

Let us now use the first interpretation to this problem. As the wire moves out, the flux enclosed in the coil is increasing. Therefore the current should decrease the net flux into the paper. This is achieved by producing a field in the opposite direc-

tion. As you can verify, a current in the direction B to A produces such a field.

EXAMPLE 22.1

A coil of wire having an area 0.02 m^2 is in a magnetic field with induction $B = 1.2$ T. If the field changes to 0.6 T in a time of 0.02 s, calculate the

Figure 22.6. The induced current has a direction such as to oppose the change producing it. Here the magnetic force is opposite to the force maintaining the velocity.

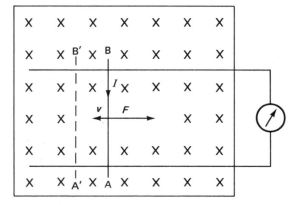

induced EMF. The coil has 50 turns and it is perpendicular to the field.

The initial flux passing through the coil is

$$BA = (1.2 \text{ T})(0.02 \text{ m}^2)$$

$$= 0.024 \text{ Wb}.$$

The final flux through the coil is

$$(0.6 \text{ T})(0.02 \text{ m}^2) = 0.012 \text{ Wb}.$$

Change in flus = final flux − initial flux

$$= 0.012 \text{ Wb} - 0.024 \text{ Wb}$$

$$= -0.012 \text{ Wb}.$$

$$\text{Induced EMF} = -N\frac{\Delta \Phi}{\Delta t} = \frac{50 \times 0.012 \text{ Wb}}{0.02 \text{ s}} = 30 \text{ V}.$$

EXAMPLE 22.2

A rectangular coil having an area 0.05 m^2 is in a magnetic field with induction $B = 0.8$ T. The coil which is initially parallel to the field is flipped to the perpendicular position (Fig. 22.7) in $\frac{1}{120}$ s. Calculate the induced EMF. The number of turns is 100.

The flux passing through the coil in its original position is zero. The flux through the coil in the

Figure 22.7. Example 22.2.

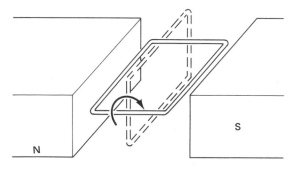

perpendicular position is

$$BA = (0.8 \text{ T}) \times (0.05 \text{ m}^2)$$

$$= 0.04 \text{ Wb}.$$

$$\text{Induced EMF} = -N\frac{\Delta \Phi}{\Delta t} = 100\frac{(0.04 \text{ Wb})}{1/120 \text{ s}} = 480 \text{ V}.$$

EXAMPLE 22.3

The moving wire in Fig. 22.4 is 0.1 m long. The field has an induction of 0.9 T. If the velocity of the wire is 10 m/s, calculate the induced voltage.

$$\text{Induced EMF} = BLv$$

$$= 0.9 \text{ T} \times 0.1 \text{ m} \times 10 \text{ m/s}$$

$$= 0.9 \text{ V}.$$

22.2 ELECTRIC GENERATOR

Electric generators convert mechanical energy to electrical energy. Mechanical energy rotates a coil in a magnetic field. The flux through the coil changes as it rotates, inducing an EMF in the coil.

The essential parts of a generator are a magnet and a coil (Fig. 22.8). In addition, there are two slip rings connected to each end of the coil. These rings are in contact with the brushes that carry the induced current to the external circuit. The coil is called the armature. The flux through the coil changes as it rotates. It is a maximum, BA, when the coil is perpendicular to the field and zero when it is parallel to the field. When the coil makes an angle θ with the field, the flux passing through are fewer than when it is perpendicular because the effective area projected by the coil perpendicular to the field is smaller (Fig. 22.9). The flux in this case is

$$\Phi = BA\sin\theta. \tag{22.5}$$

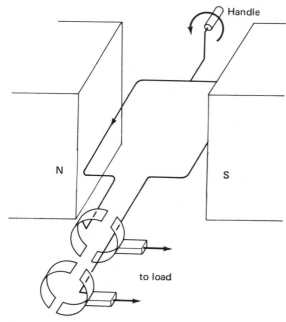

Figure 22.8. Electric generator. As the coil rotates in the magnetic field, the flux within the coil changes inducing an EMF.

If the coil is rotating with an angular velocity ω, where $\omega = \theta/t$, Eq. (22.5) can be written in the form

$$\Phi = BA \sin \omega t. \tag{22.6}$$

This equation gives us the variation of flux as a function of time.

The variation of the flux as a function of time is shown in Fig. 22.10. The flux is negative in the second half of each cycle. As the coil rotates, the side of the coil [side 1 in Fig. 22.9(a)] initially facing the north pole faces the south pole in the second half of the cycle [Fig. 22.9(d) and (e)]. This accounts for the reversal of the field in the coil.

The induced EMF is proportional to the rate at which flux changes. The changes in flux for small time intervals are marked at different points on the curve (Fig. 22.2). It is clear that the change at point 1 is zero and at point 2 the change is maximum. The rate of flux change or the slope may be calculated by dividing the flux change by the time interval. Instead of following this tedious procedure, let us use the fact that the slope of a sine function is a cosine function $[(\Delta \sin \omega t)/\Delta t = \omega \cos \omega t$ for $\Delta t \to 0]$. We then get

$$\mathcal{E} = -N\frac{\Delta \Phi}{\Delta t} = -N\frac{\Delta(BA \sin \omega t)}{\Delta t}$$

$$= NBA\,\omega \cos \omega t. \tag{22.7}$$

Figure 22.9. Flux passing through the coil as a function of its orientation.

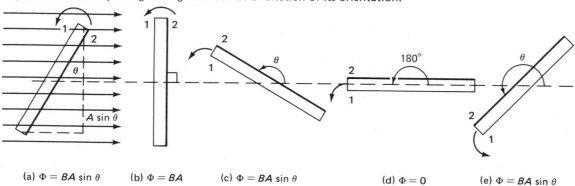

(a) $\Phi = BA \sin \theta$ (b) $\Phi = BA$ (c) $\Phi = BA \sin \theta$ (d) $\Phi = 0$ (e) $\Phi = BA \sin \theta$

(Note that the side of the coil facing the field has changed.)

From Eq. (22.7) and Fig. 22.10, we have the following.

1. The maximum value of the induced EMF is

$$\mathscr{E}_{\max} = NBA\omega \ \text{ for } \omega t = 0, \pi, \cdots$$

2. Induced EMF reverses its polarity twice in every cycle.

3. The voltage change has a period equal to the period of rotation of the coil.

The alternating voltage, with the above characteristics is called an ac (alternating current) voltage. The generator that produces ac voltage is called an ac generator. Alternating current is further discussed in the following chapter.

If the ends of the coil are connected to a split-ring commutator, instead of two slip rings, the direction of the output EMF through the external circuit remains the same. Each time the direction of induced voltage changes, the split-rings reverse the connection to the brushes (Fig. 22.11). The EMF of such a generator varies with time, but remains in the same direction. If more than one coil is used, one can obtain approximately a constant voltage output. Such generators are called dc (direct current) generators. The direction of current through an external circuit from a dc generator is the same, even though the magnitude of the voltage may sometimes change.

22.3 BACK EMF

In an electric motor, the coil is rotated by the torque produced on the current-carrying coil by the

Figure 22.10. Variation of flux and EMF with time for a rotating coil. The EMF is zero when the flux through the coil is maximum and the EMF is maximum when the flux is zero. The EMF is proportional to the rate of change of flux.

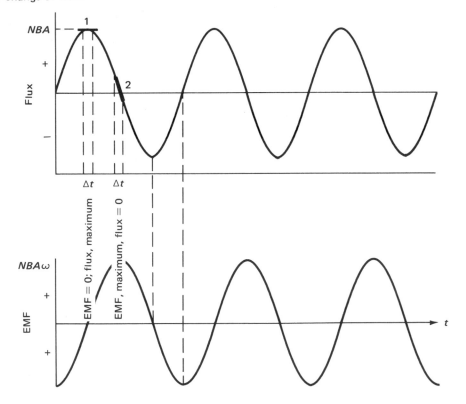

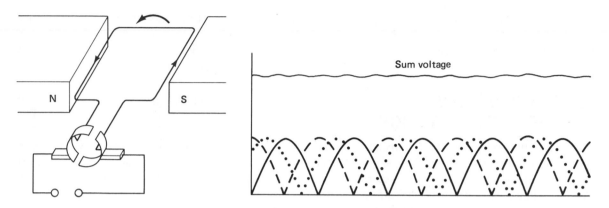

Voltage output from three coils with split rings.

Figure 22.11. A dc electric generator. Voltage output from three coils with split rings is approximately constant with time.

magnetic field. The rotating coil will generate an EMF in the opposite direction to the applied voltage that causes the rotation (Lenz's law). This induced EMF is called the back EMF or the counter EMF. The current flowing through the coil is determined by the difference between the applied EMF and the back EMF (Fig. 22.12). This current is

$$I = \frac{\mathscr{E}_a - \mathscr{E}_b}{R}, \qquad (22.8)$$

where R is the resistance of the coil. In effect, the back EMF reduces the current flowing through the coil.

The back EMF is a function of the angular speed ω, with zero at the start of the motor. Therefore, current I is extremely high at the instance of start-

Figure 22.12. Back EMF (induced EMF) opposes the applied EMF of an electric motor.

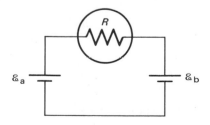

ing the motor. Special precautions, such as introducing a resistance in the coil circuit, are usually taken to reduce the starting current. Similarly an increase in current occurs if the motor gets stuck, which may result in the overheating and sometimes the burning of the coil.

Back EMF also helps in regulating the speed of the motor. Suppose the current I increases slightly, thereby increasing the torque and angular speed. The increase in ω increases the back EMF which in turn reduces the current. The opposite chain of events happens if current decreases. Thus the back EMF maintains a steady current and uniform angular speed.

EXAMPLE 22.4

The armature of a simple ac generator consists of a coil with 100 turns and an area of 0.1 m^2. The armature is rotated at a frequency of 60 Hz in a constant magnetic field having an induction of 0.8 T. Calculate the maximum voltage generated.

The maximum voltage generated is given by

$$\mathscr{E}_{max} = NBA\omega$$

We are given $N = 100, B = 0.8 \text{ T}, A = 0.1 \text{ m}^2$, and

$\omega = 2\pi f = 120\pi$ rad. Hence,

$\mathscr{E}_{max} = 100(0.8 \text{ T})(0.1 \text{ m}^2)(120\pi \text{ rad}) = 3016 \text{ V}.$

22.4 MUTUAL AND SELF-INDUCTANCE

Changes in flux due to current changes in a coil produce an induced EMF in a second coil located close to the first coil. The mutual inductance expresses the relationship between EMF induced in the second coil and the current changes in the first coil. To understand this relationship, let us consider two solenoids; the second one is wound over the first one (Fig. 22.13). The length of the first coil is l; its cross-sectional area is A and its total number of turns is N_1. The number of turns in the second coil is N_2. A current I passing through the first coil sets up a magnetic field inside the solenoid given by

$$B = \frac{\mu_0 N_1 I}{l}.$$

The total flux passing through the coil is

$$\Phi = BA = \frac{\mu_0 N_1 A I}{l}.$$

The same flux passes through the second coil. If this flux changes, a voltage $\mathscr{E}_2$ will be induced in

Figure 22.13. Mutual inductor. Current variations in the first coil induced an EMF in the second coil.

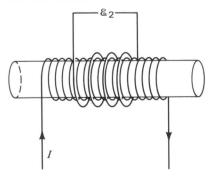

the second coil. The induced voltage $\mathscr{E}_2$ is

$$\mathscr{E}_2 = -\frac{N_2 \Delta\Phi}{\Delta t}$$

$$= -\frac{\mu_0 N_1 N_2 A}{l} \frac{\Delta I}{\Delta t}. \qquad (22.9)$$

Note that μ_0, N_1, N_2, A, and l are constants and only I changes with time. We define a new quantity called mutual inductance M by the equation

$$M = \frac{\mu_0 N_1 N_2 A}{l}, \qquad (22.10)$$

which now contains all the constants in Eq. (22.9). Equation (22.9) may now be rewritten in the form

$$\mathscr{E}_2 = -M \frac{\Delta I}{\Delta t}. \qquad (22.11)$$

The unit of mutual inductance M is called the henry (H). If a current change of 1 A in 1 s produces 1 V, the mutual inductance is 1 H. The coil system is called a mutual inductor.

Similar arguments lead us to the conclusion that when current flowing through a single coil changes with time an opposing EMF results. The inductance associated with a single coil is called self-inductance (L), which can be obtained in terms of the parameters of the coil—N the number of turns, l the length of the coil, A the area—as follows:

Field in the interior of the coil, $B = \frac{\mu_0 N I}{l}$;

Voltage induced $= -N \frac{\Delta\Phi}{\Delta t} = -\frac{N \Delta(BA)}{\Delta t}$

$$= -\frac{\mu_0 N^2 A}{l} \frac{\Delta I}{\Delta t} = -L \frac{\Delta I}{\Delta t};$$

where

Self inductance, $L = \frac{\mu_0 N^2 A}{l}$; $\qquad (22.12)$

Induced EMF $= L \dfrac{\Delta I}{\Delta t}.$ (22.13)

From Eqs. (22.10) and (22.13) it is evident that both mutual inductance and self-inductance can be increased by (a) having more turns per unit length (N/l), that is, having closely spaced windings (also having more turns in the second coil for mutual inductance); (b) increasing the area of the coil; (c) increasing the permeability constant by filling the interior of the coil with magnetic materials.

When we attempt to increase the current in a coil, it is opposed by the induced voltage. Hence work is done in increasing the current, which is stored as energy (energy associated with the magnetic field surrounding the coil). Energy stored in an inductor can be shown to be given by

Energy $= \tfrac{1}{2}LI^2.$ (22.14)

EXAMPLE 22.5

A 5 cm long solenoid contains 100 turns. The radius of its cross-section is 2 cm. (a) Calculate its self-inductance. (b) If 10 turns of a second coil is wound over this solenoid, what is its mutual inductance?

(a) Area of cross-section $= \pi r^2$

$= 3.14 \times (0.02\text{m})^2 = 0.0012 \text{ m}^2.$

$L = \dfrac{\mu_0 N^2 A}{l}$

$= \dfrac{(4\pi \times 10^{-7} \text{ T} \cdot \text{m/A})(100^2)(0.0012 \text{ m}^2)}{0.05 \text{ m}}$

$= 3.015 \times 10^{-4} \text{ H}$

$= 0.31 \text{ mH}$

(b) $M = \dfrac{\mu_0 N_1 N_2 A}{l} = 0.031 \text{ mH}.$

As this example shows, mH (millihenry) is a more practical unit than H (henry).

22.5 TRANSFORMERS

The mutual inductor, which we discussed above, gives an output voltage at the terminals of the second coil (secondary) when the current input in the first coil (primary) varies. A current variation is obtained by connecting an ac voltage ($\mathcal{E}_p$) at the primary. For a sinusoidal primary voltage, secondary voltage ($\mathcal{E}_s$) is also sinusoidal. The working principle of a transformer is similar to the mutual inductor. To increase the flux, the primary coil and the secondary coil are wound over a magnetic core (Fig. 22.14). For a transformer the ratio of the secondary voltage to primary voltage is equal to the ratio of the number turns in the respective coils:

$$\dfrac{\mathcal{E}_s}{\mathcal{E}_p} = \dfrac{N_s}{N_p}.$$ (22.15)

If $N_s < N_p$, we have $\mathcal{E}_s < \mathcal{E}_p$. Then the transformer is called a step-down transformer. If $N_s > N_p$, we get $\mathcal{E}_s > \mathcal{E}_p$. Then the transformer is called a step-up transformer.

Figure 22.14. Transformer. Voltage output of the secondary coil is proportional to voltage of the primary coil multiplied by the ratio of the number of secondary turns to the number of primary turns of wire.

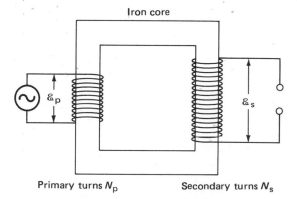

In an ideal transformer (100% efficient with no energy loss), the power output is equal to the power input

$$\mathcal{E}_s I_s = \mathcal{E}_p I_p \qquad (22.16)$$

or

$$\frac{I_p}{I_s} = \frac{\mathcal{E}_s}{\mathcal{E}_p} = \frac{N_s}{N_p} \qquad (22.17)$$

so that increasing the output voltage means a proportionate decrease in current and vice versa. Efficiency of a transformer is defined by

$$\text{Efficiency } \eta = \frac{\text{output power}}{\text{input power}}$$

$$= \frac{\text{output power}}{\text{output power} + \text{power losses}}$$

where the losses include core losses—both hysteresis losses and eddy current losses—and winding losses—ohmic or heating losses—in both primary and secondary coils.

A brief description of eddy currents is given below. When conducting material like iron is used as the core, induced currents in a plane perpendicular to the flux lines will be set up as the flux changes. These are known as eddy currents. The heating produced by these currents takes away part of the input power. To reduce eddy currents, the core is made of laminated iron sheets coated with nonconducting paint.

EXAMPLE 22.6

A step-down transformer is used to provide 6.3 V for the filament of a tube from a 110 V line source. Calculate the ratio of the primary to secondary coils. If the current at the input is ½ A, what is the maximum current we can get at the secondary coil?

We are given $\mathcal{E}_p = 110$ V and $\mathcal{E}_s = 6.3$ V. The ratio of the coils is

$$\frac{N_p}{N_s} = \frac{\mathcal{E}_p}{\mathcal{E}_s} = \frac{110}{6.3} = 17.5.$$

Assuming the transformer is 100% efficient, we get the secondary current as

$$I_s = I_p \times \frac{N_p}{N_s} = \tfrac{1}{2} \text{ A} \times 17.5 = 8.75 \text{ A}.$$

SUMMARY   Faraday's law states that flux changes induce an EMF in a coil

$$\mathcal{E} = -\frac{\Delta \Phi}{\Delta t}$$

Lenz's law states that the direction of the induced current is in such a direction as to oppose the change.

An electric generator converts mechanical energy to electrical energy. The EMF from a generator is given by

$$\mathcal{E} = NBA\omega \cos\omega t,$$

where $NBA\omega = \mathcal{E}_{max}$ and ω is the angular speed of the coil.

The EMF induced in a mutual inductance and self-inductance are

$$\mathcal{E} = M\frac{\Delta I}{\Delta t} \quad \text{and} \quad \mathcal{E} = L\frac{\Delta I}{\Delta t}.$$

A transformer steps up or steps down the input EMF. The ratio of output to input EMF is

$$\frac{\mathcal{E}_s}{\mathcal{E}_p} = \frac{N_s}{N_p}.$$

QUESTIONS AND PROBLEMS	
22.1	The primary coil of a transformer is connected to a 110 V dc source. What is the output voltage?
22.2	If you try to pull a coil out of a magnetic field, you experience an opposing force. Will such a force act on a copper sheet when it is pulled out? What will happen if the sheet of copper is slotted?
22.3	A bar magnet is dropped through a metallic hoop. Explain how the direction of the induced current changes as it drops through the loop.
22.4	The maximum voltage of a generator is $\mathcal{E}_{max} = NBA\omega$. Can increases in any or all of the quantities N, B, A, and ω, increase $\mathcal{E}_{max}$? What are the limitations you see in each case?
22.5	A coil having an area of 0.02 m^2 and a resistance of 10 Ω is kept in a magnetic field with an induction $B = 0.9$ T. The field is reversed to -0.9 T in a time of 0.1 s. Calculate (a) the voltage induced and (b) the average current in the coil.
22.6	A coil has 500 turns and an area of 0.03 m^2. The coil is placed in a magnetic field of 1.2 T and rotated from a position where the coil is parallel to the field to a perpendicular position in 0.02 s. (a) Obtain the voltage induced in the coil. (b) If the coil rotates through 180° at the same rate, what is the induced voltage in the coil?
22.7	A wire is pulled out of the magnetic field with a velocity of 2 m/s (Fig. 22.4). The length of the wire in the field is 0.5 m. The magnetic induction of the field is 0.8 T. Calculate (a) the induced voltage and (b) the average force acting on the wire. Resistance of the wire is 0.2 Ω.
22.8	An armature of an ac generator consists of 600 turns, each having an area of 0.1 m^2. The coil rotates at 3600 rev/min in a field of flux density 0.8 T. (a) What is the frequency of the induced EMF? (b) Calculate the maximum voltage induced. (c) Calculate the instantaneous EMF 0.0025 s after the coil passes a position parallel to the field.
22.9	An ac generator armature has 200 turns, each with an area of 0.06 m^2. It is rotating at a speed of 60 rev/s in a field of flux density 1 mT. Calculate the maximum induced EMF.
22.10	The starting motor of an automobile is driven by a 12 V battery. Its armature coil resistance is 0.04 Ω and the operating current is 150 A. Calculate (a) the back EMF and (b) the starting current.
22.11	A 110 V power source operates a motor at a constant current of 22 A. Its starting current is 110 A. Calculate (a) the resistance of the armature coil and (b) the back EMF.

22.12 A solenoid has 100 turns, its cross-sectional area is 0.005 m^2 and its length is 0.1 m. Calculate its self-inductance.

22.13 If the coil of Problem 22.12 is filled with iron with a relative permeability of 1200, calculate its self-inductance.

22.14 A second coil of 20 turns is wound over the coil of Problem 22.13. Calculate its mutual inductance.

22.15 A current of 2.0 A passes through a closed circuit. If the circuit is opened so that the current drops to zero in 0.5 s and the average voltage induced is 60 V, what is the self-inductance of the circuit?

22.16 When the switch is closed, the current in an inductive circuit becomes 2 A in 0.05 s. If the inductance is 150 mH, what is the induced voltage?

22.17 A current of 2.5 mA creates a flux of 0.6 × 10^{-3} Wb in a coil of 2000 turns. What is the self-inductance of the coil?

22.18 Two adjacent coils have mutual inductance of 1.5 H. (a) If the current in the primary coil changes by 50 A/s, what is the average induced EMF in the secondary coil? (b) If the resistance of the secondary coil is 10 Ω, what is the current flowing in it when connected to a 20 Ω external resistance?

22.19 A secondary coil has an EMF of 0.3 V induced in it when the primary current changes from 2.5 A to 1 A in 0.3 s. What is the mutual inductance of the coils?

22.20 If the secondary coil in Problem 22.19 has 200 turns, what is the change in flux?

22.21 A step-up transformer has 1500 turns in its primary coil and 30,000 turns in its secondary coil. If the primary is connected to a 115 V ac line, what will be the voltage induced in the secondary?

22.22 A step-up transformer with 200 turns in its primary coil is connected to a 120 V ac line. The secondary coil delivers 10,000 V and 20 mA. (a) How many turns does the secondary have? (b) What is the current in the primary?

22.23 A step-down transformer provides 20 A at 6 V at the secondary for a 110 V input at the primary. If the secondary has 100 turns, assuming 100% efficiency, calculate (a) the number of turns in the primary and (b) the current in the primary coil.

Alternating-Current Circuits

As we have seen in the previous chapter it is easier to generate alternating-current electricity. It can be stepped up or down as desired by transformers. A stepped-up high-voltage ac has the advantage of low power loss during transmission. Because of these advantages most of the electricity ($\sim$97%) produced is of the ac type.

23.1 DC AND AC

There are two types of voltage sources—alternating current (popularly known as ac) and direct current (dc) sources. In a dc source such as a battery, the polarity of the source and its magnitude remain constant with time (Fig. 23.1). If a resistance is connected to such a source, a constant current flows through it. On the other hand, the voltage from an ac source changes its magnitude or size and reverses its polarity repeatedly at constant time intervals. Current from such a source also shows similar changes with time (Fig. 23.2). For most of the ac sources the variation of voltage is sinusoidal with a period T. The number of complete cycles in a second, or frequency, is given by

$$f = 1/T,$$

where T is the period of the ac.

The peak voltage and the peak current are identified as V_{max} and I_{max} in Fig. 23.2. The voltage

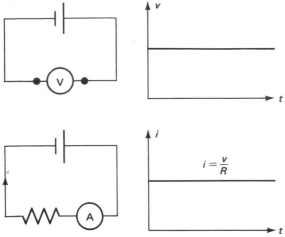

and current at any time t is given by the equations

$$v = v_{max} \sin 2\pi ft, \tag{23.1}$$

$$i = i_{max} \sin 2\pi ft, \tag{23.2}$$

with the relation $v = iR$ still holding good for current passing through a resistance.

Both voltage and current are positive during half of the cycle and negative during the other half of the cycle. Therefore, the average values of both voltage and current are zero. Does that mean there is no power output from an ac source? Our experience (for example, a toaster producing heat and a light bulb emitting heat and light when connected to an ac source) indicates that there is continuous power output from such sources. Now the question is how do we calculate the power output from an ac source? This question is answered below.

Figure 23.1. Voltage and current remain constant with time for dc sources.

Figure 23.2. (a) Variation of voltage of an ac source as a function of time. The time T is the period of oscillation. (b) When a pure resistance is connected to this source, a time-varying current results. Variation of current is in step with the voltage variation; that is, the current is maximum when the voltage is maximum and the current is zero when the voltage is zero.

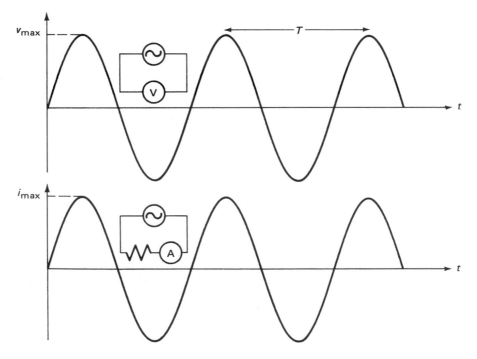

As you may recall the power loss due to heating in a resistance may be obtained from any of the following equations:

$$P = iv,$$

$$P = i^2R,$$

$$P = v^2/R.$$

Both i^2 and v^2 are positive in both halves of the cycle (Fig. 23.3) which simply means that there is a power loss, irrespective of the direction of the current or the polarity of the applied voltage. The average values of v^2 and i^2, are also shown in Fig. 23.3. The area under the curve representing v^2 is the same as area under the $(v^2)_{avg}$ curve. The power output v^2/R is equal to $(v^2)_{avg}/R$. Therefore, the power output from an ac source is equivalent to the power output from a dc source whose voltage is equal to $\sqrt{(v^2)_{avg}}$. Similarly, the area under the i^2 curve is equal to the area under the $(i^2)_{avg}$ curve. The power output i^2R is therefore equal to $(i^2)_{avg}R$. An ac source can be considered as equivalent to a dc source with a current equal to $\sqrt{(i^2)_{avg}}$. Let us designate the voltage from such an equivalent dc source by V and current by I. We have the relationships

$$V^2 = (v^2)_{avg}, \tag{23.3}$$

$$I^2 = (i^2)_{avg}, \tag{23.4}$$

or

$$V = \sqrt{(v^2)_{avg}}, \tag{23.5}$$

$$I = \sqrt{(i^2)_{avg}}. \tag{23.6}$$

The voltage V is called the effective voltage because we can use it to calculate the effective power obtainable from the ac source. Since it is equal to the square root of the average (mean) of the square of the voltage v, it is also called root mean square (rms) voltage. Similarly I is called the effective current or the rms current. Power from an ac source can be calculated using any of the following equations.

$$P = VI, \tag{23.7}$$

Figure 23.3. Plots of v^2, i^2, $(v^2)_{avg}$, and $(i^2)_{avg}$ for an ac source as functions of time.

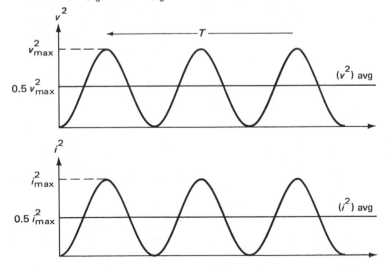

$$P = \frac{V^2}{R} = \frac{(v^2)_{avg}}{R}, \qquad (23.8)$$

$$P = I^2 R = (i^2)_{avg} R. \qquad (23.9)$$

The relationships between the effective voltage V and v_{max} and effective current I and i_{max} are

$$V^2 = (v^2)_{avg} = 0.5 \, v_{max}^2 \qquad (23.10)$$

or

$$V = \sqrt{(v^2)_{avg}} = 0.707 \, v_{max}, \qquad (23.11)$$

$$I^2 = (i^2)_{avg} = 0.5 \, i_{max}^2 \qquad (23.12)$$

or

$$I = \sqrt{(i^2)_{avg}} = 0.707 \, i_{max}. \qquad (23.13)$$

Equations (23.10) and (23.12) result from the fact that the average of $\sin^2 \omega t$ over a period T is ½. Ohms' law still can be written as

$$V = IR.$$

The values of V and I are often given to describe ac sources. The line voltage of 110 V means the effective voltage of the source is 110 V.

EXAMPLE 23.1

The line voltage of an ordinary ac source is 110 V. It is connected to a 60 Ω resistance. Calculate (a) the value of the effective current, (b) power loss, (c) peak voltage v_{max}, and (d) peak current i_{max}.

Effective current $I = \dfrac{V}{R} = \dfrac{110 \text{ V}}{60 \, \Omega} = 1.83$ A.

Power loss $= \dfrac{V^2}{R} = \dfrac{(110 \text{ V})^2}{60 \, \Omega} = 201.6$ W.

Peak voltage $v_{max} = \dfrac{V}{0.707} = \dfrac{110 \text{ V}}{0.707} = 155.6$ V.

Peak current $i_{max} = \dfrac{1.83 \text{ A}}{0.707} = 2.6$ A.

23.2 CAPACITOR AND INDUCTOR IN A DC CIRCUIT

In addition to resistance, ac circuits often contain capacitors and inductors. Before we undertake the study of these elements in ac circuits we study their effects in dc circuits for the following reasons: (a) It is comparatively easier to study them in dc circuits. (b) Such a study will provide us with valuable insight into their behavior in ac circuits. In Fig. 23.4 a capacitor (uncharged) is connected in series with a resistance R. What happens when the switch is closed, that is, when the circuit is completed? Charges will rush to the capacitor, the rate of flow being limited only by the resistance. The iniitial current is therefore $i_{t=0} = V/R$. Also at $t = 0$, the voltage on the capacitor is zero. Later the current decreases since the charges already present on the capacitor hinder other charges from accumulating on the surface of the capacitor. Eventually no more charge can reach the capacitor surface. At this point, the current becomes zero and the voltage V_C is equal to the battery voltage. The increase in voltage across the capacitor, first at a faster rate and then at a slower rate, is known as an exponential increase. Similarly the current decrease is an exponential decrease. Note that when equilibrium or the steady state is reached, there is no dc current flow in circuit containing a capacitor.

The voltage on the capacitor in an RC circuit (Fig. 23.4) increases at a faster rate if R is small or the initial current flow is high and if the value of the capacitor is small ($V_C = Q/C$). The product RC, therefore, is a measure of how fast the voltage of the capacitor increases and hence it is called the time constant of the circuit.

Time constant $\tau = RC.$ \qquad (23.14)

The time constant RC has the units of time; if R is in ohms and C is in farads, τ (pronounced tau) is in seconds. A bigger time constant means a slower

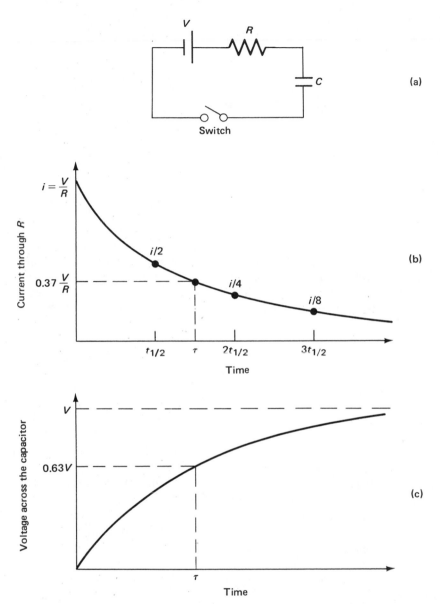

Figure 23.4. A dc circuit containing a resistor and capacitor. After the switch is closed, the current decreases slowly with time and the voltage across the capacitor increases slowly with time. In one decay time (=RC), current decreases to 37% of the maximum and the charge and voltage across the capacitor increase to 63% of their maximum values.

change in V_C and current. In one time constant V_C increases to 63% of the maximum value V. Also the current decreases by 63% of the maximum, that is, current decreases to 37% of the maximum in one time constant. This type of change may also be characterized by half-time, that is, the time taken for the current to decrease by half. The magnitude of the current decreases by half in one half-time irrespective of the starting point (Fig. 23.4).

$$V - iR - L\frac{\Delta i}{\Delta t} = 0. \tag{23.16}$$

$$\text{Half-time } t_{1/2} = 0.693 \times \tau. \tag{23.15}$$

The last term $-L\,\Delta i/\Delta t$ is the voltage induced in the inductance L. Rearranging Eq. (23.16) we get

Also note that in the RC circuit when the current through the resistance is maximum, the voltage across the capacitor is minimum and vice versa. In five time constants the capacitor gets 99.3% of the maximum charge and is considered fully charged.

$$\frac{\Delta i}{\Delta t} = \frac{V - iR}{L}. \tag{23.17}$$

As i increases from the initial zero value the slope of the current ($\Delta i/\Delta t$) decreases. Hence the current increase is exponential in this case also. The steady-state condition is reached when the current reaches the maximum value of V/R. The current through the circuit reaches the maximum value faster (a) if L is small, that is, the opposing induced voltage is less and (b) if R is high. The resistance R limits the maximum value as well as reduces the rate of increase of current. The ratio L/R is the time constant in this case.

EXAMPLE 23.2

For the circuit given in Fig. 23.4, $V = 6$ V, $R = 100\ \Omega$, and $C = 1000\ \mu$F. Calculate (a) the time constant of the circuit, (b) the maximum voltage on the capacitor, (c) the time when the voltage of the capacitor is 0.63 times the maximum voltage, and (d) the time for the current to decrease by half.

(a) Time constant of the circuit $= RC = (100\ \Omega) \times (1000 \times 10^{-6}$ F$) = 0.1$ s.

(b) Maximum voltage of the capacitor $=$ maximum voltage of the battery $= 6$ V.

(c) The voltage reaches a value of 63% of its maximum in one time constant which in this case is 0.1 s.

(d) The current through the resistance decreases by ½ in one half-time $= 0.693 \times 0.1 = 0.069$ s.

Now let us study the effect of an inductor L connected in series with a resistance R (Fig. 23.5). If the inductor is not present in the circuit the current will reach the maximum value of $i = V/R$ immediately after completing the circuit. However, the sudden increase in current results in a changing magnetic field which induces an opposing voltage in the inductor is not present in the circuit the cur- negligible initial current. As the current increases the magnetic field increases more and more slowly and reaches a constant value. At this point the induced voltage drops to zero and the current attains the maximum value.

The variation of the slope of the current can be determined by applying Kirchhoff's law to the circuit.

$$\text{Time constant } \tau = L/R \tag{23.18}$$

and the half-time is

$$t_{1/2} = 0.695\ L/R. \tag{23.19}$$

The significance of the time constant is that the higher the time constant is the slower the increase in current through the circuit. The current reaches 63% of the maximum value in one time constant (Fig. 23.5) and the voltage across the inductor decreases by 63% in the same time.

EXAMPLE 23.3

For the circuit given in Fig. 23.5, $R = 500\ \Omega$, $L = 600$ mH, $V = 10$ V. Calculate (a) the time constant of the circuit; (b) how much time it will take for the current to increase to 63% of the maximum value; (c) how much time it will take for the current to increase to 50% of the maximum; (d) the maximum value of the current.

(a) Time constant of the circuit $= L/R = 600 \times 10^{-3}$ H$/500\ \Omega = 1.2 \times 10^{-3}$ s.

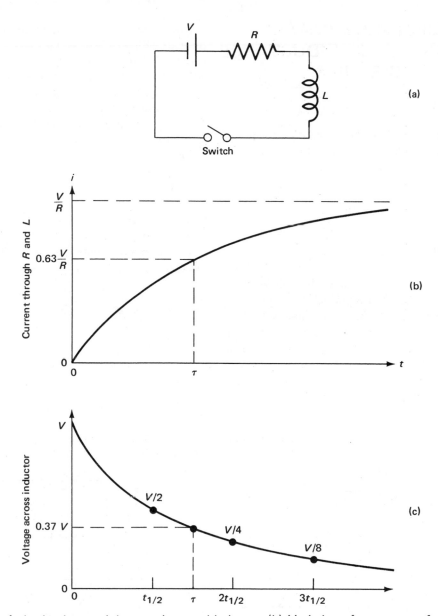

Figure 23.5. A dc circuit containing a resistor and inductor. (b) Variation of current as a function of time. (c) Variation of voltage across the inductor as a function of time. In one time constant ($= LR$) the current reaches 63% of the maximum value and voltage across the inductor decreases by 63%.

(b) The current increases to 63% of the maximum value in one time constant.

(c) Current increases to 50% of the maximum in one half-time. Half-time $= 0.695\ L/R = 8.34 \times 10^{-4}$ s.

(d) Maximum value of the current $= V/R = 10\ \text{V}/500\ \Omega = 0.02$ A.

23.3 RESISTANCE, CAPACI–
TANCE, AND INDUCTANCE
IN AC CIRCUITS

Resistance

The current i through the resistance is always in phase (in step) with the voltage v (Fig. 23.6). The effective voltage, the effective current, and the resistance R have the relation

$$I = V/R.$$

Capacitance

In a dc circuit, current flows through the capacitor for a short time during which the capacitor is charged. When an ac source is connected to the capacitor, the latter is charged and discharged alternately in opposite directions in each cycle. Therefore, the current flow through the capacitor in an ac circuit is continuous and fluctuating.

Figure 23.6. In an ac circuit containing only resistance, both current and voltage are in phase (in step).

The current and the voltage are out of phase in this case (Fig. 23.7). Voltage lags behind the current by 1/4 of a cycle (90°). The voltage maxima appear ¼T after the current maxima or the voltage maxima appear when the current is zero.

How much current flows through the circuit? The effective current I through the circuit is proportional to the effective voltage V, the relationship being

$$V = IX_C, \tag{23.20}$$

where X_C represents the opposition to the current flow due to capacitance. It is called the capacitive reactance and its units are ohms. The capacitive reactance is given by

$$X_C = \frac{1}{2\pi f C}. \tag{23.21}$$

In Eq. (23.21) if f, the frequency of the source, is expressed in Hz and C in farads, X_C is in ohms.

Figure 23.7. In an ac circuit containing only a capacitor, the voltage across the capacitor lags behind the current. The voltage maxima appear 1/4 of a cycle later than the current maxima.

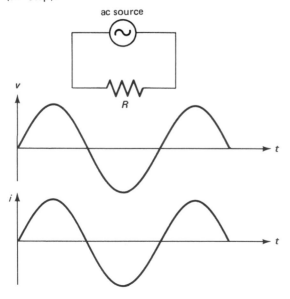

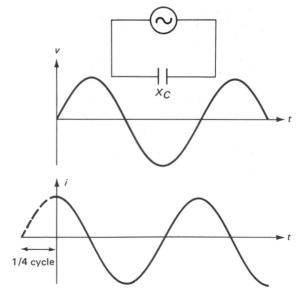

Inductor

In the case of an inductor, we also do not expect the voltage and current to be in phase. The inductor actually opposes the buildup of current and, therefore, current lags behind the voltage (voltage leads the current by 1/4 of a cycle) (Fig. 23.8). The inductive reactance X_L is given by

$$X_L = 2\pi fL. \tag{23.22}$$

If L is in henry and f in Hz, X_L is in ohms. The effective current and voltage are related by

$$V = IX_L. \tag{23.23}$$

EXAMPLE 23.4

A 10 V, 60 Hz ac source is connected to a 100 μF capacitor. Calculate (a) the reactance of the capacitor and (b) the effective current through the capacitor.

Figure 23.8. In an ac circuit containing only an inductor, current lags the voltage by 1/4 of a cycle.

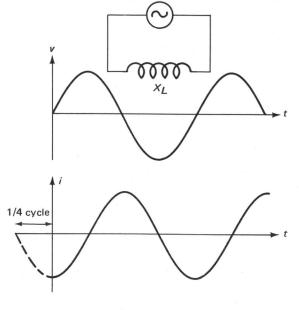

(a) $X_C = \dfrac{1}{2\pi fC} = \dfrac{1}{2\pi(60 \text{ Hz})(100 \times 10^{-6} \text{ F})}$

$\qquad = 26\,\Omega$

(b) Effective current $= \dfrac{V}{X_C} = \dfrac{10 \text{ V}}{26\,\Omega} = 0.38$ A.

EXAMPLE 23.5

There is an effective current of 10 mA in an inductance connected to a 300 Hz, 20 V source. Calculate the inductance.

$$I = \frac{V}{X_L}; \quad 10 \times 10^{-3} \text{ A} = \frac{20 \text{ V}}{X_L}; \quad X_L = 2 \times 10^3\,\Omega.$$

$$X_L = 2\pi fL; \quad L = \frac{X_L}{2\pi f} = \frac{2 \times 10^3\,\Omega}{2\pi \times 300 \text{ Hz}} = 1.06 \text{ H}.$$

Resistance, Capacitance, and Inductance (*RCL*) in Series

Resistance, capacitance, and inductance are connected in series to an ac source in Fig. 23.9. The same current is flowing through all the elements [Fig. 23.9(b)]. With respect to the current, the voltage across the resistance is in phase, the voltage across the inductance is leading, and the voltage across the capacitance is lagging. The sum of the instantaneous voltages at any time will be equal to that of the source. However, the sum of the effective voltages is not equal to the effective voltage of the source. This is because algebraic addition of effective voltages does not take into account the phase changes that exist between them. A vector addition method that takes care of the difficulty is discussed below.

The vector addition method known as the phase diagram method uses the following steps:

1. A line proportional to the voltage $V_R = IR$ across the resistor is drawn along the positive x axis (Fig. 23.10).

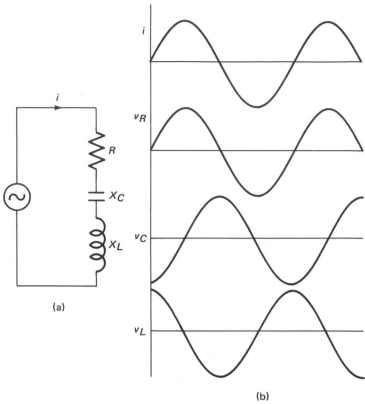

Figure 23.9. (a) An ac circuit containing a resistor, capacitor, and an inductor. (b) With respect to the current, the voltage across the resistance is in phase, the voltage across the inductance is leading, and the voltage across the capacitance is lagging.

2. A line proportional to the voltage $V_L = IX_L$ across the inductor is drawn along the positive y axis (90° to V_R).

3. A line proportional to the voltage $V_C = IX_C$ across the capacitor is drawn along the negative y axis (90° to V_R).

4. $V_L - V_C$ is then determined to obtain the sum from the reactive elements.

5. V_R and $V_L - V_C$ are then added just as we add vectors. The magnitude is given by

$$V = \sqrt{V_R^2 + (V_L - V_C)^2}. \tag{23.24}$$

6. A phase angle of the sum of the voltages is

$$\tan \phi = \frac{V_L - V_C}{V_R}. \tag{23.25}$$

If $V_C > V_L$, voltage lags current and if $V_L > V_C$, voltage leads current.

Another useful relation between the effective current I and the voltage may now be derived by recalling that

$$V_R = IR, \ V_L = IX_L, \ V_C = IX_C.$$

Upon substitution into Eq. (23.24), we have

$$V = \sqrt{I^2R^2 + I^2(X_L - X_C)^2}$$

$$= IZ. \qquad (23.26)$$

Z is called the impedance of the circuit and is given by

$$Z = \sqrt{R^2 + (X_L - X_C)^2}. \qquad (23.27)$$

The unit of impedance is the ohm.

EXAMPLE 23.6

In the circuit shown in Fig. 23.10, $X_L = 60\ \Omega$, $X_C = 30\ \Omega$, and $R = 40\ \Omega$. The effective voltage of the source is 120 V and its frequency is 60 Hz. Find (a) the impedance of the circuit, (b) the current passing through each element, and (c) the voltage across each element.

(a) Impedance of the circuit $Z = R^2$

$$= \sqrt{R^2 + (X_L - X_C)^2}$$

$$= 40^2 + (60 - 30)^2$$

$$= 50\ \Omega.$$

(b) Current I through the circuit $= \dfrac{V}{Z} = \dfrac{120\ V}{50\ \Omega}$

$$= 2.4\ A.$$

(c) Voltage across the resistance = 96 V;

Voltage across the impedance $= IX_L$

$$= 2.4\ A \times 60\ \Omega = 144\ V;$$

Voltage across the capacitance $= IX_C$

$$= 2.4\ A \times 30\ \Omega = 72\ V.$$

In the example above the algebraic sum of the voltages across the elements is larger than the source voltage, but the vector sum is equal to the source voltage as shown in Fig. 23.11.

Instruments used for ac measurements, ac voltmeters and ammeters, read effective values. Hence, if ac voltmeters are connected as shown in Fig. 23.12, in general,

$$V_C + V_L \neq V_{CL},$$

$$V_C + V_L + V_R \neq V;$$

that is, the voltage across the capacitor added to the voltage across the inductor is not equal to the voltage across the combination of capacitor and inductor. Similarly the sum across the three elements does not equal the source voltage.

Figure 23.10. Vector addition of effective voltages across each element in an *LCR* circuit.

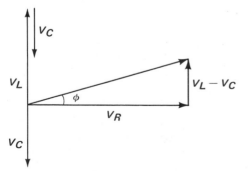

Figure 23.11. Vector addition of effective voltages of Example 23.6. Vector sum of the voltages is equal to the source voltage.

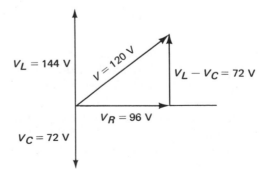

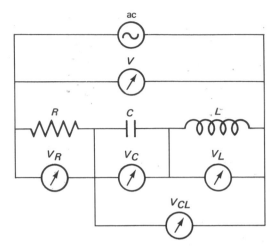

Figure 23.12. AC voltmeters read effective (rms) values. Effective voltages do not add algebraically. For example, $V_C + V_L + V_R \neq V$.

23.4 RESONANCE

The impedance Z has a minimum value when $X_L = X_C$, that is

$$2\pi f_r L = \frac{1}{2\pi f_r C}$$

or

$$f_r = \frac{1}{2\pi \sqrt{LC}}. \qquad (23.28)$$

The frequency given by Eq. (23.28) is known as the resonance frequency. For the resonant frequency the impedance is

$$Z = R.$$

The current flow in this instance is a maximum.

The response of a series RCL circuit is frequency dependent. At resonant frequency more current passes through the circuit, that is, more power is transferred from the source to the load. When the frequency of the applied voltage is farther from the resonant frequency, current flow through the circuit decreases (Fig. 23.13).

In a radio receiver, the value of a capacitor in the input circuit is adjusted so that the input circuit is in resonance with the desired frequency. Such a "tuned" resonant circuit selects the signals with resonant frequency in favor of other signals of different frequencies. The frequencies assigned to radio stations are separated from each other sufficiently that when a receiver is tuned to one station the radio waves from other stations produce very little effect (almost zero current) in the receiving circuit.

23.5 POWER FACTOR

In Fig. 23.14, v and i and vi are plotted for a pure resistance and a pure capacitance. In the case of resistance, vi is positive for both halves of the cycle, but in the case of capacitance they are alternately positive and negative. The net power loss is zero. The reason for this is that the work done on the capacitor during charging is released during discharging. Similarly there is no power loss in an inductor. If the circuit contains resistance and reactance, the power loss is only across the resistance and is given by IV_R.

The voltage V_R is the component of V along the resistance axis (Fig. 23.9), that is, the voltage component in phase with the current. Thus the effective power in a circuit is given by

$$P = IV \cos\phi. \qquad (23.29)$$

Figure 23.13. Variation of current in a series RCL circuit with frequency. The current is a maximum at the resonance frequence.

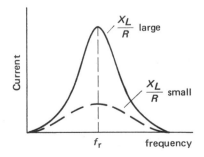

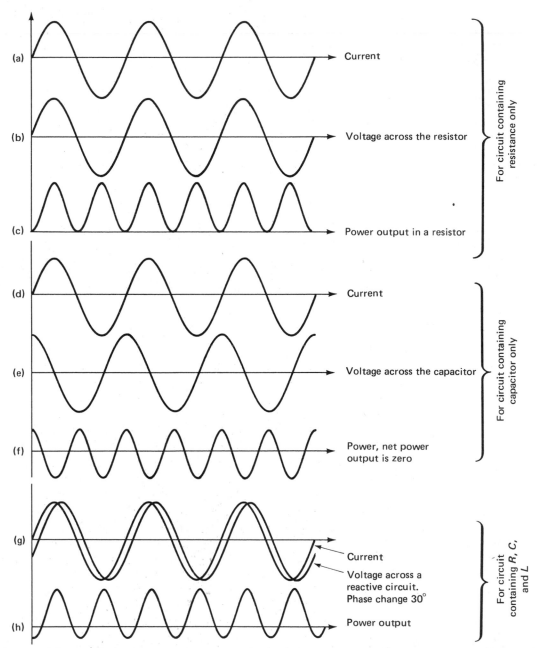

Figure 23.14. (a) and (b) Voltage is in phase with current when the load consists of only a resistance. (c) the instantaneous power *iv* in this case is positive in each half cycle. (d) and (e) Voltage is out of phase with current when the load is purely reactive. (f) Instantaneous power *iv* is alternately positive and negative in each 1/4 cycle and hence the sum of the power loss in each cycle is zero. (g) In an *RCL* circuit, the voltage is out of phase with the current by an amount determined by the values of *R, C, L,* and frequency. (h) There is a net power loss in this case.

The quantity $\cos\,\phi$ is called the power factor of the circuit. Note that $\cos\phi$ can vary from 1 for a circuit containing pure resistance to 0 for a circuit containing pure reactance. From Fig. 23.10, we obtain

$$\cos\phi = \frac{V_R}{V_Z} = \frac{R}{Z}. \qquad (23.30)$$

Substituting this value into Eq. (23.29), we get

$$P = IV\frac{R}{Z} = I^2 R. \qquad (23.31)$$

In this form the expression for power in ac circuits corresponds to that in dc circuits.

SUMMARY   An ac source voltage change is

$$v = v_{\max}\sin 2\pi ft.$$

The effective voltage is

$$V = 0.707\,v_{\max}.$$

Similarly the relation between $i_{\max}$ and effective current I is

$$I = 0.707\,i_{\max}.$$

The time constant is $\tau = RC$, for a circuit containing resistance and capacitance and $\tau = L/R$, for a circuit containing resistance and inductance.

The relations between I and V for circuits containing resistance, capacitance, and inductance are

$$V_R = IR,$$

$$V_C = IX_C,\ \ X_C = \frac{1}{2\pi fC}\ \ \text{(capacitive reactance)};$$

$$V_L = IX_L,\ \ X_L = 2\pi fL\ \ \text{(inductive reactance)};$$

$$V_Z = IZ,\ \ Z = \sqrt{R^2 + (X_L - X_C)^2}\ \ \text{(impedance)}.$$

The phase angle between current and voltage is given by

$$\tan\phi = \frac{V_L - V_C}{V_R}.$$

The circuit is in resonance when

$$X_L = X_C\ \text{or}\ f_{\mathrm{r}} = \frac{1}{2\pi LC}.$$

Power in an ac circuit is

$$P = VI\cos\phi = I^2R.$$

The term $\cos\phi$, called the power factor, is

$$\cos\phi = R/Z.$$

QUESTIONS AND PROBLEMS

23.1 Indicate which of the following statements are true. (a) There is no power loss in a capacitor or an inductor. (b) If the circuit containing resistance, capacitance, and inductance is in resonance, there is no power loss in the circuit. (c) The impedance of a circuit always increases with increasing value of the capacitance. (d) The impedance of a circuit always increases with increasing value of the inductance. (e) Impedance does not depend on frequency. (f) The sum of the effective voltages in a circuit containing L, C, and R is always equal to the effective voltage of the source.

23.2 A dc voltmeter is connected to a variable frequency ac source through a resistance. What will you observe as the frequency is slowly increased from 0 to 100 kHz?

23.3 If a capacitor is included in a dc circuit, what is the steady current?

23.4 How does the value of an inductance included in a dc circuit affect the steady current?

23.5 Light from a fluorescent bulb appears to flicker, especially when an object is moved across it. No such effect is seen in light from an incandescent bulb. Why?

23.6 A 2 μF capacitor charged to 10 V is connected to a 10 Ω resistance. What happens?

23.7 A 100 μF capacitor charged to 10 V is connected to a 100 mH inductance. What happens?

23.8 A 60 Ω resistance is connected to 110 V ac source. Calculate (a) the maximum voltage, (b) maximum current, (c) effective current, and (d) the power loss in the resistor.

23.9 A 100 W heater coil (pure resistance) is connected to a 110 V ac source. Calculate the effective current and the maximum current through the resistance.

23.10 A capacitor is rated for a maximum voltage of 1 kV. Obtain the effective voltage of an ac source to which the capacitor can be directly connected.

23.11 A 10 μF capacitor is connected to a 60 Hz source of 120 V. Calculate (a) the reactance of the capacitor and (b) the current through the capacitor.

23.12 An ac source of 60 V effective is connected to an inductor of 10 mH. The current through the inductor is 1 A. Calculate the reactance of the inductor and the frequency of the source.

23.13 An ac circuit consists of a 10 Ω resistance and a 50 μF capacitance. If the power source in the circuit has a frequency of 100 Hz, calculate the impedance of the circuit.

23.14 A 60 Hz source is connected to a circuit comprised of a 150 Ω resistance and 50 mH inductance. Calculate the impedance of the circuit.

23.15 A series circuit consists of 100 Ω, 10 mH, and 100 μF connected to a 120 V, 60 Hz power source. Calculate (a) the inductive reactance, (b) capacitive reactance, (c) impedance, (d) current in the circuit, (e) voltage across each element, and (f) phase angle.

23.16 A circuit contains 50 Ω, 500 μF, and 0.5 H in series. Calculate (a) the resonance frequency of the circuit and (b) the current through the circuit when connected to a 300 V source at resonance frequency.

23.17 Calculate the power factor of a circuit containing 1 kΩ and 50 μF when connected to a 60 Hz ac source.

23.18 An ac voltmeter reads 60 V, 100 V, and 100 V across a resistance, capacitor, and inductor, respectively. Is the circuit in resonance? What is the value of the source voltage? If the resistance is 1000 Ω, calculate the inductive and capacitive reactances.

23.19 Verify that RC has the units of time.

23.20 Verify that L/R has the units of time.

24
Atomic Physics and Spectra

In several previous chapters we mentioned atomic structure without going into detail.˙ A fuller discussion of atoms, optical spectra, and x-ray spectra is given in this chapter. Analysis of optical spectra and x-ray spectra gives information about the elemental constitution of the source as well as an explanation of the origin of the radiation. Details of such analysis and their practical applications are also given in this chapter.

24.1 BOHR'S THEORY OF THE ATOM

As mentioned in Chapter 19, an atom consists of a heavy positive nucleus and electrons. The nuclear atom model was suggested by Rutherford in 1910 based on his alpha-particle scattering experiments. Bohr further developed an atomic theory using Rutherford's model as a foundation. The questions Bohr was trying to answer with his model were the following: How are the particles, electrons and nucleus, arranged in the atom? What is the force holding them together? Why is the atomic spectrum discrete and not continuous? What causes the release of radiation? Bohr's theory has three basic postulates; two of these are discussed below, the third one will be discussed later. The first two postulates of Bohr's theory are as follows:

1. The electrostatic attractive force between the negative electron and the positive nucleus supplies the necessary force (centripetal force) to keep the electrons in circular orbit.

2. The angular momentum of the electron in an orbit around the nucleus should be equal to an integral multiple of $h/2\pi$, where h is known as Planck's constant, which is equal to 6.625×10^{-34} J·s.

As we will see below, these postulates answer the first three questions raised above.

To understand these postulates, let us consider an electron in an atom (Fig. 24.1) assumed to be in a circular orbit. If the velocity of the electron in a circular orbit about the nucleus at the center of the orbit is v, the radius of the orbit is r, and the mass of the electron is m, then the centripetal force is mv^2/r. If the charge of the nucleus is Ze, the attractive electrical force between an electron and the nucleus is $Zee/4\pi\epsilon_0 r^2$. Hence the first postulate

can be written as

$$\frac{mv^2}{r} = \frac{1}{4\pi\epsilon_0} \frac{Ze^2}{r^2}. \qquad (24.1)$$

Since mvr gives the angular momentum, the second postulate can be written in the following form:

$$mvr = nh/2\pi, \qquad (24.2)$$

where n is any integer (1, 2, 3, $\cdots$). Postulate 2 tells us that the electron should select the orbit such that its angular momentum is a multiple of $h/2\pi$ and that no other orbits are allowed. This is contrary to our experience based on mechanical systems we encounter in our daily life. For example, there is nothing in classical mechanics that restricts the value of the angular momentum of a spinning wheel to discrete values. The angular momentum of a spinning wheel can have any value from 0 to a maximum value determined by the internal forces holding the wheel together.

Another consequence of the second postulate is that, since v is discrete, the energy of an electron will have discrete values in the different allowed orbits.

The total energy of the electron E can now be obtained. It is the sum of the potential energy $[- (1/4\pi\epsilon_0)(Ze^2/r)]$ and kinetic energy ($\frac{1}{2} mv^2$). Hence with the help of Eqs. (24.1) and (24.2), we get

$$E_n = -\frac{1}{\epsilon_0^2} \frac{mZ^2 e^4}{8h^2 n^2}. \qquad (24.3)$$

From Eq.(24.3) it is evident that the energy depends inversely on n^2 and hence the integer n, all other terms in the equation being constants. The values of the constants are

Planck's constant $h = 6.62 \times 10^{-34}$ J·s;

Charge of the electron $e = 1.60 \times 10^{-19}$ C;

Mass of the electron $m = 9.11 \times 10^{-31}$ kg;

Figure 24.1. Bohr model of the atom. The electrostatic attractive force between the two charges holds the particles together in the atom. The angular momentum of the orbiting electron is equal to a multiple of $h/2\pi$.

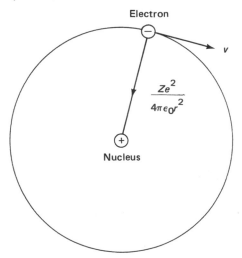

Electron

v

$\frac{Ze^2}{4\pi\epsilon_0 r^2}$

Nucleus

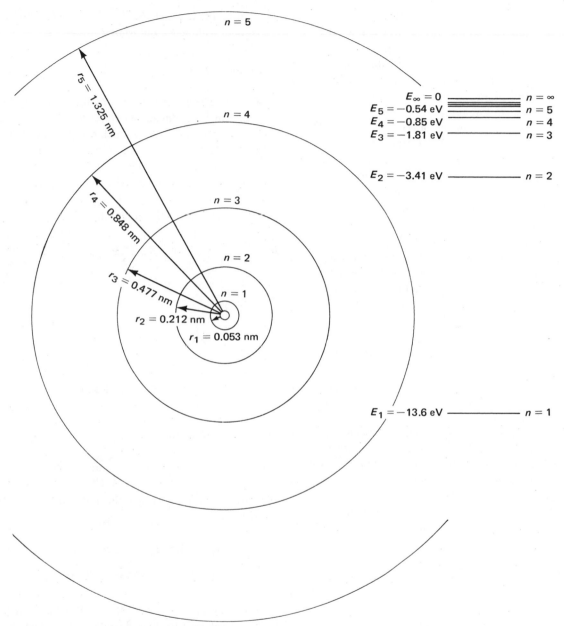

Figure 24.2. Energy levels of the electron in a hydrogen atom. Figure (left side) also shows the variation of the radius of the electron orbits as a function of the quantum number *n*.

Permittivity constant $\epsilon_0 = 8.85 \times 10^{-12}\, C^2/N \cdot m^2$. the energy of an electron in a hydrogen atom:

Substituting the values of these constants in Eq. (24.3) and putting $Z = 1$, we get the equation for

$$E_n = \frac{13.6}{n^2}\ eV. \qquad (24.4)$$

(The electron volt eV is a unit of energy used in atomic and nuclear physics. $1 \text{ eV} = 1.6 \times 10^{-19}$ J.)

Figure 24.2, known as the energy level diagram, depicts the energies for various values of n for the hydrogen atom. The number n, which determines the energy of the allowed states, is known as the principal quantum number.

The electron normally stays in the orbit with the lowest energy, i.e., the $n = 1$ orbit. The energy level for the $n = 1$ orbit is known as the ground level or the ground state. In the case of hydrogen, the energy of the ground state E_1 is -13.6 eV and for the $n = 2$ level, the energy E_2 is -3.4 eV ($13.6/2^2$). To take an electron from its ground state ($n = 1$) to the next higher level ($n = 2$), one has to supply energy equal to the difference between the states, $E_2 - E_1$ ($= 9.2$ eV). Similarly, appropriate amounts of energy must be applied to take the electron from an initial lower state to a final state of higher quantum number.

Taking an electron to a state of higher quantum number is known as excitation and the upper levels are known as the excited states.

As the quantum number n increases, the spacing between the energy states decreases, as seen in Fig. 24.2. For $n = \infty$, the energy is zero, and beyond this state, the electron has energy which is positive and continuous. The electron with positive energy is no longer bound to the atom. In this case, we say that the atom is ionized. *Ionization means removal of an electron from the atom.*

The radius of the orbit, as shown in Fig. 24.2, also increases with increasing quantum number. In the case of the hydrogen atom, the radius in nanometers is given by the following equation:

$$r_n = 0.053n^2 \text{ nm.} \qquad (24.5)$$

From Eq. (24.5), we get the radius of the first orbit equal to 0.053 nm. It is interesting to note that the radius of the first electronic orbit is about 50,000 times bigger than the radius of the nucleus. If one magnifies the atom so that the nuclear cross-section is of the size of the period at the end of this sentence, the electronic orbit will be at a distance of

~ 5 m from the nucleus. For this magnification a basketball will have a radius of 2 miles.

EXAMPLE 24.1

Calculate the energies and radii of the orbits of an electron in the quantum states $n = 2$ and $n = 3$ of a hydrogen atom. How much energy is needed to raise an electron from the $n = 2$ to the $n = 3$ state? How much energy should be supplied to ionize a hydrogen atom with its electron in the $n = 2$ orbit?

The equation for the energy is

$$E_n = \frac{-13.6}{n^2} \text{ eV.}$$

Therefore,

$$E_2 = \frac{-13.6}{2^2} = -3.4 \text{ eV,}$$

$$E_3 = \frac{-13.6}{3^2} = -1.51 \text{ eV.}$$

The equation for the radius is

$$r = 0.0529n^2 \text{ nm.}$$

Substituting for n, we get the radii for the orbits, namely,

$$r_2 = 0.0529 \times 2^2 = 0.21 \text{ nm,}$$
$$r_3 = 0.0529 \times 3^2 = 0.47 \text{ nm.}$$

The energy required to raise the electron from the $n = 2$ state to the $n = 3$ state = final energy − initial energy

$$E_3 - E_2 = -1.51 \text{ eV} - (-3.4 \text{ eV})$$
$$= 1.89 \text{ eV.}$$

Ionization means removing the electron from the

atom. Ionization energy is the minimum required to take the electron to the $n = \infty$ ($E_\infty = 0$) state. Therefore, the energy to be supplied to ionize the atom with the electron in the $n = 2$ state is $E_\infty - E_2$.

Ionization energy in this case $= 0 - (-3.4 \text{ eV})$

$$= 3.4 \text{ eV}.$$

According to Eq. (24.3), the total energy of the electron in the atom is negative. Why? What is its significance? The total energy is the sum of two terms, the electrostatic potential energy and the kinetic energy. The potential energy is negative since the force between the electron and the nucleus is attractive. Using Eqs. (24.1) and (24.2), we can show that the magnitude of the potential energy is twice the kinetic energy. This means that the bound electron is in an electrostatic potential well of the nucleus. Its kinetic energy is not sufficient to overcome the negative electrostatic potential well or to escape from the atom.

The electron in an atom is in a somewhat analogous situation that a baby finds itself in Fig. 24.3. The baby (technically speaking) is in a gravitational potential well. The potential energy of the baby is lower than when it is at the ground level or it is negative if we assume the ground level potential is zero. To raise the baby up the ladder, we have to supply energy to the baby. It is clear the baby can stay in one of the excited states defined by the steps of the ladder. It is important to remember that the excited states are not quite stable. (Consequences of instability of the excited states in the atomic case will be discussed later.) When the baby falls down from an excited state to the ground state, part of the energy difference appears as sound energy. If sufficient energy is supplied, the baby can be completely removed from the well. The baby taken out is free to move around. This process is similar to "ionization" in the atomic case.

24.2 EMISSION SPECTRA

As mentioned in the last section, an electron in the excited state is in an unstable state and it will fall back eventually to the ground level. The time for this transition is less than a nanosecond. What happens to the energy as the electron jumps from an excited level to a lower quantum number state? The third postulate of Bohr answers this question.

Figure 24.3. Energy states in a gravitational potential well analogous to electron energy states in an atom: (a) ground state; (b) excited state, usually unstable; (c) transition to ground state; (d) escaping the potential well, equivalent to ionization of electron from the atom.

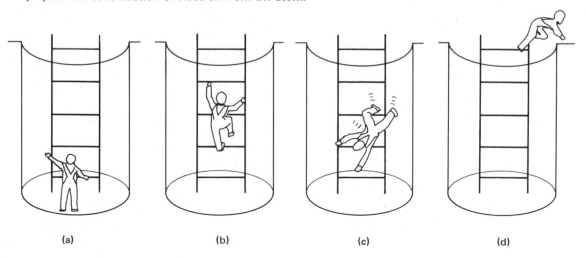

(a)   (b)   (c)   (d)

The third postulate states: *when an electron in an excited state falls to a lower state, the energy difference is radiated in the form of electromagnetic radiation, whose frequency is given by the following equation*

$$hf = E_i - E_f, \qquad (24.6)$$

where h is Planck's constant, E_i the energy of the initial state (excited state), and E_f the final state. This small wave packet that carries this energy $E_i - E_f$ is called a "photon," a quantum of energy.

EXAMPLE 24.2

Calculate the energy of each photon, frequency, and wavelength of radiation emitted during transition of an electron from the $n = 3$ state to the $n = 2$ state of the hydrogen atom. The energies corresponding to these states are calculated in Example 24.1.

$$E_i = E_3 = -1.51 \text{ eV},$$

$$E_f = E_2 = -3.40 \text{ eV}.$$

Energy of radiation $= E_i - E_f$

$$= -1.51 \text{ eV} - (-3.40 \text{ eV})$$

$$= 1.89 \text{ eV}$$

$$1 \text{ eV} = 1.6 \times 10^{-19} \text{ J}.$$

Therefore,

Energy of radiation $= 1.89 \text{ eV} \times 1.6 \times 10^{-19} \text{ J/eV}$

$$= 3.02 \times 10^{-19} \text{ J}.$$

Frequency of radiation $= \dfrac{E}{h} = \dfrac{3.02 \times 10^{-19} \text{ J}}{6.62 \times 10^{-34} \text{ J} \cdot \text{s}}$

$$= 4.57 \times 10^{14} \text{ Hz}.$$

Wavelength $= \dfrac{c}{f} = \dfrac{3 \times 10^8 \text{ m/s}}{4.57 \times 10^{14} \text{ Hz}} = 656.3 \text{ nm}.$

The radiation emitted in this case is in the visible range.

Let us consider an electric discharge tube containing hydrogen atoms. During electric discharge, atoms are excited to upper energy levels and some are even ionized. As a consequence of deexcitations, transitions from upper levels to lower levels occur and hence radiation due to all possible transitions is emitted. The transitions to the ground state produce a series of spectral lines in the uv range, known as the Lyman series, and the transitions to the $n = 2$ state produce radiations in the visible range, known as the Balmer series. Other transitions are named in Fig. 24.4. One can clearly see that the frequency separation between the lines decreases as the frequency increases as shown in Fig. 24.5 because of the way energy levels change with n. Thus the Bohr theory, for the first time, was able to account for the experimental observations. For atoms containing more than one electron, the energy levels are not so simple as those for the hydrogen atom. However, each chemical element or atom has its own characteristic set of energy levels and hence a characteristic emission spectrum. An emission spectrum, therefore, is a "fingerprint" which will help us in identifying elements in a sample.

24.3 ABSORPTION SPECTRUM

The absorption of light is a commonly observed phenomenon. As an example, consider the passage of white light through a piece of blue glass. The glass absorbs light of all colors except blue and, consequently, the emerging light appears blue. In effect, absorption substracts light in certain frequency ranges from the continuous spectrum of the incident light. The spectrum of the transmitted light is called an absorption spectrum. If the absorbing material is a gas or vapor, the absorption spectrum contains sharply defined and sometimes extremely narrow dark lines.

To understand the formation of these lines, consider light photons of 10.2 eV falling on an absorption cell containing hydrogen gas. Since the

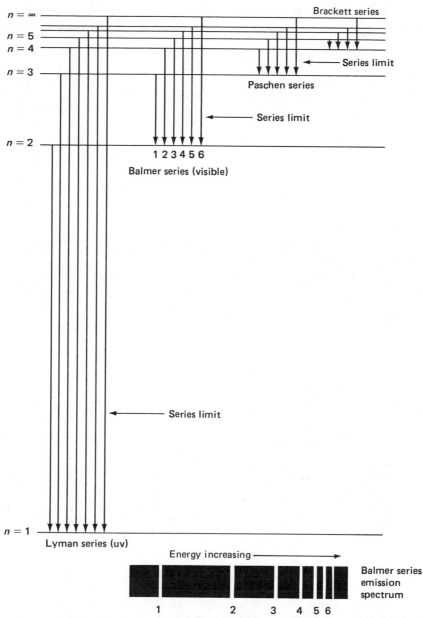

Figure 24.4. Transitions between different levels in a hydrogen atom. Transitions to the $n = 2$ state produce electromagnetic waves in the visible range. This series is known as the Balmer series. The inset at the bottom shows the emission spectrum in the visible region of the hydrogen atom. The lines are numbered according to increasing frequency.

energy of the incident photons is equal to the energy needed to excite the electron in the hydrogen atom from the $n = 1$ to $n = 2$ level, some of the photons will be absorbed by the hydrogen atoms. Similarly, if white light from an incandescent bulb passes through this gas photons with

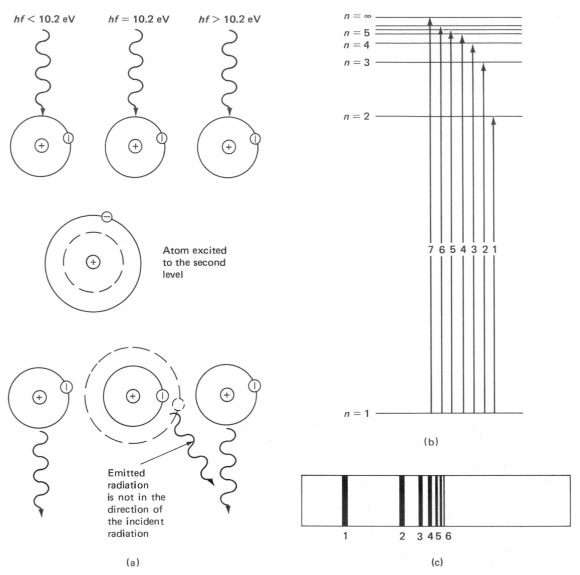

Figure 24.5. (a) When photons of the right energy are incident on the atoms, they are absorbed. Since the direction of re-emitted light is not the same as the direction of the incident light, on analysis of light passing through a gas (hydrogen in this case), dark lines are seen. (b) Possible excitations from the ground level to upper levels of hydrogen atom. (c) Absorption spectrum of hydrogen atoms. Dark lines (1, 2, 3, etc.) appear at energies at which absorption can take place.

frequencies characteristic of hydrogen will be absorbed, exciting the atoms to higher energy states in addition to the $n = 2$ states (Fig. 24.5). The transmitted light in this case will have dark lines at frequencies determined by the absorption

energies.

The emission spectrum of gaseous sources contains bright lines in a dark background while the absorption spectrum contains dark lines in a bright background. The position of the dark lines in a spectrum can give us information about the absorbing atoms.

An example of an absorption spectrum is the Fraunhofer lines in sunlight [Fig. 24.6(c)]. The hot interior of the sun emits white light (continuous spectrum). The gases in the outer regions being cooler absorb light of characteristic energies and re-emit them in all directions, thereby producing an absorption spectrum. Helium was first detected in the sun by this method. The presence of hydrogen in the sun is evident in Fig. 24.6(c).

24.4 SPECTROMETERS

Analysis of emission and absorption spectra are routinely used to determine the presence of elements and their quantities in a given sample. Such analysis is performed with instruments known as spectrometers which contain a dispersing element, a prism or a diffraction grating.

A spectrometer suitable for analysis of emission spectra is illustrated in Fig. 24.7. The lens L1 collimates the light, that is, a parallel beam of light is obtained after passing through the lens. The grating (G) disperses the light, resulting in colored beams traveling in slightly different directions. These colored beams are separately focused by the second lens (L2). At the focal plane of the second lens, we see colored lines which are colored images of the slits. The widths of these spectral lines are determined by the widths of the slits and also by the energy spread of the radiation. Better separation between the lines can be obtained if the slit is narrow, except that as the slit width approaches the wavelength of the radiation (1000 nm), the image of the slit is broadened due to diffraction. If a photographic plate is kept at the focal plane, we obtain spectra similar to the ones shown in Fig. 24.6. Figure 24.6(a) is for a line spectrum emitted by sodium; Fig. 24.6(b) is for a line spectrum emitted by hydrogen; and Fig. 24.6(c) is the spectrum obtained using sunlight.

Figure 24.6. (a) Sodium spectrum; (b) line spectrum, hydrogen source; (c) Fraunhofer absorption lines in sunlight. The presence of hydrogen in the sun is evident from (b) and (c).

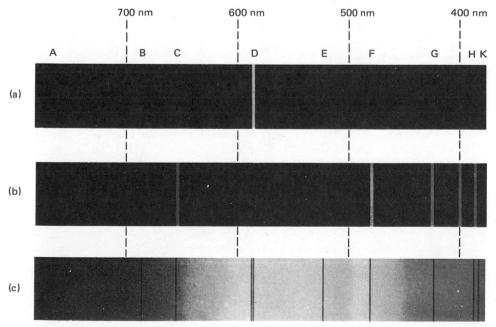

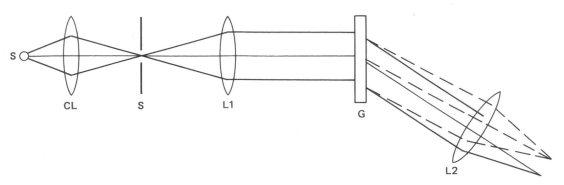

Figure 24.7. Spectrometer for analysis of light from source S. CL is a condensing lens that focuses light to the slip S. The lens L1 collimates the light, the grating G disperses the light, and lens L2 focuses the light. The spectra we see are the colored images (which appear as lines of the slit).

A simple spectrometer is shown in Fig. 24.8. It has three parts: a collimator, containing an adjustable slit and lens; a rotatable table with a vernier scale on which the prism or the grating is fixed; and a telescope to view the spectrum. The vernier scale moves over the main scale which moves with the telescope. To determine the angular position of the spectral line, the telescope is rotated so that the line is at the center of the cross-wire kept at the real image plane in the telescope.

An absorption spectrum can be obtained by a system like the one shown in Fig. 24.9. The source S gives a continuous spectrum. By rotating the grating, different parts of the spectrum are allowed to pass through slit S and fall on the sample. The detector most often used is a photoelectric detector which produces a current proportional to the intensity of light falling on the detector. As the wavelength of the light falling on the absorber varies, the current produced by the photocell varies. This variation in current is amplified and recorded on chart paper. From the absorption spectrometer, we get a record similar to the one shown in Fig. 24.10.

Figure 24.8. A simple spectrometer. A prism or a grating can be used to disperse the light.

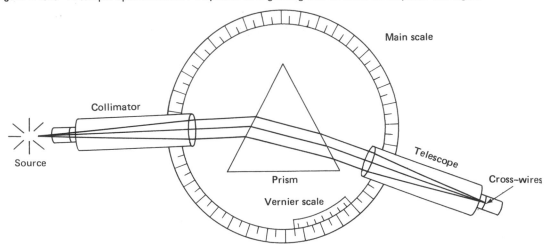

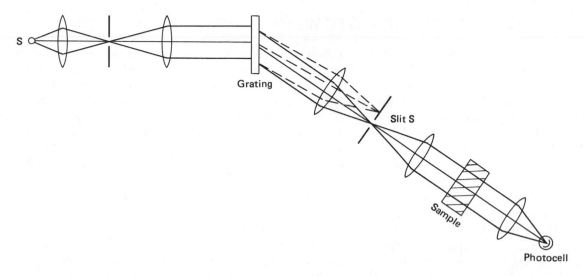

Figure 24.9. Absorption spectrometer. The photoelectric cell produces current proportional to the intensity of light falling on it. The current variation is amplified and recorded (Fig. 24.10).

Figure 24.10. Absorption spectrum of compound $CDCl_3$.

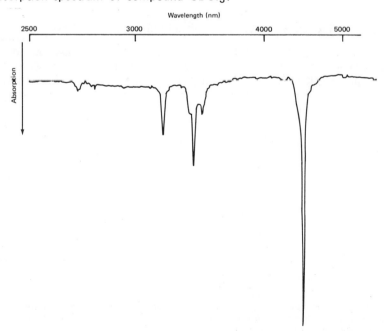

24.5　MANY–ELECTRON ATOMS

In a surprisingly simple manner, Bohr's theory was able to explain the experimental data regarding the hydrogen spectrum. However, several questions remained unanswered. How are the electrons distributed in atoms containing several electrons? What causes the variation in intensity between different spectral lines? The answers to these came with the development of quantum mechanical theory. On applying this theory to atomic systems, it became evident that each orbit cannot contain more than a certain number of electrons. Table 24.1 summarizes the results regarding the distribution of electrons.

Let us apply the rules summarized in Table 24.1. The helium atom has two positive charges (two protons) in the nucleus, and hence it should have two electrons. The two electrons can occupy the $n = 1$ shell and they do so when not excited. The next atom, lithium, has three electrons. Two of these electrons as in the helium atom occupy the $n = 1$ shell when not excited. The third electron occupies the $n = 2$ shell, which is its lowest allowed energy state. The electron distribution of a few other atoms are shown in Table 24.2.

From this table we can see some correlation between chemical properties and electronic structure. All the alkaline metals (Li, Na, K) have an electron outside their completed shells. They are easily ionized by losing their outermost electron and chemical compounds containing these positive ions are stable because of electric attraction with negative ions. Atoms such as chlorine lack one electron to complete the shell. They easily accommodate an extra electron and form negative ions. These negative ions form stable compounds with positive ions such as sodium. The compound thus formed is sodium chloride. Since the electrostatic attractive force between the ions is responsible for their stability, such compounds are called ionic compounds. In sodium chloride, sodium and chlorine exist as ions shown in Fig. 24.11. Some elements such as the ones belonging to the inert gases, have completely filled subshells and are in general not chemically reactive.

TABLE 24.1　Electron Distribution in Atomic Orbits

SYMBOLS FOR QUANTUM NUMBERS	PRINCIPAL QUANTUM NUMBERS	SUBSHELL SYMBOLS	ANGULAR MOMENTUM QUANTUM NUMBERS	MAXIMUM NUMBER OF ELECTRONS IN SUBSHELLS	TOTAL
K	1	s	0	2	2
L	2	s	0	2	8
		p	1	6	
M	3	s	0	2	
		p	1	6	18
		d	2	10	
N	4	s	0	2	
		p	1	6	32
		d	2	10	
		f	3	14	
*	n	$l = 0, 1, 2, \cdots, n-1$		$2(2l + 1)$	$2n^2$

*The last line in the table gives the relationship between the orbital angular momentum number l and the principal quantum number n, the number of electrons in the subshells of quantum number l and the maximum number of electrons in a shell of principal quantum number n.

TABLE 24.2 Electron Distribution in Selected Atoms

NAME, SYMBOL	$n=1$	$n=2$		$n=3$			COMMENTS
		s	p	s	p	d	
Hydrogen, H	1						
Helium, He	2						Shell filled
Lithium, Li	2	1					Completed shell plus one electron
Fluorine, F	2	2	5				One electron needed to complete second shell
Neon, Ne	2	2	6				Completed shells
Sodium, Na	2	2	6	1			Completed shells + one electron
Chlorine, Cl	2	2	6	2	5		One electron needed to complete $3p$ subshell

24.6 X-RAYS

The electron distribution in the copper atom is shown in Fig. 24.12. The nucleus has a positive charge of 29 and, therefore, the ordinary or neutral atom has 29 electrons. Let us consider the energy of one of the electrons in the $n = 1$ orbit. Since the force between the nucleus and the electron is approximately 29 times larger than in the case of the hydrogen atom, the electron is much closer to the atom resulting in a more negative potential energy. To a good approximation (neglecting the repulsion among electrons and other factors), the energy is given by

$$E_1 = Z^2 \times 13.6\,\text{eV} = 29^2 \times 13.6\,\text{eV} = 11.4\,\text{keV}.$$

The actual value of the energy is 8.996 keV. Therefore, if the $n = 1$ state is empty and if an electron from the upper states falls into this

Figure 24.11. Electrostatic attractive force keeps the sodium ions (+) and chlorine ions (−) together in the NaCl crystal.

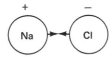

state, the radiation released will have energy in the keV range. Photons in the keV range emitted by atoms are called x-rays. These x-rays emitted during transitions to the $n = 1$ state are called K x-rays and those emitted during transitions to the $n = 2$ state are called L x-rays, etc. (Fig. 24.12).

In heavy elements, since the lower levels are filled, it is not possible to excite electrons from the $n = 1$ to the next higher level. Energy has to be supplied to take the electron to states where there are no electrons.

The common method of producing x-rays is illustrated in Fig. 24.13. Electrons from the hot filament are accelerated and allowed to strike a target. The collision between an electron and an atom in the target may result in the excitation of inner electrons and the consequent production of characteristic x-rays. In addition to the characteristic x-rays, we also get what is known as continuous x-rays. These are produced whenever there is a change in velocity of electrons during collision or deceleration near the nuclei of the target. The spectrum produced by an x-ray tube is shown in Fig. 24.14.

Roentgen's paper announcing the discovery of x-rays, published in 1895, contained a dramatic x-ray picture of his wife's hand which showed the

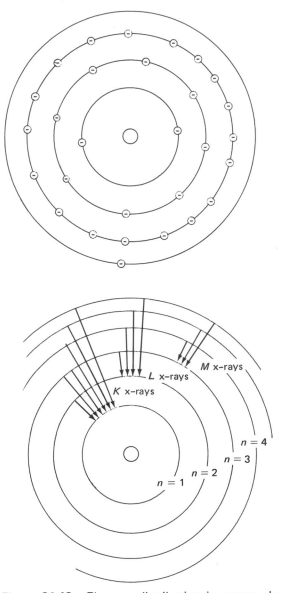

scientists who were privileged to see his paper and almost immediately x-ray machines were put to use for this purpose. The denser materials such as bones absorb x-rays, while the absorption in tissues is much less. Consequently, bones are opaque to x-rays and hence produce shadows while the tissues are transparent. Recently the merger of x-ray and computer technologies is enabling the physician to see every point in an organ, such as the brain, and project the details of the organ layer by layer. Such CAT (Computer Assisted Tomography) scanning systems can detect small tumors located deep in the organs. X-rays have found other medical and biological uses. X-rays and gamma rays (photons emitted by nuclei, with energy sometimes in the range of atomic x-rays and higher) can destroy living cells. This property is often successfully used to kill unwanted cancerous growth. X-rays are also used in genetic mutation studies. Mutations are generally harmful (see Sec. 25.8 in Chapter 25), but beneficial results have been obtained on rare occasions.

The penetrating power of x-rays increases as the energy of the x-rays increases. A good fraction of 2 MeV x-rays can penetrate steel castings up to 8 in. thick. High energy x-rays in the MeV range are, therefore, routinely used in inspecting castings, forgings, and machine parts under stress. Crystal symmetry and lattice spacing can be studied by x-ray diffraction. This is possible because the spacing between the atoms in a crystal is of the order of the wavelength of x-rays. The atoms act like a three-dimensional grating. The x-rays scattered undergo constructive interference for certain angles of scattering. The maxima in this case appear as spots (see Fig. 24.15), in contrast to lines produced by the two-dimensional optical gratings.

The lines of the x-ray spectra produced by different atoms vary from atom to atom depending on transitions to and from characteristic energy levels. The x-ray spectrum, like the optical spectrum, is thus characteristic of the atom and so this spectrum is used to analyze the chemical composition of materials and also to give values for the energy levels of the atom. An experimental arrangement for such an analysis is shown in Fig. 24.16. The

Figure 24.12. Electron distribution in copper. In the heavy elements transitions involving inner electrons produce characteristic x-rays. Electrons falling to the $n=1$ shell produce K x-rays, $n=2$ shells L x-rays, and $n=3$ shells M x-rays.

details of bones with a lighter outline of the less dense fleshy parts of the hand. Diagnostic application of x-rays was immediately obvious to the

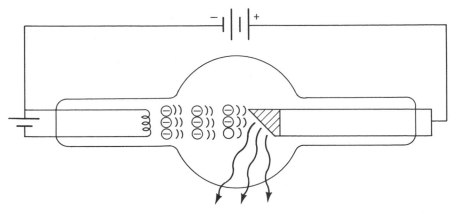

Figure 24.13. An x-ray tube. When the electrons strike the cathode, x-rays are produced.

detector, a semiconductor detector, is capable of giving information about the energy of x-rays falling on it. Typical data obtained from a stainless steel sample are shown in Fig. 24.17. There is no difficulty in detecting the presence of nickel and iron simultaneously even though they are neighboring atoms.

Determination of chemical composition by studying the x-ray spectra from the sample using equipment shown in Fig. 24.16 is known as x-ray fluorescence analysis. The atoms in the sample are excited by high energy x-rays, gamma rays, or charged particle radiation. This is one of the most popular nondestructive analysis methods. This method finds application in different fields like the analysis of paintings to uncover forgeries. The advantage is that the study can be made without damaging the paintings in any way.

Figure 24.15. X-ray diffraction pattern from a hexagonal crystal. Sixfold symmetry is evident. A perfect single crystal produces circular spots. The noncircular spots indicate that the sample contains several crystals and that the crystals are not perfectly aligned in the same direction.

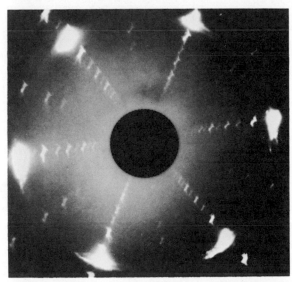

Figure 24.14. Intensity distribution as a function of wavelength of x-rays produced in a tube similar to the one given in Fig. 24.13.

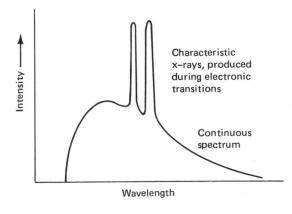

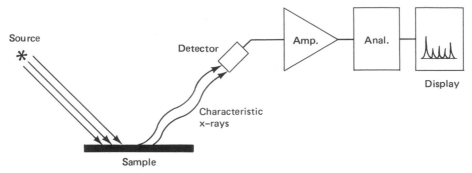

Figure 24.16. Pictorial representation of the experimental arrangement of x-ray fluorescence analysis.

24.7 LASER: LIGHT AMPLIFI-CATION BY STIMULATED EMISSION OF RADIATION

An atomic electron in an excited state (E_1) spontaneously decays to its ground state (E_0) with the emission of a photon. In spontaneous emission the direction of the emitted photons is random. A second type of decay which is triggered by another photon is stimulated emission. In this case a photon of energy $hf = E_1 - E_0$ passing near the excited atom stimulates the decay of the atom. The emitted photon travels with the incident photon and they remain in phase. If a large number of excited atoms are present a coherent beam of light builds up in the direction of the incident light (Fig. 24.18).

To obtain "lasing" action, two conditions have to be satisfied: (a) a large number of excited atoms of moderately long lifetime should be present; (b) some arrangement should be made so that light

Figure 24.17. X-rays produced by a steel target. Presence of manganese and chromium are evident from the presence of their characteristic spectrum. The atoms are excited by 14 keV gamma rays (on the right).

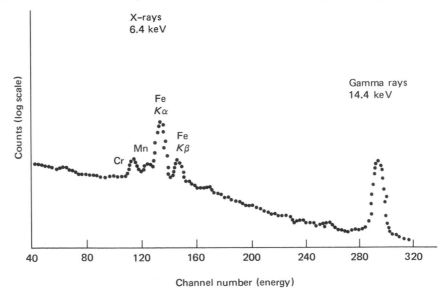

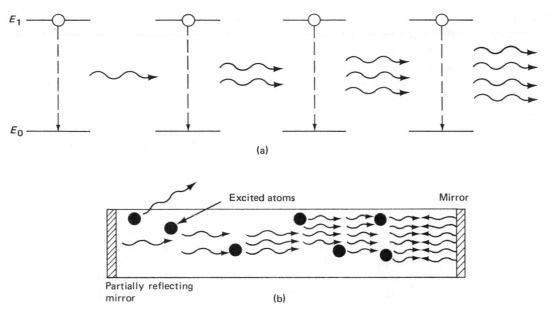

(a)

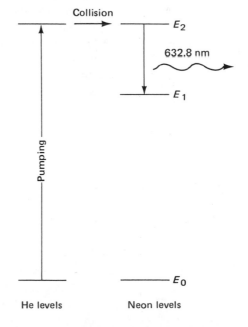

Figure 24.18. (a) Stimulated emission. In this type of emission an incident photon passing near an excited atom triggers the decay, producing a second photon in phase with the incident photon. (b) Reflecting mirrors at the ends of a laser help in the buildup of photons along the axis of the laser.

buildup takes place in one direction. Ordinarily most of the atoms are in their ground state; increasing the number of atoms in the excited state is known as population inversion. Light buildup is ordinarily obtained along the axis of the tube by placing mirrors at both ends of the laser. Light emitted in directions other than parallel to the axis eventually escape the tube, but light parallel to the axis builds up by repeated reflections back and forth.

We discuss here the helium–neon gas laser. The gas, a mixture of helium and neon in the respective ratio of 10:1, is in a discharge tube. The electric discharge is initiated in the gas by radio frequency voltage. Helium atoms are first excited by electron collisions and before they de-excite to their ground state most of them transfer their energy to neon atoms through collisions between them. The transfer is facilitated by the fact that their upper levels are of the same energy (Fig. 24.19) and the fact that helium states have a long lifetime. The excited state of neon has a sufficiently long lifetime, so that the

Figure 24.19. A simplified energy level diagram of helium and neon.

electrons do not immediately decay to the ground state. This delay helps in producing the population inversion necessary for lasing action. In the transition from the E_1 state to the E_2 state, radiation of several wavelengths is emitted by neon, the most prominent in the visible region being the red light (633.8 nm). As mentioned earlier, the mirrors at both ends of the tube reflect light back and forth producing a buildup along the axis. Laser light escapes the tube through a partially reflecting mirror at one end (Fig. 24.18).

Lasers produce intense, coherent, almost monochromatic light. Radiant energy per unit area in a laser beam is much higher than from any other type of source. Because of the extremely high energy density, lasers find applications in a wide range of areas such as in medicine and industry. The most well-known medical application is the welding of a detached retina by laser light. In industry lasers are used for microwelding, drilling precision holes, etc. Holography which involves taking and projecting three-dimensional pictures is another application.

SUMMARY

The atom contains a positively charged nucleus and electrons distributed in orbits surrounding it.

Bohr's theory has three postulates: (a) The electrostatic force between the electron and the nucleus supplies the necessary centripetal force to keep the electron in its orbit. (b) In allowed orbits the electron angular momentum (mvr) is equal to a multiple of $h/2\pi$. (c) When the electron jumps from an upper state to a lower state, the energy difference is emitted in the form of electromagnetic waves. The frequency of the waves is given by

$$E_i - E_f = hf$$

Excitation means taking an electron from its ground state to an upper state. Ionization means removing the electron out of the atom.

In many electron atoms, electron distribution follows certain rules (Table 24.1) derived from quantum mechanics. When there is a vacancy in one of the inner shells in a heavy atom, electrons from upper states fall into this vacancy and the energy difference appears as x-rays. These x-rays are called characteristic x-rays. X-rays are used in several fields such as medicine, industry, and science. The student should be familiar with some of these uses.

QUESTIONS AND PROBLEMS

24.1 The helium atom is smaller than the hydrogen atom even though helium is four times heavier than hydrogen. Explain the size difference.

24.2 The color of light from different sources varies. Explain. How many sources with distinct color compositions can you name?

24.3 An emission spectrum and an absorption spectrum have some common lines; but the emission spectrum usually has more lines. Give the reasons.

24.4 Halogen atoms form acids such as HC1 and HBr with hydrogen. Explain the formation of these compounds.

24.5 The energy of a photon is 3 eV. Calculate its frequency and wavelength.

24.6 A photon is 300 nm in wavelength. Calculate its energy.

24.7 Calculate the speed of an electron in the $n = 1$ and $n = 2$ orbits of hydrogen.

24.8 Calculate the wavelengths of the first two lines and of the series limit of the Balmer series.

24.9 Obtain the frequency of light emitted during a transition from the $n = 5$ to $n = 3$ orbit in the hydrogen atom.

24.10 Calculate the energy levels for a singly ionized helium atom (a helium atom containing one electron only).

24.11 Draw the energy level diagram for singly ionized helium using the results of Problem 24.10. Compare it with the energy levels for the hydrogen atom.

24.12 Calculate the energy required to ionize a hydrogen atom in the $n = 5$ state.

24.13 Obtain the radius of the $n = 5$ orbit of the hydrogen atom.

24.14 Obtain the radius of the $n = 1$ orbit of the helium atom.

24.15 A 1 m long gaseous discharge tube contains hydrogen gas. The average distance of collisions (mean free path) between electrons and hydrogen atoms is 0.13 m. Obtain the potential that should be applied between the ends of the tube to sustain continuous electric discharge.

24.16 How are the electrons distributed in oxygen and sodium atoms (use Table 24.2)?

24.17 Obtain in table form the electron distribution in carbon.

24.18 The x-ray diffraction pattern from a randomly aligned polycrystalline sample consists of rings instead of circular spots. Why? (See Fig. 24.15.)

25
The Nucleus and Nuclear Energy

An atomic model in which the heavy nucleus is located in the center and the electrons in shells surrounding the nucleus seems to explain most of the experimental data. We have also seen that some of the chemical properties can be explained by the distribution of electrons in the atom. Now we turn our attention to the study of the nucleus. We will attempt to discuss processes by which energy can be generated from the nucleus. This source of energy is becoming increasingly important since other sources of energy are being depleted at a very fast rate.

25.1 NUCLEAR STRUCTURE

What are the constituents of the nucleus? As mentioned earlier, the nucleus contains neutrons and protons (Chapter 19). These particles are called nucleons. The number of protons, Z, is called the atomic number. It is equal to the positive charge of the nucleus and also to the number of electrons in the atom if the atom is neutral. The number of neutrons inside the nucleus is designated by N. The number A, equal to $N + Z$, is called the atomic mass number. Since the masses of the neutron and the proton are close to one in atomic mass units, the atomic mass number is the closest integer to the mass of the atom, in atomic units. It is important to remember that the atomic mass includes the mass of the electrons; however, the contribution of electrons to the atomic mass is comparatively small.

A few atoms and their chemical symbols, the number of protons and neutrons in the nuclei of each atom, their atomic masses (in atomic mass unit, u), etc. are given in Table 25.1.

In Table 25.1 there are three atoms of hydrogen. The difference between these atoms is in the number of neutrons, since the proton number is the same. Therefore, the number of electrons in an un-ionized atom of each of the three types of atoms is the same. Since the chemical properties are determined mainly by the number and distribution of the electrons, the atoms ($_1^1$H, $_1^2$H, and $_1^3$H) are chemically similar. Such chemically similar atoms having a different number of neutrons in their nuclei are called isotopes. The isotope of hydrogen $_1^2$H is known as deuterium and $_1^3$H as tritium; the nuclei of these atoms are referred to as deuterons and tritons, respectively. All the elements have isotopes. Only in the case of hydrogen are they called by special names. The difference between the hydrogen and helium isotopes are illustrated in Fig. 25.1.

TABLE 25.1 Atomic Number (Z), Neutron* Number (N), Mass Number (A), and Atomic Mass of Selected Nuclei.

ATOM	CHEMICAL SYMBOL	Z	N	A	ATOMIC MASS (u)	COMMENTS
Hydrogen	$_1^1$H	1	0	1	1.007825	
	$_1^2$H	1	1	2	2.014102	An isotope of hydrogen, deuterium, contains a proton and a neutron
	$_1^3$H	1	2	3	3.016049	Another isotope of hydrogen called tritium is radioactive (β^-, $T_{1/2}$ = 12.2 yr)
Helium	$_2^3$He	2	1	3	3.016030	Two stable isotopes of helium.
	$_2^4$He	2	2	4	4.002603	
Carbon	$_6^{12}$C	6	6	12	12.000000	One atomic mass unit (u) is 1/12th the mass of ^{12}C
	$_6^{13}$C	6	7	13	13.003354	
	$_6^{14}$C	6	8	14	14.003241	Radioactive isotope of ^{14}C produced in atmosphere by cosmic rays; used for ^{14}C dating ($T_{1/2}$ = 5770 yr)
Uranium	$_{92}^{235}$U	92	143	235	235.043933	Alpha radioactive ($T_{1/2}$ = 7.13 × 10^8 yr)
	$_{92}^{236}$U	92	144	236	236.045733	Alpha radioactive (2.39 × 10^7 yr)
	$_{92}^{237}$U	92	145	237	237.048581	Alpha radioactive (6.75 days).

* Mass of neutron is 1.008665 u.

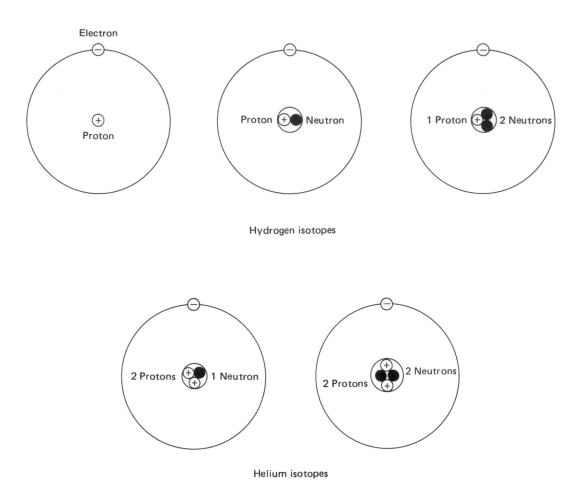

Figure 25.1. Isotopes of hydrogen and helium.

What holds the particles together inside the nucleus? This question is very significant especially since we are packing a number of positively charged protons that repel each other into a very small volume. Let us consider two protons inside a nucleus. The distance between them in an average nucleus is $\sim 10^{-14}$ m. The electrostatic repulsive force between two protons is given by

$$F = \frac{1}{4\pi\epsilon_0} \frac{e^2}{r^2}$$

$$= \left(9 \times 10^9 \ \frac{\text{N·m}^2}{\text{C}^2}\right) \frac{(1.6 \times 10^{-19} \ \text{C})^2}{(10^{-14} \ \text{m})^2} = 2.3 \ \text{N}.$$

The amount of work in bringing two protons within such a short distance is the potential energy given by

$$V = \frac{1}{4\pi\epsilon_0} \frac{e^2}{r} = \left(9 \times 10^9 \frac{\text{N} \cdot \text{m}^2}{\text{C}^2}\right) \frac{(1.6 \times 10^{-19} \text{ C})^2}{10^{-14} \text{ m}}$$

$$= 2.34 \times 10^{-14} \text{ J} = 144 \text{ keV}.$$

It is obvious that there has to be an attractive force to overcome the electrostatic repulsive force to keep the particles together. This nuclear force that holds the nucleons together has the following properties: (a) it is attractive; (b) it is stronger than the electrostatic force; (c) it acts only over very short distances ($\sim 10^{-15}$ m); (d) it is independent of the nature of the particles, that is, the nuclear forces between neutron and proton, between neutron and neutron, and between proton and proton are all equal except for the electrostatic force in the latter case.

As the number of protons increases, the electrostatic force which acts between protons increases proportionally with the square of the number of protons. To overcome this increase, the number of neutrons increases at a faster rate than the protons as the atomic number increases. This is illustrated in Fig. 25.2. One can also see that the stable nuclei lie near a line called the line of stability in this graph. Nuclei removed from this line are radioactive, nuclei on the right of the line have more neutrons than needed for stability, and nuclei on the left have less neutrons than needed for stability.

25.2 RADIOACTIVITY

What is meant by radioactivity? Some nuclei, without any external help, change into more stable nuclei by releasing energy. Such spontaneous transformations accompanied by emission of radiation are known as radioactive decay. The three different types of decay we encounter are illustrated below.

1 Alpha decay: An example of alpha activity is shown by the following equation:

$$^{226}_{90}\text{Th} \rightarrow \, ^{222}_{88}\text{Ra} + \, ^{4}_{2}\text{He}$$

In this equation $^{4}_{2}\text{He}$ represents the alpha particle; an alpha particle is the nucleus of a helium atom. Thorium is the radioactive parent nucleus and radium is the daughter nucleus. Figure 25.3 pictorially represents this alpha decay. As shown in the figure, sometimes the daughter nucleus is produced in one of its excited states.

2 Beta decay: There are three types of beta decay as shown below.

Electron decay (β^- decay),

$$^{60}\text{Co} \rightarrow \, ^{60}\text{Ni} + \, ^{0}_{-1}e$$

A negatively charged electron is emitted in this decay. The atomic number of the daughter nucleus increases by 1 in this case. One can consider this decay as a transformation where a neutron changes to a proton.

Positron decay (β^+ decay),

$$^{22}\text{Na} \rightarrow \, ^{22}\text{Ne} + \, ^{0}_{1}e$$

The positron is a particle identical to the electron except that its charge is positive. In this decay the atomic number of the daughter is less than that of the parent nucleus.

Electron capture,

$$^{55}\text{Fe} + \, ^{0}_{-1}e \rightarrow \, ^{55}\text{Mn}$$

During this transition an electron from one of the atomic shells is captured by the nucleus. In all the above beta decays another particle, the neutrino, is also emitted.

3 Gamma decay: the daughter nucleus in the

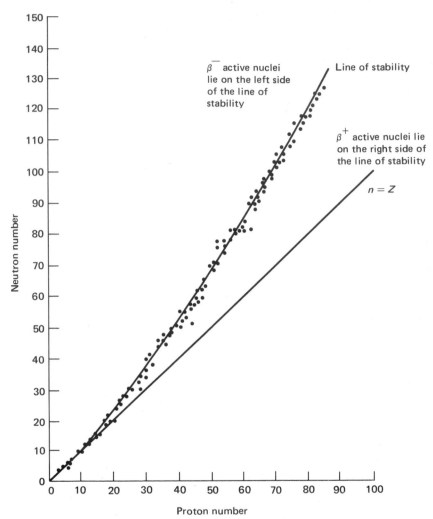

Figure 25.2. Number of neutrons versus protons in stable nuclei. Radioactive nuclei are not shown in this figure.

25.3 HALF-LIFE AND ACTIVITY

example for alpha decay is produced in an excited state. During transitions from the excited state to lower states, gamma rays are emitted. Such transitions occur in a number of radioactive decays. Gamma rays are electromagnetic waves of very high frequency.

Radioactive decay is a completely random process; it is not possible to predict when a radioactive nucleus will decay, but it is possible to predict how many will decay from a large number of radioactive nuclei. The number of nuclei that decay in a unit time is found to be proportional to the total number of nuclei present. This means that the

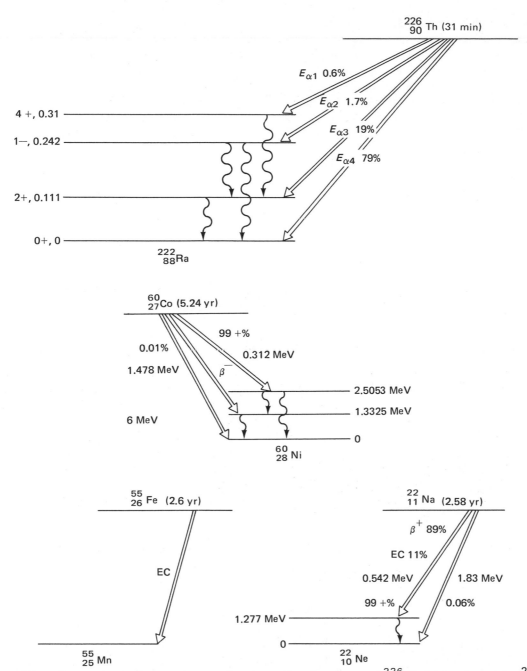

Figure 25.3. Energy level diagram representing the radioactive decay of ^{226}Th (alpha decay) to ^{222}Ra, ^{60}Co (electron decay) to ^{60}Ni. ^{55}Fe (electron capture) to ^{55}Mn, and ^{22}Na (positron decay 89%, electron capture 11% to ^{22}Ne. In most radioactive decays the daughter nucleus is produced in an excited state. In thorium decay, radium is produced in three excited states. The daughter nuclei eventually decay to the ground state emitting gamma rays.

number decaying per second, called the activity of the sample, decreases with time because the number of nuclei is decreasing. Let us take an example where we start with 1000 radioactive nuclei and 100 nuclei decay in the first second. In the next second only 90 will decay because at the end of the first second there were only 900 radioactive nuclei. At the end of the second second, we are left with 810

radioactive nuclei and 81 nuclei will decay in the third second. If we continue this calculation and plot the results we get two curves, the number of nuclei versus time and activity versus time, which are both similar [Fig. 25.4(a) and 25.4(b)]. Both curves are exponential curves.

Using Fig. 25.4 we now define half-life ($T_{1/2}$). Half-life is the time in which the number of nuclei

Figure 25.4. Number of radioactive nuclei as a function of time. The number of radioactive nuclei decreases by half in one half-life. The inset shows how the activity changes with time. Both curves show exponential decrease.

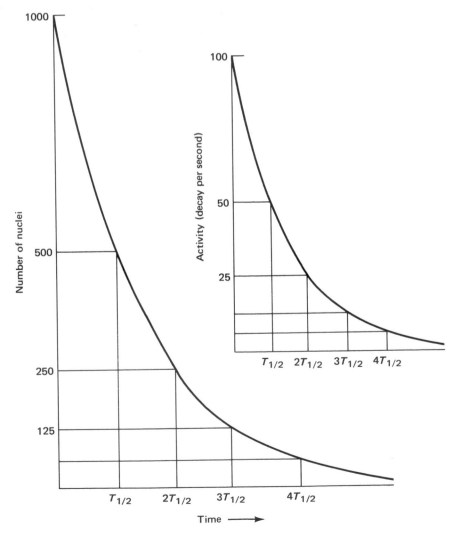

decreases by half. It is also the time in which the activity decreases by half. Experimentally half-life is determined by measuring the decrease in activity with time. An experimental setup is shown in Fig. 25.5. Radiation emitted by the source is detected by a Geiger counter. Since radiation is emitted in every decay, the amount of radiation detected will be proportional to the number of radioactive decays. When the results are plotted, we get a curve similar to Fig. 25.4(b) and from this plot the half-life can be determined.

The half-life varies from one type of nucleus to another. There are radioactive nuclei with half-lives equal to thousands of years or more ($^{232}_{90}$Th, 1.39 $\times 10^{10}$ yr), and others with half-lives of less than a second ($^{5}_{2}$He, 2×10^{-21} s). A smaller half-life means the decay is faster (Fig. 25.6); i.e., the rate of decay is inversely proportional to half-life. The activity or the number of nuclei decaying per unit time can be calculated, if the half-life is known, by the following equation

$$a = \frac{0.693}{T_{1/2}} N, \qquad (25.1)$$

Figure 25.5. Experimental setup to measure the half-life of a radioactive sample. This arrangement is suitable for a half-life of the order of hours.

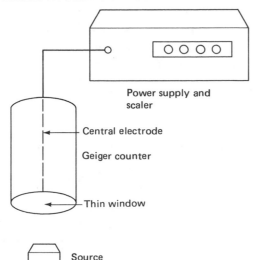

Power supply and scaler

Central electrode

Geiger counter

Thin window

Source

where a stands for activity—the number decaying per unit time—and N for the number of radioactive nuclei. One of the units of activity is the becquerel: 1 Bq = 1 disintegration per second (dps). This unit was recently adopted in honor of Henri Becquerel who discovered radioactivity. An older unit still in use, named after Madame Curie a pioneer in the study of radioactivity, is the curie (Ci)

$$1 \text{ Ci} = 3.7 \times 10^{10} \text{ dps.}$$

Multiples of this unit such as millicurie (10^{-3} Ci) are more convenient to use.

EXAMPLE 25.1

A radioactive sample has an activity of 2 mCi. Its half-life is 5 min. Calculate the number of radioactive nuclei in the sample.

Activity of the sample

$$= 2 \text{ mCi} = 2 \times 3.7 \times 10^7 \text{ dps}$$

$$= 7.4 \times 10^7 \text{ dps.}$$

Half-life = 5 min = 300 s.

Substituting the values of activity and half-life into Eq. (25.1) we get

$$7.5 \times 10^7 \text{ dps} = (0.693/300 \text{ s})N$$

or

$$N = \frac{7.4 \times 10^7 \text{ dps} \times 300 \text{ s}}{0.693}.$$

Note that since the activity is in disintegrations/second, the half-life must also be expressed in seconds to make units on both sides of the equation consistent.

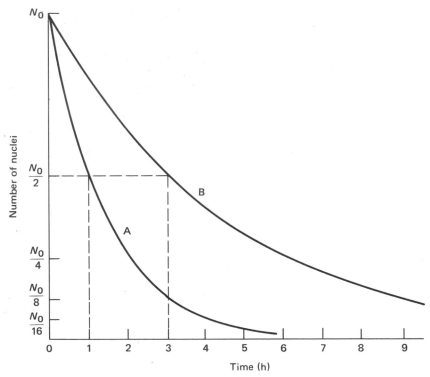

Figure 25.6. Exponential decrease in the number of radioactive nuclei. Curve A is for a source having a half-life of 1 h and curve B for a source having a half-life of 3 h. Decay is faster for the source with a shorter half-life.

25.4   BINDING ENERGY

On combining a free electron with a proton to form a hydrogen atom, radiation (light) of at least 13.6 eV is emitted. The energy of the final bound system, the hydrogen atom, is less than that of the free particles by 13.6 eV. This energy we call ionization energy. The term binding energy, used in nuclear physics, has a similar meaning to that of the ionization energy in atomic physics. To illustrate this similarity, let us consider the formation of a deuteron from a neutron and a proton. When the two particles combine, an amount of energy of 2.23 MeV in the form of gamma rays (Fig. 25.7) is released. From the masses given in Table 25.1, it is clear that the mass of the deuteron is less than the combined masses of the proton and the neutron.

To understand the relationship between the energy released and the loss of mass let us apply Einstein's mass–energy relationship

$$E = mc^2, \tag{25.2}$$

which means that a mass m, if converted completely to energy produces energy equal to mc^2. If the mass is given in kg and, if it is multiplied by the square of the velocity of light in $(m/s)^2$, we obtain the energy in joules. If the mass is given in atomic mass units, we obtain the energy in MeV by multiplying the mass by 931 or we can say 1 u is equivalent to 931 MeV.

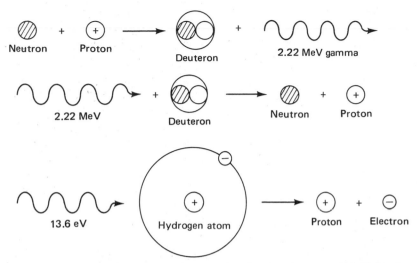

Figure 25.7. When a neutron and proton are combined to form a deuteron, 2.2 MeV of energy is released. An energy of 2.22 MeV is required to separate a neutron and proton from a deuteron. This energy is called the binding energy. In a similar situation, an energy of 13.6 eV separates the electron from a hydrogen atom.

EXAMPLE 25.2

A substance of mass 1 mg is converted completely to energy. How much energy is released? We are given $m = 1$ mg $= 1 \times 10^{-6}$ kg.

$$E = m \text{ (kg)} \times c^2 \text{ (m/s)}^2$$
$$= (1 \times 10^{-6} \text{ kg}) (3 \times 10^{10} \text{ m/s})^2$$
$$= 9 \times 10^{14} \text{ J.}$$

Applying the mass–energy relationship, we can now calculate the energy released during the formation of a deuteron from the difference in mass between the deuteron and the free particles.

Mass difference = (Mass of proton + Mass of neutron)

— Mass of deuteron.

Since nuclear masses are not ordinarily given in tables, we have to use the atomic masses that include the mass of the electron. A change from nuclear mass to atomic mass can be made by using the following equations:

Mass of hydrogen = Mass of proton

+ Mass of electron;

Mass of deuterium = Mass of deuteron

+ Mass of electron.

We get

Mass difference

= (Mass of hydrogen + Mass of neutron)

— Mass of deuterium.

Using the masses given in Table 25.1, we have

Mass difference = 1.007825 u + 1.008665 u

$$- 2.014102 \text{ u}$$

$$= 0.002388 \text{ u}.$$

Now applying Einstein's mass–energy relationship, we get

Energy released = Mass difference × 931

$$= 0.002388 \text{ u} \times 931 \text{ MeV/u}$$

$$= 2.22 \text{ MeV}.$$

The energy calculated is equal to the energy released in the form of gamma rays during the formation of the deuteron. To separate the particles, this energy has to be put back into the deuteron (Fig. 25.7). This energy is called the binding energy of the nucleus.

EXAMPLE 25.3

Calculate the binding energy of a helium nucleus.
To obtain the binding energy, first we calculate the mass difference between the helium nucleus and the masses of its constituents. The helium nucleus is made up of two protons and two neutrons.

Mass difference = 2 × Mass of hydrogen

$$+ \ 2 \times \text{Mass of neutron}$$

$$- \ \text{Mass of helium}.$$

$$= 0.030377$$

(using masses in Table 25.1).

Binding energy of helium nucleus

$$= \text{Mass difference (in u)} \times 931 \text{ MeV/u}$$

$$= 0.030377 \text{ u} \times 931 \text{ MeV/u}$$

$$= 28.3 \text{ MeV}.$$

Binding energy can be defined as the energy required to separate the nucleons in a nucleus. This can always be calculated by calculating the mass difference between the total mass of the free nucleons and the mass of the nucleus. In practice we take the mass difference between the atomic masses as illustrated in Examples 25.2 and 25.3. For a nucleus $^A_Z X$ which has Z protons and $A - Z$ neutrons, the binding energy (BE) is given in terms of the atomic masses by

$$BE = [ZM_\text{H} + (A - Z)m_n - M_\text{X}] \ 931 \text{ MeV}, \quad \textbf{(25.3)}$$

where M_H is the mass of hydrogen, m_n the mass of a neutron, and M_X the mass of the X atom, all in atomic mass units (u). It is important to remember that the binding energy is a measure of the stability of a nucleus. Binding energy means the energy required to separate the nucleons. Therefore, a higher binding energy means a higher energy input is needed to separate the particles.

The binding energy per nucleon (BE/A) is plotted in Fig. 25.8 as a function of the atomic number. For most of the nuclei BE/A is close to 8 MeV. In two regions, the low Z and high Z regions, the binding energy per nucleon is much lower than 8 MeV. To understand the low Z side consider Fig. 25.9. As the number of nucleons increase, the attractive forces acting between the nucleons increase and so does the binding energy. On the other hand, the increasing repulsive force between the protons decreases the stability and the binding energy for the higher atomic number nuclei.

25.5 NUCLEAR ENERGY

Radiation (alpha, beta, and gamma rays) carries energy in the range from 1 to 10 MeV as it is released from a radioactive source. Among the three, alpha and beta particles lose energy by

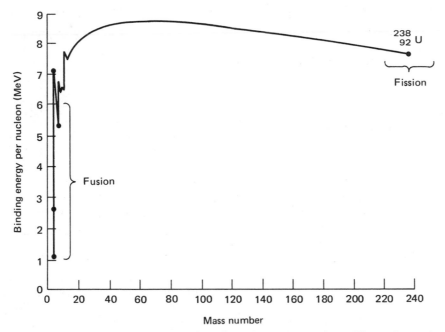

Figure 25.8. Variation of binding energy as a function of atomic number. The regions where fission and fusion may produce energy are also indicated.

collisions with atoms as they pass through matter. The range of alpha particles is very short (an alpha particle from a radioactive source can be stopped by a piece of paper) while beta particles travels a longer distance in matter. The energy lost by the radiation appears as heat energy in the material stopping the particles. A simple and direct method of using "nuclear energy" is to convert the energy of radiation into heat and this heat into electricity.

Figure 25.9. As the number of nucleons increases, the force acting on a nucleon increases.

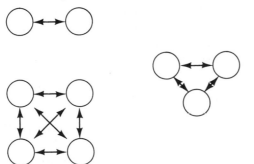

EXAMPLE 25.4

10,000 Ci of strontium–90 is kept in a copper container weighing 0.01 kg. Assuming all the beta rays emitted by the source are stopped by the container, calculate the increase in temperature in 10 min. The average energy of beta rays emitted by ^{90}Sr is 0.182 MeV.

The amount of radiation (beta rays) emitted by the source in 10 min is

$$10,000 \text{ Ci} \times 3.7 \times 10^{10} \, \frac{\text{radiation}}{\text{s·Ci}} \times \frac{60 \text{ s}}{\text{min}} \times 10 \text{ min}$$

$$= 2.22 \times 10^{17}$$

The total energy of the radiation emitted in 10 min is

$$2.22 \times 10^{17} \times 0.182 \text{ MeV} = 4.04 \times 10^{16} \text{ MeV}.$$

$$1 \text{ MeV} = 3.82 \times 10^{-14} \text{ cal}.$$

Therefore, the energy of radiation emitted in 10 min is

$$4.04 \times 10^{16} \times 3.82 \times 10^{-14} = 1.54 \times 10^3 \text{ cal}.$$

The temperature increase can be obtained using the equation (see Chapter 11).

Energy input = Mass × Specific heat of copper

× Increase in temperature

$$1.54 \times 10^3 \text{ cal} = (0.01 \text{ kg})(0.39 \times 10^3 \text{ J/°C} \cdot \text{kg})\Delta T$$

$$\Delta T = 395°C.$$

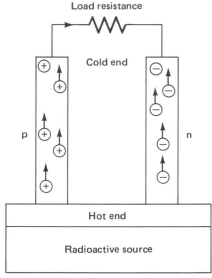

Figure 25.10. An arrangement for thermoelectric conversion of energy from a radioactive source. Charges diffuse from the hot end of the rods to the cold end maintaining an EMF.

It is worth pointing out that the energy released by radioactive sources plays an important part in maintaining the high temperatures of the earth's interior.

Now we discuss the principle of one of the simple systems used to convert heat energy to electric energy (Fig. 25.10). The two rods, designated p and n, are made of semiconducting materials like PbTe, but the difference between the two is that the conduction in one of them (p) is mainly due to positive charge carriers (holes) and in the other it is due to negative charge carriers (electrons). The charges drift from the hot to the cold end resulting in an accumulation of charges at the cold end and thus a potential difference between the cold ends of the p and n cylinders. The potential difference developed increases with temperature difference between the hot and the cold ends. Thus we achieve heat-to-electric energy conversion (thermoelectric conversion) by using this system.

A system SNAP-7E (SNAP stands for system for nuclear auxiliary power) developed for the Navy to be used as an underwater navigational beacon uses 34,000 curies of strontium-90. It has 60 pairs of lead telluride thermoelectric elements and generates 7.5 W of electricity.

Even though the efficiency of these converters are very low (see Example 25.5), they have the advantage that they work continuously with no maintenance for long periods of time. The useful lifetime of a converter depends on the half-life of the radioactive isotope used for the energy source. Strontium-90 is a suitable source because of its long half-life of 28 years.

Such power systems are used in spaceships and remote weather stations located in mid-ocean and at the poles. There are several flashing-light buoys in the sea powered by SNAP systems.

EXAMPLE 25.5

Calculate the efficiency of the conversion of heat to electricity in SNAP-7E.

To calculate the heat energy produced, we have

to first calculate the amount of radiation (beta rays) produced per second from the 124,000 Ci.

Number of beta rays emitted per second

$$= 124{,}000 \text{ Ci} \times 3.7 \times 10^{10} \text{ dps/Ci}$$

$$= 4.588 \times 10^{15} \text{ dps.}$$

Energy of the beta rays emitted per second

$$= (4.588 \times 10^{15} \text{ dps}) \times 0.182 \text{ MeV}$$

$$= 8.35 \times 10^{14} \text{ MeV/s}$$

$$= (8.35 \times 10^{14} \text{ MeV/s})(1.6 \times 10^{-13} \text{ J/MeV})$$

$$= 133.6 \text{ W.}$$

Efficiency of the generator

$$= \frac{\text{Electric power output}}{\text{Heat input}} \times 100$$

$$= \frac{7.5 \text{ W}}{133.6 \text{ W}} \times 100 = 5.6\%.$$

25.6 FISSION

The best known and widely used nuclear energy is produced from the fission process. Fission is a process in which a heavy nucleus splits into two or more parts. The fission fragments, flying apart from each other, carry a large amount of energy. To understand the energy release during fission, let us consider the BE/A versus A curve given in Fig. 25.8. The nuclei with high atomic number have less binding energy per nucleon compared to nuclei of intermediate atomic number. For example, the uranium isotope $^{236}_{92}$U has 7.6 MeV per nucleon binding energy. If this is split into two identical nuclei containing 46 protons and 72 neutrons, the resulting nuclei lie in a region of 8.5 MeV per nucleon binding energy. Because the products

have higher binding energy, energy will be released during the fission process. The difference in the binding energy per nucleon is 0.9 MeV and hence about 212 MeV will be released per fission. The following calculation will make this point clearer.

Equation (25.3) can be rearranged in the form

$$M_X = ZM_H + (A-Z)M_n - \frac{\text{BE (MeV)}}{931 \text{ (MeV/u)}}.$$

The mass of $^{236}_{92}$U, is therefore

$$M\left(^{236}_{92}\text{U}\right) = 92M_H + 144M_n$$

$$- \frac{7.6 \text{ MeV/nucleon}}{931 \text{ MeV/u}} \times 236.$$

The total mass of the product nuclei is

$$2 \times M\left(^{72}_{46}\text{X}\right) = 2\left(46M_H \times 72M_n\right.$$

$$\left. - \frac{8.5 \text{ MeV/nucleon}}{931 \text{ MeV/u}} \times 118\right).$$

The energy released is

$$\left[M\left(^{236}_{92}\text{U}\right) - 2M\left(^{72}_{46}\text{X}\right)\right]931 \text{ MeV/u}$$

$$= (8.5 - 7.6)\frac{\text{MeV}}{\text{nucleon}} \times 236$$

$$= 212 \text{ MeV.}$$

Fission, as mentioned earlier, is the splitting of a nucleus into two or more parts. In fission, in addition to two nuclei, an average of 2.5 neutrons are also emitted (Fig. 25.11).

In the example of equal splitting we discussed earlier, the product nuclei had 46 protons and 72 neutrons. Even though equal or symmetric splitting is very rare, this example points out the fact that the neutron to proton ratio (1.56) for these nuclei is higher than the ratio for the stable nuclei with $Z = 46$. (The ratio for stable nuclei is

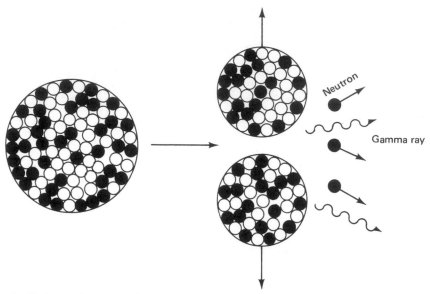

Figure 25.11. In fission, a heavy nucleus splits into two lighter nuclei. On the average about 2.5 neutrons are released in every fission. About 210 MeV of energy are released in fission. Most of this energy is carried away as kinetic energy by the nuclear product light nuclei.

~1.3 from Fig. 25.2). Therefore, the products are radioactive, decaying by a series of beta emissions. In each decay a neutron changes to a proton with electron emission and thus reduces the neutron to proton ratio. The radioactivity of the fission products is the most serious problem associated with fission reactors.

There are heavy nuclei that undergo spontaneous fission, but their rate of fission is usually very small. Thus we do not depend on spontaneous fission to produce energy. Fission initiated by a neutron,

often used in energy production, is illustrated in Fig. 25.12. The neutron, as it falls into the attractive potential well, gains some energy (the binding energy). This energy very often sets the nucleus into collective oscillation. If the energy is sufficiently large, the nucleus will undergo fission. After capture of a neutron, fission follows almost instantaneously. The neutrons produced during fission can start the next fission and thus maintain a chain reaction. To sustain the chain reaction, at least one neutron should produce the fission.

Figure 25.12. Thermal neutron fission. $^{235}_{92}$U absorbes a thermal neutron and the resulting unstable nucleus undergoes fission.

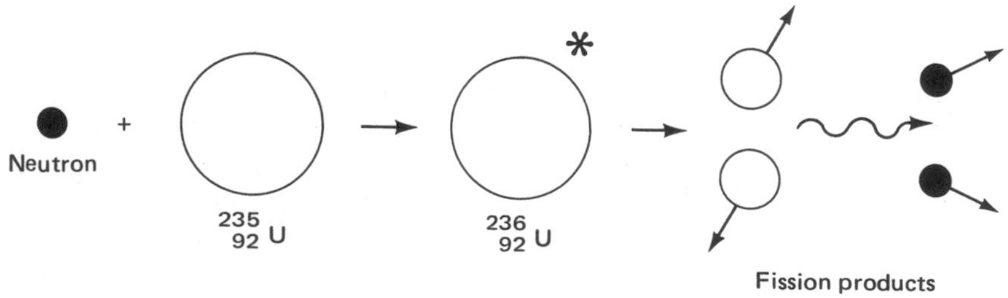

The probability of a neutron inducing fission is a function of the neutron energy. For $^{235}_{92}U$ the probability of fission is maximum when the neutron energy is of the order of 0.02 eV. This energy is equal to the average energy of atoms in a gas at room temperature and thus neutrons at this energy are known as thermal neutrons. The average energy of fission-produced neutrons is about 2 MeV and so their energy has to be reduced to increase their chance of producing fission. The process by which the energy is reduced is called thermalization or moderation.

How are the fission neutrons thermalized? As we have seen in Chapter 5, particles lose energy in elastic collisions. The amount of energy lost depends on the mass of the colliding particle; energy lost will be very small if the mass of the colliding particle is very high compared to the neutron mass and it increases as the mass of the colliding particle approaches that of the neutron mass. In the case where the masses are equal, complete transfer of energy is possible for head-on collisions. Elastic collisions between neutrons and light nuclei are used to thermalize the neutrons in a reactor. Hydrogen, or deuterium in water molecules, and graphite (carbon) are very commonly used. Because of this special function, they are called moderators.

The fission cycle where one neutron in each fission becomes available for the next fission is illustrated in Fig. 25.13. Some of the neutrons escape the reactor and others are absorbed inside the reactor by the component materials.

A reactor, where absorption and leakage are properly controlled by selection of the components, size of the core, etc., and where one neutron from every fission becomes available for a second fission, is called a critical reactor. If more than one neutron becomes available in every fission-to-fission cycle, the system will become an uncontrollable one. An atomic bomb is such a system. However, in power reactors, proper selection and distribution of its components and the use of control rods assures controllability.

There are two types of reactors in operation for power production. They are the boiling water

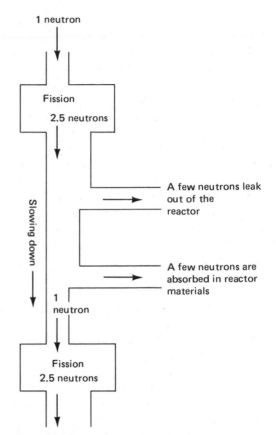

Figure 25.13. Representation of a fission-to-fission cycle. To sustain a chain reaction, after leakage and non-fission absorption is taken into account, one neutron should be available to initiate the next fission.

reactor and the pressurized water reactor (Fig. 25.14). The core structure is essentially the same in both these reactors. A reactor core contains specially shaped uranium oxide as fuel, often weighing more than 100 tons. The uranium oxide is in the shape of pellets, somewhat like an eraser at the tip of a pencil. These pellets are stacked inside a fuel rod made of stainless steel. A typical 12 ft long fuel rod may contain 200 or more of these pellets. A core assembly is made of such fuel rods. They are placed in such a way that the moderator, water, surrounds each fuel rod. Since the radius of the fuel rod is small, the chance of a fission-produced neutron to escape the rod and collide with a moderator atom is higher than its chance of meeting a uranium atom in the rod in which it is produced.

Both types of reactors use large amounts of water. This water performs two functions. First, it slows down the higher-energy fission neutrons, thus acting as the moderator. Second, it keeps the reactor core cool by absorbing its heat, which is then used to produce steam in the turbine-generator loop. Reactor cores contain control rods, made of neutron absorbing materials such as boron, cadmium, and hafnium. They help in controlling the power level in the reactor and also in shutting down

the reactors in emergency situations.

As mentioned earlier, fission fragments, flying apart from each other, carry a large amount of energy, which is converted to heat energy. One gram of uranium, when fully consumed, releases 22,770 kilowatt hours of energy (see Example 25.6) while 50 tons of coal is needed to produce an equivalent amount of energy.

EXAMPLE 25.6

Calculate the amount of heat energy released during the fission of 1 g of uranium (U-235).

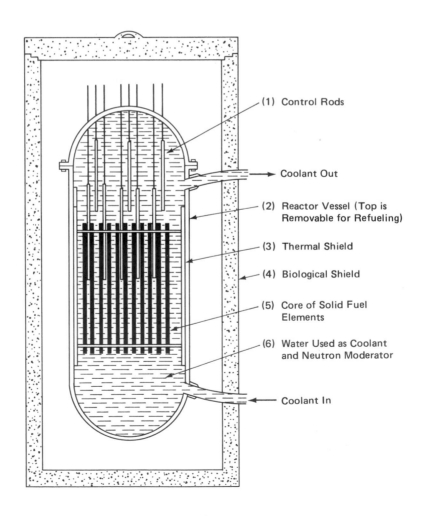

(1) Control Rods

Coolant Out

(2) Reactor Vessel (Top is Removable for Refueling)

(3) Thermal Shield

(4) Biological Shield

(5) Core of Solid Fuel Elements

(6) Water Used as Coolant and Neutron Moderator

Coolant In

Since we know the energy released in each fission is approximately 200 MeV, we have to first calculate the number of atoms in 1 g of uranium. (Note that 235 g = 1 mole for uranium-235.)

Number of atoms in 235 g of uranium

$$= 6.022 \times 10^{23} \text{ atoms/mole.}$$

Therefore, the number of atoms in 1 g is

$$\frac{6.022 \times 10^{23} \text{ atoms/mole} \times 1 \text{ g}}{235 \text{ g/mole}} = 2.56 \times 10^{21}.$$

Energy released by 1 g

$$= (\text{number of atoms}) \times (\text{energy/fission}).$$

$$= (2.56 \times 10^{21} \text{ atoms}) \times (200 \frac{\text{MeV}}{\text{atom}})$$
$$\times (1.6 \times 10^{-13} \frac{\text{J}}{\text{MeV}})$$

$$= 8.192 \times 10^{10} \text{ J.}$$

Since J/s = W, J = W · s and 3600 J = W · h.

Energy released by 1 g

$$= \frac{8.192 \times 10^{10} \text{ J}}{3600 \text{ J/W} \cdot \text{h}} = 2.27 \times 10^{7} \text{ W} \cdot \text{h.}$$

25.7 FUSION

Combining two light nuclei to produce a heavy nucleus is known as fusion. Nuclear fusion is a possible source of energy. In the low-Z region (Z < 10) the binding energy per nucleon of a nucleus formed by combining two light nuclei is higher than the binding energy per nucleon of the individual light nucleus and hence energy will be released during such a combination.

Let us consider fusion reactions between deuterium nuclei.

$$^{2}_{1}H + ^{2}_{1}H \rightarrow ^{3}_{2}He + n + 3.2 \text{ MeV}; \qquad (25.4)$$

$$^{2}_{1}H + ^{2}_{1}H \rightarrow ^{3}_{1}H + ^{1}_{1}H + 4 \text{ MeV.} \qquad (25.5)$$

These two reactions are equally probable. Half of the reactions produce tritons $(^{3}_{1}H)$ which can

Figure 25.14. Details of the core of a pressurized water reactor (left) and block diagram of a nuclear power plant (right).

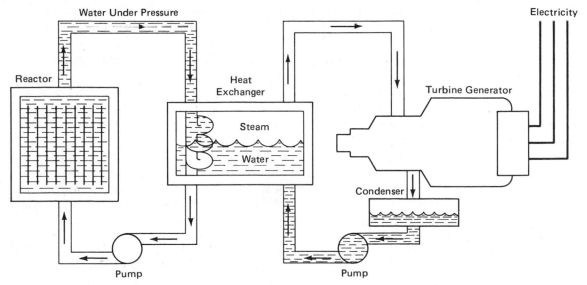

react with another deuteron as follows

$$_1^2\text{H} + _1^3\text{H} \rightarrow _2^4\text{He} + n + 17.6\,\text{MeV}. \qquad (25.6)$$

Thus, starting with five deuterons, we may end up with two helium isotopes, one proton, two neutrons and 24.8 MeV of energy. The energy output per nucleon (2.48 MeV) in these reactions is twice as much as in fission (0.9 MeV).

The fuel for the above fusion reactions is deuterium which is easily available in water. The ratio of the number of hydrogen atoms to deuterium atoms in water is 6500 to 1. Energy that could be produced by the fusion of the deuterium nuclei present in a gallon of water is enormous, almost equal to that obtainable by burning 3000 gallons of gasoline.

In spite of the above mentioned attractive features of fusion—high energy output and availability of the fuel—fusion reactors for power generation have not become feasible. The major difficulty with fusion reaction is that it is a reaction between two positively charged particles and hence the participating particles should have enough kinetic energy to overcome the electrostatic repulsion. It has been experimentally determined that, for fusion reaction to take place between deuterons with a high probability, the energy should be of the order of 10 keV. Accelerators can be used to increase the energy of deuterons to 10 keV, but the energy required to operate an accelerator is much higher than the energy released from fusion reaction. Another approach under study is to heat the deuterium ions so that the thermal energy of at least a fraction of deuterium will be enough to start the fusion reaction. Fusion reactions started by thermal energy is called thermonuclear reaction.

In the sun where fusion between hydrogen nuclei is the source of energy, fusion is maintained by thermal energy. The temperature inside the sun is about 15 million degrees Kelvin. The sun, therefore, is a fusion reactor which contains hydrogen ions, maintained at high temperature by the fusion reaction. Attempts are being made to reproduce the conditions in the sun on earth. Exploding hydrogen bombs were successful examples of such attempts. However, studies to design systems to confine ionized deuterium atoms (plasma) of sufficient density at sufficiently elevated temperatures to sustain fusion reactions for longer periods of time are still only in the experimental stage. Let us hope that these studies will produce workable systems before we run out of fossil fuels.

25.8 RADIATION HAZARDS

Before concluding this chapter, it is appropriate to discuss the effects of radiation on the human body. Specifically, attempts are made to answer the following questions

1   What are the units used in measuring the effects of radiation on the human body?
2   What are the effects of radiation on the human body?
3   What can we do to reduce exposure to radiation?

The effects induced by radiation depend on the energy it has released in a region surrounding the path of the radiation. The amount of radiation is 1 rad if the radiation released 10^{-5} J of energy in 1 g of material.

Another factor of considerable significance is the effectiveness of radiation in producing damage in a biological tissue. It is found that 1 rad of alpha particles produce 15 times more damage in the human body than 1 rad of 1 MeV x-rays. In other words, the relative biological effectiveness (RBE) of radiation should be taken into account in estimating the biological effectiveness. A unit rem (rad equivalent man) which includes the relative biological effectiveness is defined in the following way:

$$\text{rem} = \text{RBE} \times \text{rad}.$$

The human body is a complex system containing

a large number of molecules: among them the simplest and most abundant is the water molecule—perhaps the most important; also there is the very complicated DNA molecule which contains the blueprint of the living system. Molecules are made up of atoms; the electrons in the outer atomic shells can be considered as the glue holding the atoms together in a molecule. Incident radiation can directly interact with an electron and remove it, thereby producing a structural breakdown of the molecule. The type and number of products of this breakdown vary from interaction to interaction. In the case of water one may get several primary interaction products, such as electrically charged ions (H^+, OH^-, H^-, OH^+) and free radicals $(H^{\bullet}OH^{\bullet})$. These ions and free radicals are highly reactive. By reacting among themselves, they may produce hydrogen peroxide, which is extremely poisonous, and in the presence of oxygen a highly reactive HO_2 radical may be produced.

The cells in the human body contain a large number of chemicals. If these chemicals are damaged, they may be replaced by new ones within the cells as long as the control centers remain intact. However, if the molecule affected is the DNA molecule, the cell will not be able to reproduce itself and will eventually die. If, for example, the damage due to radiation is extensive in a certain tissue, and if the undamaged cells are not able to reproduce in large numbers to maintain the normal function of the tissue, the tissue will die. The death of this tissue, if sufficiently important to the organism, may lead eventually to the death of the organism. If the amount of radiation dosage received by the whole body is 10,000 rad, death is a certain consequence.

An exposure dosage in the range of 400 rem over a short period of time will result in the death of 50% of those exposed due to radiation sickness and, therefore, it is known as LD-50 (lethal dose for 50%).

Damaged cells multiply, sometimes in an uncontrolled fashion, producing more defective cells which result in cancer. A single dose of 200 rem increases the chance of leukemia threefold within 10 years after exposure. Other side effects for dosages below the LD-50 range are anemia, cardiovascular disorders, cataracts, sterility, and decrease in life expectancy.

Another type of effect, known as the genetic effect, is produced if sex cells are damaged by exposure to radiation. If the damage to a cell is not extensive enough to kill the cell, one may expect mutations in the chromosomes. Most of the time mutations are deleterious and radiation-induced mutations are adding to the naturally occurring mutations, thus increasing what is known as the "genetic load" of humanity. Increasing the genetic load will result in increasing the number of birth defects in generations to come. It is important to remember that it takes only a single radiation to damage a single cell and, therefore, to reduce genetic damage the exposure to radiation should be kept to a minimum from childhood to at least 30 years of age.

Another subtlety of genetic effects is discussed here. Quite often when a few cells in some part of the body are damaged, a sufficient number of cells will be produced to take over the functions of the damaged cells. However, such a mechanism does not work in the case of damaged sex cells. Assume that one cell out of a million sex cells is damaged. The chance of that cell getting fertilized is one in a million, but if that one is fertilized the damage will be transmitted to the next generation.

The International Commission of Radiological Protection has suggested the following upper limits of exposure: for radiation workers, 5 rem/yr and not more than 3 rem/13 week; for the general population, 0.5 rem/yr.

The above discussion is intended to make the reader aware of the dangers of radiation, both x-rays and nuclear radiation. By government regulations, when radiation sources are used in industry the public is warned of the dangers by posting the yellow radiation warning sign. However, there are some sources which the public regularly use without fully understanding the dangers, such as color television sets that produce x-rays when the electrons strike the viewing screen.

Some simple rules we should always observe when working with radiation sources are the following:

1 Keep a safe distance away from the source. Since the sources are isotropic, radiation flux and hence the rem decreases proportionally to the square of the distance ($\sim 1/r^2$).

2 Reduce the time of exposure as much as possible, that is, do not stay near a source when there is no need to stay.

3 If possible shield yourself from sources. For example, lead bricks kept in front of gamma sources reduce the intensity of radiation.

SUMMARY

A nucleus contains neutrons and protons. Nuclear forces are stronger than electrostatic forces and hence they are able to overcome the repulsion between protons in the nucleus.

Isotopes are chemically similar atoms, but the number of neutrons inside the nuclei and hence their masses are different.

Binding energy is the energy required to separate the nucleons.

Radioactivity, the emission of radiation, results from the spontaneous transformation of nuclei. There are three types of radioactive decay—alpha (α), beta (β), and gamma (λ).

The half-life is the time for half of the nuclei to decay. The activity rate of decay is the number of nuclei decaying per unit time. The units of activity are the becquerel [1 Bq = 1 dps (disintegration/s)] and the curie (1 Ci = 3.7 $\times$ 10^{10} dps).

Fission is the splitting of a nuclei into two almost equal parts; neutrons (2.5 is the average number) and gamma rays are also emitted during fission. About 200 MeV is released in fission and this energy is harnessed in nuclear reactors. Fusion is the combining of light nuclei to form heavy nuclei with an accompanying energy release. The energy of radiation that is released during radioactive decay is converted to heat and then to electricity in simple SNAP systems; however, the efficiency is very low.

Einstein's mass energy relationship

$$E = mc^2$$

is used several times in this chapter to calculate the energy released when mass changes occur. The appropriate units for the calculations are

$$E \text{ (J)} \quad = m \text{ (kg)} \ c^2 \text{ (m}^2/\text{s}^2);$$

$$E \text{ (MeV)} = m \text{ (u)} \times 931.$$

One should be familiar with the use of these equations.

25.1 It is not good to sit too close to a color television. Why?

25.2 Compare ionization energy and binding energy.

25.3 Calculate the binding energy of $^{12}_{6}C$.

25.4 Obtain the binding energy of $^{236}_{92}U$.

25.5 $^{3}_{1}H$ is radioactive. It decays to $^{3}_{2}He$. Write the equation for the radioactive decay. Calculate the maximum energy of electrons released in this decay.

25.6 A radioactive nucleus has a half-life of 2 days. Obtain the fraction of radioactive nuclei left after 8 days.

25.7 A sample has an activity of 10 mCi. If the half-life is 2 min, calculate the number of radioactive nuclei in the sample.

25.8 1000 Ci of ^{90}Sr is kept in a container thick enough to stop all the beta rays. (a) Calculate the number of strontium nuclei decaying per second. (b) How much energy is carried away by the radiation per hour? (c) Convert this energy into calories.

25.9 5000 Ci of ^{242}Cm is used in a power system. Calculate the heat energy generated per hour. ^{242}Cm is alpha active, its half-life is 8×10^{3} yr and the energy of alpha particles is 6.1 MeV.

25.10 The radioactive nuclei ^{210}Po ($T_{1/2} = 138$ days) are sometimes used for SNAP-like systems. Calculate how many curies of this source is needed to produce 1 mW of power. The energy of alpha particles from this source is 5.3 MeV.

25.11 One milligram of uranium is used to produce energy by fission. How much energy is produced?

25.12 When an electron and positron combine their mass converts to energy in the form of two gamma rays. Calculate the energy of the gamma rays. (This is called annihilation.)

25.13 The phenomenon known as pair production is a reversal of the annihilation process; that is, a gamma ray converts to an electron and positron. Calculate the minimum energy of the gamma ray that can convert to an electron and positron.

25.14 Show that the energy released in a fusion reaction represented by Eq. (25.4) is 3.2 MeV.

25.15 Show that 4 MeV is released in the fusion reaction represented by Eq. (25.5).

Electronics

It is often heard that "we live in an age of electronics." Perhaps even this statement needs further qualification today; we are actually living in an age of *solid state* electronics. For illustration look at all the electronic devices that help you in the short time between awakening and leaving for school or work: the solid state radio and timer wakes you up, the solid state microwave oven cooks your breakfast, the solid state electronic "genie" opens your garage, the solid state ignition starts your car, and the list goes on.

Technically speaking electronics means controlling the flow of electrons in a vacuum tube, a semiconductor, or any other medium to achieve a certain result such as amplification, rectification, oscillation, etc. Present day electronics uses solid state (semiconductor) components mainly while two decades ago they were curiosity items. Vacuum tubes were the major components then. We will attempt to show the difference between these two types of components and discuss some of their applications. However, the question "How do they work?" is the relevant one for the beginner. The emphasis here is to answer this question.

26.1 ELECTRONIC STRUCTURE OF SOLIDS

Electrical properties of three types of materials—conductors, semiconductors and insulators—are of interest to electrical engineers. In conductors, the electrons in the last occupied atomic orbit detach themselves and form a "sea" of free electrons. The attractive force between these electrons and the positive ions holds the atoms together in a metallic crystal (Fig. 26.1). None of these free electrons belong to an atom; instead they are shared collectively by the neighboring atomic ions. The freedom of electrons to move around easily makes metals good thermal and electrical conductors.

In most of the insulators and semiconductors, the force holding the atoms together is known as the covalent force. A simple example of such a force is one that exists between two hydrogen atoms in a hydrogen molecule. When two hydrogen atoms are brought together to form a molecule, the electrons see no distinction between the two protons. The electrons start to exchange rapidly between the two atoms. This causes an increase in the electron density in a region (Fig. 26.2) between the protons and a resulting increase in attraction for the protons towards that region. This increased attraction caused by the exchange of electrons is called the covalent force. In the representation of the covalent binding of two atoms in Fig. 22.2(b), the two lines between the atoms indicate the exchange of the two electrons. The carbon atoms in diamond (a perfect insulator) and germanium and silicon atoms (semiconductors) have four electrons to contribute to the exchange interaction between the neighboring atoms. A two-dimensional representation of the atomic arrangement in these materials is shown in Fig. 26.3. The electrons participating in the binding are called valence electrons.

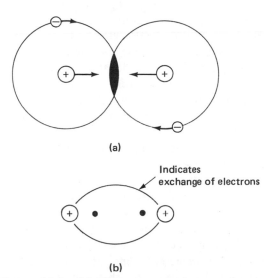

(a)

Indicates exchange of electrons

(b)

Figure 26.2. (a) Exchange of electrons between atoms of a hydrogen molecule causes an increase in electron density in the region between the atoms. Therefore the nuclei experience an increased attraction to this region. (b) Representation of an exchange interaction.

Figure 26.3. Two-dimensional representation of a germanium and silicon crystal. When an electron is raised from the valence to the conduction band, a hole (h) is created. Both contribute to conduction.

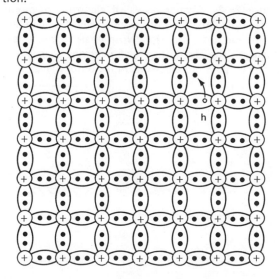

Figure 26.1. Electrons in a metal are shared collectively by neighboring atoms. These electrons are free to move around.

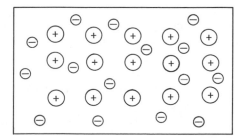

The free electrons which contribute to electric conduction are called conduction electrons.

To understand fully the electrical properties, we need to look at the modifications of the atomic energy levels that result from bringing the atoms together in a solid. As the atoms are brought together, the outer electrons start to interact with the neighboring atoms. This interaction modifies the electron energy levels into bands. Since the interaction with other atoms due to overlapping is stronger for outer electrons, these electron bands are broader than the inner electron bands (Fig. 26.4). The gap between energy bands is called the band gap or the forbidden region. Ordinarily electrons are not found in the band gap.

Schematic representations of energy bands, illustrating the differences between conductors, semiconductors, and insulators are shown in Fig. 26.5. In conductors, the conduction bands are partially filled (sometimes this happens when the valence band and conduction band overlap). Electrons in the partially filled band can accept small amounts of energy and move up to the empty part of the conduction band. This happens when a conductor is placed in an electric field. The gain in energy in the direction of the field produces electric conduction. It can be shown that the electrons in a

filled band cannot contribute to electrical conduction.

The major difference between the insulator and the semiconductor is in the magnitude of the energy gap. The energy gap of the insulator is much bigger than that of the semiconductor. For example, diamond has an energy gap of about 5 eV. The gap of a semiconductor is much smaller, for example, about 0.7 eV for germanium and 1.1 eV for silicon. Thus at room temperatures, because of their thermal energy, a large number of electrons in the upper part of the valence band can cross over the gap into the conduction band. At room temperature, about 10^{16} electrons in every cubic meter of silicon and 10^{19} electrons in every cubic meter of germanium are in the conduction band. The number of electrons in the conduction band increases with temperature. Consequently, conductivity of semiconducting materials is highly temperature dependent, increasing linearly with temperature in a range around room temperature. Such semiconductors which are relativity pure are called intrinsic semiconductors.

In addition to electrons in the conduction band, there are vacancies left in the valence band, resulting from the electrons entering the conduction band. These vacancies at the upper surface of the valence band behave like positive electrons and also contribute to the conduction as explained below. These vacancies are called "holes." Thus in the intrinsic semiconductor, both the electrons in the conduction band and the holes in the valence band contribute to the electrical conduction.

Conductivity can further be increased by adding impurity atoms into the crystals. The word "impurity" implies that these atoms are not ordinarily found in the crystal. Adding such atoms to take the place of the regular atoms is called doping. If an atom like phosphorous, with five valence electrons, is substituted for a silicon atom, one of the electrons will not participate in the binding of the atoms; only four electrons are needed for the binding. The fifth electron now is very loosely bound to the impurity atom with a very small binding energy (~0.045 eV). The electrons of the impurity

Figure 26.4. Energy levels of electrons in atoms (left). In solids the outer energy levels broaden due to interactions (right). Note that there is less change in the energy levels of electrons deep in the atom compared to electrons in the outer shells.

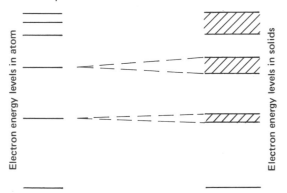

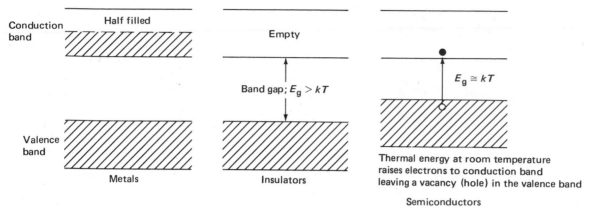

Figure 26.5. Energy bands for metals, insulators, and semiconductors.

atoms in silicon can be regarded as located at an energy level 0.045 eV below the bottom of the conduction band (Fig. 26.6). Usually at room temperature there is more than sufficient energy for the electron to jump into the conduction band from the impurity level (this can take place for temperatures below room temperature). Thus each substituted phosphorous atom can contribute an electron to this band. The atoms that donate electrons to the conduction band are called donor atoms. A semiconductor crystal doped with donor atoms is called an n-type semiconductor.

On the other hand, if an atom such as an indium atom with only three valence electrons is substituted for a silicon atom, one of the four covalent

bonds leading from the indium atom to the four silicon atoms is incomplete (Fig. 26.7). The indium atom is called an acceptor atom since it can take up an electron from the valence band, leaving a vacancy in this band. The impurity level in this case is located ~ 0.05 eV above the top of the valence band. A silicon crystal doped with indium atoms is a p-type semiconductor. The vacancy left in the valence band when an electron is excited into the acceptor impurity level is called a "hole."

If a p-type semiconductor is kept in an electric field, an electron, moving in the opposite direction to the field, can move into the hole and thus shift the hole in the direction of the field. In effect, holes move in the electric field like a positive

Figure 26.6. (a) In an n-type semiconductor, the donor atom has one more electron than needed for covalent binding with the neighboring atoms. (b) The donor electron level is close to the conduction band. At room temperature most of the donor electrons are in the conduction band, the donor atoms being positively ionized.

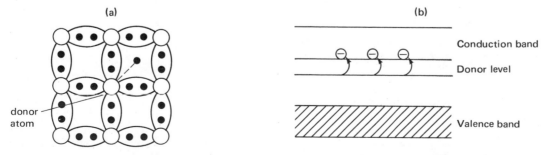

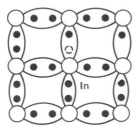
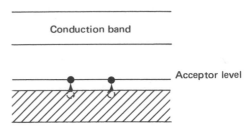

Figure 26.7. In a p-type semiconductor, an acceptor atom is substituted for a regular atom. An electron from another atom can move in to complete the binding of the acceptor atom. The energy required for this transfer of electron is very small (see the level diagram). A hole is created during this process.

charge and contribute to the conduction. In p-type materials conduction due to holes predominates over electron conduction (Fig. 26.8).

26.2 DIODES AND RECTIFIERS

So far in our discussion, we considered charge carriers—electrons and holes—inside a material. In vacuum tubes, the electrons are the charge carriers. Electrons released from the cathode travel through an evacuated region (empty space inside a glass envelope) to the plate or other electrodes. How are the electrons produced?

A metal coil called the filament, located inside the glass envelope, is heated by electric current. As the filament gets hot, the electrons gain thermal

Figure 26.8. An electron moving into a hole is equivalent to a hole moving along the electric field. A hole behaves like a positively charged particle in an electric field.

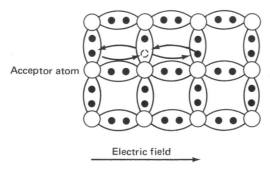

energy. Some of them will have sufficient energy to escape the metal. This process, somewhat like boiling, is called thermionic emission. In terms of the band diagram (Fig. 26.5), thermal energy is sufficient for some of the electrons to acquire enough energy (equal to or greater than $\frac{1}{2} \times$ width of the conduction band) to escape the metal. The number of electrons escaping from a given area is a function of the temperature of the filament and the properties of the material used for the filament. A pure tungsten filament has to be operated at 2100°C to get relatively good electron emission while thoriated tungsten (tungsten with thorium impurity atoms) needs to be operated at 1700°C. Electron emission can also be increased by coating the filament with oxides of barium, strontium, etc.

The electrons released by thermionic emission collect in a space near the filament, forming what is known as space charge. Negative space charge inhibits further electron emission.

The simplest method of collecting these electrons is to add another electrode as shown in Fig. 26.9. This electrode is called a plate. When the plate voltage is positive with respect to the filament, the plate attracts electrons to it. The number of electrons reaching the plate depends on the potential difference between the filament and the plate (Fig. 26.9). If the polarity of the plate is negative, the electron flow stops. From Fig. 26.9, it is evident that the diode does not obey Ohm's law. The plot of current versus voltage is not a straight line for positive voltages and the current does not flow for negative voltages. The second

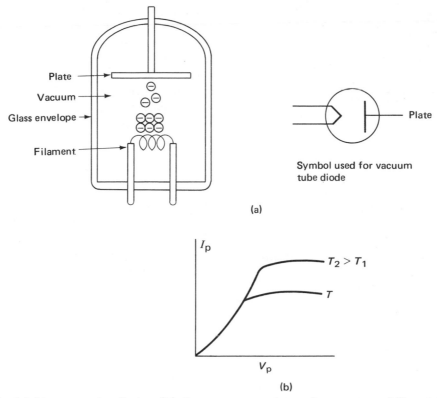

Figure 26.9. (a) Vacuum tube diode. (b) Current versus plate voltage at two different filament temperatures.

property of the diode is called rectification. Rectification means allowing current to flow in one direction only.

The juxtaposition of a region of n-type semiconducting material and a region of p-type semiconducting material forms a junction at the interface with rectifying properties. To understand the rectifying property, let us start with two pieces of n-type and p-type material (Fig. 26.10). In the n-type material the conduction electrons contributed by the donor atoms predominate. The crystal as a whole is neutral. Representation of an n-type crystal with lots of free electrons in it does not mean it is negatively charged. Similarly holes (+) contribute to the conduction in a p-type crystal, but the crystal as a whole is again neutral.

When a junction is formed, free electrons diffuse from the n-type to the p-type material and holes diffuse from the p- to n-type material. The holes diffused into the n-side and the electrons diffused into the p-side form electrically charged thin layers on both sides of the junction. The layer on the n-side is positive because of the flow of electrons *out* of that layer and the flow of holes *into* that layer. For similar reasons the layer on the p-side is negative. Let us now restrict our discussion to electrons only, keeping in mind similar effects are present for the holes also. The electrons, which are more abundant on the n-side, find it difficult to diffuse to the p-side because of the presence of this negatively charged layer. In other words, there is an electrostatic potential barrier to the continuous flow of electrons. However, one should

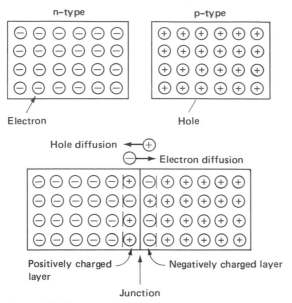

Figure 26.10. n-p Junction.

not consider it a stable situation; small diffusion currents of both charges exist in both directions but they balance each other as shown in Fig. 26.11.

Connecting a potential difference with + terminal to the n-side and − terminal to the p-side increases the electrostatic potential at the junction, thereby reducing the electron flow from the n- to p-side to zero (Fig. 26.11). This is called reverse biasing.

On the other hand, connecting a + terminal to the p-side and a − terminal to the n-side (forward biasing) reduces the electrostatic potential barrier and the current flow increases exponentially with voltage (Fig. 26.11). Thus the n-p junction has rectification properties.

Both vacuum tube diodes and semiconductor diodes are used in converting ac to dc. A single

Figure 26.11. (a) Potential seen by the electrons on both sides of a p–n junction. The current flow due to diffusion in both directions results in a zero net current. (b) Reverse biasing increases the barriers for the electrons in the n–type crystal and reduces their flow into the p–type crystal. (c) Forward biasing decreases the potential barrier. The number of electrons that can cross the barrier increases. (d) Effect of biasing on the current flow in a diode.

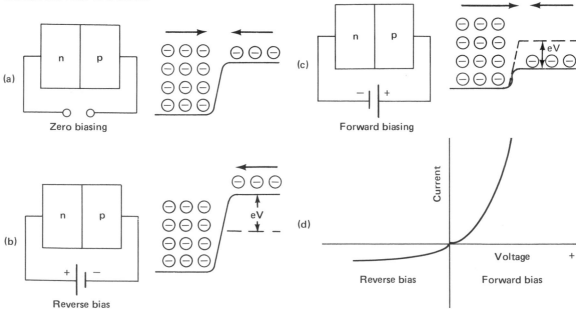

diode connected to an ac source is shown in Fig. 26.12(a); a similar circuit using a vacuum tube diode is shown in Fig. 26.12(b). In both cases voltage appears across the load resistor only during half of the cycle. Figure 26.13 shows a two diode circuit in which there is current through the load during both halves of the cycle; i.e., there is current through one diode in the first half of the cycle and current through the other diode in the second half of the cycle.

Figure 26.12. (a) An ac to dc converter using a diode. (b) An ac to dc converter using a vacuum tube diode. (c) Current flow through the load resistor as a function of time.

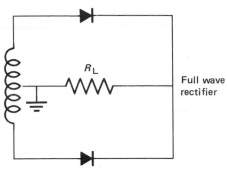

Full wave rectifier

Figure 26.13. Current flows through the load resistors during both halves of the cycle. In the first half through the top diode; in the second half through the bottom diode.

In Fig. 26.14, a bridge circuit of four diodes is used to get full wave rectification. The current flow paths in both halves are indicated in Fig. 26.14.

A pulsating voltage output from a rectifier circuit can be made smoother electronically. Adding a capacitor is one of the simplest methods. Since the capacitor is connected in parallel to the load resistor, its voltage does not decrease with the input voltage but decreases slowly, the rate of decay being decided by the time constant. Therefore a longer time constant RC will give a smoother output (Fig. 26.15).

Figure 26.14. A full wave rectifier using four diodes in a bridge circuit. The paths of the current flow in both halves of the cycle are indicated in the figure.

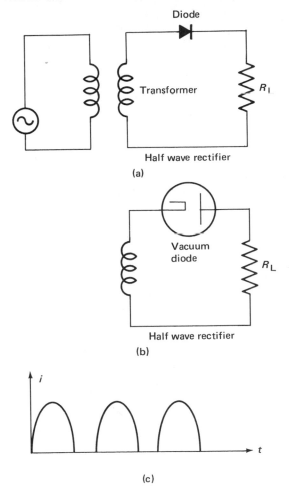

Diode

Transformer

R_1

Half wave rectifier

(a)

Vacuum diode

R_L

Half wave rectifier

(b)

(c)

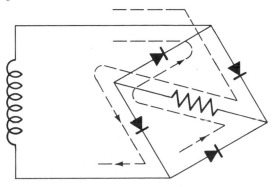

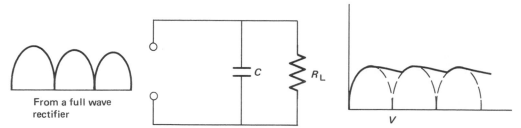

Figure 26.15. A capacitor in parallel to the load resistance reduces the ripple of the voltage output.

26.3 TRIODES AND TRANSISTORS

In a vacuum tube, a third electrode introduced between the cathode and the plate converts it from a diode to a triode. This electrode is called a grid. The grid is a helical coil with spacing between each coil to allow the electrons to pass through without hindrance to the plate (Fig. 26.16). However, because of its closeness to the filament, the grid potential controls more effectively the current flow than the plate potential. For example, if the grid is made negative by a small amount with respect to the filament, the grid repels electrons back to the filament, producing a big change in the current flow to the plate even through the plate may be highly positive.

A transistor contains three distinct parts that are analogous to a triode. The outer sections are both

the same type of material, either n or p, and the middle one is the opposite type, p or n. Therefore we have two types of transistors, designated npn or pnp in the order of the type of material in it (Fig. 26.17). The middle part, called the base, controls the current flow through the transistor just like the grid in the vacuum tube. The other two parts, called the emitter and collector, are analogous to the cathode and the plate in the tube, respectively.

To analyze the operation of a transistor let us consider an npn transistor biased as shown in Fig. 26.18. The first junction, the emitter-base junction, is forward biased and the second junction, the base-collector junction is reverse biased. The electrostatic potential as seen by an electron moving from the emitter to collector before and after

Figure 26.16. Vacuum tube triode.

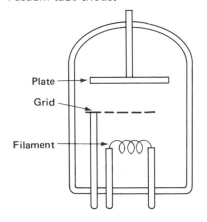

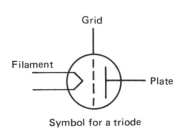

Symbol for a triode

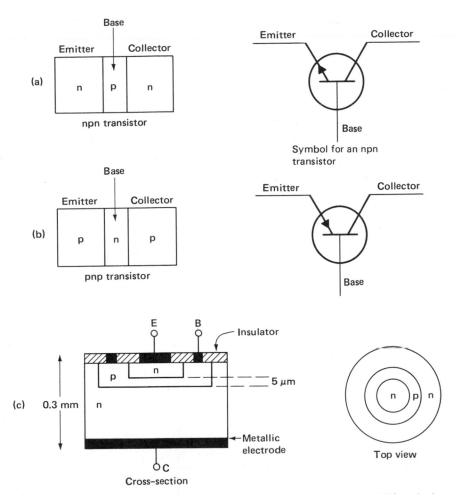

Figure 26.17. (a) An npn transistor; (b) pnp transistor; (c) structure of a diffused planar-type transistor.

biasing is shown in Fig. 26.18. Forward biasing of the first junction reduces the only barrier to electron flow. Since the potential of the collector is lower than that of the base, once the electrons reach the base, most of the electrons continue on to the collector. Therefore, the electron flow is controlled by the base. Small potential changes at the base can result in big changes in current flow.

The number of free electrons in the emitter effectively decides the maxiumum current. For this purpose, the emitter is highly doped. The base is usually made thin and lightly doped so that few electrons are lost to the base by recombination, etc. and most of the electrons reach the collector. The collector current I_c is very close to the emitter current I_e and is given by [Fig. 26.18 (a)] .

$$I_e = I_b + I_c , \qquad (26.1)$$

where I_b is the base current, I_c the collector current, and I_e the emitter current. The ratio of I_c to I_e is called alpha (α).

$$\alpha = I_c/I_e.$$

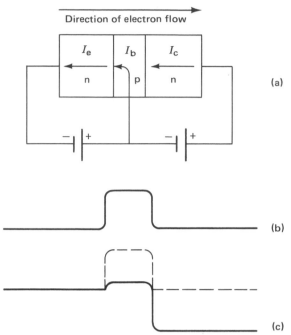

Figure 26.18. (a) Normal biasing of a transistor. (b) Electrostatic potential seen by electrons without biasing. (c) Electrostatic potential seen by electrons on biasing as shown in (a).

This number, supplied by the manufacturer for each transistor, is very close to 1, typically in the range of 0.95 to 0.99.

As mentioned in the above discussion, a signal (a voltage change) appearing at the grid or the base is amplified in the vacuum tube or the transistor. Two simple amplifier circuits, one for the tube and the other for the transistor, are shown in Fig. 26.19. In both cases the output signal is inverted. A positive increase in the base bias (grid bias) increases the current flow through the transistor (vacuum tube) and the load resistor reducing the voltage at the output point.

The small increase in the base bias increases the base current. This increase in an already small base current is extremely small, but this small change produces a big change in the collector current. The current gain, the ratio of the current in the collector circuit to the current in the base

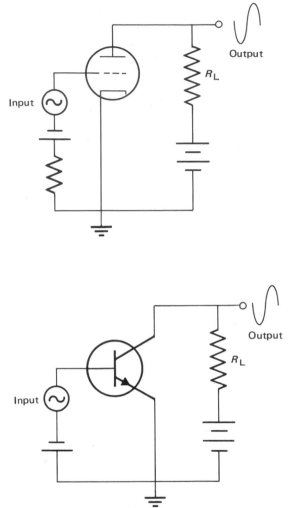

Figure 26.19. Simple amplifiers using a triode and a transistor.

circuit, can be shown to be given by

$$A_i = \frac{\alpha}{1-\alpha} , \qquad (26.2)$$

where A_i stands for the current amplification or gain.

Transistors and diodes have several advantages over their counterparts in vacuum tubes: (a) they are very compact; miniaturization has now progressed to a stage that a small "chip" the size of an ordinary stamp may contain thousands of components; (b) power consumption is very small in semiconductor devices because they do not have filaments; (c) the semiconductors operate on very small dc voltages (few volts) while the vacuum tubes require dc power supplies in the 100–300 V range for their operation. However, the transistors can not handle large power; tubes are still used in circuits for this latter purpose.

26.4 CATHODE RAY TUBE (CRT)

Wherever you see an electronic visual display—TV, oscilloscopes, computer read outs, etc.—you are looking at a cathode ray tube. A spot on the screen of a CRT is seen as bright when high-energy electrons hit that spot and convert their kinetic energy into light energy. By deflecting the electron beam up and down and/or sideways, the position of the bright spot can be moved around on the screen. Thus by controlling the position of the beam we can make the electron beam draw or write on the screen.

A cathode ray oscilloscope has the following major components to perform the functions mentioned above (Fig. 26.20): (a) An electron gun, which contains a filament and electrodes to focus and accelerate electrons to the desired velocity. In color television the electrons are accelerated to 20–35 keV. (b) Two sets of deflection plates, one for deflecting electrons in the horizontal direction and the other set for the vertical direction. The deflection in each case is proportional to the voltage applied across the plates. In some cases magnetic fields are used to deflect electrons. (c) A light emitting screen. The screen consists of a layer (behind the front face of the glass envelope) of phosphorescent material. The phosphor material is capable of producing light when hit by electrons and transmitting the light with very little absorption. Three different types of phosphor materials are used in color television screens to give three basic colors.

In oscilloscopes, we have access to the CRT through two inputs. Signals connected to the Y input (vertical deflection input) deflect the bright spot vertically while the horizontal input controls the horizontal deflection. Ordinarily we do not use the horizontal input. It is internally connected to a saw tooth pulse generated inside the oscilloscope. A saw tooth pulse (Fig. 26.21) is a linearly increasing pulse, which returns to zero voltage from a maximum value at the end of a predetermined time. The saw tooth pulse shown in Fig. 26.21 has a V_{max} of 10 V and a period of 0.1 s.

Figure 26.20. Details of a cathode ray tube used in a cathode ray oscilloscope.

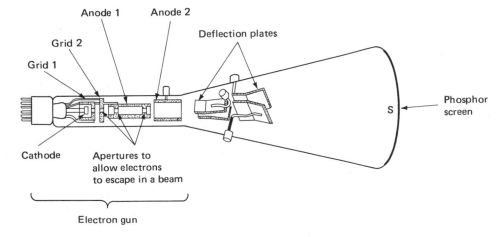

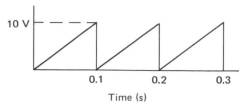

Figure 26.21. Saw tooth pulses used for horizontal sweep of electron beam in a cathode ray oscilloscope.

Let us suppose 10 V deflects the electron beam full scale horizontally. If the pulses shown in Fig. 26.21 is connected to the horizontal plates the bright spot moves from its zero position to the maximum deflection position in 0.1 s. At this point, since the voltage decreases to zero very fast, the return of the electron beam to the starting position takes place in a very short time. The sweep time can be controlled by an external knob. When the knob is turned to shorter sweep time, the saw tooth pulse generated inside has a shorter period.

The sweep time for the pulse in Fig. 26.21 is 0.1 s.

Suppose we connect a 60 cycle ac source to the vertical input and the time knob is adjusted to 0.1 s for full scale horizontal deflection. We will see the scope tracing the ac signal, the maximum deflection being proportional to the peak voltage of the ac. The number of cycles seen is given by

$$\text{Number of cycles} = \frac{\text{Sweep time}}{\text{Period of the ac}}$$

In the example we are discussing we will see 6 $[= \frac{0.1}{(1/60)}]$ complete cycles in one complete sweep. By adjusting the amplification of the input circuit, full scale vertical deflection can be obtained for several peak voltages. The knob controlling the amplification has different positions such as 0.1 V/cm, 1 V/cm, 10 V/cm, etc., which means that when the knob is set at 0.1 V/cm, an input of 0.1 V gives 1 cm deflection. The oscilloscope can thus be used to measure time-varying voltages and the periods of such voltage inputs.

SUMMARY   The following are important tems and definitions which were discussed in this chapter.

1 Intrinsic semiconductor—conductivity depends on the electrons in the conduction band which moved up from the valence band.

2 Donor atoms—atoms that have more electrons than required for binding; they readily donate electrons to the conduction band.

3 Acceptor atoms—atoms that do not have enough electrons for binding; they readily accept electrons from the valence band creating holes.

4 n-Type semiconductor—semiconductor doped with donor impurities.

5 p-Type semiconductor—semiconductor doped with acceptor atoms.

6 Rectification—current flow allowed in one direction only.

7 Thermionic emission—emission of electrons upon heating a material.

8 Rectifying junction—junction formed at the interface of an n-p semiconductor.

9 Diode (transistor) has three elements—the grid (base) controls the electron flow and the filament and collector are analogous to the filament and plate, respectively. Voltage variations at the grid (base) appear amplified at the plate (collector). A cathode ray tube uses an electron beam to produce visual display.

QUESTIONS AND PROBLEMS

26.1   Based on the band diagram of solids can you explain why light emitted by heated solid materials has a continuous spectrum while a gas emits a line spectrum.

26.2   An ac of 6.3 V is connected to a four transistor bridge. Calculate the maximum dc voltage one can obtain with this circuit.

26.3   List a few donor impurity atoms.

26.4   List a few acceptor impurity atoms.

26.5   The alpha (α) of a transistor is 0.98. Calculate the ratio of emitter current to collector current in this transistor. Calculate the current amplification factor if this transistor is used as an amplifier (Fig. 26.19).

26.6   In a transistor the collector current is 1.5 mA for a base current of 0.02 mA. Calculate α and the current amplification factor for this transistor.

26.7   The period of a saw tooth pulse used for the horizontal sweep in an oscilloscope is 1 s. How many complete oscillations will be seen on the screen if a 50 Hz ac is connected to the vertical input of an oscilloscope.

26.8   An oscilloscope is adjusted to give a 1 cm vertical deflection for 10 V. What is the maximum deflection you get when a 20 V ac is connected to it.

26.9   Explain how you can use the oscilloscope as an ac voltmeter. Do you read effective voltage in this case.

26.10 Explain how you can use the oscilloscope as an ac ammeter.

Appendix

CONVERSION EQUIVALENTS

Length

1 meter (m) = 29.37 in. = 3.281 ft = 6.214×10^{-4} mi

1 in. = 0.02540000 m; 1 ft = 0.3048 m; 1 mi = 1609 m

Area

$1 \text{ m}^2 = 10.76 \text{ ft}^2 = 1550 \text{ in.}^2$;

$1 \text{ ft}^2 = 929 \text{ cm}^2$; $1 \text{ in.}^2 = 6.452 \text{ cm}^2 = 1.273 \times 10^6$ circular mils

Volume

$1 \text{ m}^3 = 35.31 \text{ ft}^3 = 6.102 \times 10^4 \text{ in.}^3$

$1 \text{ ft}^3 = 0.02832 \text{ m}^3$; 1 U.S. gallon = 231 in.^3,

1 liter = $1.000028 \times 10^{-3} \text{ m}^3 = 61.02 \text{ in.}^3$

Time and Frequency

1 year = 365.2422 days = 8.766×10^3 h = 5.259×10^5 min = 3.156×10^7 s

Speed

1 m/s = 3.281 ft/s = 3.6 km/h = 2.237 mi/h

1 km/h = 0.2779 m/s = 0.9113 ft/s = 0.6214 mi/h

1 mi/h = 1.465 ft/s = 1.609 km/h

Mass

1 kg = 2.205 lb_m = 0.06852 slug

1 lb_m = 0.4536 kg = 0.03108 slug; 1 slug = 32.17 lb_m = 14.59 kg

Density

$1 \text{ g/cm}^3 = 1000 \text{ kg/m}^3 = 62.43 \text{ lb}_m/\text{ft}^3 = 1.940 \text{ slug/ft}^3$

$1 \text{ lb}_m/\text{ft}^3 = 0.03108 \text{ slug/ft}^3 = 16.02 \text{ kg/m}^3 = 0.01602 \text{ g/cm}^3$

Force
1 newton (N) = 0.1020 kg wt = 0.2248 lb
1 lb (force) = 4.448 N = 0.4536 kg wt

Pressure
$1 \text{ N/m}^2 = 9.869 \times 10^{-6} \text{ atm} = 1.450 \times 10^{-4} \text{ lb/in.}^2 = 0.02089 \text{ lb/ft}^2$
$= 7.501 \times 10^{-4} \text{ cm Hg} = 10^{-5} \text{ bar}$
$1 \text{ lb/in.}^2 = 144 \text{ lb/ft}^2 = 6895 \text{ N/m}^2 = 5.171 \text{ cm Hg}$
$1 \text{ atm} = 76 \text{ cm Hg} = 1.013 \times 10^5 \text{ N/m}^2 = 2116 \text{ lb/ft}^2 = 14.70 \text{ lb/in.}^2 = 760 \text{ torr}$

Work, Energy, Heat
$1 \text{ joule (J)} = 0.3289 \text{ cal} = 9.491 \times 10^{-4} \text{ Btu} = 0.7376 \text{ ft} \cdot \text{lb} = 6.242 \times 10^{18} \text{ eV}$
$1 \text{ kcal} = 4.816 \text{ joules} = 3.968 \text{ Btu} = 3087 \text{ ft} \cdot \text{lb}$
$1 \text{ eV} = 1.602 \times 10^{-19} \text{ joule}; \quad 1 \text{ unified amu} = 931.48 \text{ MeV}$
$1 \text{ kW} \cdot \text{h} = 3.6 \times 10^6 \text{ joules} = 3413 \text{ Btu} = 860.1 \text{ kcal} = 1.341 \text{ hp} \cdot \text{h}$

Power
$1 \text{ hp} = 2545 \text{ Btu/h} = 550 \text{ ft} \cdot \text{lb/s} = 745.7 \text{ watts} = 0.1792 \text{ kcal/s}$
$1 \text{ watt (W)} = 2.389 \times 10^{-4} \text{ kcal/s} = 1.341 \times 10^{-3} \text{ hp} = 0.7376 \text{ ft} \cdot \text{lb/s}$

Electric Charge
$1 \text{ faraday} = 96,487 \text{ coulombs}; \quad 1 \text{ electron charge} = 1.602 \times 10^{-19} \text{ coulomb}$

Magnetic Flux
$1 \text{ weber (Wb)} = 10^8 \text{ maxwells} = 10^8 \text{ lines}$

Magnetic Intensity
$1 \text{ tesla (T)} = 1 \text{N/A} \cdot \text{m} = 1 \text{ weber/m}^2 = 10,000 \text{ gauss}$

TRIGONOMETRIC FUNCTIONS

ANGLE (DEG)	SINE	COSINE	TANGENT	ANGLE (DEG)	SINE	COSINE	TANGENT
0°	0.000	1.000	0.000				
1°	.018	1.000	.018	46°	0.719	0.695	1.036
2°	.035	0.999	.035	47°	.731	.682	1.072
3°	.052	.999	.052	48°	.743	.669	1.111
4°	.070	.998	.070	49°	.755	.656	1.150
5°	.087	.996	.088	50°	.766	.643	1.192
6°	.105	.995	.105	51°	.777	.629	1.235
7°	.122	.993	.123	52°	.788	.616	1.280
8°	.139	.990	.141	53°	.799	.602	1.327
9°	.156	.988	.158	54°	.809	.588	1.376
10°	.174	.985	.176	55°	.819	.574	1.428
11°	.191	.982	.194	56°	.829	.559	1.483
12°	.208	.978	.213	57°	.839	.545	1.540
13°	.225	.974	.231	58°	.848	.530	1.600
14°	.242	.970	.249	59°	.857	.515	1.664
15°	.259	.966	.268	60°	.866	.500	1.732
16°	.276	.961	.287	61°	.875	.485	1.804

ANGLE (DEG)	SINE	COSINE	TANGENT	ANGLE (DEG)	SINE	COSINE	TANGENT
17°	.292	.956	.306	62°	.883	.470	1.881
18°	.309	.951	.325	63°	.891	.454	1.963
19°	.326	.946	.344	64°	.899	.438	2.050
20°	.342	.940	.364	65°	.906	.423	2.145
21°	.358	.934	.384	66°	.914	.407	2.246
22°	.375	.927	.404	67°	.921	.391	2.356
23°	.391	.921	.425	68°	.927	.375	2.475
24°	.407	.914	.445	69°	.934	.358	2.605
25°	.423	.906	.466	70°	.940	.342	2.747
26°	.438	.899	.488	71°	.946	.326	2.904
27°	.454	.891	.510	72°	.951	.309	3.078
28°	.470	.883	.532	73°	.956	.292	3.271
29°	.485	.875	.554	74°	.961	.276	3.487
30°	.500	.866	.577	75°	.966	.259	3.732
31°	.515	.857	.601	76°	.970	.242	4.011
32°	.530	.848	.625	77°	.974	.225	4.331
33°	.545	.839	.649	78°	.978	.208	4.705
34°	.559	.829	.675	79°	.982	.191	5.145
35°	.574	.819	.700	80°	.985	.174	5.671
36°	.588	.809	.727	81°	.988	.156	6.314
37°	.602	.799	.754	82°	.990	.139	7.115
38°	.616	.788	.781	83°	.993	.122	8.144
39°	.629	.777	.810	84°	.995	.105	9.514
40°	.643	.766	.839	85°	.996	.087	11.43
41°	.658	.755	.869	86°	.998	.070	14.30
42°	.669	.743	.900	87°	.999	.052	19.08
43°	.682	.731	.933	88°	.999	.035	28.64
44°	.695	.719	.966	89°	1.000	.018	57.29
45°	.707	.707	1.000	90°	1.000	.000	∞

TABLE OF THE ELEMENTS

The masses listed are based on $^{12}_{6}C = 12$ u. For an atom without any stable isotope the mass number of the long-lived isotope is given in parentheses.

ELEMENT	SYMBOL	ATOMIC NUMBER Z	AVERAGE ATOMIC MASS
Actinium	Ac	89	(227)
Aluminum	Al	13	26.9815
Americium	Am	95	(243)
Antimony	Sb	51	121.75
Argon	Ar	18	39.948
Arsenic	As	33	74.9216

ELEMENT	SYMBOL	ATOMIC NUMBER Z	AVERAGE ATOMIC MASS
Astatine	At	85	(210)
Barium	Ba	56	137.34
Berkelium	Bk	97	(247)
Beryllium	Be	4	9.0122
Bismuth	Bi	83	208.980
Boron	B	5	10.811
Bromine	Br	35	79.904
Cadmium	Cd	48	112.40
Calcium	Ca	20	40.08
Californium	Cf	98	(251)
Carbon	C	6	12.01115
Cerium	Ce	58	140.12
Cesium	Cs	55	132.905
Chlorine	Cl	17	35.453
Chromium	Cr	24	51.996
Cobalt	Co	27	58.9332
Copper	Cu	29	63.546
Curium	Cm	96	(247)
Dysprosium	Dy	66	162.50
Einsteinium	Es	99	(254)
Erbium	Er	68	167.26
Europium	Eu	63	151.96
Fermium	Fm	100	(257)
Fluorine	F	9	18.9984
Francium	Fr	87	(223)
Gadolinium	Gd	64	157.25
Gallium	Ga	31	69.72
Germanium	Ge	32	72.59
Gold	Au	79	196.967
Hafnium	Hf	72	178.49
Helium	He	2	4.0026
Holmium	Ho	67	164.930
Hydrogen	H	1	1.00797
Indium	In	49	114.82
Iodine	I	53	126.9044
Iridium	Ir	77	192.2
Iron	Fe	26	55.847
Krypton	Kr	36	83.80
Lanthanum	La	57	138.91
Lawrencium	Lr	103	(257)
Lead	Pb	82	207.19
Lithium	Li	3	6.939
Lutetium	Lu	71	174.97
Magnesium	Mg	12	24.312
Manganese	Mn	25	54.9380
Mendelevium	Md	101	(256)

ELEMENT	SYMBOL	ATOMIC NUMBER Z	AVERAGE ATOMIC MASS
Mercury	Hg	80	200.59
Molybdenum	Mo	42	95.94
Neodymium	Nd	60	144.24
Neon	Ne	10	20.183
Neptunium	Np	93	(237)
Nickel	Ni	28	58.71
Niobium	Nb	41	92.906
Nitrogen	N	7	14.0067
Nobelium	No	102	(254)
Osmium	Os	76	190.2
Oxygen	O	8	15.9994
Palladium	Pd	46	106.4
Phosphorus	P	15	30.9738
Platinum	Pt	78	195.09
Plutonium	Pu	94	(244)
Polonium	Po	84	(209)
Potassium	K	19	39.102
Praseodymium	Pr	59	140.907
Promethium	Pm	61	(145)
Protactinium	Pa	91	(231)
Radium	Ra	88	(226)
Radon	Rn	86	222
Rhenium	Re	75	186.2
Rhodium	Rh	45	102.905
Rubidium	Rb	37	85.47
Ruthenium	Ru	44	101.07
Samarium	Sm	62	150.35
Scandium	Sc	21	44.956
Selenium	Se	34	78.96
Silicon	Si	14	28.086
Silver	Ag	47	107.868
Sodium	Na	11	22.9898
Strontium	Sr	38	87.62
Sulfur	S	16	32.064
Tantalum	Ta	73	180.948
Technetium	Tc	43	(97)
Tellurium	Te	52	127.60
Terbium	Tb	65	158.924
Thallium	Tl	81	204.37
Thorium	Th	90	232.0381
Thulium	Tm	69	168.934
Tin	Sn	50	118.69
Titanium	Ti	22	47.90
Tungsten	W	74	183.85
Uranium	U	92	238.03
Vanadium	V	23	50.942

ELEMENT	SYMBOL	ATOMIC NUMBER Z	AVERAGE ATOMIC MASS
Xenon	Xe	54	131.30
Ytterbium	Yb	70	173.04
Yttrium	Y	39	88.905
Zinc	Zn	30	65.37
Zirconium	Zr	40	91.22

SELECTED ASTRONOMICAL CONSTANTS

	MASS, kg	DIAMETER, km	DISTANCE FROM SUN, km	SURFACE GRAVITY, m/s^2
sun	1.99×10^{30}	1.39×10^6		274.5862
earth	5.977×10^{24}	1.27×10^4	1.49×10^8	9.8066
moon	7.36×10^{22}	3.48×10^3		1.5447

Earth–moon distance $= 3.80 \times 10^5$ km

SI UNITS

QUANTITY	NAME OF UNIT	SYMBOL
	Base units	
length	meter	m
mass	kilogram	kg
time	second	s
electric current	ampere	A
thermodynamic temperature	kelvin	K
luminous intensity	candela	cd
	Derived units	
area	square meter	m^2
volume	cubic meter	m^3
frequency	hertz	Hz, s^{-1}
mass density (density)	kilogram per cubic meter	kg/m^3
speed, velocity	meter per second	m/s
angular velocity	radian per second	rad/s
acceleration	meter per second squared	m/s^2
angular acceleration	radian per second squared	rad/s^2
force	newton	N $(kg \cdot m/s^2)$
pressure (mechanical stress)	pascal	Pa (N/m^2)
kinematic viscosity	square meter per second	m^2/s
dynamic viscosity	newton-second per square meter	$N \cdot s/m^2$

QUANTITY	NAME OF UNIT	SYMBOL	
work, energy, quantity of heat	joule	J	(N·m)
power	watt	W	(J/s)
quantity of electricity	coulomb	C	
potential difference, electromotive force	volt	V	(W/A)
electric field strength	volt per meter	V/m	
electric resistance	ohm	Ω	(V/A)
capacitance	farad	F	(A·s/V)
magnetic flux	weber	Wb	(V·s)
inductance	henry	H	(V·s/A)
magnetic flux density	tesla	T	(Wb/m^2)
magnetic field strength	ampere per meter	A/m	
luminous flux	lumen	lm	(cd·sr)
luminance	candela per square meter	cd/m^2	
illuminance	lux	lx	(lm/m^2)
specific heat capacity	joule per kilogram kelvin	J/kg·K	
thermal conductivity	watt per meter kelvin	W/m·K	
radiant intensity	watt per steradian	W/sr	
activity (of a radioactive source)	1 per second	s$^{-1.}$	

Elementary units

plane angle	radian	rad
solid angle	steradian	sr

Answers to Problems

Chapter 1

1.4 9.6 ft, 20.3 ft, 97.4 ft^2 **1.5** 0.52, 0.86, 0.60   **1.6** 30.0 m, 36.6 m, 55°   **1.7** 230.9 ft   **1.8** 11.3°   **1.9** 57.7 ft, 57.7%   **1.10** 23 ft   **1.11** 150.38 ft^2 **1.12** 125.7 ft^3 **1.14** D_1 and D_2 in the same direction   **1.17** 32.02 m, $\theta = 2.17°$ **1.18** 15.8 mi/h, 18.4°   **1.19** 18.4°

Chapter 2

2.1 132 ft/s   **2.2** 33.14 mi/h, 0   **2.3** 40 mi/h, 53.3 mi/h, 46.7 mi/h, 40 mi/h, 46.7 mi/h, 33.3 mi/h   **2.4** 2.5 h   **2.5** 0.67 mi   **2.6** 6 m/s   **2.7** 25 mi/h, 183.3 ft   **2.8** 10.3 s, 1057.5 ft   **2.9** 1.8 m   **2.10** 6.4 s   **2.11** 2.45 s, 78.4 ft/s   **2.12** −16 ft/s   **2.13** 31.3 m/s, 3.2 s   **2.14** 50.6 ft/s   **2.15** 56.25 ft, 1.87 s, 3.74 s   **2.16** 15 ft, 0.5 s   **2.17** 2317 ft   **2.18** 43.3 m/s, 25 m/s, 2.55 s, 32 m, 220.8 m   **2.19** Velocity is a function of the angle. For 45° velocity is 176.7 ft/s.   **2.20** 62 ft/s, 30 ft   **2.21** 64 ft/s, 66.6 ft/s   **2.22** 154 m/s   **2.23** 36.1 m/s   **2.24** 22.8 m/s, 26.5 m

Chapter 3

3.3 0.625 slug, 64 ft/s^2; 160 lb, 20 lb; 3.125 slug, 62.5 lb; 49 N, 6 m/s^2; 100 kg, 2.5 m/s^2; 294 N, 150 N; 8.16 N, 833 ^{kg}; 384 lb, 12 slugs   **3.4** 140 lb, 102.5 lb, 125.64 lb   **3.5** 1.5 m/s^2, 1.4 m/s^2, 1.6 m/s^2 **3.6** 1.4 N, 14 m/s^2 **3.7** 243.8 lb   **3.8** 2.5×10^3 N   **3.9** 1.6 m/s^2, 1.6 m/s^2, 4 N   **3.10** 0.47 m/s^2, 0.47 m/s^2, 20.54 N   **3.11** 0.076 m/s^2, 3.4 N   **3.12** 4.9 m/s^2 **3.13** 4.14 m/s^2 **3.14** (a) 14.5 ft/s^2, (b) 13.7 ft/s^2 **3.15** (a) 9.8 N, 0.52 m/s^2, (b) 9.3 N, 0.48 m/s^2, (c) 10.95 N, 0.54 m/s^2 **3.16** 15 lb, 10 lb   **3.17** (a) 88.2 N, (b) 58.8 N   **3.18** 250 lb, 200 lb, 150 lb   **3.19** 9.8 m/s^2 **3.20** 1.6 m/s^2 **3.21** 19.8×10^{19} N

Chapter 4

4.1 1250 J   **4.2** kinetic energy   **4.3** 735 J, kinetic energy
4.4 2.7×10^5 ft·lb   **4.5** 1.08×10^3 N   **4.6** (a) 800 ft·lb,
(b) 16 ft/s   **4.7** 8300 ft·lb, 18600 lb   **4.8** 1.13×10^{12} J/h, 3.14
$\times 10^8$ W   **4.9** 2×10^9 ft·lb, 7.25×10^5 hp   **4.10** 1.12×10^4 W
4.11 2.5×10^3 ft·lb/s   **4.12** 400 J, 400 J   **4.13** 14.3 m/s
4.14 1.96 J, 0, 1.96 J, 0, 6.25 m/s, 0   **4.15** 49 m   **4.16** 2.94×10^3
N, 58.8 J, 58.8 J   **4.17** 128 ft/s, 114.9 ft/s   **4.18** 422 lb
4.19 5.35 N, 463 J   **4.20** 160 J, 160 J, 980 J

Chapter 5

5.6 1.1×10^4 slug·ft/s, 2.2×10^5 ft/s   **5.7** 2.82 kg·m/s, 2.82 kg·m/s,
1.88×10^3 N   **5.8** 9562 lb   **5.9** 125 lb   **5.10** 3, 3
5.11 1.5 m/s   **5.12** Inelastic, 85.3 ft/s   **5.13** 4.98 m/s
5.14 −27.5 ft/s, 2.57 ft/s   **5.15** 7.67 m/s, 2.57 m/s, 10.2 m/s
5.16 $\sim$ −100 ft/s, 0.4 ft/s, yes   **5.17** 4.3 m/s, 2.5 m/s   **5.19** 18.36
m/s, 36.5°   **5.20** 40,000 lb

Chapter 6

6.8 282.74 rad/min, 4.7 rad/s, 16964.4 rad/h, 1.6×10^4 deg/min, 269.3 deg/s,
9.7×10^5 deg/h   **6.9** 4.7 ft/s   **6.10** 314 ft/min   **6.11** 2 rev/s^2,
225 rev   **6.12** 29.3 rad/s, −0.049 rad/s^2, 8790 rad, 4.99 mi
6.15 9 kg·m^2, 12 kg·m^2 **6.16** 0.039 slug·ft^2 **6.17** 0.05 slug·ft^2
6.18 3.54 N·m, CCW, 0; 7.5 N·m, CW; 0; 13 N·m, CCW; 25 N·m,
CW; 15.96 N·m, CW; yes, CW   **6.20** 30 lb   **6.21** 39.5 N
6.22 7.17 ft/s, 76.1 lb   **6.23** 16 ft/s or 10.9 mi/h   **6.24** 13.3 ft/s, 0,
40 lb   **6.25** 1.72×10^4 mi/h, 1.52 h   **6.26** 2.49×10^4 ft/s
6.27 31.2°   **6.28** 73 mi/h   **6.29** 20 N·m, 1.25 kg·m^2, 16 rad/s^2,
80 rad/s   **6.30** 6.087 slug·ft^2, 15 rev/s^2, 8.2 slug·ft/s^2, 1.55×10^5 ft·lb
6.31 3.30×10^4 ft·lb, 56.6 ft·lb/s

Chapter 7

7.1 30.5 lb   **7.2** 73.1 lb   **7.4** 7.07 lb   **7.5** $\mu > 0.87$
7.6 60 lb   **7.7** 100 lb, 141 lb, 113 lb, 80 lb   **7.8** 72.5 lb, 197.5 lb
7.9 0.2 lb   **7.10** 50 lb, 14.4 lb, 14.4 lb   **7.11** 0.33   **7.12** 518
lb, 254 lb, 366 lb   **7.13** 0.88 ft   **7.14** 130 lb   **7.15** 65.3 lb,
52.8 lb, 38.3°   **7.16** 1833 lb, 3167 lb

Chapter 8

8.4 2.92, 124, 8 lb, 1.6, 3644 ft·lb, 1646 ft·lb   **8.5** 0.13, 2, 1.63, 82%
8.7 327 N; 3, 3, 16.7 lb, 3, 3, 250 lb, 0.4, 0.4   **8.8** 120, 100, 83%, 120
ft·lb   **8.9** 4   **8.10** 15.6 N   **8.11** 4800 rev/min, 1200 rev/min
8.12 17, 0.059   **8.13** 0.16 N·m   **8.14** 1000 lb, 1000 lb
8.15 23.6 ft·lb

Chapter 9

9.3 0.9   **9.4** 5.8 in   **9.5** 19,300 kg/m^3, 1204 lb/ft^3 **9.6** 0.7
9.7 12 kg   **9.8** 7.8, 0.79   **9.9** 2.0×10^7 Pa   **9.10** 759.8 lb/ft^3

9.11 10.34 m   **9.12** High vapor pressure   **9.14** 1.53
9.15 1000 lb, 25   **9.16** 125 lb   **9.17** 3.9×10^8 N/m^2
9.18 1×10^{-4} **9.19** 3.9×10^{12} **9.20** 5 cm   **9.21** 0.3 in.,
6.3×10^4 lb, 1.26×10^4 lb

Chapter 10
10.10 37°C   **10.11** -73.33°C, -35.6°C, 0°C, 37.8°C, 260°C
10.12 -148°F, -58°F, 122°F, 212°F, 1832°F   **10.13** 2.8 in.
10.14 299°C   **10.15** 3.1×10^{-2} cm, 2.6×10^{-2} cm; 1.5×10^{-2} cm;
1.2×10^{-2} cm   **10.16** 1.5 cm^2, 0.36 cm^2, 0.36 cm^2 **10.17** 2.8 cm^3
10.19 2.5×10^{-5}/°C, 49.975 cm, -4°C   **10.20** 0.34 in.
10.21 0.347 cm^3 **10.22** 5900 cm^3 **10.23** 8.64×10^4 N
10.24 2.88 g/cm^3 **10.25** 0.66 m^3 **10.26** 40.2 lb/in.^{2}
10.27 0.775 ft   **10.28** 43.5 kPa   **10.29** 16.1 liters

Chapter 11
11.12 4°C, 34.8°C, 18.2°C   **11.13** 25°F, 758°F, 446°F
11.14 15°C   **11.15** 1.13×10^5 Btu   **11.16** 0.28 kcal/kg·°C
11.17 22.1°C   **11.18** 0.26   **11.19** 10.5°C   **11.20** 544 cal/g
11.21 5.12×10^5 Btu/h   **11.22** 1.5 lb/s   **11.23** 0.0013 mm/s
11.24 6.4×10^{-4} kcal   **11.25** 94 kcal/s   **11.26** 3.75×10^{-2} kcal/s
11.27 0.61 kcal   **11.28** 6.68×10^{-3} kcal/s   **11.29** 3.5 times
11.30 1:1.025

Chapter 12
12.10 6.53 Btu   **12.11** 4152 J/kcal   **12.12** 1.49×10^4 ft·lb
12.13 3.7×10^9 ft·lb   **12.14** 2.85 lb   **12.15** 9.6×10^3 J
12.16 0, temperature and internal energy changes by ½.   **12.17** 4.8×10^3
12.18 4.8×10^3 J   **12.19** 62%   **12.20** 1940 J
12.21 5781 J   **12.22** 2.88×10^5 Btu/day   **12.23** 2093 J
12.24 0.8   **12.25** 7.8

Chapter 13
13.13 3.92 N/m   **13.14** 2 lb/ft, 2 lb, 4 lb/ft   **13.15** 0.001 s
13.16 0.5 Hz, 4.93 N/m, 0.0247 J, 0.0247 J, 0.246 N   **13.17** 0.635 s
13.18 308 lb/ft, 4.32 s   **13.19** 0.25 m   **13.20** 2.43 s   **13.21** 12
13.22 3.5 s   **13.23** 3750 m/s   **13.24** 1.45×10^3 m/s
13.25 1.45 m   **13.26** 1000 m/s   **13.27** 6×10^4 m   **13.28** 19.0
m/s   **13.29** 0.15 m   **13.30** 0.1 m, 20 m/s, LN   **13.31** 0.11 kg/m

Chapter 14
14.3 313.1 m/s   **14.4** 361.9 m/s   **14.5** 266.7 Hz, 533.31 Hz, 800 Hz,
1066.7 Hz   **14.6** 9680 N   **14.7** 58.67 ft/s, 949.4 Hz   **14.8** 1112.7
Hz   **14.9** 2.4×10^{-7} W   **14.10** 110 dB   **14.11** 1 Hz
14.12 352.5 Hz, 347.5 Hz   **14.13** 2, 30°   **14.14** 14,550 ft
14.15 28.2 ft

Chapter 15

15.1 9.46×10^{15} m, 9.46×10^{16} m   **15.2** 2.56 s, 26.9 d   **15.3** 5.45 $\times 10^{14}$ Hz, 1.83×10^{-15} s, 4.57×10^{14} Hz, 2.19×10^{-15} s   **15.4** 6.67 $\times 10^{14}$ Hz, 450 nm   **15.5** 605.78 nm   **15.6** 486 nm, $0.03°$, 1 mm **15.7** 0.4 mm   **15.8** 110 nm   **15.9** 250 nm   **15.12** $12.2°$, yes **15.13** $28.7°, 73.8°$ **15.14** 9.94×10^{-19} J   **15.15** 1.01×10^{20} **15.16** 2.4×10^{14} Hz, 9.34×10^{-19} J

Chapter 16

16.12 $30°$ **16.14** $0°, 27°, 58°, 48.75°$, yes   **16.15** $24.4°$ **16.16** $43.23°$, yes, yes, no   **16.18** yes   **16.19** $21°, 61.7°, 31.7°$ **16.20** 1.3   **16.21** no, $71°$ **16.22** 20 cm   **16.23** $\sim$20 cm **16.24** 21.4, real, magnification < 1, inverted   **16.25** 17.1 cm **16.26** 15 cm, 20 cm, -0.33, 1 cm, real and inverted, yes   **16.27** 60 cm **16.29** -25 cm   **16.30** -29 cm, virtual, erect; -15.4 virtual, erect; -7.1 cm, virtual, erect   **16.31** yes   **16.32** 1.4 m   **16.33** 20 cm **16.34** 24 cm, -150 cm, -200 cm, 31.25 cm, -40 cm, -40 cm **16.35** 93.9 cm   **16.36** 0.67 m, real; 1 m, real; -0.1 m virtual **16.37** -5 m, -7.5 m, -10 m, -15 m   **16.38** 27.3 ft   **16.39** -24 cm

Chapter 17

17.4 0.64 mm   **17.5** 90 mm^2 **17.6** 7.8, 1/200 s   **17.7** -300 cm, 21.4 cm   **17.8** 45.8 cm   **17.9** 150 cm, 300 cm   **17.10** 6 **17.11** 0.59 cm, 2.27 cm, 250   **17.12** 250   **17.13** 13.3 **17.14** 3 cm   **17.15** 9.8 cm   **17.16** 5.94

Chapter 18

18.1 0.19 sr   **18.2** 31 lm   **18.3** 1.1×10^4 lx   **18.4** 54.6 lx **18.5** 625 cd   **18.6** 500 lm   **18.7** 16.77 lx   **18.8** 312.5 cd **18.9** 385 lx

Chapter 19

19.4 0.356 kg   **19.5** 8.2×10^{-8} N   **19.5** 2.3×10^{12} N **19.7** 2.25×10^{11} N   **19.8** 0   **19.9** 50.9 N   **19.10** 9×10^{11} V/m, 9×10^{13} V/m, 9×10^{15} V/m, 9×10^{17} V/m   **19.11** 1.13×10^7 V/m   **19.12** 1.13×10^4 V   **19.13** 88.54 pF   **19.14** 1.8 $\times 10^{-12}$ N   **19.15** 1.6×10^{-16} J, 1.88×10^7 m/s   **19.16** 5.32 $\times 10^{-10}$ s, 3.2×10^{-15} N, 0.495 mm   **19.17** 10 J   **19.18** $1.14 \times$ 10^{-14} kg   **19.19** 885 pF, 4799 pF   **19.20** 1×10^{-5} C **19.21** 5 V   **19.22** 34 μF   **19.23** 1.54 μF   **19.24** 5 μF **19.25** 10 V, 10 V, 10 V

Chapter 20

20.4 3.1×10^{18}, 1.1×10^{22}, 2.64×10^{23} **20.5** 2.25×10^{22}, 1.79 $\times 10^{-25}$ kg, 4.05 g   **20.6** 3.13×10^{18} **20.7** 6.25×10^{16} **20.8** 2 A, 240 W   **20.9** 0.07 A, 7 V, no   **20.10** 20 A, 5.5 Ω, 2.2 kW **20.11** 40   **20.12** 100 CM, 16.9 Ω **20.13** 14 Ω **20.14** 1 A, 0.5 A; 0.75 A, 0.63 A, 0.32 A; 1.2 A, 0.4 A, 0.8 A   **20.15** 5 V, 4.75 V, 2.4 V   **20.16** 0.66 A, 0.44 A, 0.22 A; 0.25 A, 1.5 A, 1.25 A; 0.244 A, 0.09 A, 0.15 A

Chapter 21

21.13 7×10^{-12} N south　　**21.14** 3.2×10^{-14} N, 1.1 cm, 10.8 ns
21.15 8 mm　　**21.16** 1.08×10^7 m/s, 1.45×10^{-7} s　　**21.17** 0, 4.77
N out of the paper, 6.75 N　　**21.18** 0.198 T out of the paper
21.19 0.8 T out of the paper　　**21.20** 1.5×10^{-6} N·m　　**21.21** 1 rad
21.22 950 Ω, 4950 Ω, 9950 Ω　　**21.23** 0.05 Ω, 0.01 Ω, 0.05 Ω
21.24 6.7×10^{-5} T, 4×10^{-5} T, 2×10^{-5} T　　**21.25** 3.33×10^{-4} N
21.26 1.57×10^{-3} T　　**21.27** 1.26×10^{-3} T　　**21.28** 1.89 T, 1.89 T
21.29 1000 A/m, 1.5×10^6 A/m

Chapter 22

22.5 0.36 V, 0.036 A　　**22.6** 900 V, -900 V　　**22.7** 0.8 V, 1.6 N
22.8 60 Hz, 1.8×10^4 V　　**22.9** 4.5 V　　**22.10** 6 V, 300 A
22.11 1 Ω, 88 V　　**22.12** 6.3×10^{-4} H　　**22.13** 6.76 H
22.14 15.1 mH　　**22.15** 15 H　　**22.16** 6 V　　**22.17** 48 H
22.18 175 V, 2.5 A　　**22.19** 0.067 H　　**22.20** 4.5×10^{-4} Wb
22.21 2300 V　　**22.22** 1.67×10^4, 1.67 A　　**22.23** 1833, 1.1 A

Chapter 23

23.6 20×10^{-6} s　　**23.7** 5×10^{-5} s　　**23.8** 155.6 V, 2.6 A, 1.83 A,
201 W　　**23.9** 0.91 A, 1.3 A　　**23.10** 707 V　　**23.11** 265 Ω, 0.45 A
23.12 60 Ω, 955 Hz　　**23.13** 33.4 Ω　　**23.14** 151 Ω　　**23.15** 3.77
Ω, 26.5 Ω, 102.61 Ω, 1.17 A, 117 V, 31 V, 4.4 V, $-12.8°$　　**23.16** 10.1 Hz,
6 A　　**23.17** 0.9985　　**23.18** Yes, 0.06 A. 1666.7 Ω, 1666.7 Ω

Chapter 24

24.5 7.25×10^4 Hz, 4.13×10^{-7} m　　**24.6** 1×10^{15} Hz, 4.14 eV
24.7 2.2×10^6 m/s, 1.1×10^6 m/s　　**24.8** 656 nm, 487 nm, 365 nm
24.9 2.33×10^{14} Hz　　**24.10** -54.4 eV/n^2　　**24.11** 0.544 eV
24.12 1.32 nm　　**24.13** 0.026 nm　　**24.14** 104.6 V
24.15 $1s^2 2s^2 2p^4$; $1s^2 2s^2 sp^6 3s$　　**24.16** 1s state 2 electrons; 2s state 2
electrons and 2p state 2 electrons

Chapter 25

25.3 92.1 MeV　　**25.4** 1789 MeV　　**25.5** 0.018 MeV　　**25.6** 1/16
25.7 6.4×10^{10}　　**25.8** 2.7×10^{10}/s, 6.7×10^{12} MeV, 9.22×10^2 cal
25.9 15.6×10^4 cal　　**25.10** 3.2×10^4 Ci　　**25.11** 868×10^7 J
25.12 1.02 MeV　　**25.13** 1.02 MeV

Chapter 26

26.2 8.9 V　　**26.5** 1.02, 49　　**26.6** 76　　**26.7** 50
26.8 2.8 cm